Mathematical and Computer Programming Techniques for Computer Graphics

Peter Comninos

Mathematical and Computer Programming Techniques for Computer Graphics

With 311 Figures

Peter Comninos, Dip (Comp. Prog.), BSc (Hons) (Comp. Sc.), PhD (Comp. Sc.)
The National Centre for Computer Animation
Weymouth House
Bournemouth University
Poole BH12 5BB
United Kingdom
peterc@bournemouth.ac.uk

British Library Cataloguing in Publication Data
A catalogue record for this book is available from the British Library

ISBN-13: 978-1-84996-954-3
e-ISBN-13: 978-1-84628-292-8

Printed on acid-free paper

Printed in the United States of America (SPI/MVY)

9 8 7 6 5 4 3 2 1

Springer Science+Business Media
springeronline.com

"The knowledge of which geometry aims is the knowledge of the eternal."

Plato, *Republic*, VII, 52.

"There is geometry in the humming of the strings."

Pythagoras

"Mathematics is the most beautiful and most powerful creation of the human spirit. Mathematics is as old as Man."

Stefan Banach

"The mathematician's best work is art, a high perfect art, as daring as the most secret dreams of imagination, clear and limpid. Mathematical genius and artistic genius touch one another."

Gösta Mittag-Leffler

"It is not only important, it is essential!"

Dr. Strangelove

Preface

This book introduces undergraduate and postgraduate students to mathematics and related computer programming techniques used in Computer Graphics. In a gradual approach, the book exposes students to the underlying mathematical ideas and leads them towards a level of sufficient understanding of detail to be able to implement libraries and programs for 2D and 3D graphics. Through the use of numerous code examples, the students are encouraged to explore and experiment with data structures and computer programs (in the C programming language) and to master the related mathematical techniques.

This book is meant for students with a minimum prerequisite knowledge of mathematics. It assumes very little and any high school graduate should be able to follow this book. The intended reader is expected to have had some basic exposure to topics such as functions, trigonometric functions, elementary geometry and number theory, and elements of set theory. The reader is also expected to have some familiarity with some computer programming language such as C, although any algorithmic language will serve the purpose.

The book includes a simple but effective set of routines, organised as a library, that covers both 2D and 3D graphics. This parallel approach of exposing the students to the mathematical theory and showing them how to incorporate it into example programs is the major strength of this book. It both demystifies the mathematics and it demonstrates its relevance to 2D and 3D computer graphics, thus motivating and rewarding the reader.

This book is organised into ten chapters and four appendices. Chapters 1–4 are characterised as survival kits, as they introduce the basic mathematical concepts and techniques that are applied and are essential for a thorough understanding of the remaining six chapters. The material presented in this book has been used to teach mathematical and programming techniques to both Computer Scientists and Artists. For a Bachelor degree that covers the mathematics for computer graphics over three years, Chapter 1 would normally be taught at the end of year one, Chapters 2–9 would normally be taught in year two of the course and Chapter 10 may be taught at the end of year two or the beginning of year three.

Chapter 1 introduces readers to concepts of set theory and function theory. It assumes no prior knowledge of these topics and it is self-contained.

Chapter 2 deals with vectors and vector algebra. It introduces readers to these topics assuming no prior knowledge save a rudimentary understanding of 2D and 3D geometry and some elements of trigonometry. Once readers have mastered the material presented in this chapter, they will be able to solve complex vector algebra problems and to implement their solutions in computer programs. Appendix 1, which is associated with this chapter, presents an example implementation of a 3D vector-algebra library.

Chapter 3 deals with matrices and matrix algebra. It introduces readers to these topics assuming no prior knowledge of matrices but requiring a good understanding of vector algebra. Once readers have mastered the material presented in this chapter, they will be able to solve complex matrix algebra problems and to implement their solutions in computer programs. Appendix 2, which is associated with this chapter, presents an example implementation of a 4D matrix-algebra library.

Chapter 4 deals with vector spaces, which is one of the most abstract subjects dealt with in this book and thus one of the topics that some students find more difficult. This chapter requires a good understanding of both vector and matrix algebra. It is self-contained and, although it introduces the very important concept of the change of basis matrix, it may be omitted by the uninterested reader.

Chapters 5 and 6 deal with the concepts of 2D transformations and 2D clipping algorithms respectively, and their implementation. Appendix 3, which is associated with these two chapters, presents an example implementation of a comprehensive 2D graphics library.

Chapter 7 deals with the concepts of viewing and projection transformations, 3D clipping, and their implementation. Appendix 4, which is associated with this chapter, presents an example implementation of a comprehensive 3D graphics library.

Chapter 9 examines the data structures required to represent 3D models and some of the hidden-surface removal and rendering techniques used in the creation of computer generated images. This chapter also introduces readers to some of the simple empirical lighting and shading models used in real-time graphics.

Finally, Chapter 10 presents a much more detailed exposition of the nature of light and examines, in some detail, physically-based lighting and shading models, and rendering techniques and algorithms. The material presented in this chapter is more mathematically challenging.

Most of the material presented in this book has been designed to be accessible to B.A., B.Sc., M.A. and M.Sc. students of a computer animation, digital special effects or technical direction degree course. This book however will also be useful to computer science students studying a graphics or animation unit and to technical directors in CGI production.

The vector and matrix notation of this book is designed to appeal to both North American and International readers.

Acknowledgements

I express my unreserved gratitude to countless students at the National Centre for Computer Animation who have helped me "debug" this text and improve its readability. My sincere thanks also go to my wife Danièle and my daughter Celina. Without their support and understanding writing this book would have been impossible.

Last but not least I would like to express my gratitude to my colleague and friend Peter Hardie for allowing me to use his computer art on the cover of this book. More information on Peter's work can be found online at http://ncca.bournemouth.ac.uk/newhome/announce_phsab.html.

Contents

Some Definitions of Terms

Before we proceed with the main business of this book, let us start by presenting some definitions of terms frequently used in mathematics.

Definitions of Mathematical Terms

Proposition A statement that is to be proved.

True A statement that is rigorously known to be correct. In *prepositional logic* any statement can be true or false.

False A statement that is rigorously known not to be *true*.

Axiom A statement considered to be self-evidently true without the need for any *proof*. The term axiom is an archaic synonym for the term *postulate*. In contrast the terms *conjecture* or *hypothesis* denote a statement that is apparently true (i.e. it is consistent with the available data) but is not self-evidently true. The word axiom is derived from the Greek noun "*axioma*" which is derived from the Greek verb "*axio*" meaning to claim or to demand [that a statement is true].

Postulate A statement that is self-evidently true without the need for any *proof*. Postulates are the basic building blocks from which *lemmas* and *theorems* are derived. The entire topic of *Euclidean geometry* is based on five postulates that are known as *Euclid's postulates* (see below).

Conjecture A proposition that is consistent with the available data (known facts), but has neither been shown (proven) to be true or false. A conjecture is sometimes known as a *hypothesis*.

Hypothesis A proposition that is consistent with the available data (known facts), but has neither been shown (proven) to be true or false. A hypothesis is sometimes known as a conjecture.

Lemma A short *theorem* used as a stepping stone in proving a larger (more complex) *theorem.*

Theorem A statement that can be demonstrated to be true by a series of mathematical operations and logical arguments. A theorem is the embodiment of some general principle and forms part of a larger theory. The process of proving the correctness of a theorem is called a *proof.* It is estimated that 250,000 theorems are published each year. The word theorem is derived from the Greek noun "*theorima*" which is derived from the Greek verb "*theoro*" meaning to consider or to regard [that a statement is true].

Proof A rigorous mathematical argument that unambiguously demonstrates the truth of a given proposition.

What normally determines the elegance of a mathematical theory is its reliance on as few ideas as possible that we take for granted and that we do not have to or we cannot prove, i.e. its reliance on as few axioms or postulates as possible to ensure its logical consistency. In order to illustrate the use of axioms or postulates in a definition of a mathematical theory, let us examine how the famous Greek mathematician Euclid based the entire formulation of his geometry on just five self-evidently true but unprovable statements.

Euclid's Postulates

The whole logical structure of Euclidean geometry is based on the following five postulates or axioms:

1. Any two points in space define a straight-line or a straight-line segment can be drawn by joining two points in space.
2. Any straight-line segment can be extended indefinitely into a straight-line.
3. Given any straight-line segment, a circle can be drawn having this segment as its radius and one of its endpoints as its centre.
4. All right angles are congruent (i.e. equal to one another).
5. If two lines are drawn which intersect a third in such a way that the sum of the two interior angles on the same side of the third line is less than two right angles, then these two lines, if extended infinitely, must intersect each other on that side of the third line. This postulate is known as the *parallel postulate.*

1

Set Theory Survival Kit

Unlike many other branches of mathematics, where the formulation of ideas and concepts occurs gradually over time and is developed by many mathematicians before it is formalised into a single theory, the formulation of set theory is almost the single-handed creation of one mathematician, namely Georg Cantor.

Georg Ferdinand Ludwig Philipp Cantor (1845–1918) was born in Russia to a Danish father and a Russian mother and spent most of his life in Germany. Between the years 1879 and 1884 Cantor published a six-part treatise on set theory (where he introduced some of the fundamental notions of this theory) followed by the publication of a two-part treatise between the years 1895 and 1897 (where he clarified and systematised what he had introduced in his first cycle of publications).

Between the years 1897 and 1902 a number of paradoxes in Cantor's set theory began to emerge. These paradoxes were discovered by Cantor himself and, among others, by the Italian mathematician Cesare Burali-Forti (1861–1931), the German mathematician Ernst Friedrich Ferdinand Zermelo (1871–1953) and the British mathematician Bertrand Arthur William Russell (1872–1970).

In 1908, Zermelo was the first to attempt to introduce an axiomatic approach to the study of set theory. Since then, many mathematicians proved influential in the further development of set theory. Among these are the German mathematician Adolf Abraham Halevi Fraenkel (1891–1965), the Hungarian mathematician and computer scientist John von Neumann (1903–1957), the Swiss mathematician Paul Isaac Bernays (1888–1977) and the Czech mathematician Kurt Gödel (1906–1978).

Since its introduction, set theory has proved to be of great importance to the modern formulation of many topics of pure mathematics. In current mathematical practice, such topics as numbers, relations, intervals, functions and transformations are defined in terms of sets. In our study of computer graphics we will frequently use sets to explain a number of other mathematical concepts. Thus, it is important to gain a good understanding of sets and set theory.

1.1 Some Basic Notations and Definitions

1.1.1 Sets and Elements

The concept of the *set* is one of the basic concepts of mathematics and is fundamental to most branches of modern mathematics. Thus, we start our discussion by defining the terms *set* and *element* or *member*. A *set* is any *well-defined* list, collection or class of objects, in which the order and multiplicity of these objects has no significance and is ignored. These objects are called the *elements* or *members* of the set. The phrase *well-defined* means that there is a clear and unambiguous way of defining the elements of a set, i.e. of determining if a given element is a member of a given set.

Sets may be *finite* or *infinite* depending on the number of their elements.

Set theory is the branch of mathematics that concerns the study of sets and their properties.

1.1.2 Notation and Set Specification

Usually sets are denoted by upper-case bold italic characters such as $\boldsymbol{A}$, $\boldsymbol{B}$, $\boldsymbol{S}_1$ or $\boldsymbol{S}_2$, while their elements are denoted by non-bold italic characters such as a, b, e_1 or e_2.

We may define a particular set in two distinct ways. We may define a set by listing its elements. For instance:

$$\boldsymbol{A} = \{2,\ 3,\ 6,\ 8\}$$

We call such a definition the *tabular form* of the set.

Alternatively, we may define a set by stating one or more properties that its elements must satisfy in order to belong to this set. For instance:

$$\boldsymbol{B} = \{x \mid x\ is\ an\ odd\ integer\} \text{ or } \boldsymbol{B} = \{x : x\ is\ an\ odd\ integer\}$$

We call such a definition the *set-builder form* or the *set-comprehension form* of the set. Here the symbols "|" and ":" are read as "where".

Consider the following examples of set definitions:

$$\boldsymbol{S}_1 = \{John,\ Paul,\ George,\ Ringo\}$$
$$\boldsymbol{S}_2 = \{x \mid x\ is\ a\ person\ living\ in\ Europe\}$$
$$\boldsymbol{S}_3 = \{1,\ 3,\ 5,\ 7,\ \ldots\}$$
$$\boldsymbol{S}_4 = \{cyan,\ magenta,\ yellow\}$$
$$\boldsymbol{S}_5 = \{magenta,\ yellow,\ cyan\}$$
$$\boldsymbol{S}_6 = \{cyan,\ cyan,\ magenta,\ yellow,\ yellow\}$$
$$\boldsymbol{S}_7 = \{x \mid x\ is\ a\ primary\ colour\ of\ the\ subtractive\ colour\ system\}$$

Here, set $\boldsymbol{S}_1$ represents the members of the sixties popular group "The Beatles", set $\boldsymbol{S}_2$ is a very large set containing every person living in Europe at this instance in time, and S_3 is the set of all odd integers which is identical to the set B defined above.

Also, the alternative definitions S_4 to S_7 specify the same set (i.e. the primary colours of the subtractive colour system). Observe that the order of the elements and the repetition of elements in a set definition are irrelevant and are ignored.

A more general form of the set-builder form of a set can be written as:

$$S = \{x \mid \wp(x)\}$$

which denotes the set of all the entities (objects) for which the *condition* (*proposition*) $\wp(x)$ holds true. For instance, the definition:

$$S = \{x \mid x \text{ is a dog with blue eyes}\} \text{ or } S = \{x \mid \wp(x)\}$$

denotes the set of all dogs with blue eyes when the proposition $\wp(x) = x$ *is a dog with blue eyes*.

Let us consider some variations on the theme of the set-builder form of set definitions. In the following examples, $\mathbb{Z}$ denotes the set of all integers.

- $S_1 = \{x \in S \mid \wp(x)\}$ denotes the set of all elements that belong to set S and satisfy the proposition $\wp(x)$. For instance, $S_1 = \{x \in \mathbb{Z} \mid \wp(x)\}$ (where $\wp(x) = x$ *is odd*) denotes the set of odd integers.
- $S_2 = \{f(x) \mid x \in S\}$ denotes the set of elements obtained by applying the function f to the elements of set S. For instance, the set definition $S_2 = \{f(x) \mid x \in \mathbb{Z}\}$ (where $f(x) = 2x$) denotes the set of all even integers.
- $S_3 = \{f(x) \mid \wp(x)\}$ denotes the set of all elements obtained by applying the function f to all the objects that satisfy the proposition $\wp$. For instance, the definition $S_3 = \{x^2 \mid x \text{ is a member of the set } \{-3, -2, -1, 1, 2, 3\}\}$ denotes the set $\{1, 4, 9\}$. Here $f(x) = x^2$ and $\wp(x) = x$ *is a member of the set* $\{-3, -2, -1, 1, 2, 3\}$.

1.1.3 Set Membership

If an object x is a member of a set A, i.e. if A *contains* x as one of its elements, then we denote this relationship as:

$$x \in A$$

which reads x *belongs to* A, x *is a member of* A or x *is in* A.

If an object x is not a member of a set A, then we denote this relationship as:

$$x \notin A$$

which reads x *does not belong to* A, x *is not a member of* A or x *is not in* A.

The symbol "$\in$" was introduced by the Italian mathematician Giuseppe Peano in 1888 and is derived from the first letter of the Greek word "εναι" meaning "is".

1.1.4 Finite and Infinite Sets

We say that a *set is finite* if it consists of a specific number of different elements, i.e. if the process of counting its elements can terminate. Otherwise, we say that the *set is infinite*. For instance:

- If $\boldsymbol{D}$ is the set of the days of the week, then $\boldsymbol{D}$ is a finite set.
- If $\boldsymbol{O} = \{1, 3, 5, 7, \ldots\}$, then $\boldsymbol{O}$ is an infinite set.
- If $\boldsymbol{M} = \{x \mid x \text{ is a mountain of this planet}\}$, then $\boldsymbol{M}$ is a finite set, even though it may be very difficult to count all the mountains of this planet.

If a set $\boldsymbol{S}$ has n elements (where n is a non-negative integer), then we say that $\boldsymbol{S}$ has *cardinality* n.

1.2 Equality of Sets

A set $\boldsymbol{A}$ is said to be equal to a set $\boldsymbol{B}$, if both sets have the same members, i.e. if every element of $\boldsymbol{A}$ also belongs to $\boldsymbol{B}$ and if every element of $\boldsymbol{B}$ also belongs to $\boldsymbol{A}$. We denote this equality as $\boldsymbol{A} = \boldsymbol{B}$. If the two sets are not equal, then we write $\boldsymbol{A} \neq \boldsymbol{B}$. For instance:

- If $\boldsymbol{A} = \{1, 2, 3, 4\}$ and $\boldsymbol{B} = \{3, 1, 4, 2\}$, then $\boldsymbol{A} = \boldsymbol{B}$ (as a set does not change if its elements are rearranged).
- If $\boldsymbol{C} = \{5, 6, 5, 7\}$ and $\boldsymbol{D} = \{7, 5, 7, 6\}$, then $\boldsymbol{C} = \boldsymbol{D}$ (as a set does not change if its elements are repeated).

1.3 The Null Set or Empty Set

A set that contains no elements is called a *null set* or an *empty set* and is denoted by the symbol "Ø"or by two empty braces "{}". For instance:

- If $\boldsymbol{A}$ is the set of all people in the world who are older than 500 years, then $\boldsymbol{A}$ is the empty set, i.e. $\boldsymbol{A} = \emptyset$.
- If $\boldsymbol{B} = \left\{x \mid x^2 = 4 \wedge x \; is \; an \; odd \; integer\right\}$, then $\boldsymbol{B} = \emptyset$. In this definition the symbol "$\wedge$" is read as "and".
- Empty sets have zero cardinality.

1.4 Subsets

If every element of a set $\boldsymbol{A}$ is also an element of a set $\boldsymbol{B}$, then set $\boldsymbol{A}$ is called a *subset* of set $\boldsymbol{B}$. This relationship is denoted as $\boldsymbol{A} \subseteq \boldsymbol{B}$ which reads "$\boldsymbol{A}$ *is a subset of* $\boldsymbol{B}$" or "$\boldsymbol{A}$ *is contained in* $\boldsymbol{B}$". Thus, given two sets $\boldsymbol{A}$ and $\boldsymbol{B}$:

$$\boldsymbol{A} \subseteq \boldsymbol{B}, \text{ if } x \in \boldsymbol{A} \;\Rightarrow\; x \in \boldsymbol{B}$$

For instance:

- If $\boldsymbol{C} = \{1, 3, 5\}$ and $\boldsymbol{D} = \{5, 4, 3, 2, 1\}$, then $\boldsymbol{C} \subseteq \boldsymbol{D}$.
- If $\boldsymbol{E} = \{2, 4, 6\}$ and $\boldsymbol{F} = \{6, 4, 2\}$, then $\boldsymbol{E} \subseteq \boldsymbol{F}$.

Two sets $\boldsymbol{A}$ and $\boldsymbol{B}$ are said to be equal if and only if $\boldsymbol{A} \subseteq \boldsymbol{B}$ and $\boldsymbol{B} \subseteq \boldsymbol{A}$ (i.e. $\boldsymbol{A} = \boldsymbol{B} \Leftrightarrow \boldsymbol{A} \subseteq \boldsymbol{B} \wedge \boldsymbol{B} \subseteq \boldsymbol{A}$).

1.5 Supersets

If set $\boldsymbol{A}$ is a subset of set $\boldsymbol{B}$ (i.e. $\boldsymbol{A} \subseteq \boldsymbol{B}$), then we can also denote this as $\boldsymbol{B} \supseteq \boldsymbol{A}$, which reads ***B** is a superset of **A*** or ***B** contains **A***.

If set $\boldsymbol{A}$ is not a subset of set $\boldsymbol{B}$, then we can denote this as $\boldsymbol{A} \nsubseteq \boldsymbol{B}$ or $\boldsymbol{B} \nsupseteq \boldsymbol{A}$.

Observe that:

- The null set $\emptyset$ is the subset of every set.
- If $\boldsymbol{A} \nsubseteq \boldsymbol{B}$, then this means that there is at least one element of set $\boldsymbol{A}$ that is not a member of set $\boldsymbol{B}$.

1.6 Proper Subsets and Supersets

A set $\boldsymbol{A}$ is called a *proper subset* of a set $\boldsymbol{B}$ if $\boldsymbol{A}$ is a subset of $\boldsymbol{B}$ and $\boldsymbol{A}$ is not equal to $\boldsymbol{B}$, i.e.

$$\boldsymbol{A} \subset \boldsymbol{B}, \text{ if } \boldsymbol{A} \subseteq \boldsymbol{B} \text{ and } \boldsymbol{A} \neq \boldsymbol{B}$$

If set $\boldsymbol{A}$ is a proper subset of set $\boldsymbol{B}$ (i.e. $\boldsymbol{A} \subset \boldsymbol{B}$), then we can also denote this as $\boldsymbol{B} \supset \boldsymbol{A}$, which reads "***B** is a proper superset of **A***", i.e.

$$\boldsymbol{B} \supset \boldsymbol{A}, \text{ if } \boldsymbol{A} \subset \boldsymbol{B}$$

If set $\boldsymbol{A}$ is not a proper subset of set $\boldsymbol{B}$, then we can denote this as $\boldsymbol{A} \not\subset \boldsymbol{B}$ (which reads ***A** is not a proper subset of **B***) or $\boldsymbol{B} \not\supset \boldsymbol{A}$ (which reads ***B** is not a proper superset of **A***).

The use of the symbols "$\subseteq$" and "$\subset$" to represent the ordinary and proper subset operators is symmetrical to the use of the symbols "$\leq$" and "$<$" to represent the *less than or equal* and *less than* scalar operators. Similarly, the use of the symbols "$\supseteq$" and "$\supset$" to represent the ordinary and proper superset operators is symmetrical to the use of the symbols "$\geq$" and "$>$" to represent the *greater than or equal* and *greater than* scalar operators. In some literature there is no distinction made between the ordinary and proper subset/superset operators which are represented by the symbols "$\subset$" and "$\supset$", respectively.

1.7 Comparable Sets

Two sets $\boldsymbol{A}$ and $\boldsymbol{B}$ are said to be *comparable* if $\boldsymbol{A} \subset \boldsymbol{B}$ or $\boldsymbol{B} \subset \boldsymbol{A}$, that is if one of the two sets is a subset of the other set. Conversely, two sets $\boldsymbol{A}$ and $\boldsymbol{B}$ are said to be *non-comparable* (*incomparable*) if $\boldsymbol{A} \not\subset \boldsymbol{B}$ and $\boldsymbol{B} \not\subset \boldsymbol{A}$.

If a set $\boldsymbol{A}$ is not comparable to a set $\boldsymbol{B}$, then there is an element of $\boldsymbol{A}$ which is not in $\boldsymbol{B}$ and an element in $\boldsymbol{B}$ which is not in $\boldsymbol{A}$. For instance:

- If $\boldsymbol{A} = \{a, b\}$ and $\boldsymbol{B} = \{a, b, c\}$, then $\boldsymbol{A}$ is comparable to $\boldsymbol{B}$ (since $\boldsymbol{A} \subset \boldsymbol{B}$).
- If $\boldsymbol{R} = \{a, b\}$ and $\boldsymbol{S} = \{b, c\}$, then these sets are non-comparable (since $a \in \boldsymbol{R}$ and $a \notin \boldsymbol{S}$, and $c \in \boldsymbol{S}$ and $c \notin \boldsymbol{R}$).

1.8 The Universal Set

In any application of set theory, all the sets under investigation are likely to be subsets of a fixed set. We call this set the *universal set* or the *universe of discourse* and we denote it by the capital letter $\boldsymbol{U}$. Any set can act as the universal set, provided that we are investigating this particular set and its subsets. For instance, if we are investigating the set of real numbers and their subsets, then the real number set $\mathbb{R}$ can be taken to be the universal set for our investigation (i.e. in this case, $\boldsymbol{U} = \mathbb{R}$).

The universal set is only defined in the context of our investigation and it is not an absolute concept. Thus, we can not speak of an absolute universal set that contains everything.

1.9 Disjoint Sets

If two sets $\boldsymbol{A}$ and $\boldsymbol{B}$ have no elements in common (i.e. if no element of $\boldsymbol{A}$ is in $\boldsymbol{B}$ and no element of $\boldsymbol{B}$ is in $\boldsymbol{A}$), then we say that $\boldsymbol{A}$ and $\boldsymbol{B}$ are *disjoint sets*. For instance:

- If $\boldsymbol{A} = \{1, 3, 7, 8\}$ and $\boldsymbol{B} = \{2, 4, 7, 9\}$, then $\boldsymbol{A}$ and $\boldsymbol{B}$ are not disjoint (as $7 \in \boldsymbol{A}$ and $7 \in \boldsymbol{B}$).
- If $\boldsymbol{A} = \{1, 2, 3\}$ and $\mathbf{B} = \{4, 5, 6\}$, then $\boldsymbol{A}$ and $\boldsymbol{B}$ are disjoint.
- If $\boldsymbol{A} = \{x \mid x > 0\}$ and $\boldsymbol{B} = \{x \mid x < 0\}$, then $\boldsymbol{A}$ and $\boldsymbol{B}$ are disjoint.

1.10 Venn-Euler Diagrams

A Venn-Euler diagram is a pictorial representation of specific sets and their relationships, using sets of points on the plane to represent them.

These diagrams were invented by Leonhard Euler (1707–1783) and about one hundred years later by John Venn (1834–1923). Euler and Venn diagrams are identical in their appearance and are only differentiated by their domain of application. Euler used his diagrams in an attempt to illustrate specific sets and their subsets, while Venn used his diagrams to illustrate all possible relationships between specific sets. These diagrams are commonly known as Venn diagrams since they represent the only major contribution of Venn to mathematics, whereas Euler is remembered for his many contributions to the field.

In a Venn diagram, the universal set is represented by a region of the plane described by a rectangle and the other sets under investigation are represented by regions of the plane described by ellipses or closed curves. For instance, if the universal set U represents all animals, the region labelled C represents the set of all camels, the region labelled B represents the set of all birds and the region labelled A represents the set of all albatrosses, then the Venn diagram shown in Fig. 1.1 represents the relationship of these sets.

Figures 1.2–1.5 illustrate the use of Venn diagrams to represent various relationships between two sets A and B. Figure 1.2 represents the relationship "set A is a proper subset of set B", i.e. $A \subset B$. Figure 1.3 represents the relationship "set A

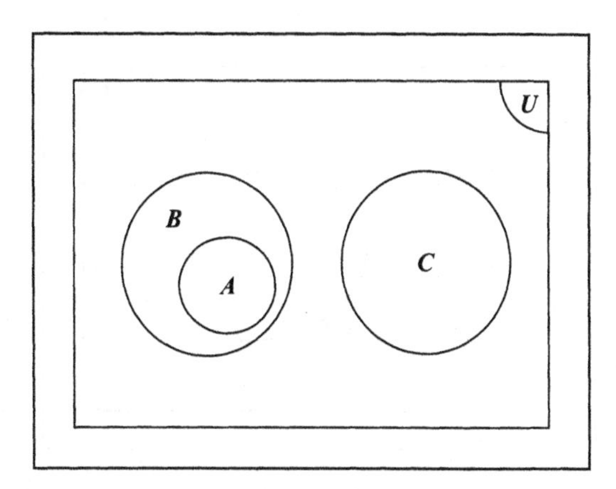

FIGURE 1.1. The relationship of camels, birds and albatrosses.

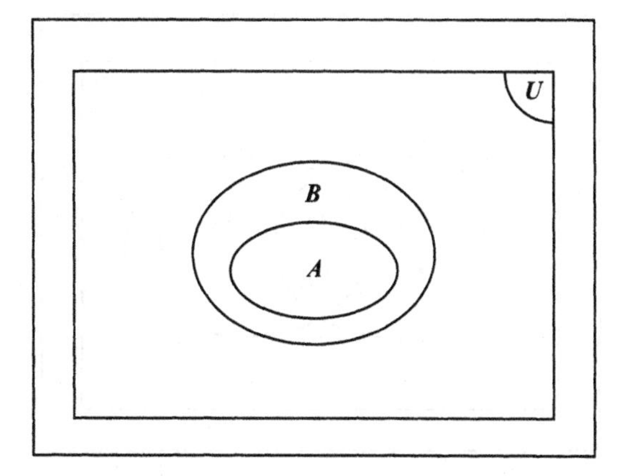

FIGURE 1.2. The Venn diagram of the set relationship $A \subset B$.

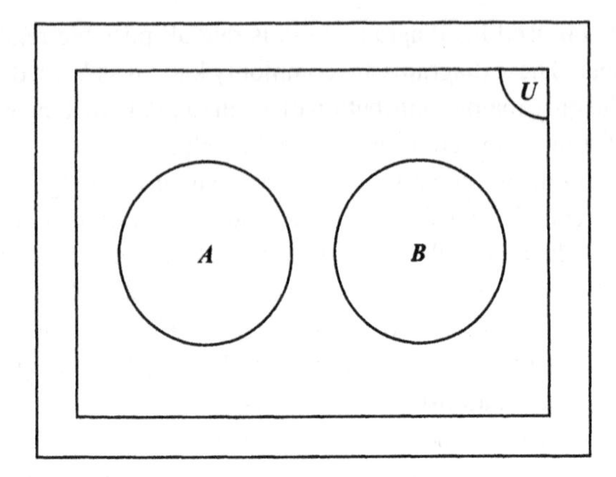

FIGURE 1.3. The Venn diagram of two disjoint sets.

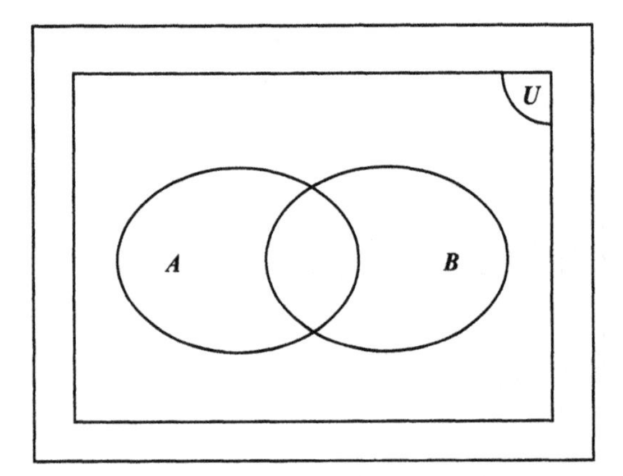

FIGURE 1.4. The Venn diagram of two incomparable sets.

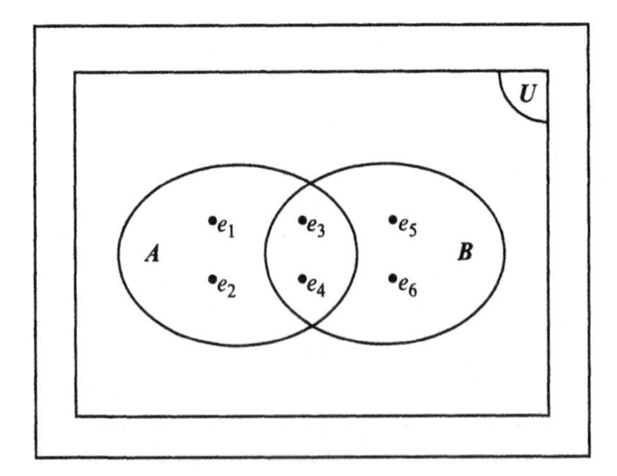

FIGURE 1.5. The Venn diagram of two sets that share some common elements.

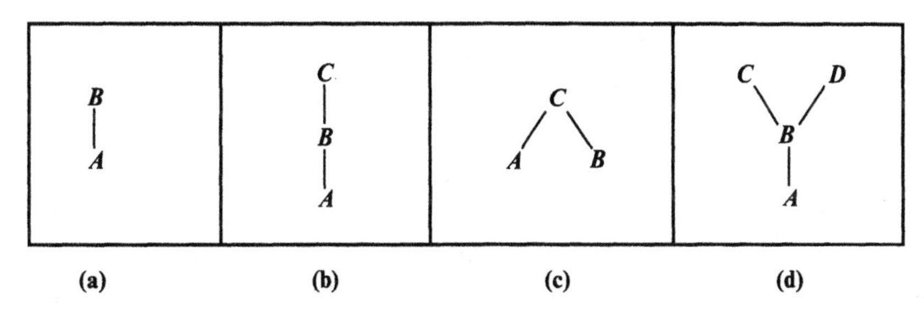

FIGURE 1.6. (a), (b), (c), (d)

is disjoint from set $\boldsymbol{B}$". Figure 1.4 represents the relationship "sets $\boldsymbol{A}$ and $\boldsymbol{B}$ are incomparable". Figure 1.5 represents the relationship between set $\boldsymbol{A} = \{e_1, e_2, e_3, e_4\}$ and set $\boldsymbol{B} = \{e_3, e_4, e_5, e_6\}$.

1.11 Line Diagrams

Another insightful and instructive way of representing the relationships that exist between sets is by using *line diagrams*. The following examples help illustrate how line diagrams can be used to represent the relationship between two or more sets:

- The set relationship $\boldsymbol{A} \subseteq \boldsymbol{B}$ is illustrated by the line diagram of Fig. 1.6a.
- The set relationship $\boldsymbol{A} \subseteq \boldsymbol{B}$ and $\boldsymbol{B} \subseteq \boldsymbol{C}$ is illustrated by the line diagram of Fig. 1.6b.
- If $\boldsymbol{A} = \{e_1\}$, $\boldsymbol{B} = \{e_2\}$ and $\boldsymbol{C} = \{e_1, e_2\}$, then the relationship of these sets is illustrated by the line diagram of Fig. 1.6c.
- If $\boldsymbol{A} = \{x\}$, $\boldsymbol{B} = \{x, y\}$, $\boldsymbol{C} = \{x, y, z\}$ and $\boldsymbol{D} = \{x, y, w\}$, then the relationship of these sets is illustrated by the line diagram of Fig. 1.6d.

1.12 Basic Set Operations

1.12.1 Set Union

The *union* of sets $\boldsymbol{A}$ and $\boldsymbol{B}$ is the set of elements that belong to set $\boldsymbol{A}$ or to set $\boldsymbol{B}$ or to both sets. We denote the union of sets $\boldsymbol{A}$ and $\boldsymbol{B}$ by $\boldsymbol{A} \cup \boldsymbol{B}$, which reads $\boldsymbol{A}$ *union* $\boldsymbol{B}$. The formal definition of the union of sets $\boldsymbol{A}$ and $\boldsymbol{B}$ is given by:

$$\boldsymbol{A} \cup \boldsymbol{B} = \{x \mid x \in \boldsymbol{A} \vee x \in \boldsymbol{B}\} \tag{1.1}$$

In this definition the symbol "$\vee$" is read as "or".

The Venn diagram representing the union of two sets is shown in Fig. 1.7. In this diagram, the shaded area represents the elements of the union $\boldsymbol{A} \cup \boldsymbol{B}$.

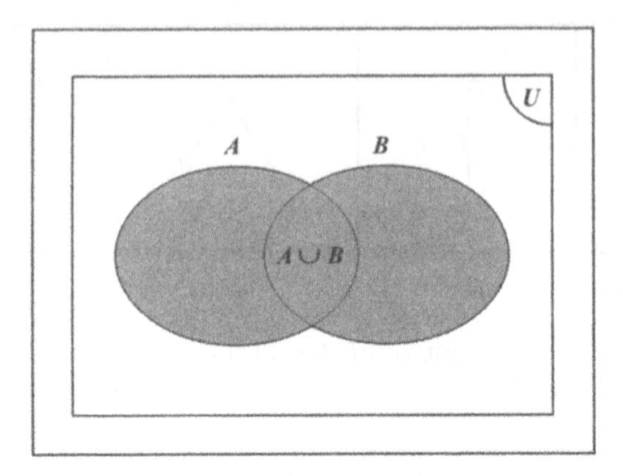

FIGURE 1.7. The Venn diagram of the union $\boldsymbol{A} \cup \boldsymbol{B}$.

Consider the following examples:

- If $\boldsymbol{A} = \{a, b, c, d\}$ and $\boldsymbol{B} = \{c, d, e, f\}$, then $\boldsymbol{A} \cup \boldsymbol{B}$ $\{a, b, c, d, e, f\}$.
- If $\boldsymbol{A} = \{x \mid x \in \mathbb{R} \wedge x > 0\}$ and $\boldsymbol{B} = \{x \mid x \in \mathbb{R} \wedge x < 0\}$, then $\boldsymbol{A} \cup \boldsymbol{B} = \{x \mid x \in \mathbb{R} \wedge x \neq 0\}$.

Given two sets $\boldsymbol{A}$ and $\boldsymbol{B}$ and their union $\boldsymbol{A} \cup \boldsymbol{B}$, we make the following observations:

- The union operation is commutative, i.e. $\boldsymbol{A} \cup \boldsymbol{B} = \boldsymbol{B} \cup \boldsymbol{A}$.
- Both sets are subsets of their union, i.e. $\boldsymbol{A} \subseteq (\boldsymbol{A} \cup \boldsymbol{B})$ and $\boldsymbol{B} \subseteq (\boldsymbol{A} \cup \boldsymbol{B})$.

In some literature the union of sets $\boldsymbol{A}$ and $\boldsymbol{B}$ is denoted as $\boldsymbol{A} + \boldsymbol{B}$ and it is called *the set-theoretic sum of* ***A*** *and* ***B***.

1.12.2 Set Intersection

The *intersection* of sets $\boldsymbol{A}$ and $\boldsymbol{B}$ is the set of elements that are common to both sets, i.e. the elements that belong both to set $\boldsymbol{A}$ and to set $\boldsymbol{B}$. We denote the intersection of set $\boldsymbol{A}$ and $\boldsymbol{B}$ by $\boldsymbol{A} \cap \boldsymbol{B}$, which reads ***A*** *intersection* ***B***. The formal definition of the intersection of sets $\boldsymbol{A}$ and $\boldsymbol{B}$ is given by:

$$\boldsymbol{A} \cap \boldsymbol{B} = \{x \mid x \in \boldsymbol{A} \wedge x \in \boldsymbol{B}\} \tag{1.2}$$

The Venn diagram representing the intersection of two sets is shown in Fig. 1.8. In this diagram, the shaded area represents the elements of the intersection $\boldsymbol{A} \cap \boldsymbol{B}$.

Consider the following examples:

- If $\boldsymbol{A} = \{a, b, c, d\}$ and $\boldsymbol{B} = \{c, d, e, f\}$, then $\boldsymbol{A} \cap \boldsymbol{B} = \{c, d\}$.
- If $\boldsymbol{A} = \{x \mid x \in \mathbb{R} \wedge x \geq 0\}$ and $\boldsymbol{B} = \{x \mid x \in \mathbb{R} \wedge x \leq 0\}$, then $\boldsymbol{A} \cap \boldsymbol{B} = \{x \mid x \in \mathbb{R} \wedge x = 0\} = \{0\}$.

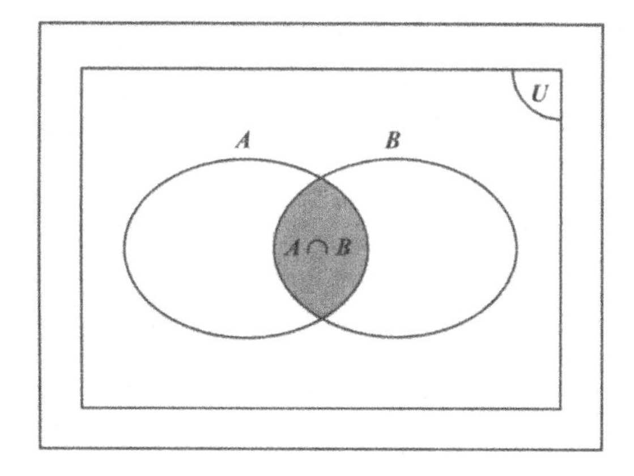

FIGURE 1.8. The Venn diagram of the intersection $\boldsymbol{A} \cap \boldsymbol{B}$.

Given two sets $\boldsymbol{A}$ and $\boldsymbol{B}$ and their intersection $\boldsymbol{A} \cap \boldsymbol{B}$, we make the following observations:

- The intersection operation is commutative, i.e. $\boldsymbol{A} \cap \boldsymbol{B} = \boldsymbol{B} \cap \boldsymbol{A}$.
- The intersection of two sets is a subset of both sets, i.e. $(\boldsymbol{A} \cap \boldsymbol{B}) \subseteq \boldsymbol{A}$ and $(\boldsymbol{A} \cap \boldsymbol{B}) \subseteq \boldsymbol{B}$.
- If two set are disjoint, then their intersection is the empty set, i.e. $\boldsymbol{A} \cap \boldsymbol{B} = \emptyset$.

In some literature the intersection of sets $\boldsymbol{A}$ and $\boldsymbol{B}$ is denoted as $\boldsymbol{A} \cdot \boldsymbol{B}$ and it is called *the set-theoretic product of* ***A*** *and* ***B***.

1.12.3 Set Difference

The *difference* of sets $\boldsymbol{A}$ and $\boldsymbol{B}$ is the set of elements that belong to set $\boldsymbol{A}$ and do not belong to set $\boldsymbol{B}$. We denote the difference of sets $\boldsymbol{A}$ and $\boldsymbol{B}$ by $\boldsymbol{A} - \boldsymbol{B}$, which reads $\boldsymbol{A}$ *difference* $\boldsymbol{B}$ or $\boldsymbol{A}$ *minus* $\boldsymbol{B}$. The formal definition of the difference of sets $\boldsymbol{A}$ and $\boldsymbol{B}$ is given by:

$$\boldsymbol{A} - \boldsymbol{B} = \{x \mid x \in \boldsymbol{A} \wedge x \notin \boldsymbol{B}\} \tag{1.3}$$

The Venn diagram representing the difference of two sets is shown in Fig. 1.9. In this diagram, the shaded area represents the elements of the difference $\boldsymbol{A} - \boldsymbol{B}$.

Consider the following examples:

- If $\boldsymbol{A} = \{a, b, c, d\}$ and $\boldsymbol{B} = \{c, d, e, f\}$, then $\boldsymbol{A} - \boldsymbol{B} = \{a, b\}$.
- If $\boldsymbol{A} = \{x \mid x \in \mathbb{R}\}$ and $\boldsymbol{B} = \{x \mid x \in \mathbb{R} \wedge x < 0\}$, then $\boldsymbol{A} - \boldsymbol{B} = \{x \mid x \in \mathbb{R} \wedge x \geq 0\}$.

Given two sets $\boldsymbol{A}$ and $\boldsymbol{B}$ and their difference $\boldsymbol{A} - \boldsymbol{B}$, we make the following observations:

- The difference operation is not commutative, i.e. $\boldsymbol{A} - \boldsymbol{B} \neq \boldsymbol{B} - \boldsymbol{A}$.
- The difference of sets $\boldsymbol{A}$ and $\boldsymbol{B}$ is a subset of set $\boldsymbol{A}$, i.e. $(\boldsymbol{A} - \boldsymbol{B}) \subseteq \boldsymbol{A}$.

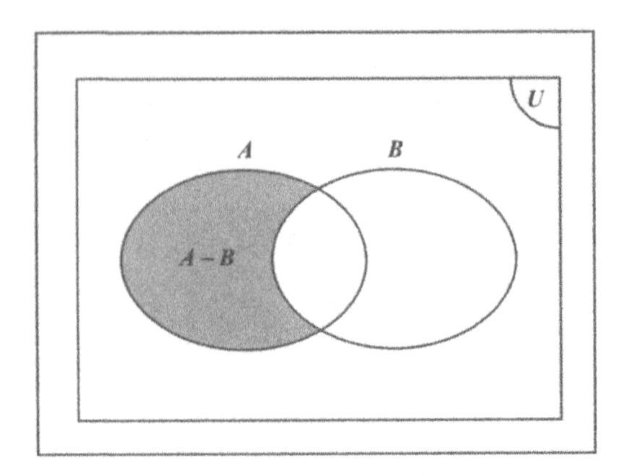

FIGURE 1.9. The Venn diagram of the difference $\boldsymbol{A} - \boldsymbol{B}$.

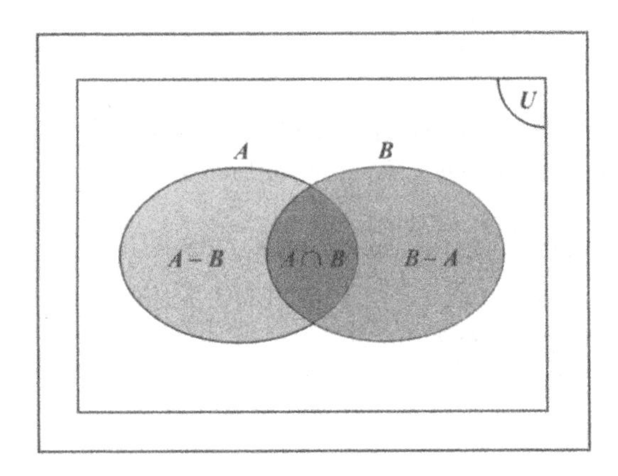

FIGURE 1.10. The Venn diagram of the sets $(\boldsymbol{A} - \boldsymbol{B})$, $(\boldsymbol{A} \cap \boldsymbol{B})$ and $(\boldsymbol{B} - \boldsymbol{A})$.

- The sets $(\boldsymbol{A} - \boldsymbol{B})$, $(\boldsymbol{A} \cap \boldsymbol{B})$ and $(\boldsymbol{B} - \boldsymbol{A})$ are *mutually disjoint*, i.e. $(\boldsymbol{A} - \boldsymbol{B}) \cap (\boldsymbol{A} \cap \boldsymbol{B}) = \emptyset$, $(\boldsymbol{A} - \boldsymbol{B}) \cap (\boldsymbol{B} - \boldsymbol{A}) = \emptyset$ and $(\boldsymbol{A} \cap \boldsymbol{B}) \cap (\boldsymbol{B} - \boldsymbol{A}) = \emptyset$. (See Fig. 1.10.)

In some literature the difference of sets $\boldsymbol{A}$ and $\boldsymbol{B}$ is denoted as $\boldsymbol{A} \backslash \boldsymbol{B}$ and is called *the set-theoretic difference of* **A** *and* **B**.

1.12.4 The Symmetric Difference of Two Sets

The *symmetric difference* of two sets $\boldsymbol{A}$ and $\boldsymbol{B}$ is the set of elements that belong to set $\boldsymbol{A}$ but not to set $\boldsymbol{B}$ and the elements that belong to set $\boldsymbol{B}$ but not to set $\boldsymbol{A}$, i.e. the elements that belong to set $\boldsymbol{A}$ or to set $\boldsymbol{B}$ but not to both sets. We denote the symmetric difference of two sets $\boldsymbol{A}$ and $\boldsymbol{B}$ by $\boldsymbol{A} \,\Delta\, \boldsymbol{B}$, which reads $\boldsymbol{A}$ *symmetric difference* $\boldsymbol{B}$. This operation is the set theoretic equivalent to the *exclusive or* (XOR) operation in Boolean algebra. The formal definition of the symmetric difference of two sets A and B is given by:

$$\begin{aligned} A \,\Delta\, B &= \{x \mid (x \in A \wedge x \notin B) \vee (x \in B \wedge x \notin A)\} \quad \text{or} \\ A \,\Delta\, B &= \{x \mid x \in A + x \in B\} \end{aligned} \tag{1.4}$$

In this definition the symbol "+" is read as "exclusive or" or "XOR" and represents the logical non-equivalence relationship.

The Venn diagram representing the symmetric difference of two sets is shown in Fig. 1.11. In this diagram, the shaded area represents the elements of the difference $A \,\Delta\, B$.

There are three additional ways of defining the symmetric difference of two sets, which derive from the above definition. The symmetric difference can be defined as the union of the intersections $(A \cap B')$ and $(A' \cap B)$, i.e.

$A \,\Delta\, B = (A \cap B') \cup (A' \cap B)$ (See Fig. 1.12.)

It can be defined as the union of the differences $(A - B)$ and $(B - A)$, i.e.

$A \,\Delta\, B = (A - B) \cup (B - A)$ (See Fig. 1.13.)

It can also be defined as the difference of the union and the intersection of the two sets, i.e.

$A \,\Delta\, B = (A \cup B) - (A \cap B)$ (See Fig. 1.14.)

Consider the following examples:

- If $A = \{a, b, c, d\}$ and $B = \{c, d, e, f\}$, then $A \,\Delta\, B = \{a, b, e, f\}$.
- If $A = \{x \mid x \in \mathbb{R} \wedge x \leq 0\}$ and $B = \{x \mid x \in \mathbb{R} \wedge x \geq 0\}$, then $A \,\Delta\, B = \{x \mid x \in \mathbb{R} \wedge x \neq 0\}$.

Given two sets A and B and their symmetric difference $A \,\Delta\, B$, we make the following observations:

- The symmetric difference operation is commutative, i.e. $A \,\Delta\, B = B \,\Delta\, A$.
- The symmetric difference operation is associative, i.e. $(A \,\Delta\, B) \,\Delta\, C = A \,\Delta\, (B \,\Delta\, C)$. (See Fig. 1.15)

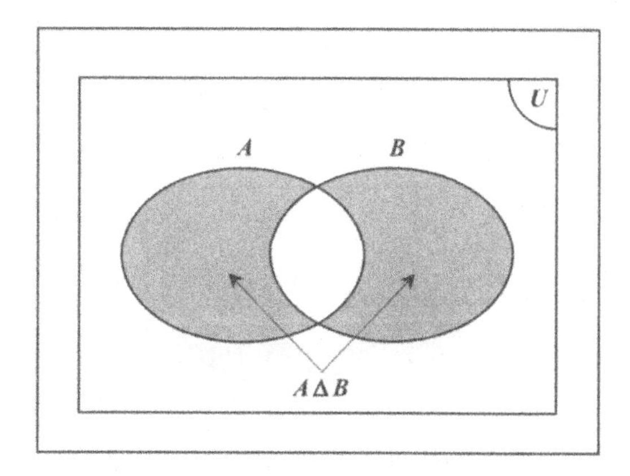

FIGURE 1.11. The Venn diagram of the symmetric difference $A \,\Delta\, B$.

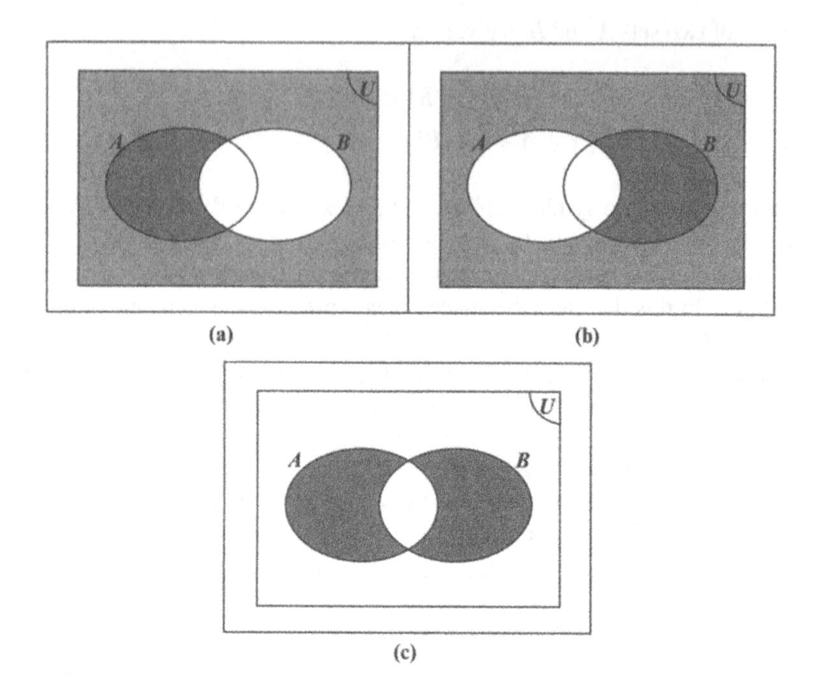

FIGURE 1.12. (a) $\boldsymbol{A} \cap \boldsymbol{B}'$. (b) $\boldsymbol{A}' \cap \boldsymbol{B}$. (c) $\boldsymbol{A} \,\Delta\, \boldsymbol{B} = (\boldsymbol{A} \cap \boldsymbol{B}') \cup (\boldsymbol{A}' - \boldsymbol{B})$.

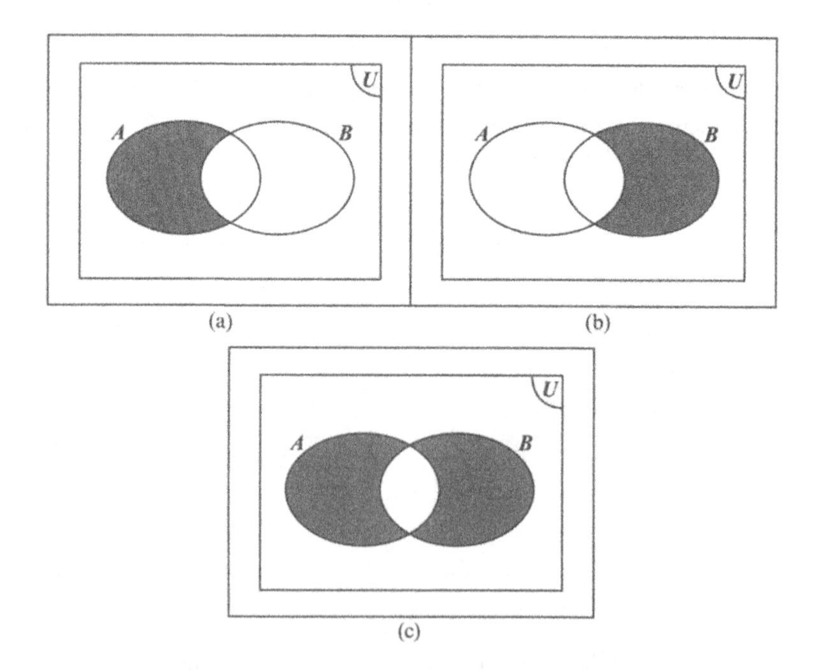

FIGURE 1.13. (a) $\boldsymbol{A} - \boldsymbol{B}$. (b) $\boldsymbol{B} - \boldsymbol{A}$. (c) $\boldsymbol{A} \,\Delta\, \boldsymbol{B} = (\boldsymbol{A} - \boldsymbol{B}) \cup (\boldsymbol{B} - \boldsymbol{A})$.

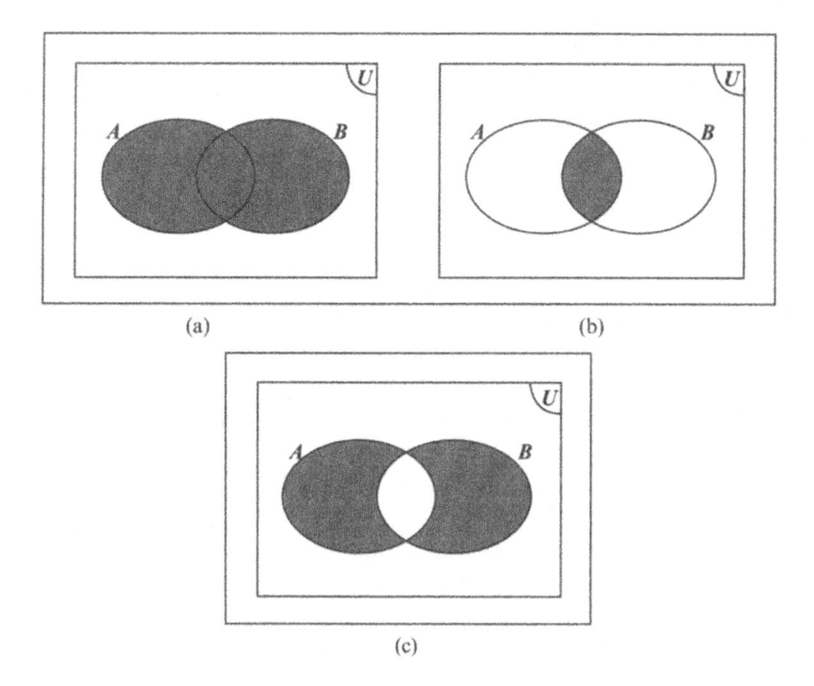

FIGURE 1.14. (a) $\boldsymbol{A \cup B}$. (b) $\boldsymbol{A \cap B}$. (c) $\boldsymbol{A \triangle B = (A \cup B) \cup (B \cap A)}$.

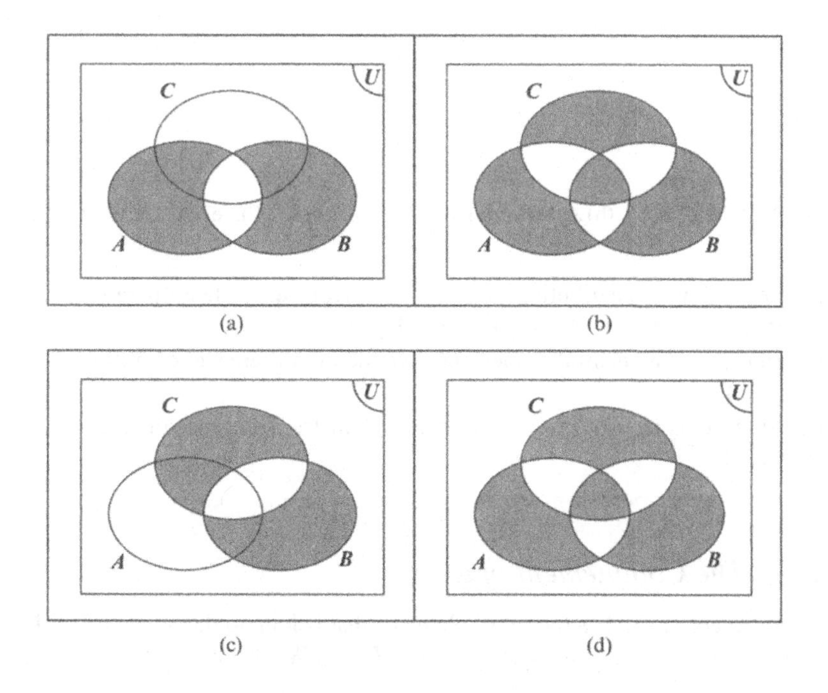

FIGURE 1.15. (a) $\boldsymbol{(A \triangle B)}$. (b) $\boldsymbol{(A \triangle B) \triangle C}$. (c) $\boldsymbol{(B \triangle C)}$. (d) $\boldsymbol{A \triangle (B \triangle C)}$.

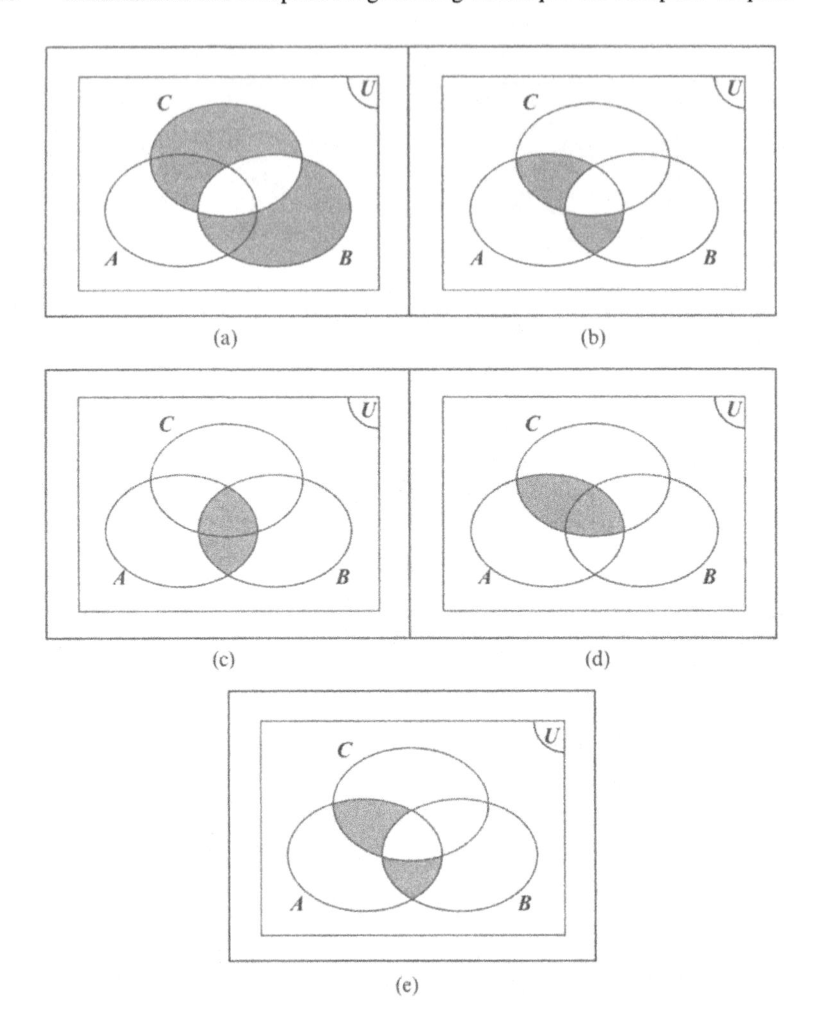

FIGURE 1.16 (a) $(\boldsymbol{B} \Delta \boldsymbol{C})$. (b) $\boldsymbol{A} \cap (\boldsymbol{B} \Delta \boldsymbol{C})$. (c) $(\boldsymbol{A} \cap \boldsymbol{B})$. (d) $(\boldsymbol{A} \cap \boldsymbol{C})$. (e) $(\boldsymbol{A} \cap \boldsymbol{B}) \Delta (\boldsymbol{A} \cap \boldsymbol{C})$.

- Set intersection is distributive over the symmetric difference operation, i.e. $\boldsymbol{A} \cap (\boldsymbol{B} \Delta \boldsymbol{C}) = (\boldsymbol{A} \cap \boldsymbol{B}) \Delta (\boldsymbol{A} \cap \boldsymbol{C})$. (See Fig. 1.16)
- The empty set is neutral under the symmetric difference operation, i.e. $\boldsymbol{A} \Delta \emptyset = \boldsymbol{A}$.
- The empty set is its own inverse under the symmetric difference operation, i.e. $\boldsymbol{A} \Delta \boldsymbol{A} = \emptyset$.

1.12.5 The Complement of a Set

The *complement* of set $\boldsymbol{A}$ is the set of elements that belong to the universal set $\boldsymbol{U}$ but do not belong to set $\boldsymbol{A}$, i.e. the difference $\boldsymbol{U} - \boldsymbol{A}$. We denote the complement

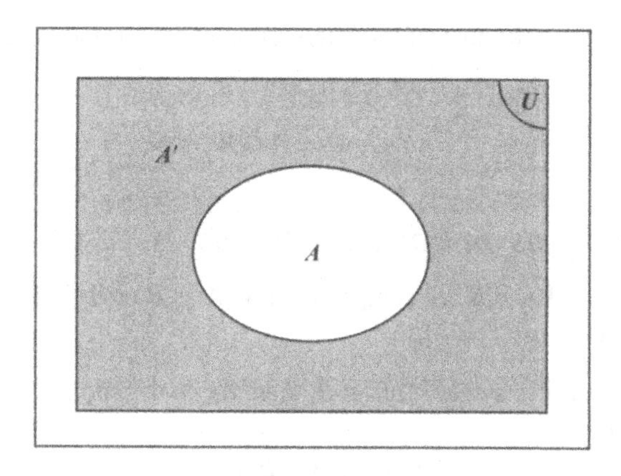

FIGURE 1.17. The Venn diagram of the complement A'.

of set A by A' or sometimes by A^c. The formal definition of the complement of a set A is given by:

$$A' = \{x \mid x \in U \wedge x \notin A\} \tag{1.5}$$

The Venn diagram representing the complement of a set is shown in Fig. 1.17. In this diagram, the shaded area represents the elements of the complement A'.

Consider the following examples:

- If $U = \{a, \ldots, z\}$ and $A = \{a, b, c, d\}$, then $A' = \{e, \ldots, z\}$.
- If $U = \{1, 2, 3, \ldots\}$ and $A = \{2, 4, 6, \ldots\}$, then $A' = \{1, 3, 5, \ldots\}$.

Given a set A and its complement A', we make the following observations:

- The union of a set A and its complement A' is the universal set, i.e. $A \cup A' = U$.
- The intersection of a set A and its complement A' is the empty set, i.e. $A \cap A' = \emptyset$.
- Any set A and its complement A' are mutually disjoint, i.e. $A \cap A' = \emptyset$.
- The complement of the universal set U is the empty set $\emptyset$ and vice versa, i.e. $U' = \emptyset$ and $\emptyset' = U$.

Theorem 1.1 *The difference of sets* A *and* B *is equal to the intersection of set* A *and the complement of set* B*, i.e.*

$$A - B = A \cap B'. \tag{T1.1}$$

Proof: We may prove the theorem as follows. By the definition of the set difference (Eq. 1.3), we have:

$$A - B = \{x \mid x \in A \wedge x \notin B\}$$

Which can be rewritten as:

$$A - B = \{x \mid x \in A \wedge x \in B'\}$$

Which in turn is the definition of the intersection of sets $\boldsymbol{A}$ and $\mathbf{B}'$. (See Eq. 1.2.) Thus,

$$\boldsymbol{A} - \boldsymbol{B} = \boldsymbol{A} \cap \boldsymbol{B}'. \qquad \square$$

1.12.6 Theorems on Comparable Sets

Given that two sets $\boldsymbol{A}$ and $\boldsymbol{B}$ are comparable (i.e. $\boldsymbol{A} \subseteq \boldsymbol{B}$ or $\boldsymbol{B} \subseteq \boldsymbol{A}$), we can prove the following theorems.

Theorem 1.2 *If set **A** is a subset of set **B**, then the intersection of sets **A** and **B** is equal to set **A**, i.e.*

$$\boldsymbol{A} \subseteq \boldsymbol{B} \Rightarrow \boldsymbol{A} \cap \boldsymbol{B} = \boldsymbol{A}. \tag{T1.2}$$

Proof: We can prove this theorem using the Venn diagram shown in Fig. 1.18. As can be seen from this diagram, the shaded area represents both set $\boldsymbol{A}$ and set $\boldsymbol{A} \cap \boldsymbol{B}$. $\square$

Theorem 1.3 *If set **A** is a subset of set **B**, then the union of sets **A** and **B** is equal to set **B**, i.e.*

$$\boldsymbol{A} \subseteq \boldsymbol{B} \Rightarrow \boldsymbol{A} \cup \boldsymbol{B} = \boldsymbol{B}. \tag{T1.3}$$

Proof: We can prove this theorem using the Venn diagram shown in Fig. 1.19. As can be seen from this diagram, the shaded area represents both set $\boldsymbol{B}$ and set $\boldsymbol{A} \cup \boldsymbol{B}$. $\square$

Theorem 1.4 *If set **A** is a subset of set **B**, then the complement of set **B** is a subset of the complement of set **A**, i.e.*

$$\boldsymbol{A} \subseteq \boldsymbol{B} \Rightarrow \boldsymbol{B}' \subseteq \boldsymbol{A}'. \tag{T1.4}$$

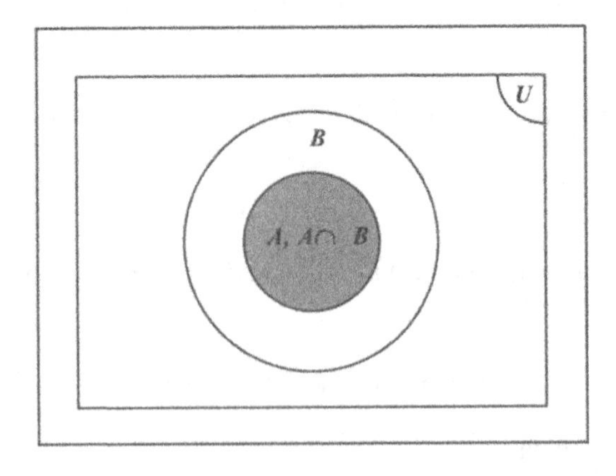

FIGURE 1.18. The graphical proof of the theorem: $\boldsymbol{A} \subseteq \boldsymbol{B} \Rightarrow \boldsymbol{A} \cap \boldsymbol{B} = \boldsymbol{A}$

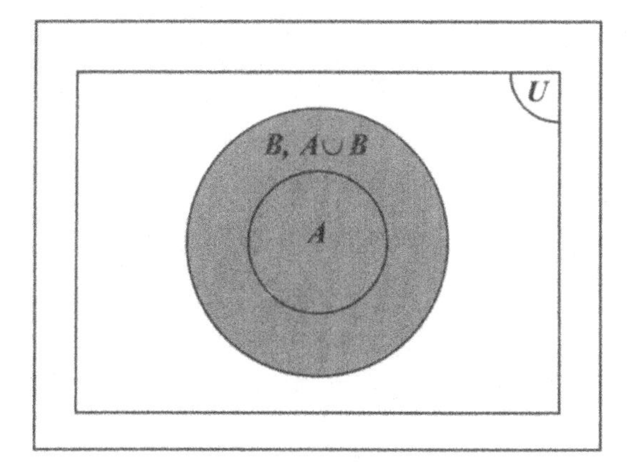

FIGURE 1.19. The graphical proof of the theorem: $\boldsymbol{A} \subseteq \boldsymbol{B} \Rightarrow \boldsymbol{A} \cup \boldsymbol{B} = \boldsymbol{B}$.

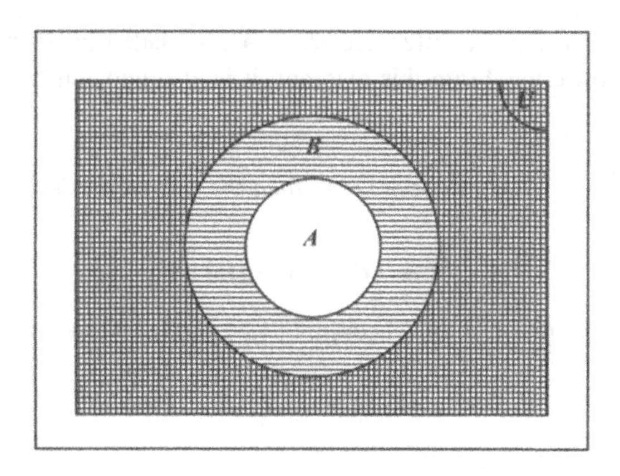

FIGURE 1.20. The graphical proof of the theorem: $\boldsymbol{A} \subseteq \boldsymbol{B} \Rightarrow \boldsymbol{B}' \subseteq \boldsymbol{A}'$.

Proof: We can prove this theorem using the Venn diagram shown in Fig. 1.20. In this diagram, the complement $\boldsymbol{A}'$ is represented by the area of $\boldsymbol{U}$ that is outside the area of $\boldsymbol{A}$ (which is shaded with a horizontal line pattern) and the complement $\boldsymbol{B}'$ is represented by the area of $\boldsymbol{U}$ that is outside the area of $\boldsymbol{B}$ (which is shaded with a vertical line pattern). The area of $\boldsymbol{U}$ that is shaded by both vertical and horizontal line patterns represents the intersection $\boldsymbol{A}' \cap \boldsymbol{B}'$. □

Theorem 1.5 *If set* **A** *is a subset of set* **B**, *then the union of sets* **A** *and* **(B − A)** *is equal to set* **B**, *i.e.*

$$\boldsymbol{A} \subseteq \boldsymbol{B} \Rightarrow \boldsymbol{A} \cup (\boldsymbol{B} - \boldsymbol{A}) = \boldsymbol{B}. \tag{T1.5}$$

Proof: We can prove this theorem using the Venn diagram shown in Fig. 1.21. In this diagram, the set $\boldsymbol{B}$ is represented by the area that is shaded with a horizontal line pattern and the set $\boldsymbol{A}'$ is represented by the area that is shaded with a vertical

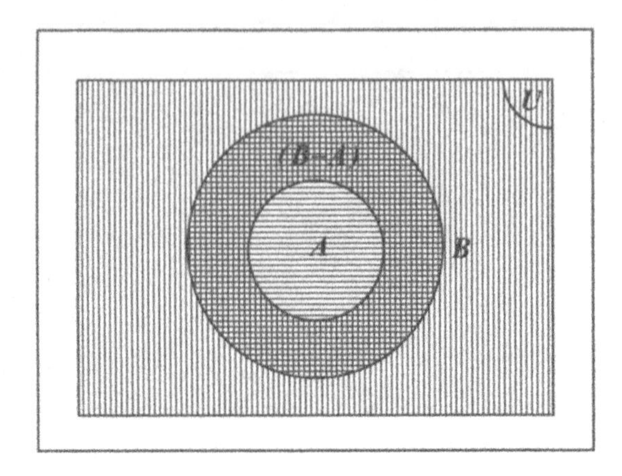

FIGURE 1.21. The graphical proof of the theorem: $\boldsymbol{A} \subseteq \boldsymbol{B} \Rightarrow \boldsymbol{A} \cup (\boldsymbol{B} - \boldsymbol{A}) = \boldsymbol{B}$.

line pattern. Thus, the set difference $(\boldsymbol{B} - \boldsymbol{A})$ is shaded by both vertical and horizontal line patterns. From this diagram, it is also apparent that $(\boldsymbol{B} - \boldsymbol{A}) = (\boldsymbol{B} \cap \boldsymbol{A}')$. □

Theorem 1.6 *If set* **A** *is a subset of set* **B**, *then the sets* $(\mathbf{B} - \mathbf{A})$ *and* $(\mathbf{B} \cap \mathbf{A}')$ *are equal, i.e.*

$$\boldsymbol{A} \subseteq \boldsymbol{B} \Rightarrow (\boldsymbol{B} - \boldsymbol{A}) = \left(\boldsymbol{B} \cap \boldsymbol{A}'\right). \tag{T1.6}$$

Proof: Refer to the proof of the Theorem (T1.5) and the Venn diagram shown in Fig. 1.21. □

1.13 The Algebra of Sets

Having examined the basic notions and definitions of sets in an intuitive and non-rigorous fashion, let us systematise what we have learned thus far by introducing a collection of rules (identities) for the algebra of sets. These rules can be used to provide rigorous mathematical proofs for any theorem of set theory.

1.13.1 The Rules of the Algebra of Sets

Given any universal set $\boldsymbol{U}$ and its subsets $\boldsymbol{A}$, $\boldsymbol{B}$, $\boldsymbol{C}$, taken in conjunction with the operations of union, intersection and complementation of sets, the following rules of the algebra of sets apply.

The Identity Rules:

$$\boldsymbol{A} \cup \emptyset = \boldsymbol{A} \tag{R1.1a}$$
$$\boldsymbol{A} \cap \boldsymbol{U} = \boldsymbol{A} \tag{R1.1b}$$
$$\boldsymbol{A} \cup \boldsymbol{U} = \boldsymbol{U} \tag{R1.2a}$$
$$\boldsymbol{A} \cap \emptyset = \emptyset \tag{R1.2b}$$

The Complement Rules:	$A \cup A' = U$	(R1.3a)
	$A \cap A' = \varnothing$	(R1.3b)
	$U' = \varnothing$	(R1.4a)
	$\varnothing' = U$	(R1.4b)
	$(A')' = A$	(R1.5)
The Idempotent Rules:	$A \cup A = A$	(R1.6a)
	$A \cap A = A$	(R1.6b)
The Commutative Rules:	$A \cup B = B \cup A$	(R1.7a)
	$A \cap B = B \cap A$	(R1.7b)
The Associative Rules:	$(A \cup B) \cup C = A \cup (B \cup C)$	(R1.8a)
	$(A \cap B) \cap C = A \cap (B \cap C)$	(R1.8b)
The Distributive Rules:	$A \cup (B \cap C) = (A \cup B) \cap (A \cup C)$	(R1.9a)
	$A \cap (B \cup C) = (A \cap B) \cup (A \cap C)$	(R1.9b)
The De Morgan Rules:	$(A \cup B)' = A' \cap B'$	(R1.10a)
	$(A \cap B)' = A' \cup B'$	(R1.10b)

In the above list of rules an *idempotent element* (or *idempotent* for short) is an entity that when operated upon by itself, yields (results in) itself.

Observe that these rules do not cover the concept of an element belonging to a set (i.e. $a \in A$). Also, that the concepts of the subset and proper subset are not covered by these rules.

In this algebra of sets, these concepts are defined as follows:

$$\text{Subset/Superset:} \quad A \subseteq B \vee B \supseteq A \Rightarrow A \cap B = A \tag{1.6}$$

$$\text{Proper Subset/Superset:} \quad A \subset B \vee B \supset A \Rightarrow A \cap B = A \wedge B - A \neq \varnothing \tag{1.7}$$

Let us now use the rules of the algebra of sets to prove some theorems.

Theorem 1.7 *Given any two sets* A *and* B*, we say that:*

$$A, B \subseteq U \Rightarrow (A \cap B) \cup (A \cap B') = A. \tag{T1.7}$$

Proof: Starting with the left-hand side of this equality and using the distributive rule (R1.9b) we get:

$$(A \cap B) \cup (A \cap B') = A \cap (B \cup B')$$

Using the complement rule (R1.3a) we get:

$$(A \cap B) \cup (A \cap B') = A \cap U$$

Finally, using the identity rule (R1.1b) we get:

$$(A \cap B) \cup (A \cap B') = A.$$

□

Theorem 1.8 *Given that* $A \subseteq B$ *and* $B \subseteq C$*, then* $A \subseteq C$*, i.e.*

$$A \subseteq B \wedge B \subseteq C \Rightarrow A \subseteq C. \tag{T1.8}$$

Proof: From the definition of a subset (1.6) we have:

$$A = A \cap B \tag{1.8}$$

and

$$B = B \cap C \tag{1.9}$$

By substituting (1.9) into the right-hand side of equality (1.8) we get:

$$A = A \cap (B \cap C)$$

Using the associative rule (R1.8b) we get:

$$A = (A \cap B) \cap C$$

Using Eq. (1.8) we get:

$$A = A \cap C$$

Which is the definition of $A \subseteq C$. □

1.13.2 The Duality Principle

In mathematics, a *dual* is a pair of identities or a grouping of two identities. As it applies to set theory, the *duality principle* can be expressed as follows. Given a set identity (i.e. an expression of set relationships) I, we can construct its dual identity I^* by interchanging each occurrence of the symbols $\cup$ and $\cap$, and the symbols $\emptyset$ and U in the original identity. For instance:

- The dual of the identity $A \cup (B \cap C) = (A \cup B) \cap (A \cup C)$ is the identity $A \cap (B \cup C) = (A \cap B) \cup (A \cap C)$.
- The dual of the identity $A \cup A' = U$ is the identity $A \cap A' = \emptyset$.
- The dual of the identity $(A \cap B) \cup (A \cap B') = A$ is the identity $(A \cup B) \cap (A \cup B') = A$.

This is precisely why, in the above table of set algebraic rules, we have arranged these rules in pairs (duals). The principle of duality implies that if we use a sequence of axioms or rules to prove the validity of an identity, then we can use the duals of these axioms and rules to prove the validity of the dual of the original identity. For instance in Theorem (T1.7) we proved that $A, B \subseteq U \Rightarrow (A \cap B) \cup (A \cap B') = A$. We can now employ the principle of duality to prove the dual of this identity.

Theorem 1.9 *Given any two sets A and B, we say that:*

$$A, B \subseteq U \Rightarrow (A \cup B) \cap (A \cup B') = A. \tag{T1.9}$$

Proof: Starting with the left-hand side of this equality and using the distributive rule (R1.9a), which is the dual of rule (R1.9b) that we have used in the first step of the proof of Theorem (T1.7), we get:

$$(A \cup B) \cap (A \cup B') = A \cup (B \cap B')$$

Using the complement rule (R1.3b), which is the dual of rule (R1.3a) that we have used in the second step of the proof of Theorem (T1.7), we get:

$$(\boldsymbol{A} \cup \boldsymbol{B}) \cap (\boldsymbol{A} \cup \boldsymbol{B}') = \boldsymbol{A} \cup \emptyset$$

Finally, using the identity rule (R1.1a), which is the dual of rule (R1.1b) that we have used in the third step of the proof of Theorem (T1.7), we get:

$$(\boldsymbol{A} \cup \boldsymbol{B}) \cap (\boldsymbol{A} \cup \boldsymbol{B}') = \boldsymbol{A}. \qquad \square$$

1.14 Numbers and Sets

1.14.1 Classes of Numbers

A number is an abstract entity that can be used to represent a quantity (i.e. a measurement, a count or an amount). There are many different classes of numbers and all numbers belonging to each class can be considered as elements belonging to a particular set. One of the earliest examples of these number classes is that of the *natural numbers*, which are represented by the set $\mathbb{N} = \{0, 1, 2, 3, \ldots\}$. Since the dawn of civilisation this class of numbers has been used for counting. The natural numbers $\{1, 2, 3, \ldots\}$ are sometimes referred to as *positive whole numbers*. If we augment the set of natural numbers by the set of *negative whole numbers*, we get the set of *integer numbers* which is represented as $\mathbb{Z} = \{\ldots, -3, -2, -1, 0, 1, 2, 3, \ldots\}$.

The ratio of integers, where the divisor is non-zero, is called a *rational number* or a *fraction*. The set of all rational numbers is represented as $\mathbb{Q} = \{x \mid x = a/b \wedge a, b \in \mathbb{Z} \wedge b \neq 0\}$. A characteristic of rational numbers is that their decimal expansion is periodic in nature.

If we augment the set of rational numbers with the set of all other numbers that have a non-periodic decimal expansion, we get the set of real numbers that are represented by the symbol $\mathbb{R}$. The set of all the real numbers that are non-rational (i.e. these numbers that can not be represented as the ratio of two integers, where the divisor is non-zero) are called *irrational numbers* and are represented by the symbol $\mathbb{Q}'$.

From the above we can observe that $\mathbb{N} \subset \mathbb{Z} \subset \mathbb{Q} \subset \mathbb{R}$. Thus, for the purposes of our discussion we will assume that the set of real numbers is our universal set, i.e. $\boldsymbol{U} = \mathbb{R}$.

In the discussion that follows we will make use of the concept of the closure of sets. So, let us begin by defining this concept.

1.14.2 Closure

Given a set $\boldsymbol{S}$, we say that this set is closed under a binary operation "$\times$", if and only if the result of this operation on two elements of the set is also an element of this set (i.e. ***iff*** $\forall\, s_1, s_2 \in \boldsymbol{S} \Rightarrow s_1 \times s_2 \in \boldsymbol{S}$). Thus, for instance, the set of natural numbers is closed under the binary operation of addition "+", since by adding two natural numbers together results in a natural number (i.e. $\forall s_1, s_2 \in \mathbb{N} \Rightarrow s_1 + s_2 \in \mathbb{N}$). Similarly, natural numbers are closed under the binary operation of multiplication "$*$", as the product of two natural numbers results in a natural number (i.e. $\forall s_1, s_2 \in \mathbb{N} \Rightarrow s_1 * s_2 \in \mathbb{N}$). Alternatively, natural numbers are not closed under the binary operation of subtraction "$-$", as the difference of two natural numbers may be a negative integer rather than a natural number (i.e. $\forall s_1, s_2 \in \mathbb{N} \not\Rightarrow /s_1 - s_2 \in \mathbb{N}$). Set $\mathbb{N}$ is also not closed under division, as the quotient of two naturals may be a real.

In general the closure of a set $\boldsymbol{S}$ is the set $C(\boldsymbol{S})$ which is the smallest superset of set $\boldsymbol{S}$.

1.14.3 The Set of Real Numbers $\mathbb{R}$

Intuitively the set of real numbers can be defined as the class of numbers whose members have a *one to one correspondence* with the set of points lying on a straight line. Loosely speaking, a one to one correspondence is a relationship between two sets which allows us to match each element of the first set with one and only one element of the second set and vice versa. Later we will define this relationship in a more rigorous way. Thus, each real number can be represented by one and only one point of the straight line and each point of the straight line represents only one real number. To do this mapping we select an arbitrary point on the line to represent the number zero. This point is called the *origin* of the *real number line*. Another point to the right of the origin is chosen to represent the number 1. Now that we have established the length of a unit on the real number line, we can represent all positive real numbers to the right of the origin and all negative numbers to the left of the origin, as shown in Fig. 1.22.

The set of real numbers is closed under the binary operations of addition, subtraction, multiplication and division.

1.14.4 The Set of Rational Numbers $\mathbb{Q}$

A rational number or fraction is a number that can be expressed as the ratio of two integers, where the divisor is non-zero. The set of rational numbers is represented as $\mathbb{Q} = \{x \mid x = a/b \wedge a, b \in \mathbb{Z} \wedge b \neq 0\}$. Each rational number can be written in an infinite number of ways, for instance: $\frac{1}{3} = \frac{2}{6} = \frac{3}{9} = \cdots = \frac{n}{3n}$ where n

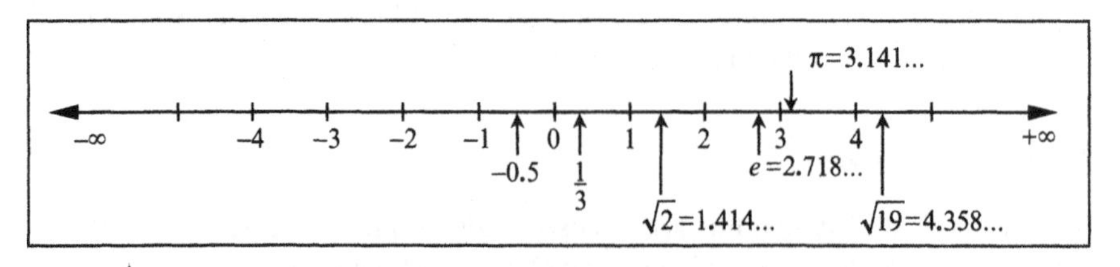

FIGURE 1.22. The real number line.

can be any integer not equal to zero. One important characteristic of any rational number is that its decimal expansion is periodic or recurring in nature. This means that its representation, following the decimal point, may start with a non-repeating sequence of digits but eventually continues with a sequence of digits that recur (repeat) ad infinitum. For instance:

- $\frac{1}{2} = 0.5\dot{0}$, where the final decimal digit recurs (i.e. $\frac{1}{2} = 0.5000\cdots$);
- $\frac{1}{3} = 0.\dot{3}$, where the final decimal digit recurs (i.e. $\frac{1}{3} = 0.333\cdots$);
- $\frac{1}{6} = 0.1\dot{6}$, where the final decimal digit recurs (i.e. $\frac{1}{6} = 0.1666\cdots$);
- $\frac{1}{7} = 0.\dot{1}4285\dot{7}$, where the sequence of decimal digits $\dot{1}4285\dot{7}$ recurs (i.e. $\frac{1}{7} = 0.142857142857142857\ldots$).

It is worth observing that, immaterial of the number base (decimal, binary, octal, etc.) that we choose to represent a rational number in, its expansion is always periodic in nature.

The set of rational numbers is closed under the binary operations of addition, subtraction, multiplication and division (except division by zero). The set of rational numbers is a proper subset of the set of real numbers, i.e. $\mathbb{Q} \subset \mathbb{R}$.

1.14.5 The Set of Irrational Numbers $\mathbb{Q}'$

An irrational number is any real number that is not a rational number. In other words any real number that cannot be expressed as the ratio of two integers, where the devisor is not zero. The decimal expansion of an irrational number is not periodic or recurring. The set of irrational numbers is defined as the complement of the set of rational numbers, when the universal set is taken to be the set of real numbers, i.e. $\mathbb{Q}' = (\mathbb{R} - \mathbb{Q})$.

We observe that $\boldsymbol{U} = \mathbb{R} \Leftrightarrow \mathbb{R} = \mathbb{Q} \cup \mathbb{Q}'$ and that $\mathbb{Q}' \subset \mathbb{R}$.

Examples of irrational numbers are $\sqrt{2}, \sqrt{3}, \sqrt[3]{5}, \pi = 3.14159\cdots$, $\mathrm{e} = 2.71828\cdots$.

1.14.6 The Set of Natural Numbers $\mathbb{N}$

As we have seen above, the set of natural numbers is defined as $\mathbb{N} = \{0, 1, 2, 3, \ldots\}$. This is the definition given by the DIN 5473 standard "Logic and Set Theory, Symbols and Concepts" published in 1992 by DIN (the Deutsches Institute für Normung-the German Institute for Standardisation). In some other literature the number zero is not included in the set of natural numbers. The natural numbers $\{1, 2, 3, \ldots\}$ are sometimes referred to as *positive whole numbers* or *cardinal numbers* and are defined as $\mathbb{N}^* = \mathbb{N} - \{0\} = \{1, 2, 3, \ldots\}$.

The set of natural numbers is closed under the binary operations of addition and multiplication but not subtraction and division, as the difference of two natural numbers may be a negative integer and the quotient of two natural numbers may be a real number.

We observe that $\mathbb{N} \subset \mathbb{R}$.

1.14.7 The Set of Integer Numbers $\mathbb{Z}$

The set of integers is the union of the set of natural numbers $\mathbb{N} = \{0, 1, 2, 3, \ldots\}$ and the set of negative whole numbers $\{-1, -2, -3, \ldots\}$. Thus, the set of integers is defined as $\mathbb{Z} = \{\ldots, -3, -2, -1, 0, 1, 2, 3, \ldots\}$ or $\mathbb{Z} = \{0, \pm 1, \pm 2, \pm 3, \ldots\}$. The name of this set originates from the German word Zahlen (meaning numbers).

The set of integers is closed under the binary operations of addition, subtraction and multiplication but not division, as the quotient of two integers may be a real number.

We observe that $\mathbb{N} \subset \mathbb{Z} \subset \mathbb{R}$.

1.14.8 Other useful Sets of Numbers

1.14.8.1 The Set of Complex Numbers $\mathbb{C}$

A superset of the set of real numbers $\mathbb{R}$ is the set of *complex numbers* $\mathbb{C}$ in which all non-constant polynomials have real roots. Complex numbers have wide application in many field of mathematics, such as: differential equations, control theory, quantum mechanics, relativity, fluid dynamics, fractals, etc. Complex numbers are defined as $\mathbb{C} = \left\{x \mid x = a + bi \wedge a, b \in \mathbb{R} \wedge i = \sqrt{-1}\right\}$. Thus, a complex number x consists of a *real part a* and an *imaginary part bi* and requires the *imaginary unit* $i = \sqrt{-1}$ (i.e. $i^2 = -1$) for its definition.

We observe that $\mathbb{R} \subset \mathbb{C}$.

1.14.8.2 The Set of Algebraic Numbers $\mathbb{A}$

An *algebraic number* is any real number that is the solution of a polynomial equation with integer coefficients of the form:

$$c_n x^n + c_{n-1} x^{n-1} + c_{n-2} x^{n-2} + \cdots + c_2 x^2 + c_1 x^1 + c_0 x^0 = 0 \quad \text{for} \quad n > 0$$

where c_i are integer coefficients and $c_n \neq 0$.

From the definition of rational numbers we know that any rational number is of the form $x = a/b$. Thus, any rational number can be expressed in the form $bx - a = 0$ (with the root $x = a/b$), which in turn means that every rational number is an algebraic number. Some irrational numbers are also algebraic numbers. For instance, the irrational number $\sqrt{2}$ can also be expressed as a polynomial with integer coefficients, $1x^2 - 2 = 0$, with the root $x = \sqrt{2}$. Thus, $\sqrt{2}$ is both irrational and algebraic.

1.14.8.3 The Set of Transcendental Numbers $\mathbb{T}$

A *transcendental number* is any irrational number that is not the solution of any polynomial equation with integer coefficients. The French mathematician Joseph Liouville (1809-1882) first established the existence of transcendental numbers in an 1851 paper. Examples of transcendental numbers are $\pi = 3.14159\cdots$, $e = 2.71828\cdots$, $\sin(\pi)$, $\cos(2\pi)$, and $\tan\left(\frac{\pi}{4}\right)$.

We observe that the set of real numbers is the union of the algebraic and transcendental numbers, i.e. $\mathbb{R} = \mathbb{A} \cup \mathbb{T}$.

1.14.9 Ordering Relations or Inequalities

An inequality is an assertion about the relative size or order of two quantities, i.e. a statement of their *relation*. Given two real numbers a and b, they may have the following ordering relations:

- $a = b$ which signifies that a is equal to b (i.e. that the numbers a and b correspond to the same point on the real number line);
- $a < b$ which signifies that a is less than b (i.e. that number a corresponds to a point, on the real line, that lies to the left of the point corresponding to the number b);
- $a \leq b$ which signifies that a is less than or equal to b (i.e. that number a corresponds to a point, on the real line, that lies on or to the left of the point corresponding to the number b);
- $a > b$ which signifies that a is greater than b (i.e. that number a corresponds to a point, on the real line, that lies to the right of the point corresponding to the number b);
- $a \geq b$ which signifies that a is greater than or equal to b (i.e. that number a corresponds to a point, on the real line, that lies on or to the right of the point corresponding to the number b).

The inequalities of real numbers exhibit the following properties:

The Reflexivity Property
Given any real number a, $a \leq a$.

The Trichotomy Property
Given any two real numbers a and b, only one of the following relations is true: $a < b$, $a = b$ or $a > b$.

The Antisymmetry Property
Given any two real numbers a and b, then $a \leq b \wedge b \leq a \Rightarrow a = b$.

The Transitivity Property
Given any three real numbers a, b and c, then $a \leq b \wedge b \leq c \Rightarrow a \leq c$.

The Addition and Subtraction Property
Given the inequalities $a \leq b$ and $a \geq b$ (between two real numbers a and b), we may add or subtract to or from both sides of these inequalities the same real number c without changing the sense of the inequalities, i.e.

- $a \leq b \Rightarrow a + c \leq b + c$
- $a \leq b \Rightarrow a - c \leq b - c$
- $a \geq b \Rightarrow a + c \geq b + c$
- $a \geq b \Rightarrow a - c \geq b - c$

The Multiplication and Division Property

Given the inequalities $a \leq b$ and $a \geq b$ (between two real numbers a and b), we may multiply or divide both sides of these inequalities by the same non-negative real number c without changing the sense of the inequalities. If, however, we multiply or divide by a negative real number, the sense of the inequalities will be reversed. Thus:

- $a \leq b \wedge c \geq 0 \Rightarrow a \cdot c \leq b \cdot c$
- $a \leq b \wedge c \geq 0 \Rightarrow a/c \leq b/c$
- $a \leq b \wedge c < 0 \Rightarrow a \cdot c \geq b \cdot c$
- $a \leq b \wedge c < 0 \Rightarrow a/c \geq b/c$
- $a \geq b \wedge c \geq 0 \Rightarrow a \cdot c \geq b \cdot c$
- $a \geq b \wedge c \geq 0 \Rightarrow a/c \geq b/c$
- $a \geq b \wedge c < 0 \Rightarrow a \cdot c \leq b \cdot c$
- $a \geq b \wedge c < 0 \Rightarrow a/c \leq b/c$

Examples

Consider the following examples of inequalities:

- $x < 6$ which signifies that the value of x is less than 6 (i.e. that number x corresponds to a point, on the real line, that lies to the left of the point corresponding to the number 6);
- $1 < x < 6$ which signifies that the value of x is greater than 1 and less than 6 (i.e. that number x corresponds to a point, on the real line, that lies to the right of the point corresponding to the number 1 and to the left of the point corresponding to the number 6);
- $1 \leq x \leq 6$ which signifies that the value of x is greater than or equal to 1 and less than or equal to 6 (i.e. that number x corresponds to a point, on the real line, that lies on or to the right of the point corresponding to the number 1 and on or to the left of the point corresponding to the number 6).

1.14.10 The Absolute Value or Modulus of a Number

The *absolute value* or *modulus* of a real number x represents its size or magnitude and is defined as:

$$|x| = \begin{cases} x, & x \geq 0 \\ -x, & \textbf{\textit{otherwise}} \end{cases}$$

We observe that $|x| \geq 0$ for all $x \in \mathbb{R}$. In geometric terms, $|x_i|$ represents the distance between the point corresponding to the value of x_i on the real number line and the origin (i.e. the point corresponding to the number 0). See Fig. 1.23.

Given two real numbers x_1 and x_2 the distance between the two points, on the real number line, corresponding to these numbers is given by $|x_2 - x_1| = |x_1 - x_2|$. See Fig. 1.24.

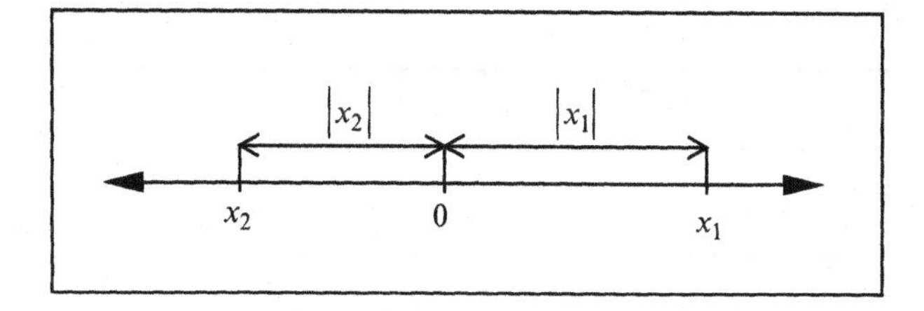

FIGURE 1.23. The absolute value of x_i.

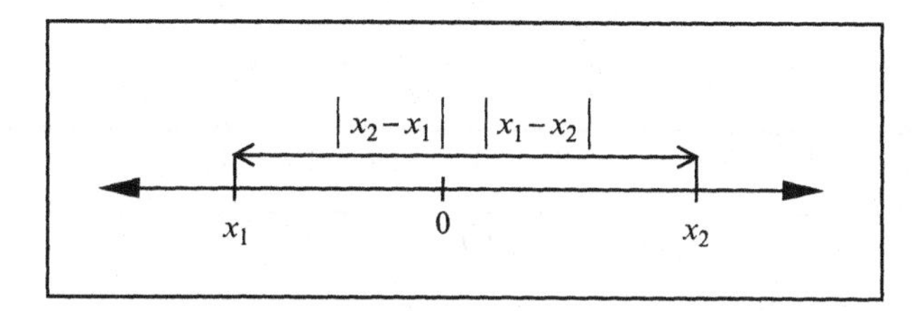

FIGURE 1.24. The distance between points x_1 and x_2.

Consider the following inequalities (illustrated in Figs. 1.25a–1.25d, respectively):

- $|x| \leq 3$ which means $-3 < x < 3$ and signifies that the value of x is greater than -3 and less than $+3$ (i.e. the number x corresponds to a point, on the real line, that lies to the right of the point corresponding to the number -3 and to the left of the point corresponding to the number $+3$);
- $|x| \leq 3$ which means $-3 < x < 3$ and signifies that the value of x is greater than or equal to -3 and less than or equal to $+3$ (i.e. the number x corresponds to a point, on the real line, that lies on or to the right of the point corresponding to the number -3 and on or to the left of the point corresponding to the number $+3$);
- $|x| > 3$ which means $x < -3 \vee x > 3$ and signifies that the value of x is either less than -3 or greater than $+3$ (i.e. the number x corresponds to a point, on the real line, that lies either to the left of the point corresponding to the number -3 or to the right of the point corresponding to the number $+3$);
- $|x| \geq 3$ which means $x \leq -3 \vee x \geq 3$ and signifies that the value of x is either less than or equal to -3, or greater than or equal to $+3$ (i.e. the number x corresponds to a point, on the real line, that lies either on or to the left of the point corresponding to the number -3, or alternatively, on or to the right of the point corresponding to the number $+3$).

Figures 1.25a-1.25d depict the graphical representation of the inequalities $|x| < 3$, $|x| \leq 3$ and $|x| \geq 3$, respectively. The darker parts of the real number line represent the points that correspond to the range of real values that satisfy these inequalities. The black and white circles indicate that the real value corresponding to that point is respectively included or excluded from the range of values that satisfy the given inequality.

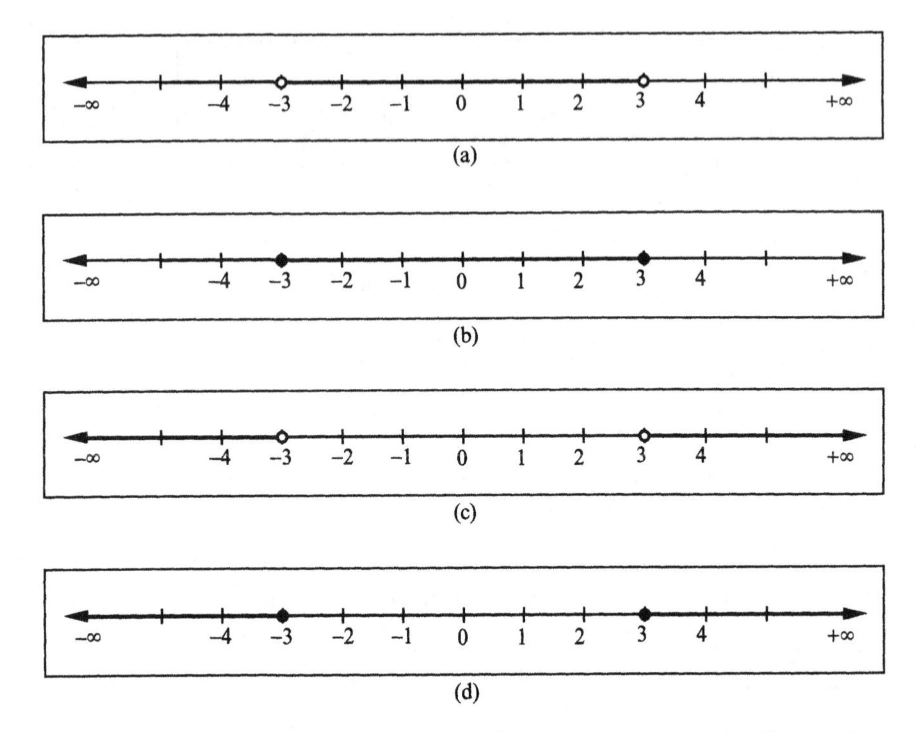

FIGURE 1.25 (a) The graphical representation of the inequality $|x| < 3$. (b) The graphical representation of the inequality $|x| \leq 3$. (c) The graphical representation of the inequality $|x| > 3$. (d) The graphical representation of the inequality $|x| \geq 3$.

1.14.11 Real Number Intervals

A real number *interval* is a subset of the real number set $\mathbb{R}$ that is bounded (limited) by two endpoints on the real number line which correspond to the real values a and b, where $a \leq b$. These two values are known as the *lower bound* or *lower limit* and the *upper bound* or *upper limit* of the interval.

Real number intervals can be *finite* or *infinite* intervals. In infinite intervals either or both of their bounds a and b can be replaced by $-\infty$ and $+\infty$, respectively. Depending on whether the values of the bounds a and b are included in or excluded from the interval, we characterise the interval either as being an *open interval* or a *closed interval*. If the value of the lower bound is included in the interval but the value of the upper bound is excluded, then the interval is known as a *closed-open interval*. While, if the value of the lower bound is excluded from the interval but the value of the upper bound is included, then the interval is known as an *open-closed interval*. Closed-open and open-closed intervals are sometimes referred to as *half-open intervals*.

Given $a, b \in \mathbb{R}$ (where $a \leq b$) we may define the following intervals:

- $I_1 = (a, b) = \{x \mid a < x < b\}$, which is an open interval depicted in Fig. 1.26a;
- $I_2 = [a, b] = \{x \mid a \leq x \leq b\}$, which is a closed interval depicted in Fig. 1.26b;

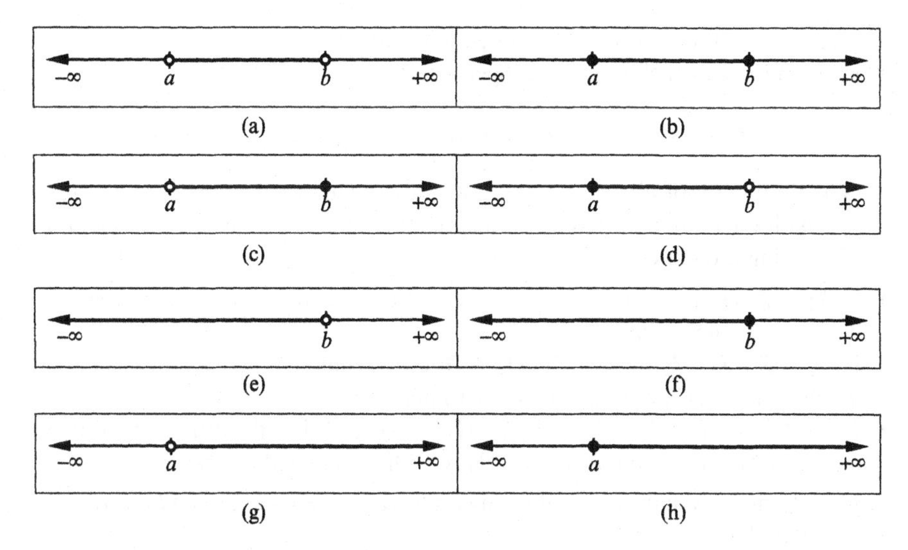

FIGURE 1.26 (a) The interval I_1. (b) The interval I_2. (c) The interval I_3. (d) The interval I_4. (e) The interval I_5. (f) The interval I_6. (g) The interval I_7. (h) The interval I_8.

- $I_3 = (a, b] = \{x \mid a < x \leq b\}$, which is a open-closed interval depicted in Fig. 1.26c;
- $I_4 = [a, b) = \{x \mid a \leq x < b\}$, which is a closed-open interval depicted in Fig. 1.26d;
- $I_5 = (-\infty, b) = \{x \mid x < b\}$, which is an open infinite interval depicted in Fig. 1.26e;
- $I_6 = (-\infty, b] = \{x \mid x \leq b\}$, which is an open-closed infinite interval depicted in Fig. 1.26f;
- $I_7 = (a, +\infty) = \{x \mid a < x\}$, which is an open infinite interval depicted in Fig. 1.26g;
- $I_8 = [a, +\infty) = \{x \mid a \leq x\}$, which is an closed-open infinite interval depicted in Fig. 1.26h;
- $I_9 = (-\infty, +\infty) = \{x \mid x \in \mathbb{R}\}$, which is an open infinite interval equal to $\mathbb{R}$.

There are also two degenerate intervals:

- $I_{10} = \{a\}$, which consists only of one value and is known as a *singleton*;
- $I_{11} = \emptyset$, which is a *zero-length interval.*

Intervals $I_1, I_2, I_3, I_4, I_{10}$ and I_{11} are known as *bounded intervals*, while intervals I_5, I_6, I_7, I_8 and I_9 are known as *unbounded intervals*.

In some European literature a slightly different notation is used to represent open and closed intervals. Instead of using the round brackets "(" and ")" to indicate the open ends of an interval we use square brackets pointing outwards, i.e. we use the symbol "]" instead of "(" and the symbol "[" instead of the symbol ")". Thus, the following intervals are equivalent in the two representations:

- $]a, b[\equiv (a, b)$, which is an open interval;
- $[a, b] \equiv [a, b]$, which is a closed interval;

- $]a, b[\equiv (a, b]$, which is a open-closed interval;
- $[a, b[\equiv [a, b)$, which is a closed-open interval.

1.14.12 Properties of Real Number Intervals

If $\mathfrak{F}$ is the family of all real number intervals (including the singleton and the zero-length interval), then given two real number intervals $\boldsymbol{I}_1$ and $\boldsymbol{I}_2$ they exhibit the following properties:

- $\boldsymbol{I}_1, \boldsymbol{I}_2 \in \mathfrak{F} \Rightarrow \boldsymbol{I}_1 \cap \boldsymbol{I}_2 \in \mathfrak{F}$, i.e. the intersection of two real number intervals is itself a real number interval;
- $\boldsymbol{I}_1, \boldsymbol{I}_2 \in \mathfrak{F} \wedge \boldsymbol{I}_1 \cap \boldsymbol{I}_2 \neq \emptyset \Rightarrow \boldsymbol{I}_1 \cup \boldsymbol{I}_2 \in \mathfrak{F}$, i.e. the union of two non-disjoint real number intervals is itself a real number interval;
- $\boldsymbol{I}_1, \boldsymbol{I}_2 \in \mathfrak{F} \wedge \boldsymbol{I}_1 \not\subset \boldsymbol{I}_2 \wedge \boldsymbol{I}_2 \not\subset \boldsymbol{I}_1 \Rightarrow (\boldsymbol{I}_1 - \boldsymbol{I}_2) \in \mathfrak{F}$, i.e. the difference of two non-comparable real number intervals is itself a real number interval.

For instance, given the two real number intervals $\boldsymbol{I}_1 = [-3,\ 3)$ and $\boldsymbol{I}_2 = (0,\ 5]$ we observe that:

- $\boldsymbol{I}_1 \cup \boldsymbol{I}_2 = [-3,\ 5]$, i.e. the interval of all real numbers that belong to either of the two intervals $\boldsymbol{I}_1$ and $\boldsymbol{I}_2$.
- $\boldsymbol{I}_1 \cap \boldsymbol{I}_2 = (0,\ 3)$, i.e. the interval of all real numbers that belong to both intervals $\boldsymbol{I}_1$ and $\boldsymbol{I}_2$.
- $\boldsymbol{I}_1 - \boldsymbol{I}_2 = [-3,\ 0]$, i.e. the interval of all real numbers that belong to $\boldsymbol{I}_1$ but not to $\boldsymbol{I}_2$.
- $\boldsymbol{I}_2 - \boldsymbol{I}_1 = [3,\ 5]$, i.e. the interval of all real numbers that belong to $\boldsymbol{I}_2$ but not to $\boldsymbol{I}_1$.

See Fig. 1.27.

1.14.13 Real Number Interval Arithmetic

When using a computer to perform arithmetic computations we can not represent real numbers accurately. Thus the results of any such computations are only approximate and susceptible to rounding errors. Occasionally it is more important to know the range of the result, of a computation, rather than the exact result itself. On such occasions we can use *interval arithmetic* to compute the range of

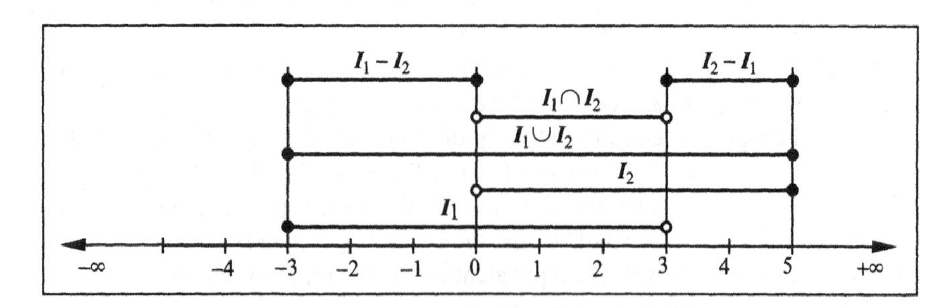

FIGURE 1.27. The graphical representation of interval set operations.

the result. Given two closed real number intervals $\boldsymbol{A} = [a_L, a_U]$ and $\boldsymbol{B} = [b_L, b_U]$, the operations of addition, subtraction, multiplication and division of real number intervals are defined as follows:

$$C = [a_L, a_U] + [b_L, b_U] = [a_L + b_L, a_U + b_U] \tag{1.10a}$$

$$C = [a_L, a_U] - [b_L, b_U] = [a_L - b_U, a_U - b_L] \tag{1.10b}$$

$$C = [a_L, a_U] \cdot [b_L, b_U] = [\min(a_L \cdot b_L, a_L \cdot b_U, a_U \cdot b_L, a_U \cdot b_U), \max(a_L \cdot b_L, a_L \cdot b_U, a_U \cdot b_L, a_U \cdot b_U)] \tag{1.10c}$$

$$C = [a_L, a_U] / [b_L, b_U] = \left[\min\left(a_L/b_L, a_L/b_U, a_U/b_L, a_U/b_U\right), \max(a_L/b_L, a_L/b_U, a_U/b_L, a_U/b_U)\right] \tag{1.10d}$$

For example, if we know that a real number a lies in the interval $\boldsymbol{A} = [-2, 2]$ and that the real number b lies in the interval $\boldsymbol{B} = [2, 3]$, then we can say with certainty that the result of their addition, subtraction, multiplication and division lie in the following intervals, respectively:

- $(a + b) \in [-2 + 2, 2 + 3] = [0, 5]$
- $(a - b) \in [-2 + 2, 2 - 3] = [-4, -2]$
- $(a \cdot b) \in [\min(-2 \cdot 2, -2 \cdot 3, 2 \cdot 2, 2 \cdot 3), \max(-2 \cdot 2, -2 \cdot 3, 2 \cdot 2, 2 \cdot 3)] = [\min(-4, -6, 4, 6), \max(-4, -6, 4, 6)] = [-6, 6]$
- $(a/b) \in \left[\min\left(\frac{-2}{2}, \frac{-2}{3}, \frac{2}{2}, \frac{2}{3}\right), \max\left(\frac{-2}{2}, \frac{-2}{3}, \frac{2}{2}, \frac{2}{3}\right)\right] = \left[\min\left(-1, -\frac{2}{3}, 1, \frac{2}{3}\right), \max\left(-1, -\frac{2}{3}, 1, \frac{2}{3}\right)\right] = [-1, 1]$

1.14.14 Bounded and Unbounded Real Number Sets

Roughly speaking, a *bounded set* is a set that has a finite size and it is thus a subset of a finite set. A set that is not bounded is called an *unbounded set.*

More precisely given a set of real numbers $\boldsymbol{S}$, if there exists a real number s_U such that $s \leq s_U$ for every $s \in \boldsymbol{S}$, then we call $\boldsymbol{S}$ an *upper bounded set* and we say that s_U is the *upper bound* of $\boldsymbol{S}$. Also if there exists a real number s_L such that $s_L \leq s$ for every $s \in \mathbf{S}$, then we call $\boldsymbol{S}$ a *lower bounded set* and we say that s_L is the *lower bound* of $\boldsymbol{S}$. If the set $\boldsymbol{S}$ is both upper and lower bounded (i.e. ***iff*** $s_L \leq s \leq s_U \forall s \in S$), then we call $\boldsymbol{S}$ a bounded set. Thus, in general a set is bounded if it is a subset of a finite interval, i.e. $\boldsymbol{S}$ is bounded ***iff*** $\boldsymbol{S} \subseteq [s_L, s_U]$.

Alternatively, we may say that a real number set $\boldsymbol{S}$ is bounded ***iff*** $|s| \leq b \forall b \geq 0 \wedge b, s \in S$, which means that $\boldsymbol{S} \subseteq [-b, b]$.

Consider the following examples:

- If $\boldsymbol{S}_1 = \left\{1, \frac{1}{2}, \frac{1}{4}, \frac{1}{8}, \ldots\right\}$, then $\boldsymbol{S}_1$ is bounded as $\boldsymbol{S}_1 \subseteq [0, 1]$.
- If $\boldsymbol{S}_2 = \{1, 3, 5, 7, \ldots\}$, then $\boldsymbol{S}_2$ is unbounded as we can not find a positive number b such that $|s| \leq b \; \forall \; s \in S_2$.
- If $\boldsymbol{S}_3 = \{33, -2, 56, 77, -22\}$, then $\boldsymbol{S}_3$ is bounded as $\boldsymbol{S}_3 \subseteq [-22, 77]$.

We observe that if a real number set is finite, then its is bounded (as is the case of set $\boldsymbol{S}_3$ above) and if it is infinite, then it may be bounded (as is the case of set $\boldsymbol{S}_1$ above) or it may be unbounded (as is the case of set $\boldsymbol{S}_2$ above).

1.15 Ordered Pairs and Ordered n-tuples

An *ordered pair* is a collection of two objects such that we can distinguish one object as being the *first element* of the ordered pair and the other as being its *second element*. An ordered pair whose first element is a and second element is b is denoted as (a, b). An ordered pair is sometimes referred to as an *ordered 2-tuple*.

A collection of three objects such that we can distinguish its first, second and third elements is called an *ordered triple*, an *ordered triplet* or an *ordered 3-tuple* and it is denoted by (a, b, c), where $a \in \boldsymbol{A}$, $b \in \boldsymbol{B}$ and $c \in \boldsymbol{C}$. An ordered triple can be defined recursively from the definition of an ordered pair, if we allow the second element of the ordered pair to be an ordered pair itself, i.e. the ordered pair $(a, (b, c))$ gives rise to the ordered triple (a, b, c).

In general, a collection of n ordered elements is called an *ordered n-tuple*. There are two main properties that distinguish an ordered n-tuple from a set with n elements. Unlike a set:

- An ordered n-tuple can contain an object more than once.
- The objects in an ordered n-tuple appear in a certain order.

Thus, two n-tuples are equal ***iff*** their corresponding elements are equal, i.e.

$$(a_1, a_2, \ldots, a_n) = (b_1, b_2, \ldots, b_n) \Leftrightarrow a_1 = b_1, a_2 = b_2, \ldots, a_n = b_n \quad (1.11)$$

Consider the following examples of ordered pairs:

- The ordered pairs $(1, 5)$ and $(5, 1)$ are not equal, as the order of their elements is not the same.
- The set $\{-1, 3\}$ is not an ordered pair as its elements are not ordered, i.e. $\{-1, 3\} = \{3, -1\}$.
- $(-1, -1)$ and $(3, 3)$ are valid ordered pairs, as an ordered pair may have duplicate elements.
- The Cartesian coordinates of point $\boldsymbol{P}$ in Fig. 1.28 form the ordered pair $(2, 3)$.

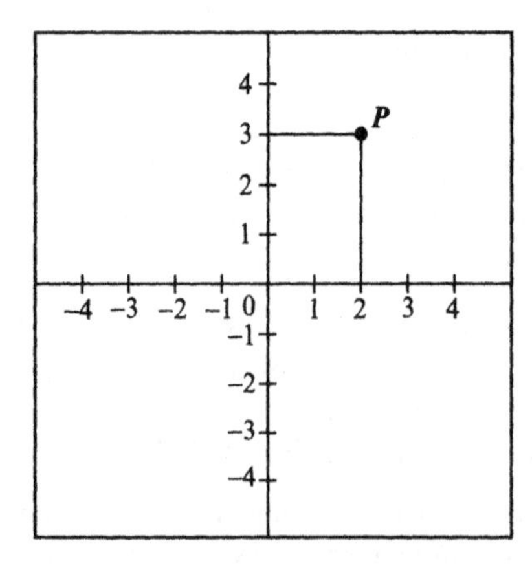

FIGURE 1.28. The Cartesian coordinates of a point P form an ordered pair(2,3).

1.16 The Cartesian Product of Sets

The set of all ordered pairs (a, b), where $a \in \boldsymbol{A}$ and $b \in \boldsymbol{B}$, is called the *Cartesian product* of sets $\boldsymbol{A}$ and $\boldsymbol{B}$ and is denoted as $\boldsymbol{A} \times \boldsymbol{B}$, which is read as $\boldsymbol{A}$ *cross* $\boldsymbol{B}$. The formal definition of the Cartesian product of two sets $\boldsymbol{A}$ and $\boldsymbol{B}$ is given by:

$$\boldsymbol{A} \times \boldsymbol{B} = \{(a, b) \mid a \in \boldsymbol{A} \wedge b \in \boldsymbol{B}\} \tag{1.12}$$

Consider the following examples:

- If $\boldsymbol{X} = \{x_1, x_2, x_3\}$ and $Y = \{y_1, y_2\}$, then their Cartesian product is given by: $\boldsymbol{X} \times \boldsymbol{Y} = \{(x_1, y_1), (x_1, y_2), (x_2, y_1), (x_2, y_2), (x_3, y_1), (x_3, y_2)\}$.
- The Cartesian plane, shown in Fig. 1.28, is the set of all ordered pairs that are elements of the Cartesian product $\mathbb{R}^2 = \mathbb{R} \times \mathbb{R}$.

  We observe that:

- If set $\boldsymbol{A}$ has n elements and set $\boldsymbol{B}$ has m elements, then their Cartesian product $\boldsymbol{A} \times \boldsymbol{B}$ has $n \times m$ elements.
- If set $\boldsymbol{A}$ or set $\boldsymbol{B}$ are empty, then their Cartesian product is the empty set, i.e. $\boldsymbol{A} = \emptyset \vee \boldsymbol{B} = \emptyset \Rightarrow \boldsymbol{A} \times \boldsymbol{B} = \emptyset$.
- If one of two sets is infinite and the other is not empty, then their Cartesian product is an infinite set.
- In general, the product of two sets $\boldsymbol{A}$ and $\boldsymbol{B}$ is not commutative, i.e. $\boldsymbol{A} \times \boldsymbol{B} \neq \boldsymbol{B} \times \boldsymbol{A}$.

The binary Cartesian product, defined in Eq. (1.12), can be extended to the n-ary Cartesian product as follows:

$$\prod_{i=1}^{n} \boldsymbol{S}_i = \boldsymbol{S}_1 \times \boldsymbol{S}_2 \times \cdots \times \boldsymbol{S}_n = \{(s_1, s_2, \cdots, s_n) \mid s_1 \in \boldsymbol{S}_1 \wedge s_2 \in \boldsymbol{S}_2 \wedge \cdots \wedge s_n \in \boldsymbol{S}_n\} \tag{1.13}$$

An example of this type of product is the three-dimensional Cartesian space, which is defined as the Cartesian product $\mathbb{R}^3 = \mathbb{R} \times \mathbb{R} \times \mathbb{R}$.

The Cartesian product is named after the French mathematician and philosopher René Descartes (1596–1650) who formulated the principles of analytical geometry and first used the product $\mathbb{R} \times \mathbb{R}$.

The Cartesian product $\boldsymbol{A} \times \boldsymbol{B}$ of two sets $\boldsymbol{A}$ and $\boldsymbol{B}$ can be represented graphically by a *coordinate diagram*. In such a diagram the horizontal and vertical axes represent the first and second sets of the Cartesian product. Subdivisions along these axes represent the respective elements of these sets and the points of intersection of vertical and horizontal lines, corresponding to these set elements, represent the ordered pairs that define the Cartesian product.

For instance, given a set $\boldsymbol{A} = \{a_1, a_2, a_3, a_4\}$ and a set $\boldsymbol{B} = \{b_1, b_2, b_3\}$ the coordinate diagram representing the Cartesian product $\boldsymbol{A} \times \boldsymbol{B}$ is shown in Fig. 1.29. In this diagram, the intersection point $\boldsymbol{P}$ represents the ordered pair (a_2, b_3). Altogether there are twelve intersection points representing the ordered pairs that define the Cartesian product $\boldsymbol{A} \times \boldsymbol{B} = \{(a_1, b_1), (a_1, b_2), (a_1, b_3), (a_2, b_1), (a_2, b_2), (a_2, b_3), (a_3, b_1), (a_3, b_2), (a_3, b_3), (a_4, b_1), (a_4, b_2), (a_4, b_3)\}$.

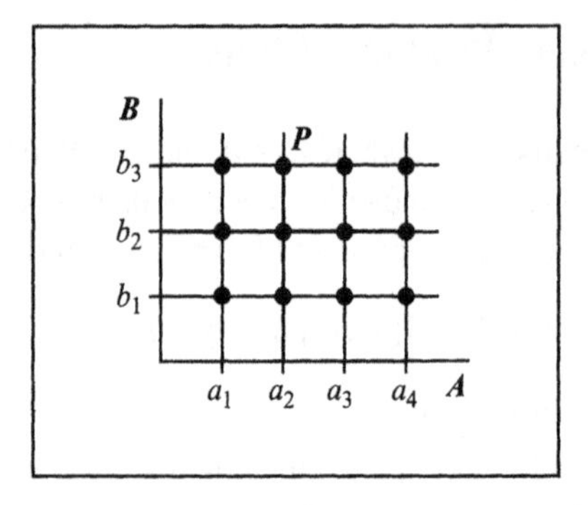

FIGURE 1.29. The coordinate diagram of the Cartesian product $\boldsymbol{A} \times \boldsymbol{B}$.

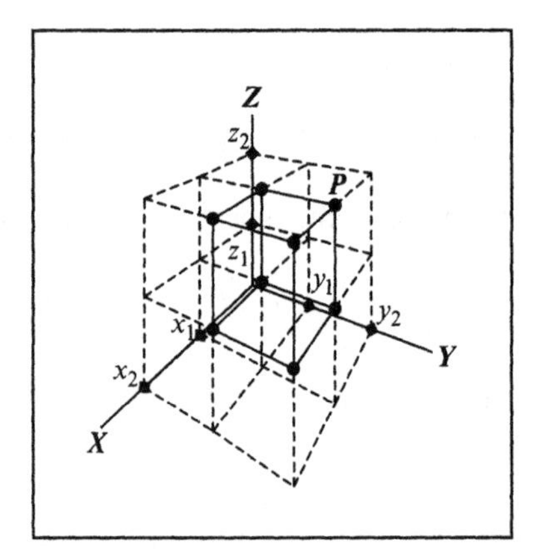

FIGURE 1.30. The coordinate diagram of the Cartesian product $\boldsymbol{X} \times \boldsymbol{Y} \times \boldsymbol{Z}$.

Similarly, given the three sets $\boldsymbol{X} = \{x_1, x_2\}, \boldsymbol{Y} = \{y_1, y_2\}$ and $\boldsymbol{Z} = \{z_1, z_2\}$ the coordinate diagram of their Cartesian product $\boldsymbol{X} \times \boldsymbol{Y} \times \boldsymbol{Z} = \{(x_1, y_1, z_1), (x_1, y_1, z_2), (x_1, y_2, z_1), (x_1, y_2, z_2), (x_2, y_1, z_1), (x_2, y_1, z_2), (x_2, y_2, z_1), (x_2, y_2, z_2)\}$ is depicted in Fig. 1.30. In this diagram, the intersection point $\boldsymbol{P}$ represents the ordered triplet (x_1, y_2, z_2).

1.17 Functions

Roughly speaking, a *function* is a relation that allows us to associate each element of an *input set* $\boldsymbol{I}$ with a unique element of an *output set* $\boldsymbol{O}$. The output set may be the same set as the input set. Thus, a function is a *deterministic procedure* or *rule* that allows us to assign a unique output to each given input. For example, given the set of members of the popular sixties group "The Beatles" $\boldsymbol{I} = \{$*John, Paul,*

George, Ringo}, the set of instruments $\boldsymbol{O}$ = {*lead guitar, base guitar, accompanying guitar, drums*} and the function f = *person i plays instrument o* (where $i \in \boldsymbol{I}$ and $o \in \boldsymbol{O}$), it is easy to see that we can use the function f to assign a unique musical instrument to each member of the group. Thus, we say that:

- $f(i_1) = o_1$, i.e. "John plays the lead guitar";
- $f(i_2) = o_2$, i.e. "Paul plays the base guitar";
- $f(i_3) = o_3$, i.e. "George plays the accompanying guitar";
- $f(i_4) = o_4$, i.e. "Ringo plays the drums".

The term function was first introduced in the late seventeenth century by the German mathematician Gottfried Leibniz (1646–1716) and it was later refined by the Swiss mathematician Leonhard Euler (1707–1783) and the German mathematician Karl Weierstrass (1815–1897). The German mathematician Peter Dirichlet (1805–1859) introduced the formal definition of functions.

1.17.1 The Formal Definition of a Function

More formally, a function can be defined as follows. Suppose that to each element i_i of an input set $\boldsymbol{I}$ we can assign a unique element o_j of an output set $\boldsymbol{O}$. We call such a relation between the sets $\boldsymbol{I}$ and $\boldsymbol{O}$ a function f. This relationship is denoted as $f : \boldsymbol{I} \to \boldsymbol{O}$, which is read as *$f$ is a function of $\boldsymbol{I}$ into $\boldsymbol{O}$* or *f maps $\boldsymbol{I}$ into $\boldsymbol{O}$*. The input set $\boldsymbol{I}$ is called the *domain of the function* and the output set $\boldsymbol{O}$ is called the *co-domain of the function.*

If i is an element of set $\boldsymbol{I}$, then the element o of set $\boldsymbol{O}$ (which is assigned to it) is known as the *image* of element i and is denoted as $f(i) = o$. Also, element i is known as the *preimage* of element o. For the relationship f to be a well-defined function, it must satisfy the following two constraints:

- f must be *total*, i.e. for all elements $i \in \boldsymbol{I}$ there must exist an $o \in \boldsymbol{O}$ such that $f(i) = o$.
- Many input values may be assigned to one output value, but only one output value may be assigned to each input value, i.e. $f(i_i) = o_i \wedge f(i_i) = o_j \Rightarrow o_i = o_j$.

The above definition can be expressed, in a more concise form, as follows. A function f from an input set $\boldsymbol{I}$ into an output set $\boldsymbol{O}$ is a subset of the Cartesian product $\boldsymbol{I} \times \boldsymbol{O}$, where for each element i in set $\boldsymbol{I}$ there exists a unique image o in set $\boldsymbol{O}$ such that the ordered pair (i, o) is in the function f.

The domain of a function $f : \boldsymbol{I} \to \boldsymbol{O}$ can be *restricted* to a subset $\boldsymbol{S} \in \boldsymbol{I}$, in which case the *restricted function* is denoted as $f \mid \boldsymbol{S} : \boldsymbol{S} \to \boldsymbol{O}$.

Consider the following examples:

- The function f maps every real number x to its cube, i.e. $f(x) = x^3$. Here, the real number set is both the domain and the co-domain of the function f, i.e. $f : \mathbb{R} \to \mathbb{R}$. The image of -2 is -8 (i.e. $f : -2 \to -8$) and the image of 2 is 8 (i.e. $f : 2 \to 8$). See Fig. 1.31.

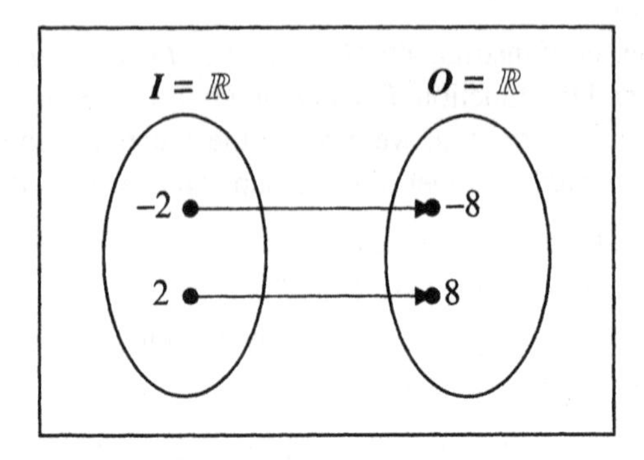

FIGURE 1.31. The mapping of the function $f(x) = x^3$.

- Given the sets $I = \{a, b, c, d\}$ and $O = \{1, 2, 3\}$, we may define a function f by the correspondence $f(a) = 1, f(b) = 2, f(c) = 1$ and $f(d) = 2$ or more concisely by:

$$f(x) = \begin{cases} 1, & x = a \vee x = c \\ 2, & x = b \vee x = d \end{cases}$$

  Observe that for f to be a function $f : I \to O$, each elements of the domain must be assigned to a unique element of the co-domain, but not all the elements of the co-domain need be assigned as images of elements of the domain. Thus, element 1 of the co-domain is the image of elements a and c of the domain, element 2 of the co-domain is the image of elements b and d of the domain, and element 3 of the co-domain has not been assigned to any of the elements of the domain. See Fig. 1.32.
- Let $D = \{0, 1\}$ and the function f assign each rational number to element 1 of set D and each irrational number to element 0 of set D. This function is known as the Dirichlet function $f : \mathbb{R} \to D$ and is denoted as:

$$f(x) = \begin{cases} 1, & x \in \mathbb{Q} \\ 0, & x \in \mathbb{Q}' \end{cases}$$

  See Fig. 1.33.

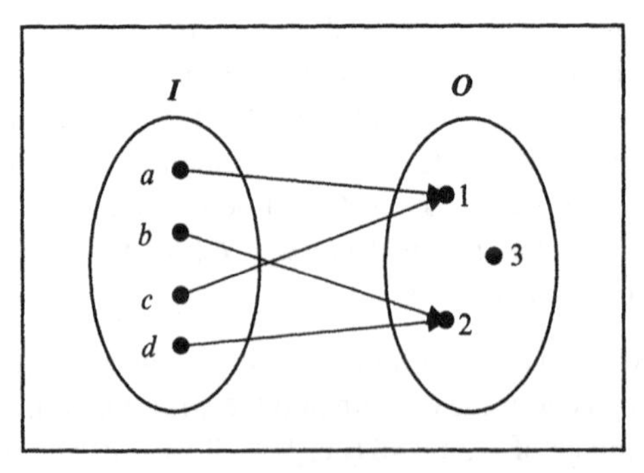

FIGURE 1.32. The mapping of the function $f(x) = \begin{cases} 1, & x = a \vee x = c \\ 2, & x = b \vee x = d \end{cases}$.

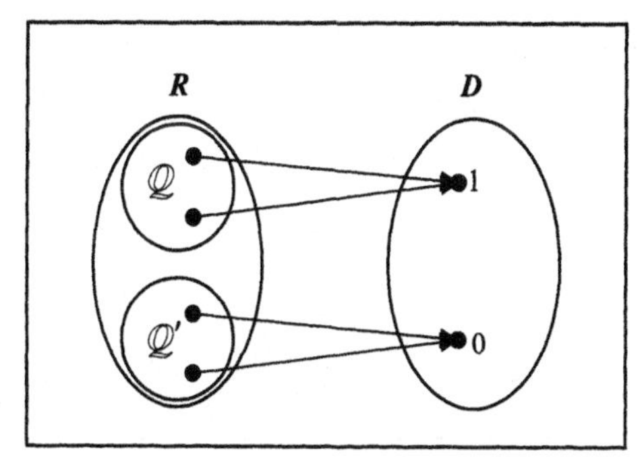

FIGURE 1.33. The mapping of the function $f(x) = \begin{cases} 1, & x = \mathbb{Q} \\ 0, & x = \mathbb{Q}' \end{cases}$.

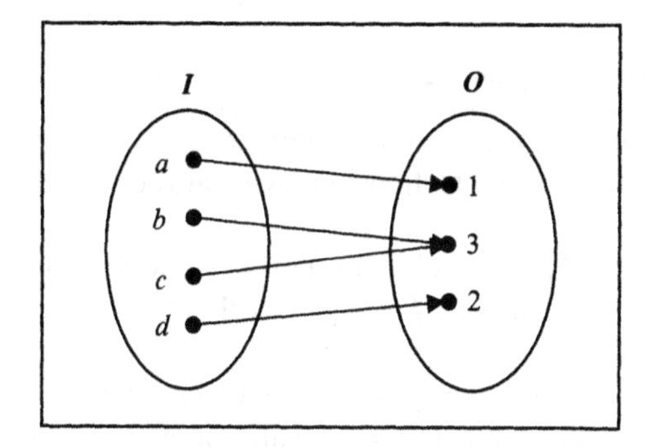

FIGURE 1.34. The mapping of a function that is defined diagrammatically.

- Given the sets $\boldsymbol{I} = \{a, b, c, d\}$ and $\boldsymbol{O} = \{1, 2, 3\}$, the function $\boldsymbol{f} : \boldsymbol{I} \to \boldsymbol{O}$ may be defined in diagrammatic form as shown in Fig. 1.34.

1.17.2 Mappings, Operators and Transformations

Given two general sets $\boldsymbol{I}$ and $\boldsymbol{O}$, the function $\boldsymbol{f}$ of $\boldsymbol{I}$ into $\boldsymbol{O}$ is frequently known as a *mapping of* $\boldsymbol{I}$ *into* $\boldsymbol{O}$ and the notation $\boldsymbol{f} : \boldsymbol{I} \to \boldsymbol{O}$ reads "$\boldsymbol{f}$ *maps* $\boldsymbol{I}$ *into* $\boldsymbol{O}$". The mapping (or function) of a set $\boldsymbol{I}$ into a set $\boldsymbol{O}$ is sometimes denoted by $\boldsymbol{I} \xrightarrow{f} \boldsymbol{O}$ or depicted in diagrammatic form as shown in Fig. 1.35.

If the same set $\boldsymbol{S}$ is both the domain and the co-domain of a function $\boldsymbol{f}$ (i.e. $\boldsymbol{f} : \boldsymbol{S} \to \boldsymbol{S}$), then this function is frequently referred to as an *operator on the set* $\boldsymbol{S}$ or a *transformation on the set* $\boldsymbol{S}$.

1.17.3 Equality of Functions

Two functions $\boldsymbol{f}_1$ and $\boldsymbol{f}_2$ are said to be *equal* (i.e. $\boldsymbol{f}_1 = \boldsymbol{f}_2$) if the following two conditions are satisfied:

- Both functions have the same domain $\boldsymbol{I}$.
- For every element $i \in \boldsymbol{I}$ the same image is assigned to it by both functions (i.e. $\boldsymbol{f}_1(i) = \boldsymbol{f}_2(i)$).

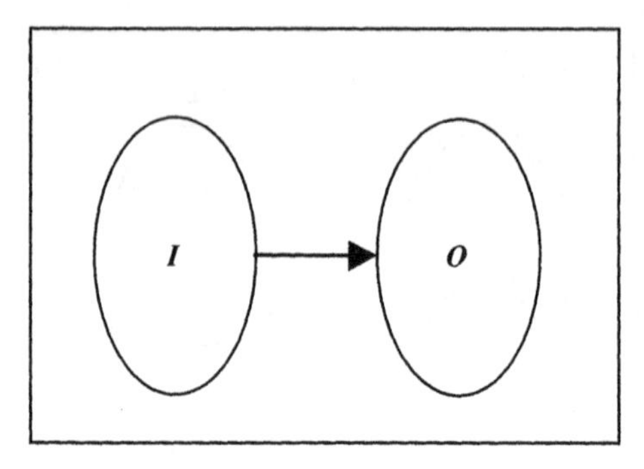

FIGURE 1.35. The mapping $I \xrightarrow{f} O$.

Consider the following examples:

- Given the two function $f_1(x) = x^3$, where $x \in \mathbb{Q}$, and $f_2(x) = x^3$, where $x \in \mathbb{Q}'$, we say that these functions are unequal (i.e. $f_1 \neq f_2$) as they have different domains.
- Given two functions $f_1(x) = x^3$ and $f_2(y) = y^3$, where x,y $\in \mathbb{R}$, we say that these functions are equal (i.e. $f_1 = f_2$) as they have the same domain and $f_1(x) = f_2(y)$ for every $x, y \in \mathbb{R}$ and $x = y$.
- Given that the function f_1 is defined by the mapping shown in Fig. 1.36 and that the function f_2 is defined as $f_2(x) = x^3$ with a domain $I_2 = \{1, 2\}$, we say that these two functions are equal, as their domains are equal (i.e. $I_1 = I_2 = \{1, 2\}$) and equal elements in their co-domains are assigned to equal elements in their domains (i.e. $f_1(1) = f_2(1) = 1$ and $f_1(2) = f_2(2) = 8$). Here, the co-domain of function f_2 is $O_2 = \{1, 8\}$. The fact that element 27 of the co-domain of function f_1 is not assigned to any element of its domain does not effect the equality of the two functions. Thus, two functions can be equal even though they have different co-domains. In this case, $O_2 \subset O_1$.

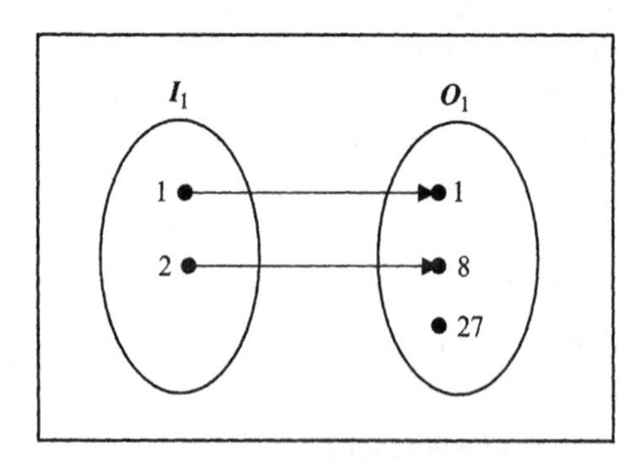

FIGURE 1.36. The diagrammatic mapping of the function f_1.

1.17.4 The Range of a Function

If f is a function of set I into set O (i.e. $f : I \to O$), then as we have seen in the above example not every element of its co-domain O need be an image of an element of its domain I. The elements of O that are the image of at least one element of I are known as the *range of the function* f. The range of a function $f : I \to O$ is denoted as $f(I)$.

A function $f : I \to O$ whose domain and range are the same (i.e. $f(I) = O$) is known as an *endofunction*.

Consider the following examples:

- Given that the function f is defined by the mapping shown in Fig. 1.36 (i.e. $f_1 : I_1 \to O_1$), the range of this function is $f_1(I_1) = \{1, 8\}$ which, as we can see, is different from its co-domain $O_1 = \{1, 8, 27\}$.
- Given the function $f_2(x) = |x|$, where $x \in \mathbb{R}$, the range of this function is $f_2(\mathbb{R}) = \{x \mid x \in \mathbb{R} \wedge x \geq 0\}$.

Thus, we observe that the co-domain of a function is the set of *possible* outputs and the range of a function is the set of *actual* outputs.

1.17.5 Different Types of Functions

We can distinguish the following four different types of functions, which we will examine in more detail.

- Many-to-one functions
- Injective functions or one-to-one functions
- Surjective functions or onto functions
- Bijective functions or one-to-one correspondences

1.17.5.1 Many-to-One Functions

A *many-to-one function* or *many-to-one mapping* is a function $f : I \to O$, which maps more than one distinct input values to one distinct output value.

More precisely, a function $f : I \to O$ is a many-to-one function if more than one elements i_i of its domain are assigned the same element o_j of its co-domain (i.e. if the correspondence $f(i_1) = f(i_2) = \cdots = f(i_n) = o_j$ is allowed under the function f).

Consider the following examples:

- The function $f(x) = |x|$ (where $x \in \mathbb{R}$) is a many-to-one function, as $f(x) = f(-x) = |x|$. See Fig. 1.37.
- The function $f(x) = x^2$ (where $x \in \mathbb{R}$) is a many-to-one function, as $f(x) = f(-x) = x^2$. See Fig. 1.38.

1.17.5.2 Injective Functions or One-to-One Functions

An *injective function* or *one-to-one function* or *injection* is a function $f : I \to O$, which maps each distinct input value to a distinct output value.

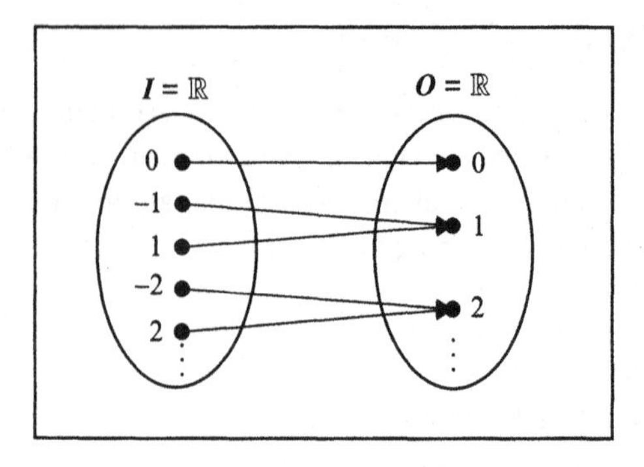

FIGURE 1.37. The mapping of the function $f(x) = |x|$.

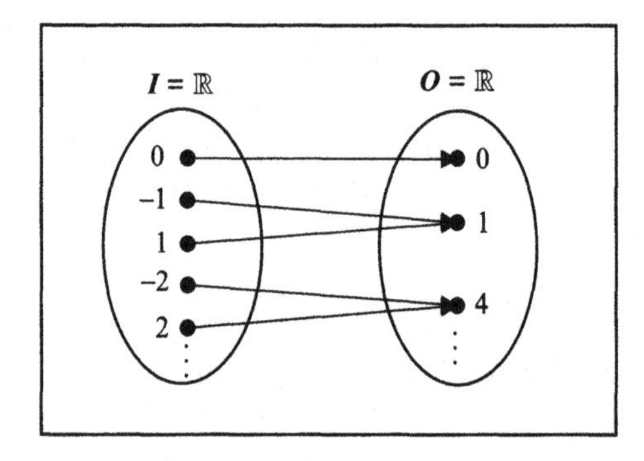

FIGURE 1.38. The mapping of the function $f(x) = x^2$.

More precisely, a function $f : I \to O$ is a one-to-one function if each distinct elements i_i of its domain is assigned a distinct element o_j of its co-domain. Alternatively, if two elements of the domain have the same image o_j in its co-domain, then the two elements must be equal (i.e. $f(i_1) = o_j \wedge f(i_2) = o_j \Rightarrow i_1 = i_2$).

Consider the following examples:

- The function $f(x) = x^3$ (where $x \in \mathbb{R}$) is *injective*, as distinct input values generate distinct output values. See Fig. 1.39.
- The function $f(x) = 3x + 2$ (where $x \in \mathbb{R}$) is also injective. See Fig. 1.40.

If a function $f : I \to O$ is injective, then its co-domain O has at least as many elements as its domain I.

1.17.5.3 Surjective Functions or Onto Functions

A *surjective function* or *onto function* or *surjection* is a function $f : I \to O$, which contains no other values in its co-domain than the images of the values of its domain.

More precisely, a function $f : I \to O$ is an onto function if every element of its co-domain is the image of at least one element of its domain. Alternatively,

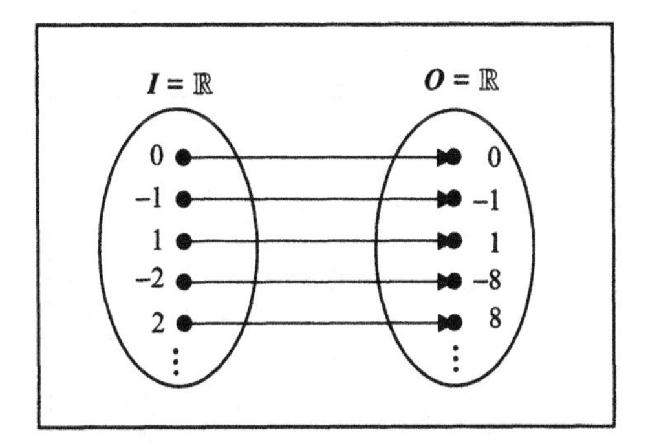

FIGURE 1.39. The mapping of the function $f(x) = x^3$.

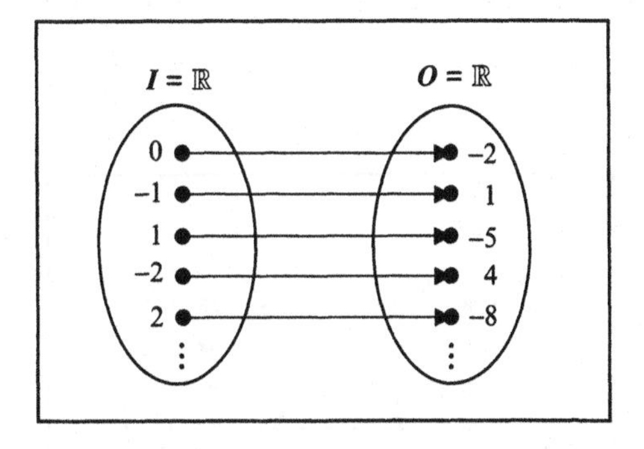

FIGURE 1.40. The mapping of the function $f(x) = 3x + 2$.

a function $f : I \to O$ is *surjective* if its range is equal to its co-domain (i.e. *iff* $f : (I) = O$). Equally, a function is surjective if for each image in its co-domain there exists a preimage in its domain.

Consider the following examples:

- The function $f : I \to O$ defined by the mapping depicted in Fig. 1.41 is surjective, as for every image o_j there exists a preimage i_i.
- The function $f(x) = 3x$ (where $x \in \mathbb{R}$) is also surjective, as for every image $3x$ there exists a preimage x. See Fig. 1.42.
- The function $f : I \to O$ defined by the mapping depicted in Fig. 1.43 is not surjective, as the image 4 has no preimage.

If a function $f : I \to O$ is surjective, then its domain I has at least as many elements as its co-domain O.

1.17.5.4 Bijective Functions or One-to-One Correspondences

A *bijective function* or *one-to-one correspondence* or *bijection* is a function $f : I \to O$, which is both injective (i.e. a one-to-one function) and surjective (i.e. an onto function). In other words, a bijection is a function that creates a correspondence that maps one input value to exactly one output value.

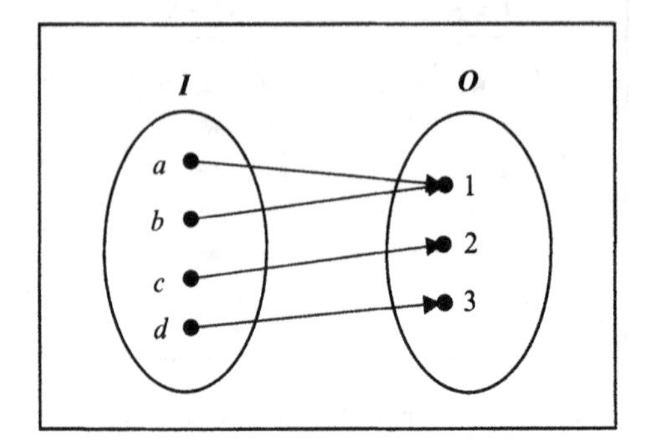

FIGURE 1.41. The mapping of surjective function.

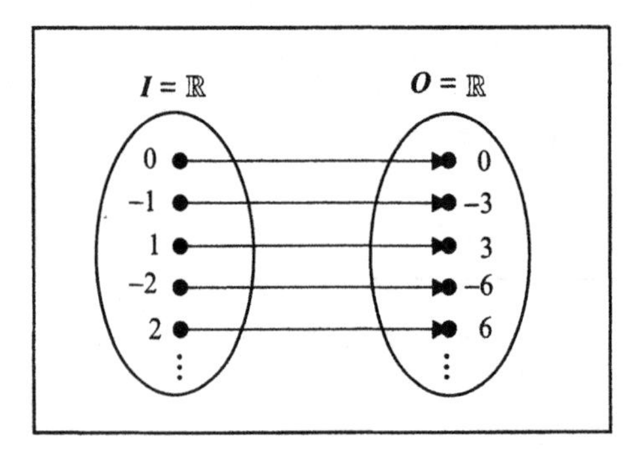

FIGURE 1.42. The mapping of the function $f(x) = 3x$.

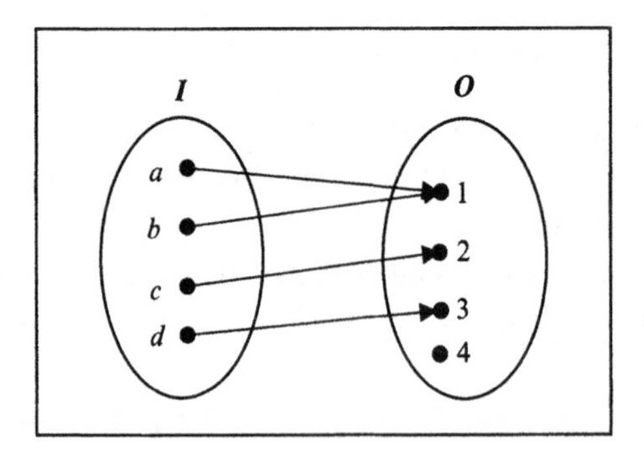

FIGURE 1.43. The mapping of non-surjective function.

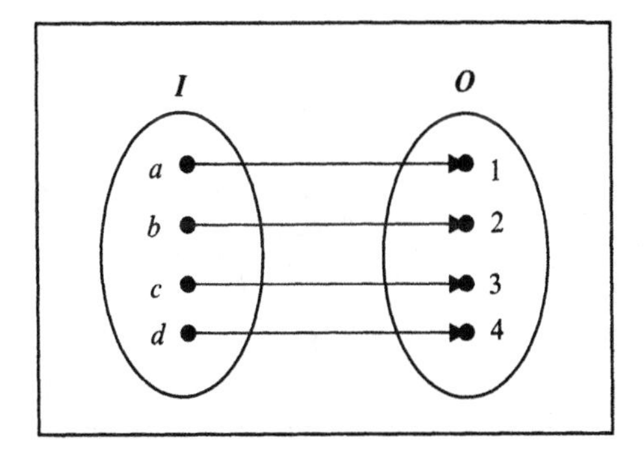

FIGURE 1.44. The mapping of bijective function.

More precisely, a function $\boldsymbol{f} : \boldsymbol{I} \to \boldsymbol{O}$ is a bijective function if every image o_j of its co-domain there is exactly one preimage i_i in its domain such that $\boldsymbol{f}(i_i) = o_j$.

Consider the following examples:

- The function $\boldsymbol{f}(x) = 3x$ (where $x \in \mathbb{R}$) is bijective, as it is both injective and subjective. See Fig. 1.42;
- The function $\boldsymbol{f} : \boldsymbol{I} \to \boldsymbol{O}$ defined by the mapping depicted in Fig. 1.44 is also bijective.

If the sets $\boldsymbol{I}$ and $\boldsymbol{O}$ are finite, then the function $\boldsymbol{f} : \boldsymbol{I} \to \boldsymbol{O}$ is bijective ***iff*** its domain and its co-domain have exactly the same number of elements.

1.17.6 Constant Functions

A *constant function* is a function $\boldsymbol{f} : \boldsymbol{I} \to \boldsymbol{O}$, which maps all the elements i_i of its domain to only one element o_j of its co-domain (i.e. $\boldsymbol{f}(i_i) = o_j \forall i_i \in I \mid o_j \in \boldsymbol{O}$). The range of a constant function is a set containing the unique element of its co-domain onto which all the elements of its domain are mapped (i.e. $\boldsymbol{f}(\boldsymbol{I}) = \{o_j\}$). Thus, the range of a constant function is a singleton.

For instance, the function $\boldsymbol{f} : \boldsymbol{I} \to \boldsymbol{O}$ defined by the mapping depicted in Fig. 1.45 is a constant function.

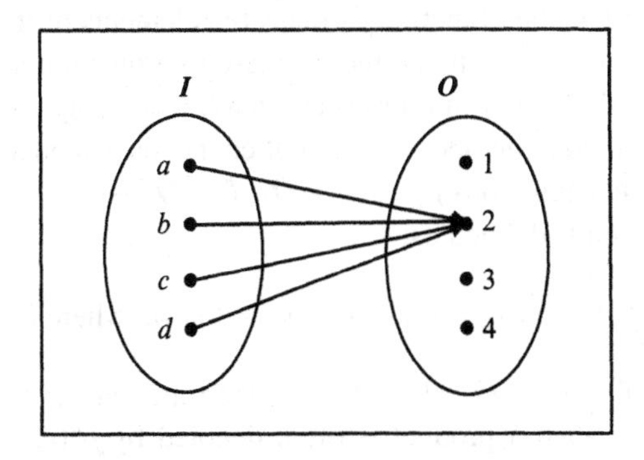

FIGURE 1.45. The mapping of constant function.

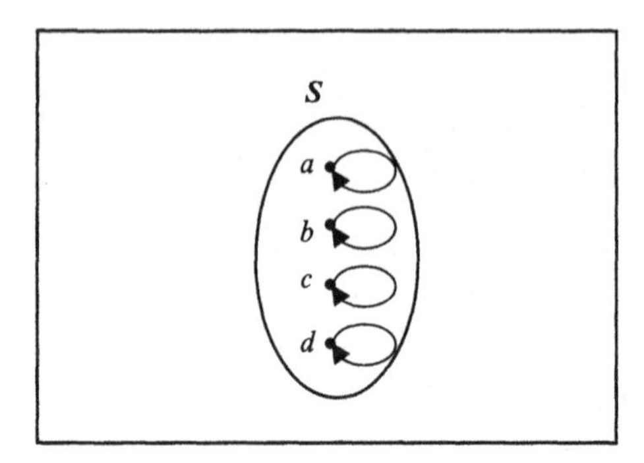

FIGURE 1.46. The mapping of an identity function.

1.17.7 The Identity Function or Identity Transformation

An *identity function* or *identity transformation* is a function $f : S \to S$, which maps each element of its domain to itself (i.e. $f(s_i) = s_i, \forall s_i \in S$). The identity function is denoted by $\mathbf{1}$ or $\mathbf{1}_s$. Figure 1.46 depicts an identity function.

1.17.8 The Composition or Product of Functions

A *composite function* or the *composition of one function on another* is the output value arrived at by using the output of one function as input into a second function.

More precisely, given two functions $f_1 : I_1 \to O_1$ and $f_2 : O_1 \to O_2$ the composition of these functions is denoted by $(f_2 \circ f_1)$ and is defined as $(f_2 \circ f_1)(i) = f_2(f_1(i)) = o$, where $i \in I_1 \wedge o \in O_2$. The operation $(f_2 \circ f_1)$ reads as f_1 *composed with* f_2, f_1 *composition* f_2, f_2 *circle* f_1 or f_2 *oh* f_1. The operator "o" itself is known as the *function product operator* or the *function composition operator*.

This new function $(f_2 \circ f_1)$ is a mapping of the domain I_1, of the first function, into the co-domain O_2 of the second function. This operation can be thought of as occurring in two steps. First, function f_1 maps the elements of its domain I_1 into its co-domain O_1, then function f_2 maps the elements of its domain O_1 into its co-domain O_2. The combined effect of these two function applications maps the elements of its domain I_1 into its co-domain O_2. See Fig. 1.47.

The function composition operation can of course be extended recursively to n levels. Thus, given n functions $f_1 : I_1 \to O_1, f_2 : O_1 \to O_2, \cdots, f_n : O_{n-1} \to O_n$ their composition is defined as:

$$(f_n \circ \cdots \circ f_2 \circ f_1)(i) = f_n(\cdots f_2(f_1(i))\cdots) = o, \text{ where } i \in I_1 \wedge o \in O_n$$

Given a function $f_1 : I \to O$, where $O \subseteq I$, the function can be composed with itself (i.e. $f \circ f$). Such a product is often denoted by $f^2(i)$ and is defined as $f^2(i) = (f \circ f)(i) = f(f(i)) = o$, where $i \in I \wedge o \in O$.

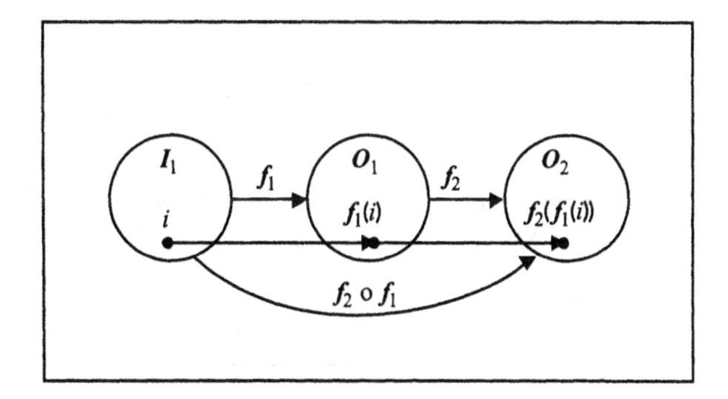

FIGURE 1.47. The mapping of function composition $(f_2 \circ f_1)$.

This notation can be generalised to n levels of function composition:

$$f^n(i) = (f \circ \cdots \circ f \circ f)(i) = f(\cdots f(f(i))\cdots) = o, \text{ where } i \in \boldsymbol{I} \wedge o \in \boldsymbol{O}.$$

In this notation the superscript is used to indicate levels of function composition except in the case of trigonometric functions, where for historical reasons the superscript is used to indicate raising the function to a power rather than function composition. For instance, $\cos^2\theta$ is defined to be $\cos\theta \cdot \cos\theta$ and not $\cos(\cos\theta)$.

Given two functions $f_1 : \boldsymbol{I}_1 \to \boldsymbol{O}_1$, and $f_2 : \boldsymbol{O}_1 \to \boldsymbol{O}_2$, in the general case the composition of these functions is not commutative (i.e. $f_2 \circ f_1 \neq f_1 \circ f_2$). To illustrate this consider the following example. Given two functions $f_1 : \mathbb{R} \to \mathbb{R}$ (which is defined as $f_1(x) = 3x + 2$) and $f_2 : \mathbb{R} \to \mathbb{R}$ (which is defined as $f_2(x) = x^2$), let us compute the function compositions $(f_2 \circ f_1)$ and $(f_1 \circ f_2)$:

$$(f_2 \circ f_1)(x) = f_2(f_1(x)) = f_2(3x + 2) = (3x + 2)^2 = 9x^2 + 12x + 4$$

and $(f_1 \circ f_2)(x) = f_1(f_2(x)) = f_1(x^2) = 3x^2 + 2$

Thus, in general $(f_2 \circ f_1) \neq (f_1 \circ f_2)$. The above two compositions could only be equal ***iff*** $9x^2 + 12x + 4 = 3x^2 + 2$ (i.e. ***iff*** $6x^2 + 12x + 2 = 0$, which in this particular case has no real solution).

If $f : \boldsymbol{I} \to \boldsymbol{O}$, then pre-multiplying this function by the identity function $\mathbf{1}_o$ on its co-domain or post-multiplying it by the identity function $\mathbf{1}_I$ on its domain does not alter the function (i.e. $\mathbf{1}_O \circ f = f$ and $f \circ \mathbf{1}_I = f$).

Given three functions $f_1 : \boldsymbol{I}_1 \to \boldsymbol{O}_1, f_2 : \boldsymbol{O}_1 \to \boldsymbol{O}_2$ and $f_3 : \boldsymbol{O}_2 \to \boldsymbol{O}_3$, the composition of these three functions $(f_3 \circ f_2 \circ f_1) : \boldsymbol{I}_1 \to \boldsymbol{O}_3$ is associative (i.e. $(f_3 \circ f_2) \circ f_1 = f_3 \circ (f_2 \circ f_1)$). See Fig. 1.48.

1.17.9 The Inverse of a Function

Let $f : \boldsymbol{I} \to \boldsymbol{O}$ be a function in its most general form. For such a function each element of its domain is mapped to an element of its co-domain and one or more elements of its domain are mapped to a given element of its co-domain

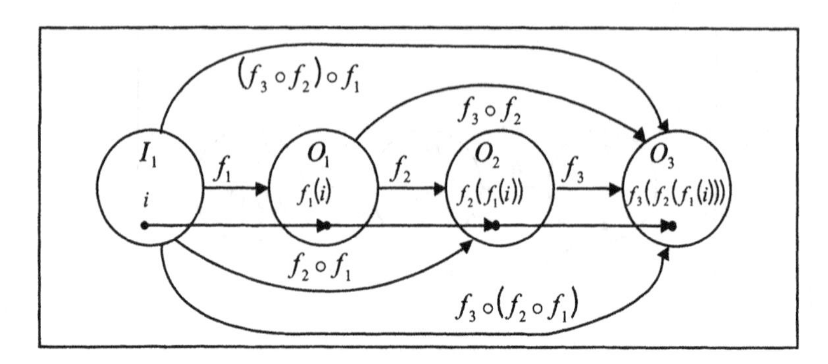

FIGURE 1.48. The mapping of function composition $(f_3 \circ f_2) \circ f_1 = f_3 \circ (f_2 \circ f_1)$.

(i.e. $\boldsymbol{f}(i_1) = \boldsymbol{f}(i_2) = \ldots = \boldsymbol{f}(i_n) = o_j$, where $i_i \in \boldsymbol{I}$ and $o_j \in \boldsymbol{I}$). Thus, each element of the co-domain of $\boldsymbol{f}$ (that is an image under this function) may have one or more preimages and the function maps each preimage i_i (in its domain) to its corresponding image o_j (in its co-domain). A function that maps each image o_j to its corresponding preimages i_i is known as the inverse of function $\boldsymbol{f}$ and is denoted by $\boldsymbol{f}^{-1}$. We read $\boldsymbol{f}^{-1}$ as *$\boldsymbol{f}$ inverse* or *the inverse of $\boldsymbol{f}$*.

1.17.9.1 Applying the Inverse of a Function to an Element of its Co-domain

If o_j is an image under the function $\boldsymbol{f} : \boldsymbol{I} \to \boldsymbol{O}$, then the inverse of this image is given by $\boldsymbol{f}^{-1}(o_j) = \{i_1, i_2, \ldots, i_n\}$ where $i_i \in \boldsymbol{I}$ and $o_j \in \boldsymbol{O}$. This set is known as the *inverse image* of o_j.

If $o_j \in \boldsymbol{f}(\boldsymbol{I})$ (i.e. if o_j is in the range of the function, which means that it is the image at least one $i_i \in \boldsymbol{I}$), then the inverse image of o_j is given by $\boldsymbol{f}^{-1}(o_j) = \{i_i \mid i_i \in \boldsymbol{I} \wedge \boldsymbol{f}(i_i) = o_j\}$. Alternatively, if $o_k \in \boldsymbol{O} \wedge o_k \notin \boldsymbol{f}(\boldsymbol{I})$ (i.e. if o_k is in the co-domain but not in the range of the function, which means that it is not the image of any $i_i \in \boldsymbol{I}$), then the inverse image of o_k is given by an empty set (i.e. $\boldsymbol{f}^{-1}(o_k) = \emptyset$). See Fig. 1.49.

For example, given the function $\boldsymbol{f} : \boldsymbol{I} \to \boldsymbol{O}$ defined by the mapping depicted in Fig. 1.50 then by applying the inverse function to the elements of the co-domain we get:

$$\boldsymbol{f}^{-1}(1) = \{a, c\}$$
$$\boldsymbol{f}^{-1}(2) = \{b, d\}$$
$$\boldsymbol{f}^{-1}(3) = \emptyset$$

Observe that the image 1 has two preimages a and c, the image 2 has two preimages b and d, and element 3 of the co-domain is not the image of any element of the domain under this function (thus its inverse image is the empty set).

1.17.9.2 Applying the Inverse of a Function to a Subset of its Co-domain

Now let us widen the definition of the inverse of a function. Suppose that we have a function $\boldsymbol{f} : \boldsymbol{I} \to \boldsymbol{O}$ and that $\boldsymbol{C}$ is a subset of its co-domain (i.e. $\boldsymbol{C} \subseteq \boldsymbol{O}$). Then

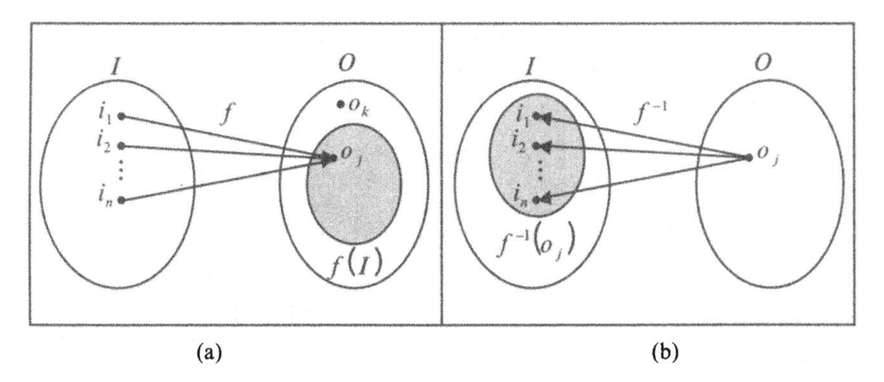

(a) (b)

FIGURE 1.49 (a) The mapping of the elements $i_1, i_2, \ldots, i_n$ onto the image o_j under the function $\boldsymbol{f}$. (b) The mapping of the image o_j to the preimages $\{i_1, i_2, \ldots, i_n\}$under the inverse function$\boldsymbol{f}^{-1}$.

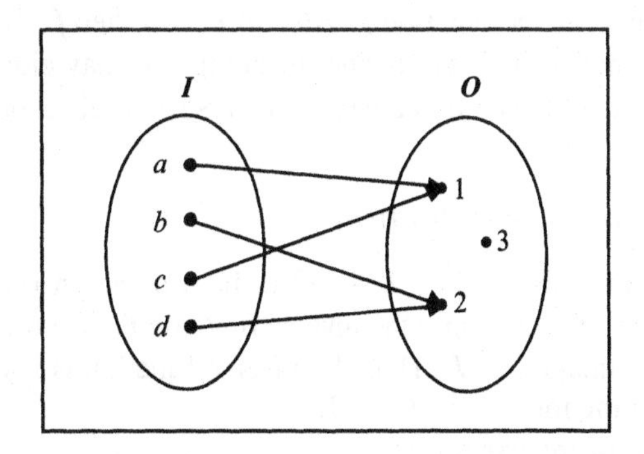

FIGURE 1.50. The mapping of a function that is defined diagrammatically.

the inverse of this subset under the mapping $\boldsymbol{f}$ is denoted by $\boldsymbol{f}^{-1}(\boldsymbol{C})$ and is defined to be the set of all the elements of the domain $\boldsymbol{I}$ that are mapped onto some image contained in set $\boldsymbol{C}$, i.e. $\boldsymbol{f}^{-1}(\boldsymbol{C}) = \{i \mid i \in \boldsymbol{I} \wedge \boldsymbol{f}(i) \in \boldsymbol{C}\}$. See Fig. 1.51.

Let us re-examine the function $\boldsymbol{f} : \boldsymbol{I} \rightarrow \boldsymbol{O}$ defined by the mapping depicted in Fig. 1.50. By applying the inverse of the function $\boldsymbol{f}$ to subsets of its co-domain we get:

$\boldsymbol{f}^{-1}(\{1, 2, 3\}) = \{a, b, c, d\}$
$\boldsymbol{f}^{-1}(\{1, 2\}) = \{a, b, c, d\}$
$\boldsymbol{f}^{-1}(\{1\}) = \{a, c\}$
$\boldsymbol{f}^{-1}(\{2\}) = \{b, d\}$
$\boldsymbol{f}^{-1}(\{3\}) = \emptyset$

In this example the range of the function is $\boldsymbol{f}(\boldsymbol{I}) = \{1, 2\}$, and as can be seen, the inverse of the range of $\boldsymbol{f}$ is $\boldsymbol{f}^{-1}(\boldsymbol{f}(\boldsymbol{I})) = \boldsymbol{I}$. From this result it should be apparent that the mapping represented by the inverse function $\boldsymbol{f}^{-1}$ is the reverse of the

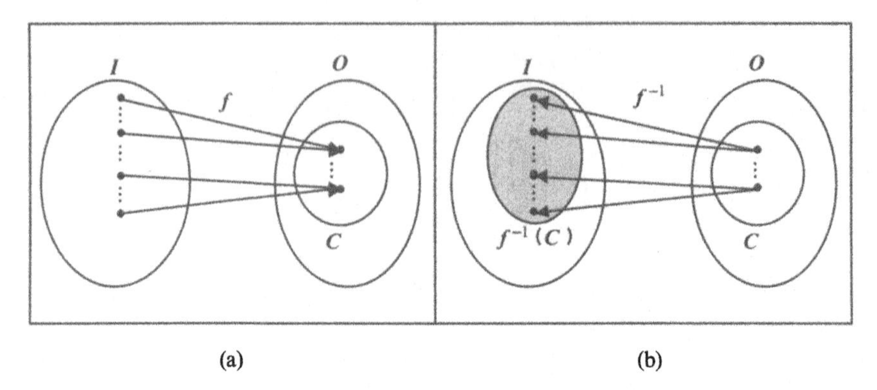

FIGURE 1.51 (a) The mapping of the elements of the domain whose images are in the subset $\boldsymbol{C}$ of the co-domain under the function $\boldsymbol{f}$. (b) The mapping of the images that are members of the subset $\boldsymbol{C}$ of the co-domain under the inverse function $\boldsymbol{f}^{-1}$.

mapping represented by the function $\boldsymbol{f}$. Also, if $o \in \boldsymbol{O}$ then $\boldsymbol{f}^{-1}(o) = \boldsymbol{f}^{-1}(\{o\})$. Thus an inverse function has two distinct meanings. It may either represent the inverse of an element of the co-domain or the inverse of a subset of the co-domain.

1.17.10 The Inverse Function

Next, we restrict the function $\boldsymbol{f} : \boldsymbol{I} \to \boldsymbol{O}$ to be a bijection (i.e. a one-to-one correspondence). If we now apply the inverse function $\boldsymbol{f}^{-1}$ to each image $o \in \boldsymbol{O}$, we get a single preimage $i \in \boldsymbol{I}$. Thus the inverse function is itself a one-to-one correspondence of the form $\boldsymbol{f}^{-1} : \boldsymbol{O} \to \boldsymbol{I}$.

Consider the following examples:

- Let the function $\boldsymbol{f} : \boldsymbol{I} \to \boldsymbol{O}$ be defined by the mapping depicted in Fig. 1.52a. From the diagram we can see that this function is a bijection (i.e. both one-to-one and onto) and thus its inverse function $\boldsymbol{f}^{-1} : \boldsymbol{O} \to \boldsymbol{I}$ exists and is defined by the mapping depicted in Fig. 1.52b.

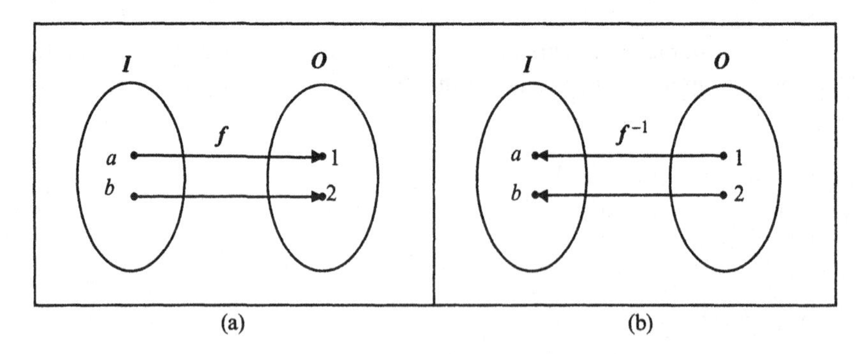

FIGURE 1.52 (a) The mapping of the function $\boldsymbol{f} : \boldsymbol{I} \to \boldsymbol{O}$. (b) The mapping of the function $\boldsymbol{f}^{-1} : \boldsymbol{I} \to \boldsymbol{O}$.

- Let the function $f : \mathbb{R} \to \mathbb{R}$ be defined as $f(x) = x^3$. This function is a bijection as it is both one-to-one and onto. Thus its inverse function $f^{-1} : \mathbb{R} \to \mathbb{R}$ exists and is defined as $f^{-1}(x) = x^{1/3} = \sqrt[3]{x}$.

1.17.11 Theorems on the Inverse Function

For functions that are bijections (i.e. one-to-one correspondences) we can prove the following two theorems.

Theorem 1.10 *If the function $f : I \to O$ is a bijection (which implies that its inverse function exists and is given by $f^{-1} : O \to I$), the composite function $(f^{-1} \circ f) : I \to I$ is equal to the identity function 1_I on the domain of f and the composite function $(f \circ f^{-1}) : O \to O$ is equal to the identity function 1_O on the co-domain of f. See Fig. 1.53. This can be expressed more concisely as:*

$$f : I \to O \wedge f^{-1} : O \to I \Rightarrow \left(f^{-1} \circ f\right) = 1_I \wedge \left(f \circ f^{-1}\right) = 1_O. \quad \text{(T1.10)}$$

Proof: To illustrate this theorem consider the following example. Given the bijective function $f : I \to O$ defined by the mapping depicted in Fig. 1.52a

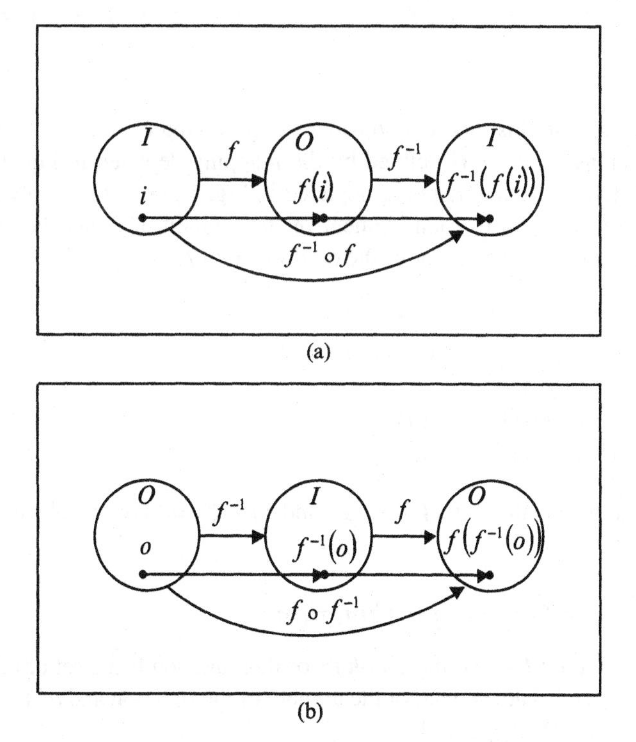

FIGURE 1.53 (a) The mapping of function composition $(f^{-1} \circ f)$. (b) The mapping of function composition $(f \circ f^{-1})$.

and its inverse $f^{-1} : \boldsymbol{O} \rightarrow \boldsymbol{I}$) defined by the mapping depicted in Fig. 1.52b, let us apply the composite function $\left(f^{-1} \circ f\right)$ to each element of the domain of f and the composite function $\left(f \circ f^{-1}\right)$ to each element of the co-domain of f:

$$\left(f^{-1} \circ f\right)(a) = f^{-1}(f(a)) = f^{-1}(1) = a$$
$$\left(f^{-1} \circ f\right)(b) = f^{-1}(f(b)) = f^{-1}(2) = b$$

and

$$\left(f \circ f^{-1}\right)(1) = f\left(f^{-1}(1)\right) = f(a) = 1$$
$$\left(f \circ f^{-1}\right)(2) = f\left(f^{-1}(2)\right) = f(b) = 2$$

Thus, it is apparent that $\left(f^{-1} \circ f\right) = \mathbf{1}_I$ and $\left(f \circ f^{-1}\right) = \mathbf{1}_O$. □

Theorem 1.11 *If the functions* $f : \boldsymbol{I} \rightarrow \boldsymbol{O}$ *and* $g : \boldsymbol{O} \rightarrow \boldsymbol{I}$ *are bijections, then* $g = f^{-1}$ *iff the composite function* $(g \circ f) : \boldsymbol{I} \rightarrow \boldsymbol{I}$ *is equal to the identity function* $\mathbf{1}_I$ *on the domain of* f *and the composite function* $(f \circ g) : \boldsymbol{O} \rightarrow \boldsymbol{O}$ *is equal to the identity function* $\mathbf{1}_o$ *on the co-domain of* f*. This can be expressed more concisely as:*

$$f : \boldsymbol{I} \rightarrow \boldsymbol{O} \wedge g : \boldsymbol{O} \rightarrow \boldsymbol{I} \Rightarrow \left(g = f^{-1} \Leftrightarrow (g \circ f) = \mathbf{1}_I \wedge \left(f \circ f^{-1}\right) = \mathbf{1}_O\right). \tag{T1.11}$$

Proof: To illustrate this theorem consider the following example. Given the bijective functions $f : \boldsymbol{I} \rightarrow \boldsymbol{O}$ defined by the mapping depicted in Fig. 1.54a and $g : \boldsymbol{O} \rightarrow \boldsymbol{I}$ defined by the mapping depicted in Fig. 1.54b, let us apply the composite function $(g \circ f)$ to each element of the domain of f and the composite function $(f \circ g)$ to each element of the co-domain of f:

$$(g \circ f)(a) = g(f(a)) = g(1) = a$$
$$(g \circ f)(b) = g(f(b)) = g(2) = b$$

and

$$(f \circ g)(1) = f(g(1)) = f(a) = 1$$
$$(f \circ g)(2) = f(g(2)) = f(b) = 2$$

Thus, it is apparent that $(g \circ f) = \mathbf{1}_I$ and $(f \circ g) = \mathbf{1}_O$, which implies that $g = f^{-1}$.

1.17.12 The Graph of a Function

Given a function $f : \boldsymbol{I} \rightarrow \boldsymbol{O}$, the *graph* f^* of this function is the set of all ordered pairs (i, o) where i is an element of the domain of the function and o is an element of its co-domain. More precisely:

$$f^* = \{(i, o) \mid i \in \boldsymbol{I} \wedge o = f(i)\}$$

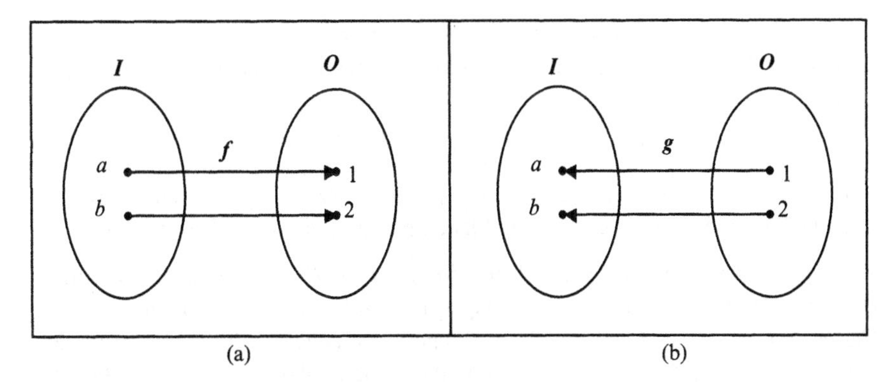

FIGURE 1.54 (a) The mapping of the function $f : I \rightarrow O$. (b) The mapping of the function $g : O \rightarrow I$.

Frequently the term *graph of a function* refers to the graphical representation of this set of ordered pairs.

Consider the following examples:

- Given the function $f : I \rightarrow O$ defined by $f(i) = \begin{cases} 1, & i = a \\ 2, & i = b \\ 3, & i = c \end{cases}$, then its graph is given by $f^* = \{(a, 1), (b, 2), (c, 3)\}$.
- Given the function $f : I \rightarrow O$ defined by the mapping depicted in Fig. 1.55, then its graph is given by $f^* = \{(a, 1), (b, 2), (c, 3)\}$.
- Given the function $f : X \rightarrow \mathbb{N}$ (where $X = \{0, 1, 2, 3\}$) defined as $f(x) = x^2 + 2$, then its graph is given by $f^* = \{(0, 2), (1, 3), (2, 6), (3, 11)\}$.
- Given the function $f : \mathbb{N} \rightarrow \mathbb{N}$ defined as $f(x) = x^2 + 2$, then the graph of this function is the infinite set $f^* = \{(0, 2), (1, 3), (2, 6), (3, 11), (4, 18), \ldots\}$.

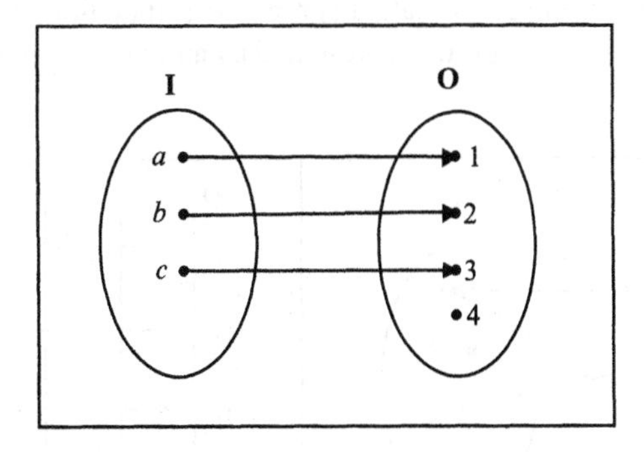

FIGURE 1.55. The mapping of the function $f : I \rightarrow O$.

Recall that, given a mapping $f : I \rightarrow O$, for f to be a function of I into O every element $i \in I$ must a have a unique image $o \in O$, but not all elements of O need to have a preimage in I. Thus we can make the following observations about the graph f^* of function f:

- For each $i \in \boldsymbol{I}$ there exists an ordered pair $(i, o) \in \boldsymbol{f}^*$.
- Each $i \in \boldsymbol{I}$ appears as the first element of only one ordered pair of the graph $\boldsymbol{f}^*$, i.e.

$$\boldsymbol{f}(i, o_1) \in \boldsymbol{f}^* \wedge \boldsymbol{f}(i, o_2) \in \boldsymbol{f}^* \Rightarrow o_1 = o_2.$$

- The graph $\boldsymbol{f}^*$, of the general function $\boldsymbol{f} : \boldsymbol{I} \to \boldsymbol{O}$, is a proper subset of the Cartesian product of its domain and its co-domain, i.e.: $\boldsymbol{f}^* \subset \boldsymbol{I} \times \boldsymbol{O}$.

Let $\boldsymbol{f}^*$ be the graph of a general function $\boldsymbol{f} : \boldsymbol{I} \to \boldsymbol{O}$. As we have seen above $\boldsymbol{f}^* \subset \boldsymbol{I} \times \boldsymbol{O}$, thus we could *graph* (display) $\boldsymbol{f}^*$ on the coordinate diagram of the Cartesian product $\boldsymbol{I} \times \boldsymbol{O}$. To illustrate this point, consider the following example. Given the function $\boldsymbol{f} : \boldsymbol{I} \to \boldsymbol{O}$ defined by the mapping depicted in Fig. 1.56a, the graph of this function is given by $\boldsymbol{f}* = \{(i_1, o_1), (i_2, o_3), (i_3, o_2), (i_4, o_1)\}$ and it is *graphed* on the coordinate diagram depicted in Fig. 1.56b.

Thus, if the graph $\boldsymbol{f}^*$ of a general function $\boldsymbol{f} : \boldsymbol{I} \to \boldsymbol{O}$ is graphed on the coordinate diagram of the Cartesian product $\boldsymbol{I} \times \boldsymbol{O}$, then we observe that:

- Each vertical line of the coordinate diagram (corresponding to an element of $\boldsymbol{I}$) contains one and only one point (represented by an ordered pair) of the graph $\boldsymbol{f}^*$.
- Some of the horizontal lines may contain more than one point, as an image in the co-domain may have more than one preimages in the domain of the function $\boldsymbol{f}$.

Let us now examine the case where the function $\boldsymbol{f}$ is a bijection (a one-to-one correspondence) by considering the following example. Given the bijective function $\boldsymbol{f} : \boldsymbol{I} \to \boldsymbol{O}$ defined by the mapping depicted in Fig. 1.57a, the graph of this function is given by $\boldsymbol{f}^* = \{(i_1, o_1), (i_2, o_3), (i_3, o_2)\}$ and it is *graphed* on the coordinate diagram depicted in Fig. 1.57b.

Thus, if the graph $\boldsymbol{f}^*$ of a bijective function $\boldsymbol{f} : \boldsymbol{I} \to \boldsymbol{O}$ is graphed on the coordinate diagram of the Cartesian product $\boldsymbol{I} \times \boldsymbol{O}$, then we observe that:

- Each vertical line of the coordinate diagram (corresponding to an element of $\boldsymbol{I}$) contains one and only one point (represented by an ordered pair) of the graph $\boldsymbol{f}^*$.

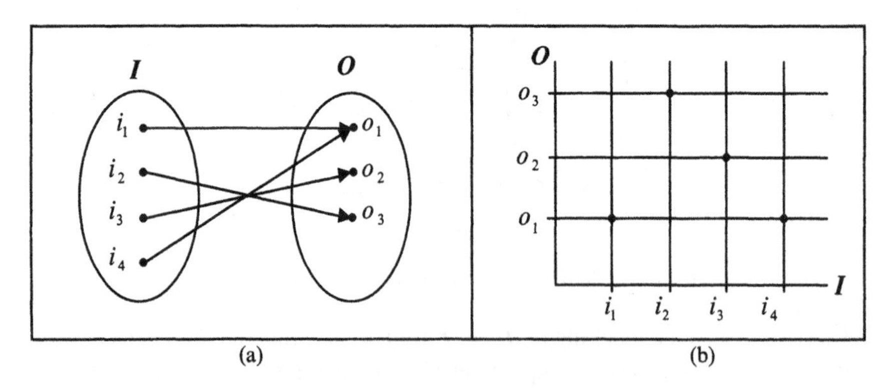

FIGURE 1.56 (a) The mapping of the function $\boldsymbol{f} : \boldsymbol{I} \to \boldsymbol{O}$. (b) The graph $\boldsymbol{f}^*$ on the coordinate diagram of $\boldsymbol{I} \times \boldsymbol{O}$.

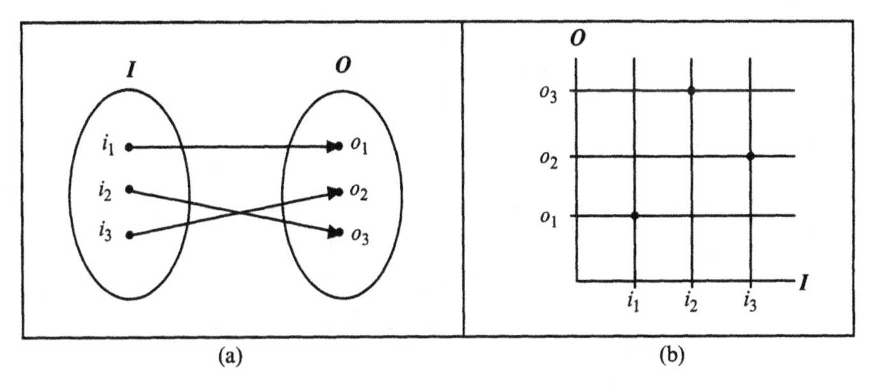

FIGURE 1.57 (a) The mapping of the function $f : I \to O$. (b) The graph f^* on the coordinate diagram of $I \times O$.

- Each horizontal line of the coordinate diagram (corresponding to an element of O) contains one and only one point (represented by an ordered pair) of the graph f^*.

To further clarify the relationship between a function f and its graph f^* displayed on the coordinate diagram of the Cartesian product $I \times O$ consider the following example. Given two sets $I = \{i_1, i_2, i_3, i_4\}$ and $O = \{o_1, o_2, o_3\}$, and the three coordinate diagrams shown in Fig. 1.58, we observe that:

- the set of points $\{(i_1, o_1), (i_2, o_3), (i_3, o_1), (i_4, o_2)\}$ of the coordinate diagram depicted in Fig. 1.58a is the graph f^* of a function $f : I \to O$, as each vertical line of the coordinate diagram contains one and only one point contained in the graph of f;
- The set of points $\{(i_1, o_1), (i_2, o_2), (i_4, o_3)\}$ of the coordinate diagram depicted in Fig. 1.58b is not the graph of a function, as the vertical line corresponding to the element i_3 of the coordinate diagram does not contain a point (which indicates that $i_3 \in I$ does not have an image in O).
- The set of points $\{(i_1, o_1), (i_2, o_2), (i_3, o_1), (i_3, o_2), (i_4, o_3)\}$ of the coordinate diagram depicted in Fig. 1.58c is not the graph of a function, as the vertical line corresponding to the element i_3 of the coordinate diagram contains two points (which indicates that $i_3 \in I$ does not have a unique image in O).

1.17.13 The Redefinition of a Function as a Set of Ordered Pairs

As we have seen above the graph f^* of a function $f : I \to O$ is a proper subset of the Cartesian product $I \times O$ and has the following properties:

- For each $i \in I$ there exists an ordered pair $(i, o) \in f^*$.
- No two different ordered pairs of f^* have the same first element.

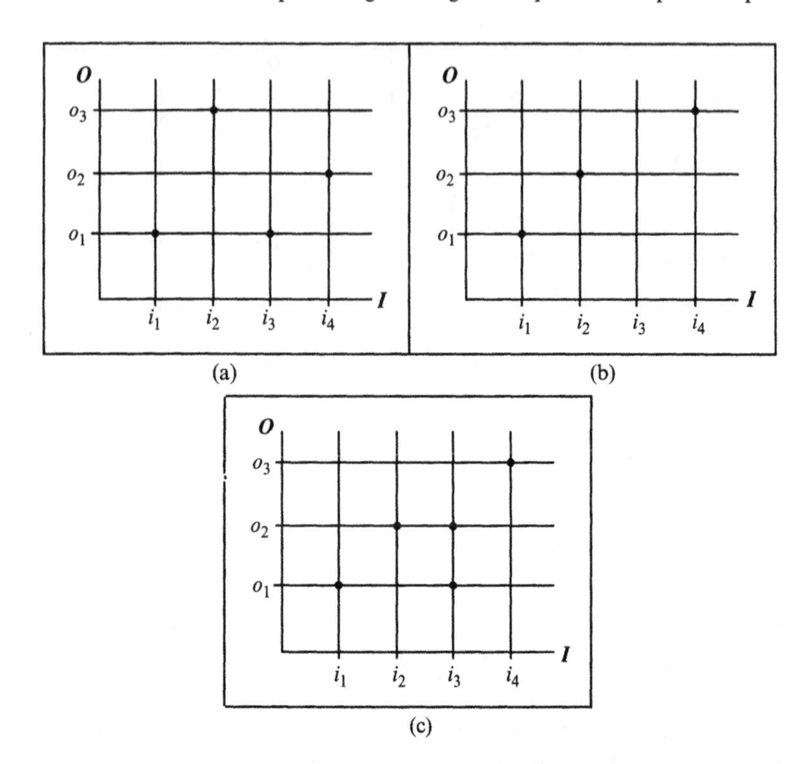

FIGURE 1.58 (a) A set of points on the coordinate diagram of $\boldsymbol{I} \times \boldsymbol{O}$. (b) A set of points on the coordinate diagram of $\boldsymbol{I} \times \boldsymbol{O}$. (c) A set of points on the coordinate diagram of $\boldsymbol{I} \times \boldsymbol{O}$.

Thus, we have a rule that assigns to each element $i \in \boldsymbol{I}$ a unique image $o \in \boldsymbol{O}$. In other words, $\boldsymbol{f}^*$ is a function of $\boldsymbol{I}$ into $\boldsymbol{O}$. This observation leads us to redefine the concept of the function as a set of ordered pairs. Thus, we say that: *a function $f : \boldsymbol{I} \rightarrow \boldsymbol{O}$ is a subset of $\boldsymbol{I} \times \boldsymbol{O}$ in which all elements $i \in \boldsymbol{I}$ appear as the first element of an ordered pair (i, o) that belongs to the set of ordered pairs $\boldsymbol{f}$.*

Given this new definition of a function as a set of ordered pairs belonging to a subset of $\boldsymbol{I} \times \boldsymbol{O}$, we can make the following observations:

- If $\boldsymbol{f}$ is a set of ordered pairs in the coordinate diagram of the Cartesian product $\boldsymbol{I} \times \boldsymbol{O}$ (where each vertical line of this diagram contains one and only one point that belongs to the set $\boldsymbol{f}$), then $\boldsymbol{f}$ is a function of $\boldsymbol{I}$ into $\boldsymbol{O}$.
- If the function $\boldsymbol{f} : \boldsymbol{I} \rightarrow \boldsymbol{O}$ is a one-to-one correspondence (bijection), then the inverse function $\boldsymbol{f}^{-1}$ is defined to be the set of ordered pairs that define the function $\boldsymbol{f}$ taken in the reverse order, i.e. $\boldsymbol{f}^{-1} = \{(o, i) \mid (i, o) \in \boldsymbol{f}\}$.

1.18 Families of Indexed Sets

A set $\boldsymbol{I}$ whose elements $i \in \boldsymbol{I}$ *index* (are used to refer to) members of another set $\boldsymbol{S}$ is known as an *index set.* As an example suppose that we have the sets $\boldsymbol{S}_1 = \{s_{1,1}, s_{1,2}, \ldots, s_{1,k}\}$, $\boldsymbol{S}_2 = \{s_{2,1}, s_{2,2}, \ldots, s_{2,l}\}$, $\boldsymbol{S}_3 = \{s_{3,1}, s_{3,2}, \ldots, s_{3,m}\}$

and a set $\boldsymbol{I} = \{1, 2, 3\}$, i.e. for each element $i \in \boldsymbol{I}$ there exists a set $\boldsymbol{S}_i$. Then, the set $\boldsymbol{I}$ is known as the *index set*, the element $i \in \boldsymbol{I}$ is known as the *index* of set $\boldsymbol{S}_i$, the sets $\boldsymbol{S}_1$, $\boldsymbol{S}_2$ and $\boldsymbol{S}_3$ are known as the *indexed sets*, and $\boldsymbol{S} = \{\boldsymbol{S}_1, \boldsymbol{S}_2, \boldsymbol{S}_3\}$ is the set of all indexed sets $\boldsymbol{S}_i$. Such a collection of sets is often referred to as an *indexed family of sets* and is denoted as $\mathfrak{S} = \{\boldsymbol{S}_i\}_{i \in I}$.

Families and sets are alternative ways of representing collections of entities. These two representations only differ formally, as an indexed family of sets $\mathfrak{S} = \{\boldsymbol{S}_i\}_{i \in I}$ can give rise to set of indexed sets $\boldsymbol{S} = \{\boldsymbol{S}_i \mid i \in \boldsymbol{I}\}$ and vice-versa a set of indexed sets $\boldsymbol{S} = \{\boldsymbol{S}_i \mid i \in \boldsymbol{I}\}$ can give rise to an indexed family of sets $\mathfrak{S} = \{\boldsymbol{S}_i\}_{i \in I}$. The main difference between sets and families is that the repetition of an element in a set definition is ignored and does not alter its definition, while the repetition of an element in a family definition alters its definition.

Alternatively, we may view an indexed family of sets $\mathfrak{S} = \{\boldsymbol{S}_i\}_{i \in I}$ as a function $f : \boldsymbol{I} \rightarrow \mathfrak{S}$ that assigns each element $i \in \boldsymbol{I}$ of its domain (which is the index set $\boldsymbol{I}$ of the indexed family of sets $\mathfrak{S}$) to a unique element $\boldsymbol{S}_i$ of its range (which is the indexed family of sets $\mathfrak{S}$ itself).

Consider the following examples:

- Suppose that we have the interval definition $\boldsymbol{I}_j = \{i \mid \mathrm{j} \leq i \leq \mathrm{j} + 1\}$, where $j \in \boldsymbol{J}$ and $\boldsymbol{J} = \{0, 1, 2, 3\}$. This definition gives rise to the intervals $\boldsymbol{I}_0 = [0, 1]$, $\boldsymbol{I}_1 = [1, 2]$, $\boldsymbol{I}_2 = [2, 3]$, $\boldsymbol{I}_3 = [3, 4]$ and to the indexed family of intervals $\mathfrak{I} = \left\{I_j\right\}_{j \in J}$.
- Suppose that we have the set definition $\boldsymbol{S}_i = \left\{s \mid s = i^n \forall \mathrm{n} \in \mathbb{N}^+\right\}$, where $i \in \mathbb{N}$ and where the symbol $\mathbb{N}^+$ denotes the set of the positive natural numbers. This definition gives rise to the sets:

$$\begin{aligned} S_0 &= \left\{0^1, 0^2, 0^3, 0^4, \ldots\right\} = \{0, 0, 0, 0, \ldots\}, \\ S_1 &= \left\{1^1, 1^2, 1^3, 1^4, \ldots\right\} = \{1, 1, 1, 1, \ldots\}, \\ S_2 &= \left\{2^1, 2^2, 2^3, 2^4, \ldots\right\} = \{2, 4, 8, 16, \ldots\}, \\ &\vdots \end{aligned}$$

  and to the indexed family of sets $\mathfrak{S} = \{\boldsymbol{S}_i\}_{i \in \mathbb{N}}$.
- Suppose that we have the set definition $\boldsymbol{S}_i = \{s \mid s = i \cdot n \ \forall n \in \mathbb{N}\}$, where $i \in \mathbb{N}$. This definition gives rise to the sets:

$$\begin{aligned} \boldsymbol{S}_0 &= \{0 \cdot 0, 0 \cdot 1, 0 \cdot 2, 0 \cdot 3, \ldots\} = \{0, 0, 0, 0, \ldots\}, \\ \boldsymbol{S}_1 &= \{1 \cdot 0, 1 \cdot 1, 1 \cdot 2, 1 \cdot 3, \ldots\} = \{0, 1, 2, 3, \ldots\}, \\ \boldsymbol{S}_2 &= \{2 \cdot 0, 2 \cdot 1, 2 \cdot 2, 2 \cdot 3, \ldots\} = \{0, 2, 4, 6, \ldots\}, \\ &\vdots \end{aligned}$$

  and to the indexed family of sets $\mathfrak{S} = \{\boldsymbol{S}_i\}_{i \in \mathbb{N}}$. Observe that in this example the set $\boldsymbol{S}_1$ is identical to the index set $\mathbb{N}$ and is the universal set for all indexed sets $\boldsymbol{S}_i$.

Above we have observed that any indexed family of sets $\mathfrak{S} = \{S_i\}_{i \in I}$ can be thought of as a function $f : I \to \mathfrak{S}$. Replacing the function f by the identity function on $\mathfrak{S}$ we get $1_{\mathfrak{S}} : \mathfrak{S} \to \mathfrak{S}$, which is another way of saying that $\mathbb{S} = \{S_i\}_{i \in \mathfrak{S}}$, where $S_i \in \mathfrak{S}$ and $i = S_i$ (i.e. the index of any set in the family of indexed sets $\mathfrak{S}$ is the set itself). Now, we can rewrite $\mathfrak{S}$ as $\mathfrak{S} = \{i\}_{i \in \mathfrak{S}}$ and say that *any indexed family of sets* $\mathfrak{S}$ *can be indexed by itself.*

1.19 The Generalised Set Union and Intersection Operations

Earlier in this chapter we have defined the operations of union and intersection between two sets. Now, we will extend these operations to deal with a finite number of sets. Thus, given n sets $S_1, S_2, \ldots, S_n$ we refine the definition of these set operations as follows. The union of n sets is defined as:

$$\bigcup_{i=1}^{n} S_i = S_1 \cup S_2 \cup \cdots \cup S_n \tag{1.14}$$

or

$$\bigcup_{i=1}^{n} S_i = \{s \mid (\exists i \in \{1, 2, \ldots, \mathrm{n}\}) : (s \in S_i)\} \tag{1.15}$$

Which reads "the union of the sets S_i for i from 1 to n is equal to the set of elements s such that there exists a value of i that belongs to the set $\{1, 2, \ldots, n\}$ for which element s belongs to the set S_i", i.e. the elements of the union of the sets belong to at least one of the sets S_i.

Similarly, the intersection of n sets is defined as:

$$\bigcap_{i=1}^{n} S_i = S_1 \cap S_2 \cap \cdots \cap S_n \tag{1.16}$$

or

$$\bigcap_{i=1}^{n} S_i = \{s \mid (\forall i \in \{1, 2, \ldots, n\}) : (s \in S_i)\} \tag{1.17}$$

Which reads "the intersection of the sets S_i for i from 1 to n is equal to the set of elements s such that for all values of i that belong to the set $\{1, 2, \ldots, n\}$ element s belongs to the set S_i", i.e. the elements of the intersection of the sets belong to all the sets S_i.

As we have seen before in some literature the union and intersection of two sets A and B are denoted as $A + B$ and $A \cdot B$, respectively. If this notation is used, the generalised union and intersection of n sets $S_1, S_2, \cdots, S_n$ are denoted as:

$$\bigcup_{i=1}^{n} S_i \equiv \sum_{i=1}^{n} S_i = S_1 + S_2 + \cdots + S_n \tag{1.18}$$

$$\bigcap_{i=1}^{n} S_i \equiv \prod_{i=1}^{n} S_i = S_1 \cdot S_2 \ldots S_n \tag{1.19}$$

Next, we further generalise the union and intersection operations to deal with families of indexed sets. Thus, given a family of indexed sets $\mathfrak{S} = \{S_i\}_{i \in I}$ (where I is the index set of the family) we refine the definition of the union and intersection operations as follows. The union operation of a family of indexed sets is defined as:

$$\bigcup_{i \in I} S_i = \{s \mid (\exists i \in I) : (s \in S_i)\} \tag{1.20}$$

If the index set is not ambiguous, the notation $\bigcup_{i \in I} S_i$ can be shortened to $\bigcup S_i$. Similarly, the intersection of a family of indexed sets is defined as:

$$\bigcap_{i \in I} S = \{s \mid (\forall i \in I) : (s \in S_i)\} \tag{1.21}$$

Again, If the index set is not ambiguous, the notation $\bigcap_{i \in I} S_i$ can be shortened to $\bigcap S_i$.

1.19.1 The Negation of the Generalised Set Operations

We must be particularly careful when negating a generalised union or intersection operation on a family of indexed sets. For instance, the expression $s \in \bigcup_{i \in I} S_i$ means that there is at least one value of the index i for which $s \in S_i$ and conversely the expression $s \notin \bigcup_{i \in I} S_i$ means that $s \notin S_i$ for all values of the index $i \in I$. Similarly, the expression $s \in \bigcap_{i \in I} S_i$ means that $s \in S_i$ for all values of the index $i \in I$ and conversely the expression $s \notin \bigcap_{i \in I} S_i$ means that $s \notin S_i$ for at least one value of the index i.

1.19.2 Some Algebraic Rules for the Generalised Set Operations

The distributive rules of the union over intersection (R1.9a) and of the intersection over union (R1.9b) can be extended to deal with a set A and a family of indexed sets $\mathfrak{S} = \{S_i\}_{i \in I}$ (where I is the index set of the family). Starting from the distributive rule of the union over intersection $A \cup (B \cap C) = (A \cup B) \cap (A \cup C)$ we get:

$$A \cup \left(\bigcap_{i \in I} S_i \right) = \bigcap_{i \in I} (A \cup S_i) \tag{1.22}$$

Similarly, starting from the distributive rule of the intersection over union $A \cap (B \cup C) = (A \cap B) \cup (A \cap C)$ we get:

$$A \cap \left(\bigcup_{i \in I} S_i \right) = \bigcup_{i \in I} (A \cap S_i) \tag{1.23}$$

Also, the De Morgan rules for the negation of the union of two sets (R1.10a) and the negation of the intersection of two sets (R1.10b) can be extended to deal with a family of indexed sets $\mathfrak{S} = \{S_i\}_{i \in I}$ (where I is the index set of the family). Starting from the De Morgan rule for the negation of the union of two sets $(A \cup B)' = A' \cap B'$ we get:

$$\left(\bigcup_{i \in I} S_i \right)' = \bigcap_{i \in I} S_i' \tag{1.24}$$

Similarly, starting from the De Morgan rule for the negation of the intersection of two sets $(A \cap B)' = A' \cup B'$ we get:

$$\left(\bigcap_{i \in I} S_i \right)' = \bigcup_{i \in I} S_i' \tag{1.25}$$

1.20 The Cardinality or Size of a Set

When dealing with sets we frequently wish to know what is the size of a set or indeed if two sets have the same size. With finite sets this is easy, all we have to do is to count the number of their elements. With infinite sets, however, this is more difficult (as we can not count the elements of infinite sets) and it depends on how we define the sets to have the same number of elements.

1.20.1 Equivalent Sets

Two sets A and B are said to be *equivalent*, *equipotent* or *equipollent* if there exists a bijective function (i.e. a one-to-one correspondence) f such that $f : A \rightarrow B$. The equivalence of sets A and B is denoted by $A \sim B$.

If the sets A and B are finite, then it is easy to see that these sets can only be equivalent if and only if they have the same number of elements. For instance, given the sets $A = \{0, 1, 2, 3\}$, $B = \{0, 3, 6, 9\}$ and the bijective function $f : A \rightarrow B$ defined by $f(x) = 3x$ it is easy to see that each element of the domain of f has a unique image in its co-domain and that each element in its co-domain has a unique preimage in its domain. This is inherent in the definition of a bijective function. It should also be apparent that the function f can only be defined if its domain and its co-domain have exactly the same number of elements.

On the other hand, if the two sets are infinite, one of the sets can be equivalent to a proper subset of itself. For instance, consider the sets $\mathbb{N} = \{0, 1, 2, 3, \ldots\}$, $M = \{0, 3, 6, 9, \ldots\}$ and the bijective function $f : \mathbb{N} \rightarrow M$ defined by $f(x) = 3x$. Here both sets $\mathbb{N}$ and M are infinite, the function f is bijective and its is obvious that $M \subset \mathbb{N}$. Thus, we arrive at the following definitions:

- A set is infinite if it is equivalent to a proper subset of itself.
- A set is finite if it is not equivalent to a proper subset of itself.

Given three sets $\boldsymbol{A}$, $\boldsymbol{B}$ and $\boldsymbol{C}$ we observe that:

- For any set $\boldsymbol{A}$, $\boldsymbol{A} \sim \boldsymbol{A}$ (i.e. $(\forall \boldsymbol{A}) : (\boldsymbol{A} \sim \boldsymbol{A})$).
- If $\boldsymbol{A} \sim \boldsymbol{B}$, then $\boldsymbol{B} \sim \boldsymbol{A}$ (i.e. $(\boldsymbol{A} \sim \boldsymbol{B}) \Leftrightarrow (\boldsymbol{B} \sim \boldsymbol{A})$).
- If $\boldsymbol{A} \sim \boldsymbol{B}$ and $\boldsymbol{B} \sim \boldsymbol{C}$, then $\boldsymbol{A} \sim \boldsymbol{C}$ (i.e. $(\boldsymbol{A} \sim \boldsymbol{B}) \wedge (\boldsymbol{B} \sim \boldsymbol{C}) \Rightarrow (\boldsymbol{A} \sim \boldsymbol{C})$).

1.20.2 The Cardinal Number or Cardinality of a Set

The *cardinal number*, *cardinal*, *cardinality* or *power* of a set $\boldsymbol{S}$ describes the size of the set, is denoted as $\#(\boldsymbol{S})$ or $|\boldsymbol{S}|$ and is defined as "*if a set* $\boldsymbol{S}$ *contains a finite number of unique elements, then its cardinality* $|\boldsymbol{S}|$ *is the number of elements in* $\boldsymbol{S}$". Thus:

the set $\{\}$ has cardinality 0,
the set $\{0\}$ has cardinality 1,
the set $\{0, 1\}$ has cardinality 2,
the set $\{0, 1, 2\}$ has cardinality 3,
and so on...

According to this definition the sets $\boldsymbol{S}_1 = \{1, 2, 3\}$ and $\boldsymbol{S}_2 = \{a, b, c\}$ have the same cardinality and are said to be equivalent, equipotent, equipollent or *equinumerous* (equal in their number of elements).

1.21 The Power Set of a Set

Given a set $\boldsymbol{S}$, the *power set* of $\boldsymbol{S}$ is the set of all subsets of $\boldsymbol{S}$. The power set of a set $\boldsymbol{S}$ is denoted as $\mathcal{P}(S)$ or 2^S. For instance, given the set $\boldsymbol{S} = \{1, 2, 3\}$, its power set $\mathcal{P}(S) = \{\{\}, \{1\}, \{2\}, \{3\}, \{1, 2\}, \{1, 3\} \{2, 3\}, \{1, 2, 3\}\}$.

If the set $\boldsymbol{S}$ has cardinality $|\boldsymbol{S}| = n$, then its power set has cardinality $|\mathcal{P}(\boldsymbol{S})| = 2^n$. In the above example $|\boldsymbol{S}| = 3$ and $|\mathcal{P}(\boldsymbol{S})| = 2^3 = 8$.

To represent a set with cardinality n in computer memory we would require an n-bit field, with each bit of the field corresponding to a given element of the set. The presence of a given element in the set would be indicated by setting the corresponding bit of the n-bit field to 1, while the absence of an element would be indicated by setting the corresponding bit to 0.

2

Vector Algebra Survival Kit

Vector algebra is one of the most important topics in modern mathematics. The Czech mathematician Bernard Bolzano first developed the concept of a vector in 1804. This concept was further developed by the French mathematician Jean Argand, the German mathematician August Möbius, the Irish mathematician Sir William Hamilton (who is believed to have coined the term vector) and culminated in the development of the *theory of vectors* by the Polish mathematician Hermann Grassmann in 1844. The concept of the vector continued to develop through the late nineteenth century and early twentieth century until the American mathematician Josiah Gibbs introduced the *theory of vector analysis* in 1890 and the *theory of vector spaces* in 1930.

In our study of computer graphics we will frequently use vectors to solve a variety of problems in such diverse areas as geometric modelling, transformations, projections, visibility determination, lighting, shading and texturing, and the development of curves, surfaces and deformations. Thus it is important to gain a thorough understanding of vector algebra.

2.1 Some Basic Definitions and Notation

Let us start by defining the terms *scalar* and *vector*. A scalar is a quantity that is completely determined by a single numerical value, which consists of a possibly signed *magnitude* (i.e. a *real* number). For instance, quantities such as temperature, time, length, mass, and speed are all represented by scalars. A vector, on the other hand, is a quantity that is determined by a *direction* and a *magnitude*. It is a *directed* quantity represented by an arrow. The direction and the length of this arrow determine the vector's direction and magnitude, respectively.

In everyday language it is common to use quantities such as speed and velocity interchangeably. This use however is inaccurate as the speed of a vehicle represents the magnitude of its movement alone and gives no indication of its direction of movement, while the velocity of a vehicle represents both its magnitude and its direction of movement. A scalar is a single dimensional quantity, while a vector is a multidimensional quantity.

In this book we denote vectors in bold italic notation such as $\vec{v}$, $\bar{v}$ or $\boldsymbol{v}$, while scalars are denoted by Greek letters or non-bold italic characters such as λ or l.

A three-dimensional geometric vector can be seen as a translation in three-dimensional Euclidean space $\mathcal{E}^3$. Given a point $\boldsymbol{P}$ in $\mathcal{E}^3$, we may use a vector $\boldsymbol{v}$ to move this point to a new position $\boldsymbol{P}'$, as shown in Fig. 2.1. Sometimes we call the vector $\boldsymbol{v}$ a *displacement vector*, as it displaces point $\boldsymbol{P}$ to its new position $\boldsymbol{P}'$. The distance between points $\boldsymbol{P}$ and $\boldsymbol{P}'$ is called the *magnitude* of the vector $\boldsymbol{v}$ and it is denoted by $|\boldsymbol{v}|$ or the non-bold italic version of the vector name v. This distance is the same for all points $\boldsymbol{P}$ in $\mathcal{E}^3$. Thus, the original position of point $\boldsymbol{P}$ is immaterial. For all vectors $\boldsymbol{v}$ we say that $|\boldsymbol{v}| \geq 0$.

There exists a special vector whose magnitude is zero. This vector translates every point $\boldsymbol{P}$ onto itself, i.e. the position of the point is left unchanged and the magnitude of the vector is zero. We call this vector the *zero vector* or *null vector* and we denote it by $\vec{\boldsymbol{0}}$ or $\boldsymbol{0}$ when there is no notational ambiguity (i.e. when it can not be misinterpreted as a scalar). No direction is associated with the zero vector. Thus,

$$\left|\vec{\boldsymbol{0}}\right| = 0 \tag{2.1}$$

It follows that

$$|\boldsymbol{v}| > 0 \quad \text{for every } \boldsymbol{v} \neq \vec{\boldsymbol{0}} \tag{2.2}$$

A vector $\boldsymbol{v}$ whose magnitude is equal to one (i.e. $|\boldsymbol{v}|=1$) is called a *unit vector*.

From the above discussion it should be apparent that all non-zero vectors are characterised by their *direction* and their *magnitude* and that two vectors are equal if they have the same direction and the same magnitude.

Given two points $\boldsymbol{P}_1$ and $\boldsymbol{P}_2$ in E^3, the straight line segment between $\boldsymbol{P}_1$ and $\boldsymbol{P}_2$, as well as the direction from $\boldsymbol{P}_1$ to $\boldsymbol{P}_2$, is denoted by $\overrightarrow{\boldsymbol{P}_1\boldsymbol{P}_2}$ or $\overline{\boldsymbol{P}_1\boldsymbol{P}_2}$ and is called a *directed segment*. See Fig. 2.2. The points $\boldsymbol{P}_1$ and $\boldsymbol{P}_2$ are called the *initial* and *terminal* points of the directed segment, respectively. The distance between the two points is called the *magnitude* of $\overrightarrow{\boldsymbol{P}_1\boldsymbol{P}_2}$ and is denoted by $\left|\overrightarrow{\boldsymbol{P}_1\boldsymbol{P}_2}\right|$ or $|\boldsymbol{P}_1\boldsymbol{P}_2|$. We use the notation $(\boldsymbol{P}_1\boldsymbol{P}_2)$ to denote the signed (directed) distance between the two points. This of course means that when travelling along the line defined by the

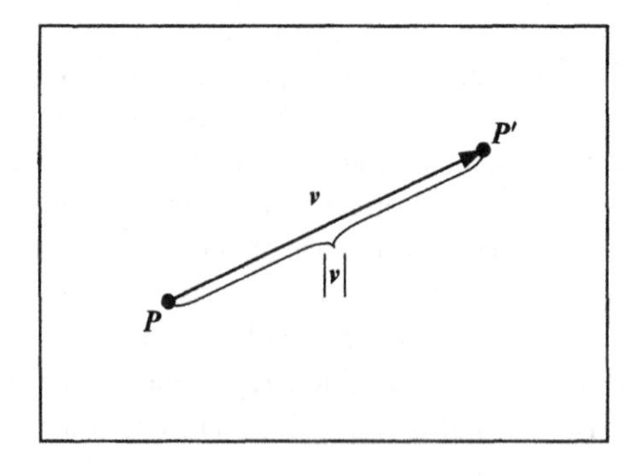

FIGURE 2.1. A displacement vector.

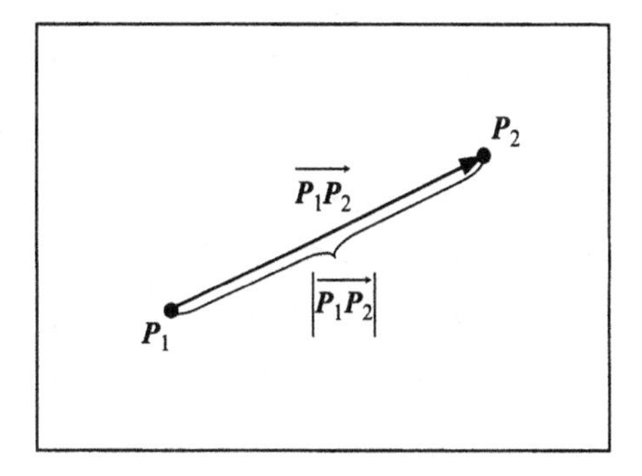

FIGURE 2.2. A directed segment.

points P_1 and P_2, distances are taken to be positive in one direction and negative in the opposite direction. Thus,

$$(P_2P_1) = -(P_1P_2) \tag{2.3}$$

From the above discussion it should be apparent that all non-zero directed segments are defined by their direction and their magnitude. Two directed segments are said to be *equivalent* if they have the same direction and magnitude. We can extend the concept of the directed segment to include the case where the initial and terminal points are the same. Thus, $\overrightarrow{PP}$ is a directed segment of zero magnitude and a non-unique direction.

If a vector $\boldsymbol{v}$ moves a point P to P', then $\boldsymbol{v}$ has the same direction and magnitude as the directed segment $\overrightarrow{PP'}$. The directed segment $\overrightarrow{PP'}$ is said to be *representative* of the vector $\boldsymbol{v}$ with initial point P. Similarly, given a point P' there is *one and only one* representative with terminating point P'. Conversely, given a directed segment $\overrightarrow{PP'}$ there is only one vector $\boldsymbol{v}$ represented by $\overrightarrow{PP'}$.

Two directed segments are said to represent the *same* vector if and only if they are equivalent.

Confusion between a vector and its representative directed segment must be avoided as it leads to serious errors in vector algebra. Despite the fact that both the vector $\boldsymbol{v}$ and its representative directed segment $\overrightarrow{PP'}$ have the same direction and magnitude, $\overrightarrow{PP'}$ has an initial point associated with it, whereas $\boldsymbol{v}$ does not. For this reason a directed segment is sometimes called a *localised vector*.

If two non-zero vectors $\boldsymbol{v}_1$ and $\boldsymbol{v}_2$ have the same direction they are said to be *equidirectional* or *parallel* and if they have an opposite direction they are said to be *opposite* or *antiparallel*. Two vectors that are either parallel or antiparallel are said to be *collinear*. Parallel vectors are denoted by $\boldsymbol{v}_1 \uparrow\uparrow \boldsymbol{v}_2$, antiparallel vectors are denoted by $\boldsymbol{v}_1 \uparrow\downarrow \boldsymbol{v}_2$ and collinear vectors are denoted by $\boldsymbol{v}_1 || \boldsymbol{v}_2$. If $\overrightarrow{P_1P'_1}$ and $\overrightarrow{P_2P'_2}$ are the representatives of $\boldsymbol{v}_1$ and $\boldsymbol{v}_2$, respectively, then the vectors are parallel if the lines $P_1P'_1$ and $P_2P'_2$ are coincident or parallel. It is convenient to think of the zero vector $\vec{\boldsymbol{0}}$ as being parallel to any vector.

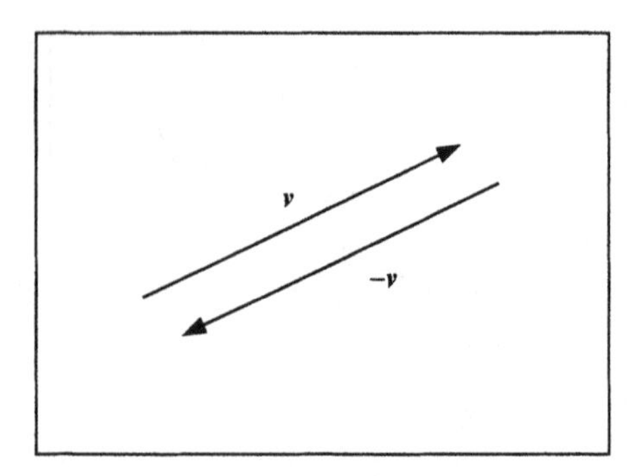

FIGURE 2.3. A vector and its inverse vector.

Given a non-zero vector $\boldsymbol{v}$, the vector with the same magnitude but an opposite direction is called the *negative* or *inverse* of $\boldsymbol{v}$ and is denoted by $-\boldsymbol{v}$. If $\overrightarrow{\boldsymbol{PP'}}$ is the representative of $\boldsymbol{v}$, then $\overrightarrow{\boldsymbol{P'P}}$ is representative of $-\boldsymbol{v}$. See Fig. 2.3. For any vector $\boldsymbol{v}$ we have:

$$-(-\boldsymbol{v}) = \boldsymbol{v} \tag{2.4}$$

2.2 Multiplication of a Vector by a Scalar

If $\boldsymbol{v}$ is a non-zero vector and α is a positive real number (scalar), then we define their product $\alpha\boldsymbol{v}$ to be the vector with the same direction as $\boldsymbol{v}$ and a magnitude $\alpha\,|\boldsymbol{v}|$ or αv. See Fig. 2.4. If α is a negative number, we define $\alpha\boldsymbol{v}$ to be $(-\alpha)(-\boldsymbol{v})$, so that $\alpha\boldsymbol{v}$ is the vector with a direction opposite to $\boldsymbol{v}$ (i.e. antiparallel to $\boldsymbol{v}$) and a magnitude $-\alpha v$. We also define $0\boldsymbol{v}$ to be the zero vector for any vector $\boldsymbol{v}$, and $\alpha\vec{\boldsymbol{0}}$ to be the zero vector for any scalar value α. The operation of forming the product $\alpha\boldsymbol{v}$ is called *scalar multiplication* of $\boldsymbol{v}$ by α.

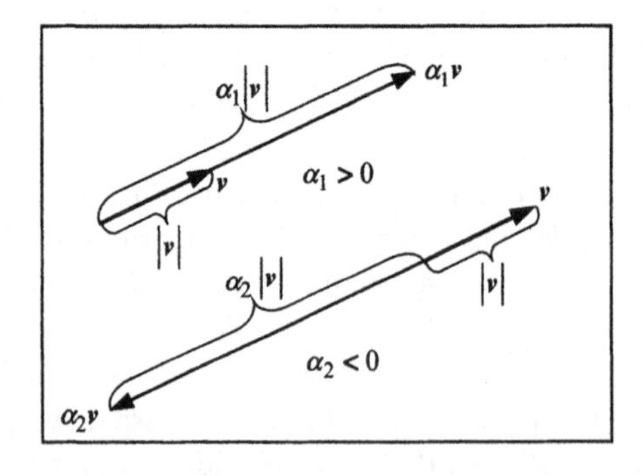

FIGURE 2.4. Scalar multiplication.

The following axioms and rules of vector algebra apply to the vector by scalar product for all vectors $\boldsymbol{v}$, $\boldsymbol{v}_1$ and $\boldsymbol{v}_2$ and all scalars α and β:

Existence of the vector by scalar product:

$\alpha\boldsymbol{v}$ is a vector (A2.1)

Existence of the zero element:

$0\boldsymbol{v} = \boldsymbol{v}0 = \vec{\boldsymbol{0}}$ (A2.2)

Existence of the neutral element:

$1\boldsymbol{v} = \boldsymbol{v}1 = \boldsymbol{v}$ (A2.3)

Associative law:

$\alpha(\beta\boldsymbol{v}) = (\alpha\beta)\boldsymbol{v}$ (R2.1)

Distributive laws:

$(\alpha + \beta)\boldsymbol{v} = \alpha\boldsymbol{v} + \beta\boldsymbol{v}$ (R2.2)

$\alpha(\boldsymbol{v}_1 + \boldsymbol{v}_2) = \alpha\boldsymbol{v}_1 + \alpha\boldsymbol{v}_2$ (R2.3)

2.3 Vector Addition

If vector $\boldsymbol{v}_1$ moves point $\boldsymbol{P}_1$ to $\boldsymbol{P}_2$ and vector $\boldsymbol{v}_2$ moves point $\boldsymbol{P}_2$ to $\boldsymbol{P}_3$, then the combined effect of $\boldsymbol{v}_1$ followed by $\boldsymbol{v}_2$ moves $\boldsymbol{P}_1$ to $\boldsymbol{P}_3$. See Fig. 2.5. The directed segment $\overrightarrow{\boldsymbol{P}_1\boldsymbol{P}_3}$ represents a vector $\boldsymbol{v}_3$, which is unaffected by the choice of $\boldsymbol{P}_1$. So we obtain a unique vector $\boldsymbol{v}_3$ by combining $\boldsymbol{v}_1$ and $\boldsymbol{v}_2$. $\boldsymbol{v}_3$ is known as the *sum* of $\boldsymbol{v}_1$ and $\boldsymbol{v}_2$:

$$\boldsymbol{v}_3 = \boldsymbol{v}_1 + \boldsymbol{v}_2 \tag{2.5}$$

The sum of vectors is sometimes referred to as the *resultant vector*. The operation forming $\boldsymbol{v}_3$ from $\boldsymbol{v}_1$ and $\boldsymbol{v}_2$ is called *vector addition*. Vector addition can also be expressed in terms of directed segments. Thus,

$$\overrightarrow{\boldsymbol{P}_1\boldsymbol{P}_3} = \overrightarrow{\boldsymbol{P}_1\boldsymbol{P}_2} + \overrightarrow{\boldsymbol{P}_2\boldsymbol{P}_3} \tag{2.6}$$

where $\overrightarrow{\boldsymbol{P}_1\boldsymbol{P}_3}$ is the representative of $\boldsymbol{v}_1 + \boldsymbol{v}_2$.

The above result is sometimes referred to as the *triangle rule*. This rule states that if two vectors $\boldsymbol{v}_1$ and $\boldsymbol{v}_2$ are represented in direction and magnitude by two

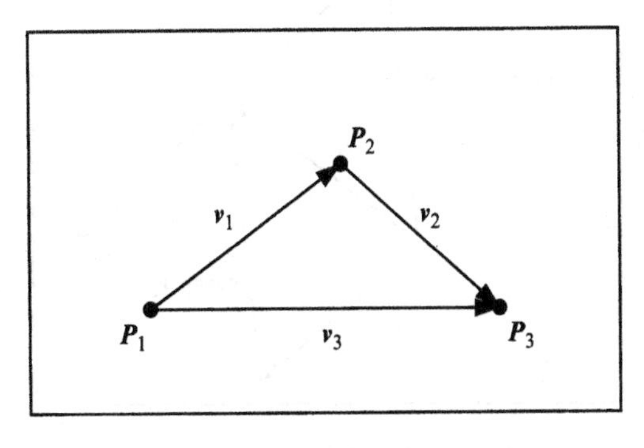

FIGURE 2.5. Vector addition.

consecutive sides of a triangle, then their sum $v_3 = v_1+v_2$ is represented in direction and magnitude by the third side (i.e. the closing side) of this triangle. The direction of the resultant vector is as shown in Fig. 2.5, i.e. from the initial point of v_1 to the terminal point of v_2.

The triangle rule can be generalised to the *polygon rule*, which deals with the summation of n vectors. In this case we arrange the vectors in such a fashion that the initial point of each vector (to be added to the sum) is placed at the terminal point of the vector preceding it in the summation, thus forming a *vector polygon*. Then the *resultant vector* is the vector with the same initial point as the first vector in the summation and the same terminal point as the last vector in the summation. The direction of the resultant vector r is as shown in Fig. 2.6a. If the vector polygon is closed, then the resultant vector is the zero vector. See Fig. 2.6b

The *difference* $v_1 - v_2$ of two vectors v_1 and v_2 is defined as the vector:

$$v_3 = v_1 - v_2 = v_1 + (-v_2) \tag{2.7}$$

The operation that forms v_3 from v_1 and v_2 is called *vector subtraction*. See Fig. 2.7.

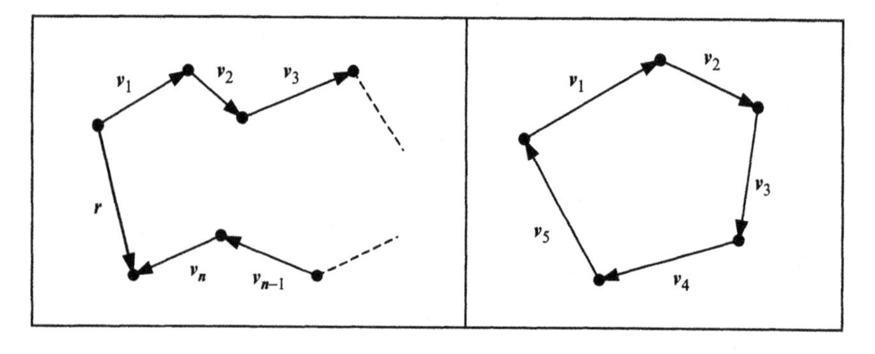

FIGURE 2.6. (a) The polygon rule. (b) A closed vector polygon.

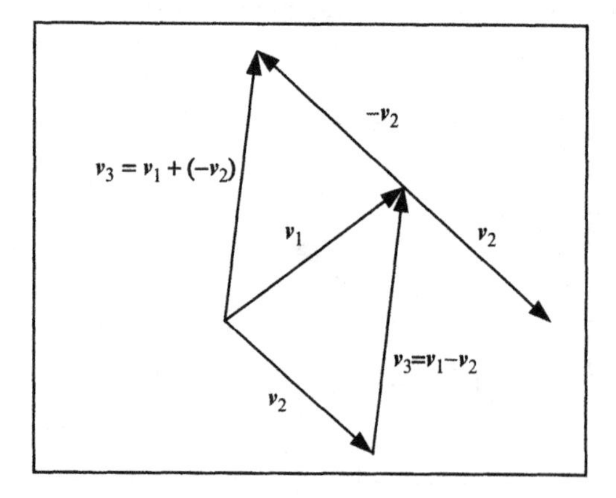

FIGURE 2.7. Vector subtraction.

The following axioms and rules of vector algebra apply to vector addition hold for all vectors $\boldsymbol{v}$, $\boldsymbol{v}_1$, $\boldsymbol{v}_2$ and $\boldsymbol{v}_3$:

Existence of the vector sum:
$\boldsymbol{v}_1 + \boldsymbol{v}_2$ is a vector (A2.4)
Existence of the neutral element:
$\vec{\boldsymbol{0}} + \boldsymbol{v} = \boldsymbol{v} + \vec{\boldsymbol{0}} = \boldsymbol{v}$ (A2.5)
Existence of the inverse element:
$\boldsymbol{v} + (-\boldsymbol{v}) = (-\boldsymbol{v}) + \boldsymbol{v} = \vec{\boldsymbol{0}}$ (A2.6)
Commutative law:
$\boldsymbol{v}_1 + \boldsymbol{v}_2 = \boldsymbol{v}_2 + \boldsymbol{v}_1$ (R2.4)
Associative law:
$(\boldsymbol{v}_1 + \boldsymbol{v}_2) + \boldsymbol{v}_3 = \boldsymbol{v}_1 + (\boldsymbol{v}_2 + \boldsymbol{v}_3)$ (R2.5)
Change of detection rule:
$-(\boldsymbol{v}_1 - \boldsymbol{v}_2) = \boldsymbol{v}_2 - \boldsymbol{v}_1$ (R2.6)

2.4 Position Vectors and Free Vectors

The *position vector* of a point $\boldsymbol{P}$ in $\mathcal{E}^3$ relative to an origin $\boldsymbol{O}$ is defined to be the vector representing the directed segment $\overrightarrow{\boldsymbol{OP}}$. The position vector of a point $\boldsymbol{P}$ is sometimes referred to as a *bound vector*, as its initial point is fixed at the origin $\boldsymbol{O}$ and its terminal point is fixed at the point $\boldsymbol{P}$. The position vector $\boldsymbol{p}$ of a point $\boldsymbol{P}$ relative to the origin $\boldsymbol{O}$ is sometimes denoted by:

$$\vec{\boldsymbol{p}}(\boldsymbol{P}) = \overrightarrow{\boldsymbol{OP}} \tag{2.8}$$

In contrast a *free vector* or an *orientation vector* is any *unbound* vector that is free to be translated in an arbitrary fashion.

2.5 The Vector Equation of a Line

Lines in $\mathcal{E}^3$ can be represented by equations involving position vectors. Let $\boldsymbol{P}_1$ and $\boldsymbol{P}_2$ be two points on a line, having position vectors $\boldsymbol{p}_1$ and $\boldsymbol{p}_2$, respectively,

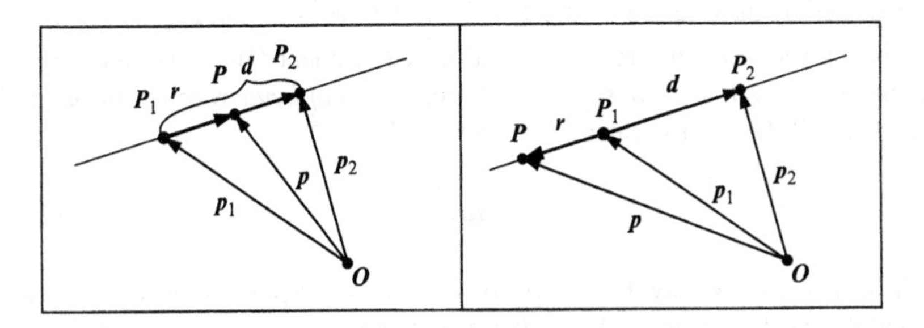

FIGURE 2.8. A line passing through two points.

relative to some origin $\boldsymbol{O}$. From Fig. 2.8 we see that:

$$\overrightarrow{P_1P_2} = \overrightarrow{P_1O} + \overrightarrow{OP_2} \tag{2.9}$$

which represents the *direction vector*:

$$\begin{aligned} \boldsymbol{d} &= (-\boldsymbol{p}_1) + \boldsymbol{p}_2 \\ &= \boldsymbol{p}_2 - \boldsymbol{p}_1 \end{aligned} \tag{2.10}$$

Now, let $\boldsymbol{P}$ be a point whose position vector relative to the origin $\boldsymbol{O}$ is $\boldsymbol{p}$. This point will lie on the line $\boldsymbol{P}_1\boldsymbol{P}_2$ *if and only if* $\overrightarrow{P_1P}$ and $\overrightarrow{P_1P_2}$ represent collinear vectors. Since $\overrightarrow{P_1P}$ represents vector $\boldsymbol{r} = \boldsymbol{p} - \boldsymbol{p}_1$, it follows that point $\boldsymbol{P}$ lies on the line $\boldsymbol{P}_1\boldsymbol{P}_2$ if and only if $\boldsymbol{p} - \boldsymbol{p}_1 = t(\boldsymbol{p}_2 - \boldsymbol{p}_1)$ for some scalar t, i.e. if and only if

$$\boldsymbol{p} = (1 - t)\boldsymbol{p}_1 + t\boldsymbol{p}_2 \tag{2.11}$$

This is the *parametric vector equation of the line* $\boldsymbol{P}_1\boldsymbol{P}_2$ relative to some origin $\boldsymbol{O}$. Here t is a parameter that can assume any real value. It is worth noting that at point $\boldsymbol{P}_1$ $t = 0$, at point $\boldsymbol{P}_2$ $t = 1$, at points preceding the directed segment $\overrightarrow{P_1P_2}$ $t < 0$ and at points following it $t > 1$.

2.6 Linear Dependence/Independence of Vectors

If $\boldsymbol{a}$ is a non-zero vector, then any vector $\boldsymbol{v}$ which is collinear to the vector $\boldsymbol{a}$ is of the form $\alpha\boldsymbol{a}$ for some scalar α. When vectors $\boldsymbol{a}$ and $\boldsymbol{v}$ are parallel then $\alpha = |\boldsymbol{v}| / |\boldsymbol{a}|$, when vectors $\boldsymbol{a}$ and $\boldsymbol{v}$ are antiparallel then $\alpha = -|\boldsymbol{v}| / |\boldsymbol{a}|$ and finally when $\boldsymbol{v} = \vec{\boldsymbol{0}}$ then $\alpha = 0$.

If vector $\boldsymbol{v}$ is of the form $\alpha\boldsymbol{a}$ (i.e. $\boldsymbol{v} = \alpha\boldsymbol{a}$), then we say that vector $\boldsymbol{v}$ is *linearly dependent* on vector $\boldsymbol{a}$. This definition remains valid even if $\boldsymbol{a} = \vec{\boldsymbol{0}}$. When $\boldsymbol{a} \neq \vec{\boldsymbol{0}}$, vector $\boldsymbol{v}$ is linearly dependent on vector $\boldsymbol{a}$ if and only if vectors $\boldsymbol{v}$ and $\boldsymbol{a}$ are collinear.

Next, assume that we have two vectors $\boldsymbol{a}$ and $\boldsymbol{b}$ that are non-collinear and are represented by the directed segments $\overrightarrow{OA}$ and $\overrightarrow{OB}$, as shown in Fig. 2.9.

Given a vector $\boldsymbol{v}$ represented by the directed segment $\overrightarrow{OV}$, where point $\boldsymbol{V}$ lies on the plane $\boldsymbol{AOB}$, then we say that the vector $\boldsymbol{v}$ is *coplanar* with vectors $\boldsymbol{a}$ and $\boldsymbol{b}$ if and only if there exist scalars α and β such that:

$$\boldsymbol{v} = \alpha\boldsymbol{a} + \beta\boldsymbol{b} \tag{2.12}$$

In such a case, we say that the vector $\boldsymbol{v}$ is *linearly dependent* on vectors $\boldsymbol{a}$ and $\boldsymbol{b}$, and that vector $\boldsymbol{v}$ is formed as a *linear combination* of vectors $\boldsymbol{a}$ and $\boldsymbol{b}$. Further we say that vectors $\boldsymbol{a}$, $\boldsymbol{b}$ and $\boldsymbol{v}$ are *coplanar vectors*. Thus, any vector lying on a

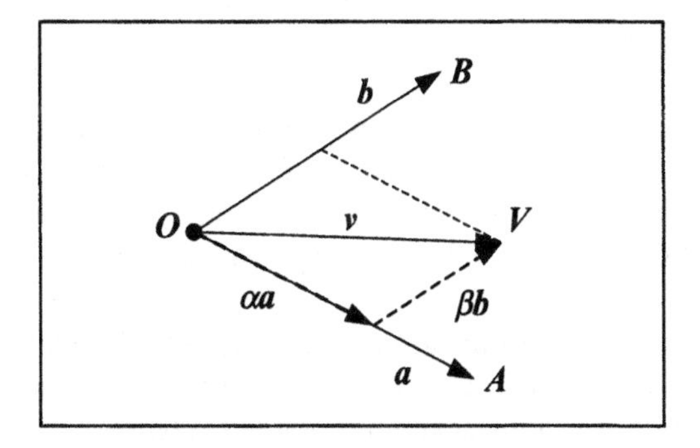

FIGURE 2.9. Linear dependence on two vectors.

plane defined by two other vectors is linearly dependent on these vectors, i.e. it can be expressed as the sum of multiples of these vectors.

Finally, assume that we have three non-coplanar vectors $\boldsymbol{a}$, $\boldsymbol{b}$, $\boldsymbol{c}$ represented by the directed segments $\overrightarrow{\boldsymbol{OA}}$, $\overrightarrow{\boldsymbol{OB}}$, $\overrightarrow{\boldsymbol{OC}}$, respectively, as shown in Fig. 2.10. The relationship of these vectors implies that points $\boldsymbol{O}, \boldsymbol{A}, \boldsymbol{B}, \boldsymbol{C}$ are not coplanar.

A vector $\boldsymbol{v}$ is said to be linearly dependent on vectors $\boldsymbol{a}, \boldsymbol{b}, \boldsymbol{c}$ if and only if there exist scalars α, β, γ such that:

$$\boldsymbol{v} = \alpha \boldsymbol{a} + \beta \boldsymbol{b} + \gamma \boldsymbol{c} \tag{2.13}$$

Thus any vector of $\mathcal{E}^3$ is linearly dependent on three non-coplanar vectors.

In general, given a set of n non-zero vectors $\{\boldsymbol{v}_i\}_{i=1}^n$ and a set of scalars $\{\alpha_i\}_{i=1}^n$, there exists a linear combination of these vectors that vanishes, i.e. equal to the null vector:

$$\alpha_1 \boldsymbol{v}_1 + \alpha_2 \boldsymbol{v}_2 + \cdots + \alpha_n \boldsymbol{v}_n = \vec{\boldsymbol{0}} \tag{2.14}$$

If this linear combination of vectors vanishes with at least one of the scalars $\alpha_i \neq 0$, then we say that these vectors are *linearly dependent*. If however this linear combination of vectors only vanishes when all scalars $\{\alpha_i = 0\}_{i=1}^n$, then we say that these vectors are *linearly independent*.

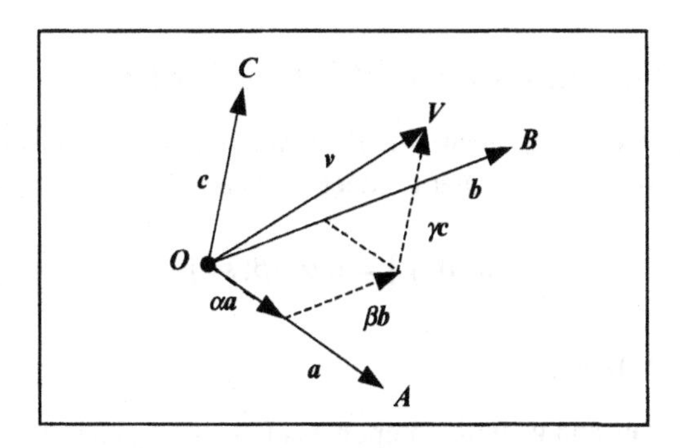

FIGURE 2.10. Linear dependence on three vectors.

2.7 Vector Bases

Given an ordered set of three non-coplanar vectors $\boldsymbol{a}, \boldsymbol{b}, \boldsymbol{c}$, we call this a *basis* of E^3. The order of the vectors is significant. If $\boldsymbol{a}, \boldsymbol{b}, \boldsymbol{c}$ is a basis of E^3, then $\boldsymbol{b}, \boldsymbol{c}, \boldsymbol{a}$ is a different basis of E^3. The vectors $\boldsymbol{a}, \boldsymbol{b}, \boldsymbol{c}$ forming the basis are called *base vectors*. Every other vector in E^3 can be expressed as a linear combination of a given basis.

In general, a basis of an n-dimensional *vector space* E^n is defined by a set of n linearly independent vectors $\{\boldsymbol{v}_i\}_{i=1}^n$ in E^n. Every other vector $\boldsymbol{v}$ in this vector space can be expressed as a linear combination of the base vectors:

$$\boldsymbol{v} = \alpha_1\boldsymbol{v}_1 + \alpha_2\boldsymbol{v}_2 + \cdots + \alpha_n\boldsymbol{v}_n \tag{2.15}$$

where $\{\alpha_i\}_{i=1}^n$ are scalars.

2.8 The Components of a Vector

Let $\boldsymbol{a}, \boldsymbol{b}, \boldsymbol{c}$ be a basis. Then vector $\boldsymbol{v}$ is linearly dependent on the base vectors $\boldsymbol{a}$, $\boldsymbol{b}, \boldsymbol{c}$ and there exist a unique triple of scalars α, β, γ such that $\boldsymbol{v} = \alpha\boldsymbol{a} + \beta\boldsymbol{b} + \gamma\boldsymbol{c}$. The scalars α, β, γ are called the *components of vector* $\boldsymbol{v}$ *relative to the basis* $\boldsymbol{a}, \boldsymbol{b}, \boldsymbol{c}$.

Let $\boldsymbol{v}$ be the set of all the vectors $\boldsymbol{v}$ and $\boldsymbol{C}$ be the set of all the triples of components $[\alpha, \beta, \gamma]$ where α, β, γ are scalars. There is a *one to one correspondence* between sets $\boldsymbol{v}$ and $\boldsymbol{C}$ associating each vector $\boldsymbol{v}$ with a triple of components $[\alpha, \beta, \gamma]$. Thus a triple of components is uniquely determined by $\boldsymbol{v}$ and given the triple $[\alpha, \beta, \gamma]$ there is only one vector v that satisfies the equation $\boldsymbol{v} = \alpha\,\boldsymbol{a} + \beta\,\boldsymbol{b} + \gamma\,\boldsymbol{c}$. Using this correspondence between vectors and the ordered triples of scalars we can prove properties of vectors by working in terms of their components rather than the vectors themselves, remembering that the basis relative to which a vector is expressed uniquely determines its components. Thus, changing the basis relative to which a vector is expressed changes its components.

Let us now see how multiplication of a vector by a scalar and vector addition can be defined in terms of the components of vectors.

2.8.1 Multiplication of a Vector by a Scalar

Given a vector $\boldsymbol{v}$ with components $[\alpha, \beta, \gamma]$ relative to some basis $\boldsymbol{a}, \boldsymbol{b}, \boldsymbol{c}$ and a scalar s, then the vector by a scalar product sv in terms of its components is given by:

$$s[\alpha, \beta, \gamma] = [s\alpha, s\beta, s\gamma] \tag{2.16}$$

2.8.2 Vector Addition

Given two vectors $\boldsymbol{v}_1$ and $\boldsymbol{v}_2$ with components $[\alpha_1, \beta_1, \gamma_1]$ and $[\alpha_2, \beta_2, \gamma_2]$ relative to some basis $\boldsymbol{a}, \boldsymbol{b}, \boldsymbol{c}$, then the vector sum $\boldsymbol{v}_1 \pm \boldsymbol{v}_2$ in terms of its components is given by:

$$[\alpha_1, \beta_1, \gamma_1] \pm [\alpha_2, \beta_2, \gamma_2] = [\alpha_1 \pm \alpha_2, \beta_1 \pm \beta_2, \gamma_1 \pm \gamma_2] \tag{2.17}$$

2.8.3 Vector Equality

Given two vectors $\boldsymbol{v}_1$ and $\boldsymbol{v}_2$ with components $[\alpha_1, \beta_1, \gamma_1]$ and $[\alpha_2, \beta_2, \gamma_2]$ relative to some basis $\boldsymbol{a}$, $\boldsymbol{b}$, $\boldsymbol{c}$, then the vectors are said to be equal if their corresponding components are equal, i.e.

$$[\alpha_1, \beta_1, \gamma_1] = [\alpha_2, \beta_2, \gamma_2] \quad \Leftrightarrow \quad \alpha_1 = \alpha_2 \wedge \beta_1 = \beta_2 \wedge \gamma_1 = \gamma_2 \tag{2.18}$$

The components $[\alpha, \beta, \gamma]$ of a vector $\boldsymbol{v}$ can either be written in row or in column form, i.e.

$$[\alpha, \beta, \gamma] \text{ or } \begin{bmatrix} \alpha \\ \beta \\ \gamma \end{bmatrix}$$

The row form is said to be the *transpose* of the column form and vice versa, i.e.

$$[\alpha, \beta, \gamma]^T = \begin{bmatrix} \alpha \\ \beta \\ \gamma \end{bmatrix} \text{ and } \begin{bmatrix} \alpha \\ \beta \\ \gamma \end{bmatrix}^T = [\alpha, \beta, \gamma]$$

These two representations are equivalent and the choice of one over the other is a matter of taste.

2.9 Orthogonal, Orthonormal and Right-Handed Vector Bases

A basis is said to be *orthogonal* if its base vectors are mutually perpendicular and it is said to be *orthonormal* if its base vectors are mutually perpendicular unit vectors.

Given an orthogonal basis $\boldsymbol{a}$, $\boldsymbol{b}$, $\boldsymbol{c}$ and any vector $\boldsymbol{v}$, the directed segments $\overrightarrow{OA}$, $\overrightarrow{OB}$, $\overrightarrow{OC}$, $\overrightarrow{OV}$ represent the $\boldsymbol{a}$, $\boldsymbol{b}$, $\boldsymbol{c}$, $\boldsymbol{v}$ vectors, respectively, as shown in Fig. 2.11.

Let $\boldsymbol{P}$ be the foot of the perpendicular from point $\boldsymbol{V}$ to the plane $\boldsymbol{AOB}$ (i.e. the projection of point $\boldsymbol{V}$ on the plane $\boldsymbol{AOB}$) and let $\boldsymbol{Q}$ be the foot of the perpendicular from point $\boldsymbol{P}$ to the line $\boldsymbol{OA}$. From this figure we see that:

$$\overrightarrow{OV} = \overrightarrow{OQ} + \overrightarrow{QP} + \overrightarrow{PV} \tag{2.19}$$

If vector $\boldsymbol{v}$ has components $[\alpha, \beta, \gamma]$ relative to the orthogonal basis $\boldsymbol{a}$, $\boldsymbol{b}$, $\boldsymbol{c}$, then $\overrightarrow{OV}$ represents the vector sum $\alpha\,\boldsymbol{a} + \beta\,\boldsymbol{b} + \gamma\,\boldsymbol{c}$. But the directed segments $\overrightarrow{OQ}$, $\overrightarrow{QP}$, $\overrightarrow{PV}$ have the same directions as $\boldsymbol{a}$, $\boldsymbol{b}$, $\boldsymbol{c}$, respectively. Thus $\overrightarrow{OQ}$, $\overrightarrow{QP}$,

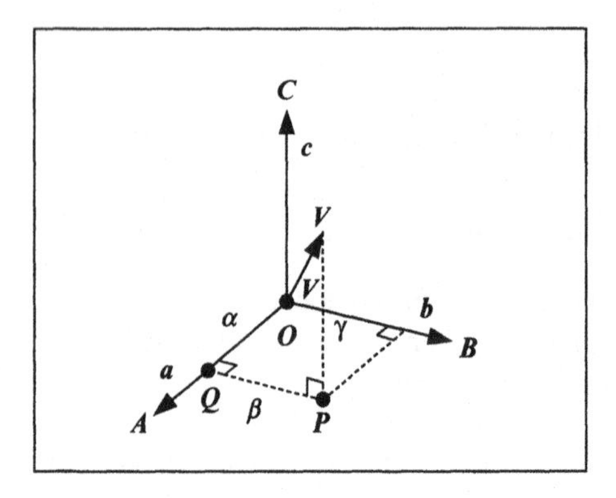

FIGURE 2.11. An orthogonal vector basis.

$\overrightarrow{PV}$ represent the vectors $\alpha\boldsymbol{a}$, $\beta\boldsymbol{b}$, $\gamma\boldsymbol{c}$, respectively. From this we can deduce that the magnitudes of the components α, β, γ are equal to the magnitudes of the orthogonal projections of the directed segment $\overrightarrow{OV}$ onto the lines $\boldsymbol{OA}$, $\boldsymbol{OB}$, $\boldsymbol{OC}$, respectively. The sign of α is positive if points $\boldsymbol{V}$ and $\boldsymbol{A}$ are on the same side of the plane $\boldsymbol{BOC}$ and negative if they are on opposite sides of this plane. Similarly for the signs of β and γ.

For a non-orthogonal basis we can not obtain the components of a vector, with respect to this basis, by orthogonal projections as we have done above. In a later section we resolve this problem using triple scalar products.

Let $\boldsymbol{a}$, $\boldsymbol{b}$, $\boldsymbol{c}$ be three non-zero non-coplanar vectors, represented by the directed segments $\overrightarrow{OA}$, $\overrightarrow{OB}$, $\overrightarrow{OC}$, respectively, form a basis, as shown in Fig. 2.12.

Suppose that there is a right-handed screw aligned with the directed segment $\overrightarrow{OC}$ and pointing in the same direction as $\overrightarrow{OC}$. A clockwise rotation of the screw will cause it to advance along $\overrightarrow{OC}$ and will also cause $\overrightarrow{OA}$ and $\overrightarrow{OB}$ to rotate so that $\overrightarrow{OA}$ rotates towards the original position of $\overrightarrow{OB}$. Alternatively, suppose that the observer's eye is situated at point $\boldsymbol{C}$ and is looking along $\overrightarrow{OC}$ towards the point $\boldsymbol{O}$.

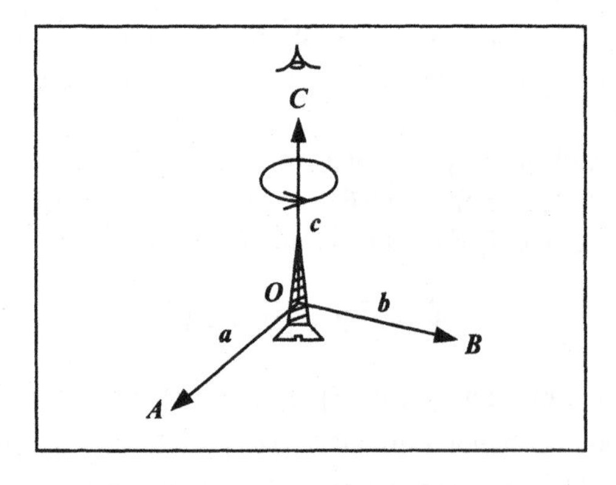

FIGURE 2.12. A right-handed basis.

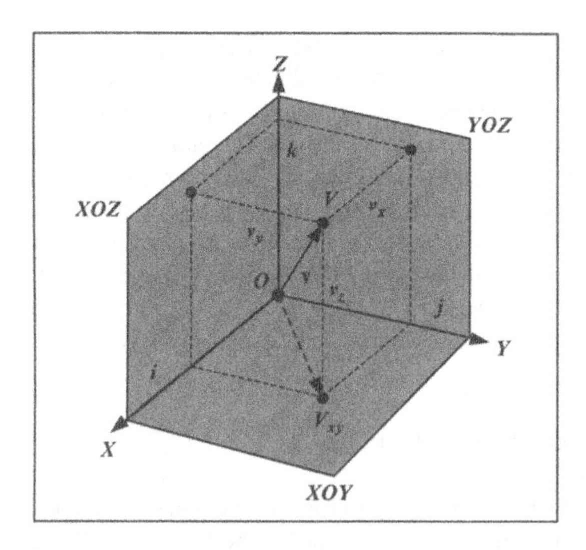

FIGURE 2.13. A rectangular Cartesian coordinate system.

A counter-clockwise rotation about $\overrightarrow{OC}$ will cause $\overrightarrow{OA}$ and $\overrightarrow{OB}$ to rotate so that $\overrightarrow{OA}$ rotates towards the original position of $\overrightarrow{OB}$. Any ordered set of vectors $\boldsymbol{a}, \boldsymbol{b}, \boldsymbol{c}$ that are arranged thus is called a *right-handed system* of vectors or a *right-handed basis*.

2.10 Cartesian Bases and Cartesian Coordinates

A basis that is both orthonormal and right-handed is called a *Cartesian basis*.

Let $\boldsymbol{i}, \boldsymbol{j}, \boldsymbol{k}$ be a Cartesian basis, let $\boldsymbol{O}$ be any fixed point and let the directed segments $\overrightarrow{OX}, \overrightarrow{OY}, \overrightarrow{OZ}$ represent the base vectors $\boldsymbol{i}, \boldsymbol{j}, \boldsymbol{k}$, respectively, as shown in Fig. 2.13.

If $\boldsymbol{v}$ is the point with position vector $\boldsymbol{v}$ relative to the origin $\boldsymbol{O}$, the components of vector $\boldsymbol{v}$ relative to the basis $\boldsymbol{i}, \boldsymbol{j}, \boldsymbol{k}$ are the perpendicular distances of point $\boldsymbol{V}$ from the planes $\boldsymbol{YOZ}, \boldsymbol{XOZ}, \boldsymbol{XOY}$, respectively, with the sign conventions described in Section 2.9.

In this section we abandon the conventions we have adopted earlier of denoting the components of vectors by Greek letters. We write v_x, v_y, v_z for the components of the vector $\boldsymbol{v}$ relative to some Cartesian basis $\boldsymbol{i}, \boldsymbol{j}, \boldsymbol{k}$. Thus,

$$\boldsymbol{v} = v_x\boldsymbol{i} + v_y\boldsymbol{j} + v_z\boldsymbol{k} \tag{2.20}$$

By expressing the position vectors of points in this way, we establish a *one to one* correspondence between vectors and ordered triples of real numbers. Here the triple $[v_x, v_y, v_z]$ corresponds to point $\boldsymbol{V}$ with position vector $\boldsymbol{v} = v_x\boldsymbol{i} + v_y\boldsymbol{j} + v_z\boldsymbol{k}$. Such a correspondence is called a *rectangular Cartesian coordinate system* and the components of the position vector are called the *coordinates* of the point to which they correspond.

The coordinates of the origin $\boldsymbol{O}$ are $[0, 0, 0]$, the coordinates of a point on the line $\boldsymbol{OX}$ are given by $[x, 0, 0]$, the coordinates of a point on the plane $\boldsymbol{XOY}$ are given by $[x, y, 0]$ and finally the coordinates of a point in space are given by $[x, y, z]$. Similar properties hold for the other *coordinate axes* and *major planes*.

It is sometimes convenient to use vector notation rather than coordinate notation. When doing so, we often write $\boldsymbol{v} = \left[v_x, v_y, v_z\right]$ to mean the same thing as $\boldsymbol{v} = v_x\boldsymbol{i} + v_y\boldsymbol{j} + v_z\boldsymbol{k}$. We frequently abbreviate the expression *the point with coordinates* $\left[v_x, v_y, v_z\right]$ by the expression *the point* $\left[v_x, v_y, v_z\right]$ and the expression *the point with position vector* $\boldsymbol{v}$ by the expression *the point* $\boldsymbol{v}$. Such abbreviations are valid so long as only one coordinate system is involved.

2.11 The Length of a Vector

Let $\left[v_x, v_y, v_z\right]$ be the components of a vector $\boldsymbol{v}$ related to some Cartesian basis $\boldsymbol{i}, \boldsymbol{j}, \boldsymbol{k}$, as shown in Fig. 2.13. Let point $\boldsymbol{V}_{xy}$ be the foot of the perpendicular of point $\boldsymbol{V}$ onto the $\boldsymbol{XOY}$ plane. From this figure we see that:

$$\begin{aligned} |\boldsymbol{v}|^2 &= \left|\overrightarrow{\boldsymbol{OV}}\right|^2 \\ &= \left|\overrightarrow{\boldsymbol{OV}_{xy}}\right|^2 + v_z^2 \end{aligned}$$

But:

$$\begin{aligned} \left|\overrightarrow{\boldsymbol{OV}_{xy}}\right|^2 &= v_x^2 + v_y^2 \\ \therefore\quad |\boldsymbol{v}|^2 &= v_x^2 + v_y^2 + v_z^2 \\ \therefore\quad |\boldsymbol{v}| &= \sqrt{v_x^2 + v_y^2 + v_z^2} \end{aligned} \tag{2.21}$$

2.12 The Scalar Product of Vectors

Let $\boldsymbol{a}$ and $\boldsymbol{b}$ be two non-zero vectors represented by the directed segments $\overrightarrow{\boldsymbol{OA}}$ and $\overrightarrow{\boldsymbol{OB}}$, respectively. The angle θ between the two vectors is the angle between their representatives, as seen in Fig. 2.14. We assume that $0 \leq \theta \leq \pi$. If $\theta = 0$, then the two vectors have the same direction, if $\theta = \pi$, then the two vectors point in opposite directions and if $\theta = \frac{\pi}{2}$, then the two vectors are perpendicular.

The *scalar product* or *dot product* or *inner product* $\boldsymbol{a} \odot \boldsymbol{b}$ of vectors $\boldsymbol{a}$ and $\boldsymbol{b}$ is defined as:

$$\boldsymbol{a} \odot \boldsymbol{b} = |\boldsymbol{a}|\,|\boldsymbol{b}| \cos\theta \tag{2.22}$$

If $a = \vec{\boldsymbol{0}}$ or $b = \vec{\boldsymbol{0}}$, then the scalar product $\boldsymbol{a} \odot \boldsymbol{b}$ is defined to be zero.

If $\boldsymbol{a}$ and $\boldsymbol{b}$ are unit vectors, then their scalar product simplifies to $\boldsymbol{a} \odot \boldsymbol{b} = \cos\theta$.

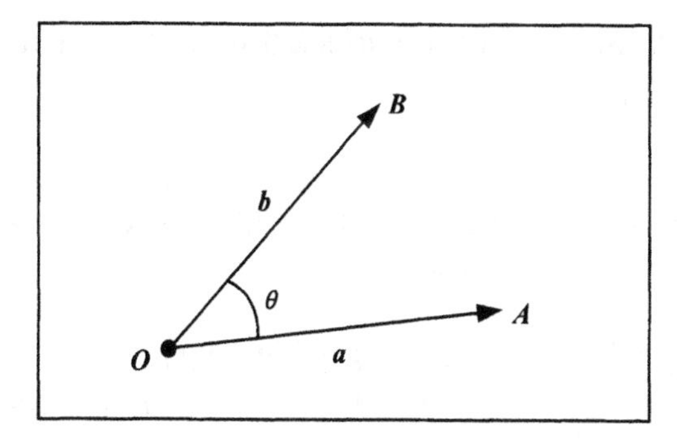

FIGURE 2.14. The angle between two vectors.

If $a \perp b$ (i.e. $\boldsymbol{a}$ and $\boldsymbol{b}$ are perpendicular), then their scalar product is $\boldsymbol{a} \odot \boldsymbol{b} = 0$, as $cos\left(\frac{\pi}{2}\right) = 0$.

From the above it should be apparent that the scalar product of two vectors is a scalar. This implies that the scalar product is only defined for two vectors. This further implies that the only power of vectors that is defined is the square power, i.e. $\boldsymbol{a}^2 = \boldsymbol{a} \odot \boldsymbol{a} = |\boldsymbol{a}|^2$.

Applying Eq. (2.22) to the base vectors of an orthonormal basis $\boldsymbol{i}, \boldsymbol{j}, \boldsymbol{k}$ we obtain:

$$\boldsymbol{i} \odot \boldsymbol{i} = \boldsymbol{j} \odot \boldsymbol{j} = \boldsymbol{k} \odot \boldsymbol{k} = 1$$

and

$$\boldsymbol{i} \odot \boldsymbol{j} = \boldsymbol{j} \odot \boldsymbol{k} = \boldsymbol{i} \odot \boldsymbol{k} = 0 \tag{2.23}$$

since $cos\,(0) = 1$ and $cos\left(\frac{\pi}{2}\right) = 0$.

The above result can be written in a more condensed form using *Kronecker's* symbol that summarises orthonormality:

$$\delta_{l,m} = \boldsymbol{v}_l \odot \boldsymbol{v}_m = \begin{cases} 1, & l = m \\ 0, & otherwise \end{cases}$$

$$\text{with} \quad l, m \in \{1, 2, 3\} \quad \text{and} \quad \text{where} \quad \boldsymbol{v}_1 = \boldsymbol{i},\ \boldsymbol{v}_2 = \boldsymbol{j},\ \boldsymbol{v}_3 = \boldsymbol{k} \tag{2.24}$$

The following axioms and vector algebra rules apply to the scalar product of vectors for all vectors $\boldsymbol{v}_1, \boldsymbol{v}_2, \boldsymbol{v}_3$ and $\boldsymbol{v}$, and all scalars α:

Existence of the scalar product:

$\boldsymbol{v}_1 \odot \boldsymbol{v}_2$ is a scalar (A2.7)

Powers of a vector:

$$\boldsymbol{v}^2 = \boldsymbol{v} \odot \boldsymbol{v} = |\boldsymbol{v}|^2 \tag{A2.8}$$

Commutative law:

$$\boldsymbol{v}_1 \odot \boldsymbol{v}_2 = \boldsymbol{v}_2 \odot \boldsymbol{v}_1 \tag{R2.7}$$

Distributive laws:

$$(\alpha \boldsymbol{v}_1) \odot \boldsymbol{v}_2 = \boldsymbol{v}_1 \odot (\alpha \boldsymbol{v}_2) = \alpha\,(\boldsymbol{v}_1 \odot \boldsymbol{v}_2) \tag{R2.8}$$

$$\boldsymbol{v}_1 \odot (\boldsymbol{v}_2 + \boldsymbol{v}_3) = \boldsymbol{v}_1 \odot \boldsymbol{v}_2 + \boldsymbol{v}_1 \odot \boldsymbol{v}_3 \tag{R2.9}$$

2.13 The Scalar Product Expressed in Terms of its Components

If $\boldsymbol{p}$, $\boldsymbol{q}$, $\boldsymbol{r}$ is a basis and if $\boldsymbol{a}$, $\boldsymbol{b}$ are vectors with components $\left[a_x, a_y, a_z\right]$ and $\left[b_x, b_y, b_z\right]$, respectively, relative to this basis, then the scalar product $\boldsymbol{a} \odot \boldsymbol{b}$ (in terms of its components) is defined as:

$$\begin{aligned}\boldsymbol{a} \odot \boldsymbol{b} &= \left(a_x\boldsymbol{p} + a_y\boldsymbol{q} + a_z\boldsymbol{r}\right) \odot \left(b_x\boldsymbol{p} + b_y\boldsymbol{q} + b_z\boldsymbol{r}\right) \\ &= a_x b_x \boldsymbol{p} \odot \boldsymbol{p} + a_y b_y \boldsymbol{q} \odot \boldsymbol{q} + a_z b_z \boldsymbol{r} \odot \boldsymbol{r} + \left(a_y b_z + a_z b_y\right) \boldsymbol{q} \odot \boldsymbol{r} \\ &\quad + \left(a_z b_x + a_x b_z\right) \boldsymbol{r} \odot \boldsymbol{p} + \left(a_x b_y + a_y b_x\right) \boldsymbol{p} \odot \boldsymbol{q}\end{aligned} \tag{2.25}$$

When dealing with an orthonormal basis $\boldsymbol{i}, \boldsymbol{j}, \boldsymbol{k}$, using equation (2.23) we can simplify the above definition to:

$$\boldsymbol{a} \odot \boldsymbol{b} = a_x b_x + a_y b_y + a_z b_z \tag{2.26}$$

2.14 Properties and Applications of the Scalar Product

In this section, unless otherwise stated or implied, we will assume that the components of all vectors used are defined with respect to a Cartesian basis $\boldsymbol{i}, \boldsymbol{j}, \boldsymbol{k}$.

2.14.1 The Magnitude of a Vector Using its Components

Using Axiom (A2.8) we can now define the *magnitude*, *length* or *norm* of a vector in terms of the scalar product as:

$$|\boldsymbol{v}| = \sqrt{\boldsymbol{v} \odot \boldsymbol{v}} = \sqrt{v_x^2 + v_y^2 + v_z^2} \tag{2.27}$$

Which accords with the definition we have given in Eq. (2.21).

2.14.2 Normalising a Vector

Given a non-zero vector $\boldsymbol{v}$ we can *normalise* this vector (i.e. cause it to become a unit-vector) by dividing the vector by its magnitude. This *normalised vector* is given by:

$$\hat{\boldsymbol{v}} = \frac{\boldsymbol{v}}{|\boldsymbol{v}|} = \left[\frac{v_x}{\sqrt{v_x^2 + v_y^2 + v_z^2}}, \frac{v_y}{\sqrt{v_x^2 + v_y^2 + v_z^2}}, \frac{v_z}{\sqrt{v_x^2 + v_y^2 + v_z^2}}\right] \tag{2.28}$$

The magnitude of a normalised vector is $\left|\hat{\boldsymbol{v}}\right| = 1$.

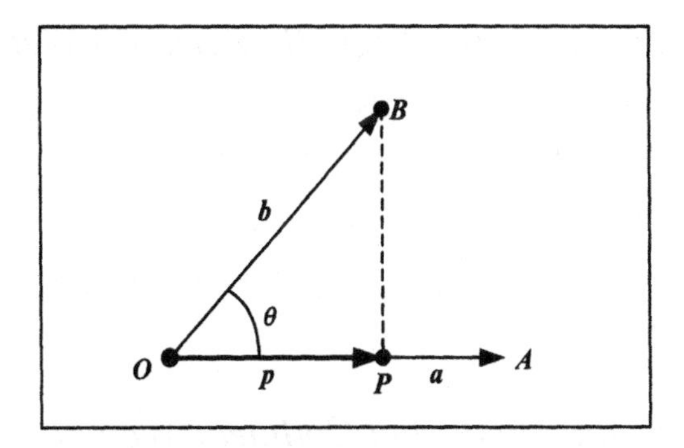

FIGURE 2.15. Projection of a vector onto another.

2.14.3 The Projection of a Vector onto Another

Suppose that $\boldsymbol{a}$, $\boldsymbol{b}$ are non-zero and non-collinear vectors represented by the directed segments $\overrightarrow{OA}$, $\overrightarrow{OB}$, respectively, as shown in Fig. 2.15. Let P be the foot of the perpendicular from point B to the line OA. Then $\overrightarrow{OP}$ is the projection of $\overrightarrow{OB}$ onto $\overrightarrow{OA}$ and vector $\boldsymbol{p}$ represents the projection of vector $\boldsymbol{b}$ onto vector $\boldsymbol{a}$. Vector $\boldsymbol{p}$ is known as the *projection vector*. From the diagram it should be apparent that the magnitude of $\boldsymbol{p}$ is not affected by the magnitude of $\boldsymbol{a}$.

Initially, let us suppose that the vectors $\boldsymbol{a}$, $\boldsymbol{b}$ are unit vectors. Using simple trigonometry we determine that in this case the magnitude of the projection vector $\boldsymbol{p}$ is $|\boldsymbol{p}| = \cos\theta$ and the vector itself is given by:

$$\boldsymbol{p} = \boldsymbol{a} \cdot \cos\theta \tag{2.29}$$

Next, let us assume that vector $\boldsymbol{a}$ is a unit vector and $\boldsymbol{b}$ is any non-zero vector. In this case the magnitude of the projection vector $\boldsymbol{p}$ is $|\boldsymbol{p}| = |\boldsymbol{b}|\cos\theta$ and the vector itself is given by:

$$\boldsymbol{p} = \boldsymbol{a} \cdot |\boldsymbol{b}|\cos\theta \tag{2.30}$$

Finally, let us relax all restrictions and assume that $\boldsymbol{a}$, $\boldsymbol{b}$ are any non-zero vectors. In this case the magnitude of the projection vector $\boldsymbol{p}$ is $|\boldsymbol{p}| = |\boldsymbol{b}|\cos\theta$ (as before) but now the vector itself is given by:

$$\boldsymbol{p} = \hat{\boldsymbol{a}} \cdot |\boldsymbol{b}|\cos\theta = \frac{\boldsymbol{a}}{|\boldsymbol{a}|}|\boldsymbol{b}|\cos\theta$$

$$\therefore \ \boldsymbol{p} = \boldsymbol{a} \cdot \frac{|\boldsymbol{b}|}{|\boldsymbol{a}|}\cos\theta \tag{2.31}$$

Observe that above we have normalised vector $\boldsymbol{a}$ before we scaled it by the magnitude of the projection vector $|\boldsymbol{p}|$.

If vectors $\boldsymbol{a}$, $\boldsymbol{b}$ are parallel, then we define the projection of $\boldsymbol{b}$ onto $\boldsymbol{a}$ to be $\boldsymbol{b}$ itself.

2.14.4 The Cosine of the Angle Between two Vectors

Given two non-zero vectors $\boldsymbol{a}$ and $\boldsymbol{b}$ their scalar product is given by $\boldsymbol{a} \odot \boldsymbol{b} = |\boldsymbol{a}|\,|\boldsymbol{b}|\cos\theta$. Solving this equation for $\cos\theta$ we obtain:

$$\cos\theta = \frac{\boldsymbol{a} \odot \boldsymbol{b}}{|\boldsymbol{a}|\,|\boldsymbol{b}|} = \frac{a_x b_x + a_y b_y + a_z b_z}{\sqrt{a_x^2 + a_y^2 + a_z^2}\sqrt{b_x^2 + b_y^2 + b_z^2}} \tag{2.32}$$

2.14.5 The Scalar Product of Collinear Vectors

Assume that we have two non-zero collinear vectors $\boldsymbol{a}$ and $\boldsymbol{b}$.

If the vectors are parallel, then their scalar product is given by:

$$\boldsymbol{a} \odot \boldsymbol{b} = |\boldsymbol{a}|\,|\boldsymbol{b}| \quad \Leftrightarrow \quad \boldsymbol{a} \uparrow\uparrow \boldsymbol{b} \tag{2.33}$$

since $\cos(0) = 1$.

If the vectors are antiparallel, then their scalar product is given by:

$$\boldsymbol{a} \odot \boldsymbol{b} = -\,|\boldsymbol{a}|\,|\boldsymbol{b}| \quad \Leftrightarrow \quad \boldsymbol{a} \uparrow\downarrow \boldsymbol{b} \tag{2.34}$$

since $\cos(\pi) = -1$.

2.14.6 The Scalar Product of Orthogonal Vectors

Given two non-zero orthogonal vectors $\boldsymbol{a}$ and $\boldsymbol{b}$, the cosine of the angle between them will be zero and thus their scalar product *vanishes* (i.e. it is zero). This is both *a necessary and sufficient condition.* This means that the converse is also true, i.e. if the scalar product of two vectors vanishes, then the vectors are orthogonal. We say that:

$$\boldsymbol{a} \odot \boldsymbol{b} = 0 \quad \Leftrightarrow \quad \boldsymbol{a} \perp \boldsymbol{b} \tag{2.35}$$

since $\cos\left(\frac{\pi}{2}\right) = 0$.

2.15 The Direction Ratios and Direction Cosines of a Vector

Let $\boldsymbol{i}, \boldsymbol{j}, \boldsymbol{k}$ be a Cartesian basis. Suppose that the vector $\boldsymbol{v}$ is a non-zero vector, with components $[v_x, v_y, v_z]$ relative to the $\boldsymbol{i}, \boldsymbol{j}, \boldsymbol{k}$ basis, that is represented by the directed segment $\overrightarrow{OV}$, as seen in Fig. 2.16. Let point V_{xy} be the foot of the perpendicular from point V to the XOY plane. Then the coordinates of point V and the components of vector $\boldsymbol{v}$ are given by the triple $[v_x, v_y, v_z]$.

The components v_x, v_y, v_z are known as the *direction ratios* of vector $\boldsymbol{v}$. They allow us to express vector $\boldsymbol{v}$ as the sum of the $v_x\boldsymbol{i}$, $v_y\boldsymbol{j}$, $v_z\boldsymbol{k}$ vectors, which are collinear to the base vectors $\boldsymbol{i}, \boldsymbol{j}, \boldsymbol{k}$ and have magnitudes v_x, v_y, v_z, respectively.

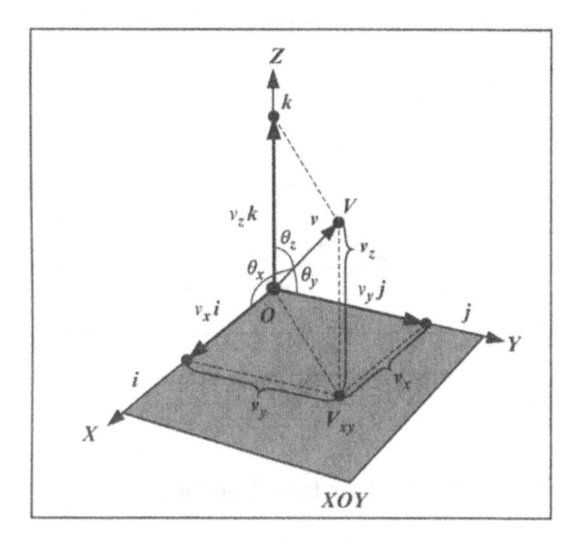

FIGURE 2.16. The direction ratios and direction cosines of a vector.

These vectors are projections of vector $\boldsymbol{v}$ onto the three base vectors. $\theta_x, \theta_y, \theta_z$ are the angles subtended by the vector $\boldsymbol{v}$ and each of the base vectors i, j, k, respectively.

If we compute the scalar product of vector $\boldsymbol{v}$ with each of the base vectors $\boldsymbol{i}, \boldsymbol{j}, \boldsymbol{k}$, respectively we obtain:

$$\begin{aligned} \boldsymbol{v} \odot i &= v_x \\ \boldsymbol{v} \odot j &= v_y \\ \boldsymbol{v} \odot k &= v_z \end{aligned} \tag{2.36}$$

Which are the lengths of the projection vectors $v_x \boldsymbol{i}$, $v_y \boldsymbol{j}$, $v_z \boldsymbol{k}$. Thus,

$$\begin{aligned} |v_x \boldsymbol{i}| &= v_x = |\boldsymbol{v}| \cos(\theta_x) \\ |v_y \boldsymbol{j}| &= v_y = |\boldsymbol{v}| \cos(\theta_y) \\ |v_z \boldsymbol{k}| &= v_z = |\boldsymbol{v}| \cos(\theta_z) \end{aligned} \tag{2.37}$$

Solving the above equation for the cosines of the angles we obtain:

$$\begin{aligned} \cos(\theta_x) &= \frac{v_x}{|\boldsymbol{v}|} = \frac{v_x}{\sqrt{v_x^2 + v_y^2 + v_z^2}} \\ \cos(\theta_y) &= \frac{v_y}{|\boldsymbol{v}|} = \frac{v_y}{\sqrt{v_x^2 + v_y^2 + v_z^2}} \\ \cos(\theta_z) &= \frac{v_z}{|\boldsymbol{v}|} = \frac{v_z}{\sqrt{v_x^2 + v_y^2 + v_z^2}} \end{aligned} \tag{2.38}$$

These three quantities are known as the *direction cosines* of the vector $\boldsymbol{v}$. Thus, the direction cosines of a vector $\boldsymbol{v}$ are the components of the normalised vector $\hat{\boldsymbol{v}}$.

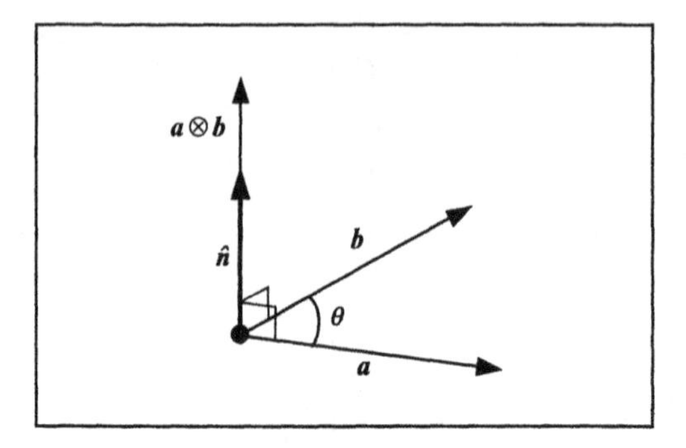

FIGURE 2.17. The vector product of two vectors.

2.16 The Vector Product of two Vectors

Let $\boldsymbol{a}, \boldsymbol{b}$ be two non-collinear vectors. Let $\hat{\boldsymbol{n}}$ be a unit vector which is perpendicular to both $\boldsymbol{a}$ and $\boldsymbol{b}$, which, when taken together with $\boldsymbol{a}$ and $\boldsymbol{b}$, forms a right-handed system $\boldsymbol{a}, \boldsymbol{b}, \hat{\boldsymbol{n}}$ in this order. The vector $\hat{\boldsymbol{n}}$ is uniquely determined by $\boldsymbol{a}$ and $\boldsymbol{b}$, as it is the only vector that satisfies all the above constraints. Let θ be the angle between vectors $\boldsymbol{a}$ and $\boldsymbol{b}$, as shown in Fig. 2.17.

The *vector product* or *cross product* or *outer product* $\boldsymbol{a} \otimes \boldsymbol{b}$ of two vectors $\boldsymbol{a}, \boldsymbol{b}$ is defined as:

$$\boldsymbol{a} \otimes \boldsymbol{b} = (|\boldsymbol{a}|\,|\boldsymbol{b}|\sin\theta) \cdot \hat{\boldsymbol{n}} \tag{2.39}$$

If $\boldsymbol{a}$ and $\boldsymbol{b}$ are collinear vectors, then their vector product is defined to be the zero vector. This includes the cases where one or both vectors are the zero vector.

If $\boldsymbol{a}$ and $\boldsymbol{b}$ are non-collinear vectors, then any vector perpendicular to both $\boldsymbol{a}$ and $\boldsymbol{b}$ is collinear to $\boldsymbol{a} \otimes \boldsymbol{b}$.

From the above discussion it should be apparent that the vector product of two vectors is defined in three-dimensional space $\mathcal{E}^3$.

The following axioms and vector algebra rules apply to the vector product for all vectors $\boldsymbol{v}_1, \boldsymbol{v}_2, \boldsymbol{v}_3$ and all scalars α:

Existence of the vector product:

$\boldsymbol{v}_1 \otimes \boldsymbol{v}_2$ is a vector with magnitude $|\boldsymbol{v}_1|\,|\boldsymbol{v}_2|\sin(\theta)$ (A2.9)

Vector product of collinear vectors:

$$\boldsymbol{v}_1 \otimes \boldsymbol{v}_2 = \vec{\boldsymbol{0}} \quad \Leftrightarrow \quad \boldsymbol{v}_1 || \boldsymbol{v}_2 \tag{A2.10}$$

Associative law (does not apply):

$$\boldsymbol{v}_1 \otimes (\boldsymbol{v}_2 \otimes \boldsymbol{v}_3) \neq (\boldsymbol{v}_1 \otimes \boldsymbol{v}_2) \otimes \boldsymbol{v}_3 \tag{R2.10}$$

Anti-commutative law:

$$\boldsymbol{v}_1 \otimes \boldsymbol{v}_2 = -(\boldsymbol{v}_2 \otimes \boldsymbol{v}_1) \tag{R2.11}$$

Distributive laws:

$$(\alpha \boldsymbol{v}_1) \otimes \boldsymbol{v}_2 = \boldsymbol{v}_1 \otimes (\alpha \boldsymbol{v}_2) = \alpha (\boldsymbol{v}_1 \otimes \boldsymbol{v}_2) \tag{R2.12}$$

$$\boldsymbol{v}_1 \otimes (\boldsymbol{v}_2 + \boldsymbol{v}_3) = \boldsymbol{v}_1 \otimes \boldsymbol{v}_2 + \boldsymbol{v}_1 \otimes \boldsymbol{v}_3 \tag{R2.13}$$

Given a Cartesian basis $\boldsymbol{i}, \boldsymbol{j}, \boldsymbol{k}$, then the vector products of combinations of its base vectors are given by:

$$\boldsymbol{i} \otimes \boldsymbol{i} = \boldsymbol{j} \otimes \boldsymbol{j} = \boldsymbol{k} \otimes \boldsymbol{k} = \vec{\mathbf{0}} \quad \text{(N.B. This result is also true for any basis)} \tag{2.40}$$

$$\boldsymbol{i} \otimes \boldsymbol{j} = \boldsymbol{k}, \quad \boldsymbol{j} \otimes \boldsymbol{k} = \boldsymbol{i}, \quad \boldsymbol{k} \otimes \boldsymbol{i} = \boldsymbol{j} \tag{2.41}$$

$$\boldsymbol{j} \otimes \boldsymbol{i} = -\boldsymbol{k}, \quad \boldsymbol{k} \otimes \boldsymbol{j} = -\boldsymbol{i}, \quad \boldsymbol{i} \otimes \boldsymbol{k} = -\boldsymbol{j} \tag{2.42}$$

2.17 The Vector Product Expressed in Terms of its Components

If $\boldsymbol{p}, \boldsymbol{q}, \boldsymbol{r}$ is a right-handed basis and if $\boldsymbol{a}, \boldsymbol{b}$ are vectors with components $\left[a_x, a_y, a_z\right]$ and $\left[b_x, b_y, b_z\right]$, respectively, relative to this basis, then the vector product $\boldsymbol{a} \otimes \boldsymbol{b}$ (in terms of its components) is defined as:

$$\begin{aligned}
\boldsymbol{a} \otimes \boldsymbol{b} &= \left(a_x\boldsymbol{p} + a_y\boldsymbol{q} + a_z\boldsymbol{r}\right) \otimes \left(b_x\boldsymbol{p} + b_y\boldsymbol{q} + b_z\boldsymbol{r}\right) \\
&= a_x\boldsymbol{p} \otimes b_x\boldsymbol{p} + a_x\boldsymbol{p} \otimes b_y\boldsymbol{q} + a_x\boldsymbol{p} \otimes b_z\boldsymbol{r} \\
&\quad + a_y\boldsymbol{q} \otimes b_x\boldsymbol{p} + a_y\boldsymbol{q} \otimes b_y\boldsymbol{q} + a_y\boldsymbol{q} \otimes b_z\boldsymbol{r} \\
&\quad + a_z\boldsymbol{r} \otimes b_x\boldsymbol{p} + a_z\boldsymbol{r} \otimes b_y\boldsymbol{q} + a_z\boldsymbol{r} \otimes b_z\boldsymbol{r} \\
&= a_x b_x\boldsymbol{p} \otimes \boldsymbol{p} + a_y b_y\boldsymbol{q} \otimes \boldsymbol{q} + a_z b_z\boldsymbol{r} \otimes \boldsymbol{r} \\
&\quad + \left(a_y b_z\boldsymbol{q} \otimes \boldsymbol{r} + a_z b_y\boldsymbol{r} \otimes \boldsymbol{q}\right) + \left(a_z b_x\boldsymbol{r} \otimes \boldsymbol{p} + a_x b_z\boldsymbol{p} \otimes \boldsymbol{r}\right) \\
&\quad + \left(a_x b_y\boldsymbol{p} \otimes \boldsymbol{q} + a_y b_x\boldsymbol{q} \otimes \boldsymbol{p}\right)
\end{aligned}$$

Using the anti-commutative law (R2.11) and since $\boldsymbol{p} \otimes \boldsymbol{p} = \vec{\mathbf{0}}$, $\boldsymbol{q} \otimes \boldsymbol{q} = \vec{\mathbf{0}}$ and $\boldsymbol{r} \otimes \boldsymbol{r} = \vec{\mathbf{0}}$ we can simplify the above expression to:

$$\boldsymbol{a} \otimes \boldsymbol{b} = \left(a_y b_z - a_z b_y\right) \boldsymbol{q} \otimes \boldsymbol{r} + \left(a_z b_x - a_x b_z\right) \boldsymbol{r} \otimes \boldsymbol{p} + \left(a_x b_y - a_y b_x\right) \boldsymbol{p} \otimes \boldsymbol{q} \tag{2.43}$$

The above definition of the vector product applies to any right-handed basis $\boldsymbol{p}, \boldsymbol{q}, \boldsymbol{r}$. When dealing with a Cartesian basis $\boldsymbol{i}, \boldsymbol{j}, \boldsymbol{k}$ using Eq. (2.41) we can further simplify the above definition to:

$$\boldsymbol{a} \otimes \boldsymbol{b} = \left(a_y b_z - a_z b_y\right) \boldsymbol{i} + \left(a_z b_x - a_x b_z\right) \boldsymbol{j} + \left(a_x b_y - a_y b_x\right) \boldsymbol{k} \tag{2.44}$$

Thus, the vector product $\boldsymbol{a} \otimes \boldsymbol{b}$ is a vector with components:

$$\boldsymbol{a} \otimes \boldsymbol{b} = \left[\left(a_y b_z - a_z b_y\right), \left(a_z b_x - a_x b_z\right), \left(a_x b_y - a_y b_x\right)\right] \tag{2.45}$$

Alternatively:

$$\boldsymbol{a} \otimes \boldsymbol{b} = \begin{bmatrix} a_x \\ a_y \\ a_z \end{bmatrix} \otimes \begin{bmatrix} b_x \\ b_y \\ b_z \end{bmatrix} = \begin{bmatrix} a_y b_z - a_z b_y \\ a_z b_x - a_x b_z \\ a_x b_y - a_y b_x \end{bmatrix} \tag{2.46}$$

Finally, the vector product can be represented as the *determinant* of a *matrix* whose first column consists of the base *vectors*:

$$\begin{aligned}
\boldsymbol{a} \otimes \boldsymbol{b} &= \begin{vmatrix} \boldsymbol{i} & a_x & b_x \\ \boldsymbol{j} & a_y & b_y \\ \boldsymbol{k} & a_z & b_z \end{vmatrix} \\
&= \boldsymbol{i} \cdot \begin{vmatrix} a_y & b_y \\ a_z & b_z \end{vmatrix} - \boldsymbol{j} \cdot \begin{vmatrix} a_x & b_x \\ a_z & b_z \end{vmatrix} + \boldsymbol{k} \cdot \begin{vmatrix} a_x & b_x \\ a_y & b_y \end{vmatrix} \\
&= \boldsymbol{i} \cdot \left(a_y b_z - a_z b_y\right) - \boldsymbol{j} \cdot \left(a_x b_z - a_z b_x\right) + \boldsymbol{k} \cdot \left(a_x b_y - a_y b_x\right) \\
&= \boldsymbol{i} \cdot \left(a_y b_z - a_z b_y\right) + \boldsymbol{j} \cdot \left(a_z b_x - a_x b_z\right) + \boldsymbol{k} \cdot \left(a_x b_y - a_y b_x\right) \qquad (2.47)
\end{aligned}$$

Such a determinant can only be evaluated symbolically, as we can not compute the value of a determinant that contains symbols. An identical result would be arrived at by transposing the rows and columns of the above determinant.

2.18 Properties of the Vector Product

2.18.1 The Geometric Interpretation of the Vector Product

Let $\boldsymbol{a}$, $\boldsymbol{b}$ be two non-collinear vectors that are represented by the directed segments $\overrightarrow{\boldsymbol{OA}}, \overrightarrow{\boldsymbol{OB}}$, respectively. Let θ be the angle between vectors $\boldsymbol{a}$ and $\boldsymbol{b}$. Let $\boldsymbol{C}$ be the fourth corner of the parallelogram that has $\boldsymbol{OA}$ and $\boldsymbol{OB}$ as two adjacent sides, as shown in Fig. 2.18. Using simple trigonometry, the area of the triangle $\boldsymbol{AOB}$ is given by $\frac{1}{2}\,|\boldsymbol{a}| \cdot h = \frac{1}{2}\,|\boldsymbol{a}|\,|\boldsymbol{b}|\,|\sin\theta|$, thus the area of the parallelogram $\boldsymbol{AOBC}$ is $|\boldsymbol{a}|\,|\boldsymbol{b}|\sin\theta$. But the magnitude of the vector product $\boldsymbol{a} \otimes \boldsymbol{b}$ is $|\boldsymbol{a} \otimes \boldsymbol{b}| = ||\boldsymbol{a}|\,|\boldsymbol{b}|\sin\theta|$. Thus, the magnitude of the vector product $\boldsymbol{a} \otimes \boldsymbol{b}$ is equal to the area of the parallelogram spanned by vectors $\boldsymbol{a}$ and $\boldsymbol{b}$. Extra care should be taken with the sign of angle θ, which is positive on the left diagram but negative on the right diagram of Fig. 2.18.

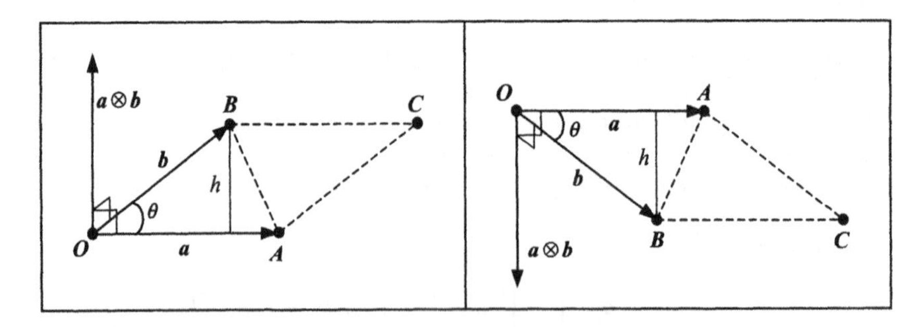

FIGURE 2.18. The magnitude of the vector product.

2.18.2 The Magnitude of the Vector Product in Terms of its Components

As we have seen in Eq. (2.45), given two non-collinear vectors $\boldsymbol{a}$ and $\boldsymbol{b}$ with components $\left[a_x, a_y, a_z\right]$ and $\left[b_x, b_y, b_z\right]$ relative to a Cartesian basis $\boldsymbol{i}, \boldsymbol{j}, \boldsymbol{k}$, the components of their vector product are:

$$\boldsymbol{a} \otimes \boldsymbol{b} = \left[\left(a_y b_z - a_z b_y\right), \left(a_z b_x - a_x b_z\right), \left(a_x b_y - a_y b_x\right)\right]$$

Thus the magnitude of the vector product $\boldsymbol{a} \otimes \boldsymbol{b}$, in terms of the components of $\boldsymbol{a}$ and $\boldsymbol{b}$, is given by:

$$|\boldsymbol{a} \otimes \boldsymbol{b}| = \sqrt{\left(a_y b_z - a_z b_y\right)^2 + \left(a_z b_x - a_x b_z\right)^2 + \left(a_x b_y - a_y b_x\right)^2} \tag{2.48}$$

2.18.3 The Square of the Magnitude of the Vector Product

Given two non-collinear vectors $\boldsymbol{a}$ and $\boldsymbol{b}$, the square of the magnitude of their vector product $\boldsymbol{a} \otimes \boldsymbol{b}$ is given by:

$$\begin{aligned} |\boldsymbol{a} \otimes \boldsymbol{b}|^2 &= |\boldsymbol{a}|^2 |\boldsymbol{b}|^2 \sin^2\theta \\ &= |\boldsymbol{a}|^2 |\boldsymbol{b}|^2 \left(1 - \cos^2\theta\right) \\ &= |\boldsymbol{a}|^2 |\boldsymbol{b}|^2 - \left(|\boldsymbol{a}| |\boldsymbol{b}| \cos\theta\right)^2 \\ &= |\boldsymbol{a}|^2 |\boldsymbol{b}|^2 - \left(\boldsymbol{a} \odot \boldsymbol{b}\right)^2 \\ \therefore\ |\boldsymbol{a} \otimes \boldsymbol{b}|^2 &= |\boldsymbol{a}|^2 |\boldsymbol{b}|^2 - \left(\boldsymbol{a} \odot \boldsymbol{b}\right)^2 \end{aligned} \tag{2.49}$$

2.18.4 The Magnitude of the Sine of the Angle between Two Vectors

Given two non-zero and non-collinear vectors $\boldsymbol{a}$ and $\boldsymbol{b}$, the magnitude of their vector product $\boldsymbol{a} \otimes \boldsymbol{b}$ is given by:

$$|\boldsymbol{a} \otimes \boldsymbol{b}| = \left||\boldsymbol{a}|\,|\boldsymbol{b}\,| \sin\theta\right|$$

Solving this equation for $|\sin\theta|$ we obtain:

$$|\sin\theta| = \frac{|\boldsymbol{a} \otimes \boldsymbol{b}|}{|\boldsymbol{a}|\,|\boldsymbol{b}|} \tag{2.50}$$

If the components of the vectors $\boldsymbol{a}$ and $\boldsymbol{b}$ are defined relative to a Cartesian basis $\boldsymbol{i}, \boldsymbol{j}, \boldsymbol{k}$, then $|\sin(\theta)|$ is given by:

$$|\sin\theta| = \frac{|\boldsymbol{a} \otimes \boldsymbol{b}|}{|\boldsymbol{a}\,|\,|\boldsymbol{b}|} = \frac{\sqrt{\left(a_y b_z - a_z b_y\right)^2 + \left(a_z b_x - a_x b_z\right)^2 + \left(a_x b_y - a_y b_x\right)^2}}{\sqrt{a_x^2 + a_y^2 + a_z^2}\sqrt{b_x^2 + b_y^2 + b_z^2}} \tag{2.51}$$

2.19 Triple Products of Vectors

Triple products of vectors combine the operations of scalar multiplication and/or vector multiplication and define products involving three or four vectors.

2.19.1 *The Triple Scalar Product*

Given any three vectors $\boldsymbol{a}, \boldsymbol{b}, \boldsymbol{c}$, the products $(\boldsymbol{a} \otimes \boldsymbol{b}) \odot \boldsymbol{c}$ and $\boldsymbol{a} \odot (\boldsymbol{b} \otimes \boldsymbol{c})$ (which are equal) are known as the *triple scalar products* of $\boldsymbol{a}, \boldsymbol{b}, \boldsymbol{c}$ in this particular order. They are denoted by $(\boldsymbol{a}, \boldsymbol{b}, \boldsymbol{c})$ or $[\boldsymbol{a}, \boldsymbol{b}, \boldsymbol{c}]$ or $\langle \boldsymbol{a}, \boldsymbol{b}, \boldsymbol{c} \rangle$ or $\det(\boldsymbol{a}, \boldsymbol{b}, \boldsymbol{c})$. Thus the triple scalar product of three vectors $\boldsymbol{a}, \boldsymbol{b}, \boldsymbol{c}$ is defined as:

$$(\boldsymbol{a}, \boldsymbol{b}, \boldsymbol{c}) = (\boldsymbol{a} \otimes \boldsymbol{b}) \odot \boldsymbol{c} = \boldsymbol{a} \odot (\boldsymbol{b} \otimes \boldsymbol{c}) \tag{2.52}$$

From this definition we see that the triple scalar product is in reality the scalar product of two vectors (one of which is the vector product of two other vectors) and thus it produces a scalar result.

If vectors $\boldsymbol{a}, \boldsymbol{b}, \boldsymbol{c}$ have components $[a_x, a_y, a_z]$,$[b_x, b_y, b_z]$,$[c_x, c_y, c_z]$ relative to a Cartesian basis $\boldsymbol{i}, \boldsymbol{j}, \boldsymbol{k}$, then the triple scalar product is defined in terms of these components as:

$$\begin{aligned}
(\boldsymbol{a}, \boldsymbol{b}, \boldsymbol{c}) &= \boldsymbol{a} \odot (\boldsymbol{b} \otimes \boldsymbol{c}) \\
&= [a_x, a_y, a_z] \odot ([b_x, b_y, b_z] \otimes [c_x, c_y, c_z]) \\
&= [a_x, a_y, a_z] \odot [(b_y c_z - b_z c_y), (b_z c_x - b_x c_z), (b_x c_y - b_y c_x)] \\
\therefore \quad (\boldsymbol{a}, \boldsymbol{b}, \boldsymbol{c}) &= a_x (b_y c_z - b_z c_y) + a_y (b_z c_x - b_x c_z) + a_z (b_x c_y - b_y c_x) \qquad (2.53)
\end{aligned}$$

Alternatively we can define the triple scalar product using the determinant of a matrix whose rows or columns are the components of $\boldsymbol{a}, \boldsymbol{b}, \boldsymbol{c}$:

$$\begin{aligned}
(\boldsymbol{a}, \boldsymbol{b}, \boldsymbol{c}) &= \boldsymbol{a} \odot (\boldsymbol{b} \otimes \boldsymbol{c}) \\
&= \begin{vmatrix} a_x & a_y & a_z \\ b_x & b_y & b_z \\ c_x & c_y & c_z \end{vmatrix} \\
&= a_x \begin{vmatrix} b_y & b_z \\ c_y & c_z \end{vmatrix} - a_y \begin{vmatrix} b_x & b_z \\ c_x & c_z \end{vmatrix} + a_z \begin{vmatrix} b_x & b_y \\ c_x & c_y \end{vmatrix} \\
&= a_x (b_y c_z - b_z c_y) - a_y (b_x c_z - b_z c_x) + a_z (b_x c_y - b_y c_x) \\
\therefore \quad (\boldsymbol{a}, \boldsymbol{b}, \boldsymbol{c}) &= a_x (b_y c_z - b_z c_y) + a_y (b_z c_x - b_x c_z) + a_z (b_x c_y - b_y c_x) \qquad (2.54)
\end{aligned}$$

The reason why the triple scalar product is of interest is because of its geometric interpretation. The triple scalar product $(\boldsymbol{a} \otimes \boldsymbol{b}) \odot \boldsymbol{c}$ is equal in magnitude to the volume of the parallelepiped having vectors $\boldsymbol{a}, \boldsymbol{b}, \boldsymbol{c}$ as concurrent sides.

Let $\boldsymbol{a}, \boldsymbol{b}, \boldsymbol{c}$ be three non-coplanar vectors, which are represented by the directed segments $\overrightarrow{OA}$, $\overrightarrow{OB}$, $\overrightarrow{OC}$, respectively. Let ***AOBD*** be the parallelogram spanning

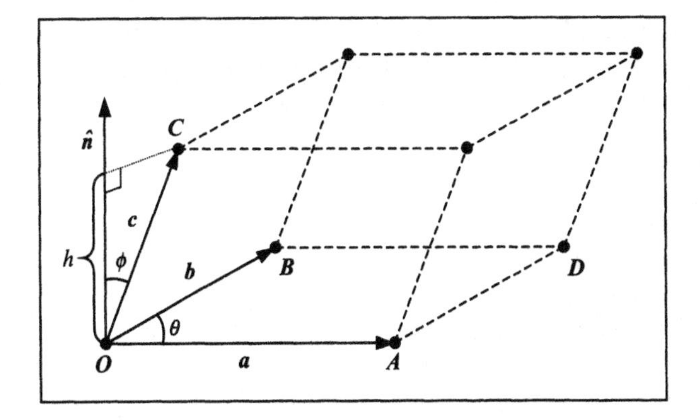

FIGURE 2.19. The parallelepiped whose volume is defined by the triple scalar product.

the vectors $\boldsymbol{a}$ and $\boldsymbol{b}$, as shown in Fig. 2.19. Let $\hat{\boldsymbol{n}}$ be the unit vector normal to the plane defined by the vectors $\boldsymbol{a}$ and $\boldsymbol{b}$, i.e.

$$\hat{\boldsymbol{n}} = \frac{\boldsymbol{a} \otimes \boldsymbol{b}}{|\boldsymbol{a} \otimes \boldsymbol{b}|}$$

Finally, let θ be the angle between vectors $\boldsymbol{a}$ and $\boldsymbol{b}$, and let ϕ be the angle between vectors $\hat{\boldsymbol{n}}$ and $\boldsymbol{c}$.

By definition the vector product $\boldsymbol{a} \otimes \boldsymbol{b}$ is given by:

$$\boldsymbol{a} \otimes \boldsymbol{b} = (|\boldsymbol{a}|\,|\boldsymbol{b}| \sin\theta) \cdot \hat{\boldsymbol{n}}$$

As we have seen in Subsection 2.18.1, the area α of the parallelogram of the base is given by:

$$\alpha = \big||\boldsymbol{a}|\,|\boldsymbol{b}| \sin\theta\big|$$

The magnitude of the vector product can now be rewritten as:

$$|\boldsymbol{a} \otimes \boldsymbol{b}| = \left|\alpha \cdot \hat{\boldsymbol{n}}\right|$$

and the magnitude of the triple scalar product $(\boldsymbol{a} \otimes \boldsymbol{b}) \odot c$ can be rewritten as:

$$|(\boldsymbol{a} \otimes \boldsymbol{b}) \odot \boldsymbol{c}| = \left|\left(\alpha \cdot \hat{\boldsymbol{n}}\right) \odot \boldsymbol{c}\right| = \alpha \cdot \left|\left(\hat{\boldsymbol{n}} \odot \boldsymbol{c}\right)\right| = \alpha \cdot \left|\hat{\boldsymbol{n}}\right| \cdot |\boldsymbol{c}| \cdot |\cos\phi| = \alpha \cdot h = \nu$$

Where $h = |c| \cdot |\cos\phi|$ represents the height of the parallelepiped and $\nu = \alpha \cdot h$ represents the volume of the parallelepiped.

Looking at the triple scalar product $(\boldsymbol{a} \otimes \boldsymbol{b}) \odot \boldsymbol{c}$ from this geometric point of view has a number of useful consequences. If any two of the vectors $\boldsymbol{a}$, $\boldsymbol{b}$, $\boldsymbol{c}$ are collinear, then the parallelepiped collapses to a plane and has zero volume, i.e. $(\boldsymbol{a} \otimes \boldsymbol{b}) \odot \boldsymbol{c} = 0$. Similarly, if any of the vectors is the zero vector or if all the vectors are coplanar, then the parallelepiped collapses and has zero volume.

Another important application of the triple scalar product is that it allows us to determine the handedness of a basis. Suppose that vectors $\boldsymbol{a}$, $\boldsymbol{b}$, $\boldsymbol{c}$ form a basis.

This implies that $(\boldsymbol{a}, \boldsymbol{b}, \boldsymbol{c}) \neq 0$. If $(\boldsymbol{a}, \boldsymbol{b}, \boldsymbol{c}) > 0$, then this basis is *right-handed*. Alternatively, if $(\boldsymbol{a}, \boldsymbol{b}, \boldsymbol{c}) < 0$, then it is *left-handed*. Thus, the basis $\boldsymbol{i}, \boldsymbol{j}, \boldsymbol{k}$ is right-handed, while the basis $\boldsymbol{i}, \boldsymbol{j}, \boldsymbol{k}$ is left-handed.

The following axioms and vector algebra rules apply to the triple scalar product for all vectors $\boldsymbol{v}_1, \boldsymbol{v}_2, \boldsymbol{v}_3$:

Existence of the triple scalar product:

$$(\boldsymbol{v}_1, \boldsymbol{v}_2, \boldsymbol{v}_3) \text{ is a scalar} \tag{A2.11}$$

Triple scalar product of coplanar vectors:

$$if\ all\ \boldsymbol{v}_1, \boldsymbol{v}_2, \boldsymbol{v}_3\ are\ coplanar \Rightarrow (\boldsymbol{v}_1, \boldsymbol{v}_2, \boldsymbol{v}_3) = 0 \tag{A2.12}$$

Triple scalar product of zero vectors:

$$\boldsymbol{v}_1 = \vec{\boldsymbol{0}} \vee \boldsymbol{v}_2 = \vec{\boldsymbol{0}} \vee \boldsymbol{v}_3 = \vec{\boldsymbol{0}} \Rightarrow (\boldsymbol{v}_1, \boldsymbol{v}_2, \boldsymbol{v}_3) = 0 \tag{A2.13}$$

Triple scalar product of collinear vectors:

$$\boldsymbol{v}_1 || \boldsymbol{v}_2 \vee \boldsymbol{v}_2 || \boldsymbol{v}_3 \vee \boldsymbol{v}_3 || \boldsymbol{v}_1 \Rightarrow (\boldsymbol{v}_1, \boldsymbol{v}_2, \boldsymbol{v}_3) = 0 \tag{A2.14}$$

Cyclic permutation rule:

$$(\boldsymbol{v}_1, \boldsymbol{v}_2, \boldsymbol{v}_3) = (\boldsymbol{v}_2, \boldsymbol{v}_3, \boldsymbol{v}_1) = (\boldsymbol{v}_3, \boldsymbol{v}_1, \boldsymbol{v}_2) \tag{R2.14}$$

$$\text{i.e., } \boldsymbol{v}_1 \odot (\boldsymbol{v}_2 \otimes \boldsymbol{v}_3) = \boldsymbol{v}_2 \odot (\boldsymbol{v}_3 \otimes \boldsymbol{v}_1) = \boldsymbol{v}_3 \odot (\boldsymbol{v}_1 \otimes \boldsymbol{v}_2)$$

$$\text{and } (\boldsymbol{v}_1 \otimes \boldsymbol{v}_2) \odot \boldsymbol{v}_3 = (\boldsymbol{v}_2 \otimes \boldsymbol{v}_3) \odot \boldsymbol{v}_1 = (\boldsymbol{v}_3 \otimes \boldsymbol{v}_1) \odot \boldsymbol{v}_2$$

Non-cyclic permutation rule:

$$(\boldsymbol{v}_2, \boldsymbol{v}_1, \boldsymbol{v}_3) = (\boldsymbol{v}_3, \boldsymbol{v}_2, \boldsymbol{v}_1) = (\boldsymbol{v}_1, \boldsymbol{v}_3, \boldsymbol{v}_2) = -(\boldsymbol{v}_1, \boldsymbol{v}_2, \boldsymbol{v}_3) \tag{R2.15}$$

2.19.2 The Triple Vector Product

Given any three vectors $\boldsymbol{a}, \boldsymbol{b}, \boldsymbol{c}$, the products $(\boldsymbol{a} \otimes \boldsymbol{b}) \otimes \boldsymbol{c}$ and $\boldsymbol{a} \otimes (\boldsymbol{b} \otimes \boldsymbol{c})$ (which are not equal in general) are known as the *triple vector products* of $\boldsymbol{a}$, $\boldsymbol{b}$, $\boldsymbol{c}$ in this particular order. If vectors $\boldsymbol{a}, \boldsymbol{b}, \boldsymbol{c}$ have components $[a_x, a_y, a_z]$, $[b_x, b_y, b_z]$, $[c_x, c_y, c_z]$ relative to a Cartesian basis $\boldsymbol{i}, \boldsymbol{j}, \boldsymbol{k}$, then the triple vector product is defined in terms of these components as:

$$\begin{aligned}(\boldsymbol{a} \otimes \boldsymbol{b}) \otimes \boldsymbol{c} &= \left([a_x, a_y, a_z] \otimes [b_x, b_y, b_z]\right) \otimes [c_x, c_y, c_z] \\ &= \left[(a_y b_z - a_z b_y), (a_z b_x - a_x b_z), (a_x b_y - a_y b_x)\right] \otimes [c_x, c_y, c_z]\end{aligned}$$

$$\therefore\ (\boldsymbol{a} \otimes \boldsymbol{b}) \otimes \boldsymbol{c} = \begin{bmatrix} (a_z b_x - a_x b_z) c_z - (a_x b_y - a_y b_x) c_y \\ (a_x b_y - a_y b_x) c_x - (a_y b_z - a_z b_y) c_z \\ (a_y b_z - a_z b_y) c_y - (a_z b_x - a_x b_z) c_x \end{bmatrix}^T \tag{2.55}$$

Expanding the first component of this vector we get:

$$\begin{aligned}(a_z b_x - a_x b_z) c_z - (a_x b_y - a_y b_x) c_y &= a_z c_z b_x - b_z c_z a_x - b_y c_y a_x + a_y c_y b_x \\ &= a_z c_z b_x - b_z c_z a_x - b_y c_y a_x + a_y c_y b_x \\ &\quad + a_x c_x b_x - b_x c_x a_x \\ &= (a_x c_x + a_y c_y + a_z c_z) b_x \\ &\quad - (b_x c_x + b_y c_y + b_z c_z) a_x \\ &= (\boldsymbol{a} \odot \boldsymbol{c}) b_x - (\boldsymbol{b} \odot \boldsymbol{c}) a_x\end{aligned}$$

Expanding its second component we get:

$$\begin{aligned}\left(a_x b_y - a_y b_x\right) c_x - \left(a_y b_z - a_z b_y\right) c_z &= a_x c_x b_y - b_x c_x a_y - b_z c_z a_y + a_z c_z b_y \\ &= a_x c_x b_y - b_x c_x a_y - b_z c_z a_y + a_z c_z b_y \\ &\quad + a_y c_y b_y - b_y c_y a_y \\ &= \left(a_x c_x + a_y c_y + a_z c_z\right) b_y \\ &\quad - \left(b_x c_x + b_y c_y + b_z c_z\right) a_y \\ &= (\boldsymbol{a} \odot \boldsymbol{c})\, b_y - (\boldsymbol{b} \odot \boldsymbol{c})\, a_y\end{aligned}$$

Expanding its third component we get:

$$\begin{aligned}\left(a_y b_z - a_z b_y\right) c_y - \left(a_z b_x - a_x b_z\right) c_x &= a_y c_y b_z - b_y c_y a_z - b_x c_x a_z + a_x c_x b_z \\ &= a_y c_y b_z - b_y c_y a_z - b_x c_x a_z + a_x c_x b_z \\ &\quad + a_z c_z b_z - b_z c_z a_z \\ &= \left(a_x c_x + a_y c_y + a_z c_z\right) b_z \\ &\quad - \left(b_x c_x + b_y c_y + b_z c_z\right) a_z \\ &= (\boldsymbol{a} \odot \boldsymbol{c})\, b_z - (\boldsymbol{b} \odot \boldsymbol{c})\, a_z\end{aligned}$$

Thus:

$$\begin{aligned}(\boldsymbol{a} \otimes \boldsymbol{b}) \otimes \boldsymbol{c} &= [((\boldsymbol{a} \odot \boldsymbol{c}) b_x - (\boldsymbol{b} \odot \boldsymbol{c}) a_x), \\ &\qquad \left((\boldsymbol{a} \odot \boldsymbol{c}) \odot b_y - (\boldsymbol{b} \odot \boldsymbol{c}) a_y\right), ((\boldsymbol{a} \odot \boldsymbol{c}) b_z - (\boldsymbol{b} \odot \boldsymbol{c}) a_z)] \\ \therefore\ (\boldsymbol{a} \otimes \boldsymbol{b}) \otimes \boldsymbol{c} &= (\boldsymbol{a} \odot \boldsymbol{c}) \boldsymbol{b} - (\boldsymbol{b} \odot \boldsymbol{c}) \boldsymbol{a} \qquad (2.56)\end{aligned}$$

Similarly we can prove that:

$$\boldsymbol{a} \otimes (\boldsymbol{b} \otimes \boldsymbol{c}) = (\boldsymbol{a} \odot \boldsymbol{c}) \boldsymbol{b} - (\boldsymbol{a} \odot \boldsymbol{b}) \boldsymbol{c}$$

The following axioms and vector algebra rules apply to the triple vector product for all vectors $\boldsymbol{v}_1, \boldsymbol{v}_2, \boldsymbol{v}_3$:

Existence of the triple vector product:

$(\boldsymbol{v}_1 \otimes \boldsymbol{v}_2) \otimes \boldsymbol{v}_3$ is a vector (A2.15)

Associative law (does not apply):

$$(\boldsymbol{v}_1 \otimes \boldsymbol{v}_2) \otimes \boldsymbol{v}_3 \neq \boldsymbol{v}_1 \otimes (\boldsymbol{v}_2 \otimes \boldsymbol{v}_3) \qquad \text{(R2.16)}$$

Permutation rule:

$$(\boldsymbol{v}_1 \otimes \boldsymbol{v}_2) \otimes \boldsymbol{v}_3 = \boldsymbol{v}_3 \otimes (\boldsymbol{v}_2 \otimes \boldsymbol{v}_1) \qquad \text{(R2.17)}$$

Expansion rules:

$$(\boldsymbol{v}_1 \otimes \boldsymbol{v}_2) \otimes \boldsymbol{v}_3 = (\boldsymbol{v}_1 \odot \boldsymbol{v}_3)\, \boldsymbol{v}_2 - (\boldsymbol{v}_2 \odot \boldsymbol{v}_3)\, \boldsymbol{v}_1 \qquad \text{(R2.18)}$$

$$\boldsymbol{v}_1 \otimes (\boldsymbol{v}_2 \otimes \boldsymbol{v}_3) = (\boldsymbol{v}_1 \odot \boldsymbol{v}_3)\, \boldsymbol{v}_2 - (\boldsymbol{v}_1 \odot \boldsymbol{v}_2)\, \boldsymbol{v}_3 \qquad \text{(R2.19)}$$

2.19.3 The Scalar Product of Two Vector Products

Given any four vectors $\boldsymbol{a}$, $\boldsymbol{b}$, $\boldsymbol{c}$, $\boldsymbol{d}$, the product $(\boldsymbol{a} \otimes \boldsymbol{b}) \odot (\boldsymbol{c} \otimes \boldsymbol{d})$ is known as the *scalar product of two vector products* or the *scalar product of four vectors*. Such a product results in a scalar.

First we use the cyclic permutation rule (R2.14) to rewrite this product:

$$(\boldsymbol{a} \otimes \boldsymbol{b}) \odot (\boldsymbol{c} \otimes \boldsymbol{d}) = \boldsymbol{c} \odot (\boldsymbol{d} \otimes (\boldsymbol{a} \otimes \boldsymbol{b}))$$

Where for the cyclic permutation $\boldsymbol{v}_1 \odot (\boldsymbol{v}_2 \otimes \boldsymbol{v}_3) = \boldsymbol{v}_2 \odot (\boldsymbol{v}_3 \otimes \boldsymbol{v}_1)$ we used mappings $(\boldsymbol{a} \otimes \boldsymbol{b}) \to \boldsymbol{v}_1, \boldsymbol{c} \to \boldsymbol{v}_2, \boldsymbol{d} \to \boldsymbol{v}_3$.

Next we apply the expansion rule (R2.19) to rewrite this result:

$$\boldsymbol{c} \odot (\boldsymbol{d} \otimes (\boldsymbol{a} \otimes \boldsymbol{b})) = \boldsymbol{c} \odot (\boldsymbol{a}(\boldsymbol{b} \odot \boldsymbol{d}) - \boldsymbol{b}(\boldsymbol{d} \odot \boldsymbol{a})) = (\boldsymbol{c} \odot \boldsymbol{a})(\boldsymbol{b} \odot \boldsymbol{d}) - (\boldsymbol{c} \odot \boldsymbol{b})(\boldsymbol{d} \odot \boldsymbol{a})$$

Where for the expansion $\boldsymbol{v}_1 \otimes (\boldsymbol{v}_2 \otimes \boldsymbol{v}_3) = (\boldsymbol{v}_1 \odot \boldsymbol{v}_3)\, \boldsymbol{v}_2 - (\boldsymbol{v}_1 \odot \boldsymbol{v}_2)\, \boldsymbol{v}_3$, we used mappings $\boldsymbol{d} \to \boldsymbol{v}_1, \boldsymbol{a} \to \boldsymbol{v}_2, \boldsymbol{b} \to \boldsymbol{v}_3$.

$$\therefore \quad (\boldsymbol{a} \otimes \boldsymbol{b}) \odot (\boldsymbol{c} \otimes \boldsymbol{d}) = (\boldsymbol{c} \odot \boldsymbol{a})(\boldsymbol{b} \odot \boldsymbol{d}) - (\boldsymbol{c} \odot \boldsymbol{b})(\boldsymbol{d} \odot \boldsymbol{a}) \tag{2.57}$$

Using this result we can calculate the square of the vector product as:

$$(\boldsymbol{a} \otimes \boldsymbol{b})^2 = (\boldsymbol{a} \odot \boldsymbol{a})(\boldsymbol{b} \odot \boldsymbol{b}) - (\boldsymbol{a} \odot \boldsymbol{b})(\boldsymbol{a} \odot \boldsymbol{b}) = |\boldsymbol{a}|^2\, |\boldsymbol{b}|^2 - (\boldsymbol{a} \odot \boldsymbol{b})^2 \tag{2.58}$$

2.19.4 The Vector Product of Two Vector Products

Given any four vectors $\boldsymbol{a}$, $\boldsymbol{b}$, $\boldsymbol{c}$, $\boldsymbol{d}$, the product $(\boldsymbol{a} \otimes \boldsymbol{b}) \otimes (\boldsymbol{c} \otimes \boldsymbol{d})$ is known as the *vector product of two vector products* or the *vector product of four vectors*. Such a product results in a vector.

Using the expansion rule (R2.18) we can rewrite this product as:

$$\begin{aligned}(\boldsymbol{a} \otimes \boldsymbol{b}) \otimes (\boldsymbol{c} \otimes \boldsymbol{d}) &= (\boldsymbol{a} \odot (\boldsymbol{c} \otimes \boldsymbol{d}))\, \boldsymbol{b} - (\boldsymbol{b} \odot (\boldsymbol{c} \otimes \boldsymbol{d}))\, \boldsymbol{a} \\ &= (\boldsymbol{a}, \boldsymbol{c}, \boldsymbol{d})\, \boldsymbol{b} - (\boldsymbol{b}, \boldsymbol{c}, \boldsymbol{d})\, \boldsymbol{a}\end{aligned} \tag{2.59}$$

Where for the expansion $(\boldsymbol{v}_1 \otimes \boldsymbol{v}_2) \otimes \boldsymbol{v}_3 = (\boldsymbol{v}_1 \odot \boldsymbol{v}_3)\, \boldsymbol{v}_2 - (\boldsymbol{v}_2 \odot \boldsymbol{v}_3)\, \boldsymbol{v}_1$, we used mappings $\boldsymbol{a} \to \boldsymbol{v}_1, \boldsymbol{b} \to \boldsymbol{v}_2$, $(\boldsymbol{c} \otimes \boldsymbol{d}) \to \boldsymbol{v}_3$.

Similarly, using the expansion rule (R2.19) we can rewrite this product as:

$$\begin{aligned}(\boldsymbol{a} \otimes \boldsymbol{b}) \otimes (\boldsymbol{c} \otimes \boldsymbol{d}) &= ((\boldsymbol{a} \otimes \boldsymbol{b}) \odot \boldsymbol{d})\, \boldsymbol{c} - ((\boldsymbol{a} \otimes \boldsymbol{b}) \odot \boldsymbol{c})\, \boldsymbol{d} \\ &= (\boldsymbol{a}, \boldsymbol{b}, \boldsymbol{d})\, \boldsymbol{c} - (\boldsymbol{a}, \boldsymbol{b}, \boldsymbol{c})\, \boldsymbol{d}\end{aligned} \tag{2.60}$$

Where for the expansion $\boldsymbol{v}_1 \otimes (\boldsymbol{v}_2 \otimes \boldsymbol{v}_3) = (\boldsymbol{v}_1 \odot \boldsymbol{v}_3)\, \boldsymbol{v}_2 - (\boldsymbol{v}_1 \odot \boldsymbol{v}_2)\, \boldsymbol{v}_3$, we used mappings $(\boldsymbol{a} \otimes \boldsymbol{b}) \to \boldsymbol{v}_1, \boldsymbol{c} \to \boldsymbol{v}_2, \boldsymbol{d} \to \boldsymbol{v}_3$.

Thus, the vector product of two vector products is defined as:

$$(\boldsymbol{a} \otimes \boldsymbol{b}) \otimes (\boldsymbol{c} \otimes \boldsymbol{d}) = (\boldsymbol{a}, \boldsymbol{c}, \boldsymbol{d})\, \boldsymbol{b} - (\boldsymbol{b}, \boldsymbol{c}, \boldsymbol{d})\, \boldsymbol{a} = (\boldsymbol{a}, \boldsymbol{b}, \boldsymbol{d})\, \boldsymbol{c} - (\boldsymbol{a}, \boldsymbol{b}, \boldsymbol{c})\, \boldsymbol{d} \tag{2.61}$$

2.20 The Components of a Vector Relative to a Non-orthogonal Basis

Let $\boldsymbol{s}$, $\boldsymbol{t}$, $\boldsymbol{u}$ be three non-coplanar and non-zero vectors that form a right-handed basis. Let the directed segments $\overrightarrow{OS}, \overrightarrow{OT}, \overrightarrow{OU}$ represent the vectors $\boldsymbol{s}, \boldsymbol{t}, \boldsymbol{u}$, respectively, as shown in Fig. 2.20. Let $\boldsymbol{v}$ be any vector and let the directed segment

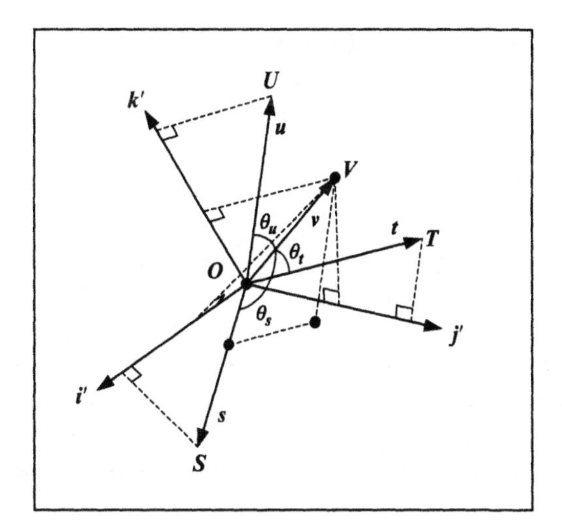

FIGURE 2.20. The non-orthonormal right-handed $\boldsymbol{s}, \boldsymbol{t}, \boldsymbol{u}$ basis.

$\overrightarrow{OV}$ represent vector $\boldsymbol{v}$. The components of the $\boldsymbol{s}, \boldsymbol{t}, \boldsymbol{u}, \boldsymbol{v}$ vectors with respect to a Cartesian basis $\boldsymbol{i}, \boldsymbol{j}, \boldsymbol{k}$ are given by $[s_x, s_y, s_z]$,$[t_x, t_y, t_z]$,$[u_x, u_y, u_z]$,$[v_x, v_y, v_z]$, respectively.

Suppose we wish to express the components of $\boldsymbol{v}$ with respect to the right-handed basis $\boldsymbol{s}, \boldsymbol{t}, \boldsymbol{u}$. That is, we wish to determine the triple of components $[v_s, v_t, v_u]$.

Had the $\boldsymbol{s}$, **t**, **u** basis been a Cartesian basis like the $\boldsymbol{i}, \boldsymbol{j}, \boldsymbol{k}$ basis, this would be simple. As we have seen in Section 2.14.3, we would project $\boldsymbol{v}$ onto each of the $\boldsymbol{s}, \boldsymbol{t}, \boldsymbol{u}$ base vectors of this basis. The magnitudes of these projection vectors would be equal to the components v_s, v_t, v_u. Thus, when $\boldsymbol{s} = \boldsymbol{i}$, $\boldsymbol{t} = \boldsymbol{j}$ and $\boldsymbol{u} = \boldsymbol{k}$ these components are given by:

$$\begin{aligned} v_s &= |v| \cdot \cos(\theta_s) = \boldsymbol{s} \odot \boldsymbol{v} \\ v_t &= |v| \cdot \cos(\theta_t) = \boldsymbol{t} \odot \boldsymbol{v} \\ v_u &= |v| \cdot \cos(\theta_u) = \boldsymbol{u} \odot \boldsymbol{v} \end{aligned}$$

where $\theta_s, \theta_t, \theta_u$ are the angles subtended by $\boldsymbol{v}$ and $\boldsymbol{s}, \boldsymbol{t}, \boldsymbol{u}$, respectively.

The above result can be generalised as follows:

$$v_s = \frac{\textit{the magnitude of the projection of } \boldsymbol{v} \textit{ onto } \boldsymbol{i}}{\textit{the magnitude of the projection of } \boldsymbol{s} \textit{ onto } \boldsymbol{i}} = \frac{\boldsymbol{i} \odot \boldsymbol{v}}{\boldsymbol{i} \odot \boldsymbol{s}}$$

$$v_t = \frac{\textit{the magnitude of the projection of } \boldsymbol{v} \textit{ onto } \boldsymbol{j}}{\textit{the magnitude of the projection of } \boldsymbol{t} \textit{ onto } \boldsymbol{j}} = \frac{\boldsymbol{j} \odot \boldsymbol{v}}{\boldsymbol{j} \odot \boldsymbol{t}}$$

$$v_u = \frac{\textit{the magnitude of the projection of } \boldsymbol{v} \textit{ onto } \boldsymbol{k}}{\textit{the magnitude of the projection of } \boldsymbol{u} \textit{ onto } \boldsymbol{k}} = \frac{\boldsymbol{k} \odot \boldsymbol{v}}{\boldsymbol{k} \odot \boldsymbol{u}}$$

Which when $\boldsymbol{s} = \boldsymbol{i}, \boldsymbol{t} = \boldsymbol{j}$ and $\boldsymbol{u} = \boldsymbol{k}$ reduces to:

$$v_s = \frac{\boldsymbol{i} \odot \boldsymbol{v}}{\boldsymbol{i} \odot \boldsymbol{i}} = \frac{\boldsymbol{s} \odot \boldsymbol{v}}{1} = \boldsymbol{s} \odot \boldsymbol{v}$$
$$v_t = \frac{\boldsymbol{j} \odot \boldsymbol{v}}{\boldsymbol{j} \odot \boldsymbol{j}} = \frac{\boldsymbol{t} \odot \boldsymbol{v}}{1} = \boldsymbol{t} \odot \boldsymbol{v}$$
$$v_u = \frac{\boldsymbol{k} \odot \boldsymbol{v}}{\boldsymbol{k} \odot \boldsymbol{k}} = \frac{\boldsymbol{u} \odot \boldsymbol{v}}{1} = \boldsymbol{u} \odot \boldsymbol{v}$$

In the general case however $\boldsymbol{s} \neq \boldsymbol{i}, \boldsymbol{t} \neq \boldsymbol{j}$ and $\boldsymbol{u} \neq \boldsymbol{k}$, we must compute a new set of vectors $\boldsymbol{i}', \boldsymbol{j}', \boldsymbol{k}'$ that correspond to base vectors $\boldsymbol{i}, \boldsymbol{j}, \boldsymbol{k}$. As vector $\boldsymbol{i}$ is normal to the plane defined by the other two vectors $\boldsymbol{j}, \boldsymbol{k}$ (of the $\boldsymbol{i}, \boldsymbol{j}, \boldsymbol{k}$ basis), so $\boldsymbol{i}'$ must be normal to the plane defined by vectors $\boldsymbol{t}, \boldsymbol{u}$ (of the $\boldsymbol{s}, \boldsymbol{t}, \boldsymbol{u}$ basis). Similarly $\boldsymbol{j}'$ must be normal to the plane defined by $\boldsymbol{u}, \boldsymbol{s}$, and $\boldsymbol{k}'$ must be normal to the plane defined by $\boldsymbol{s}, \mathbf{t}$.

Thus:

$$\boldsymbol{i}' = \boldsymbol{t} \otimes \boldsymbol{u}$$
$$\boldsymbol{j}' = \boldsymbol{u} \otimes \boldsymbol{s}$$
$$\boldsymbol{k}' = \boldsymbol{s} \otimes \boldsymbol{t}$$

Using this new set of vectors $\boldsymbol{i}', \boldsymbol{j}', \boldsymbol{k}'$ we can compute the components of vector $\boldsymbol{v}$ with respect to the right-handed basis $\boldsymbol{s}, \boldsymbol{t}, \boldsymbol{u}$:

$$\begin{aligned} v_s &= \frac{(\boldsymbol{t} \otimes \boldsymbol{u}) \odot \boldsymbol{v}}{(\boldsymbol{t} \otimes \boldsymbol{u}) \odot \boldsymbol{s}} = \frac{(\boldsymbol{t}, \boldsymbol{u}, \boldsymbol{v})}{(\boldsymbol{t}, \boldsymbol{u}, \boldsymbol{s})} \\ v_t &= \frac{(\boldsymbol{u} \otimes \boldsymbol{s}) \odot \boldsymbol{v}}{(\boldsymbol{u} \otimes \boldsymbol{s}) \odot \boldsymbol{t}} = \frac{(\boldsymbol{u}, \boldsymbol{s}, \boldsymbol{v})}{(\boldsymbol{u}, \boldsymbol{s}, \boldsymbol{t})} \\ v_u &= \frac{(\boldsymbol{s} \otimes \boldsymbol{t}) \odot \boldsymbol{v}}{(\boldsymbol{s} \otimes \boldsymbol{t}) \odot \boldsymbol{u}} = \frac{(\boldsymbol{s}, \boldsymbol{t}, \boldsymbol{v})}{(\boldsymbol{s}, \boldsymbol{t}, \boldsymbol{u})} \end{aligned} \tag{2.62}$$

Expanding the triple scalar products we obtain:

$$\begin{aligned} v_s &= \frac{(\boldsymbol{t}, \boldsymbol{u}, \boldsymbol{v})}{(\boldsymbol{t}, \boldsymbol{u}, \boldsymbol{s})} = \frac{\left(t_y u_z - t_z u_y\right) v_x + \left(t_z u_x - t_x u_z\right) v_y + \left(t_x u_y - t_y u_x\right) v_z}{\left(t_y u_z - t_z u_y\right) s_x + \left(t_z u_x - t_x u_z\right) s_y + \left(t_x u_y - t_y u_x\right) s_z} \\ v_t &= \frac{(\boldsymbol{u}, \boldsymbol{s}, \boldsymbol{v})}{(\boldsymbol{u}, \boldsymbol{s}, \boldsymbol{t})} = \frac{\left(u_y s_z - u_z s_y\right) v_x + \left(u_z s_x - u_x s_z\right) v_y + \left(u_x s_y - u_y s_x\right) v_z}{\left(u_y s_z - u_z s_y\right) t_x + \left(u_z s_x - u_x s_z\right) t_y + \left(u_x s_y - u_y s_x\right) t_z} \\ v_u &= \frac{(\boldsymbol{s}, \boldsymbol{t}, \boldsymbol{v})}{(\boldsymbol{s}, \boldsymbol{t}, \boldsymbol{u})} = \frac{\left(s_y t_z - s_z t_y\right) v_x + \left(s_z t_x - s_x t_z\right) v_y + \left(s_x t_y - s_y t_x\right) v_z}{\left(s_y t_z - s_z t_y\right) u_x + \left(s_z t_x - s_x t_z\right) u_y + \left(s_x t_y - s_y t_x\right) u_z} \end{aligned} \tag{2.63}$$

The above discussion was useful in illuminating the geometric aspects of our problem. It illustrated how geometric arguments are helpful in reasoning and solving problems in vector algebra, but it did not provide a conclusive mathematical proof of the correctness of our arguments. At some stage of our discussion we claimed that a *result could be generalised* without offering any evidence to back our claim. We will do so in the next section.

2.21 The Decomposition of a Vector According to a Basis

In this section we revisit the problem dealt with in the previous section, but we do so in a more mathematically rigorous way and provide a vector algebraic proof of our argument.

We start by restating the problem. Given four non-coplanar and non-zero vectors $\boldsymbol{a}$, $\boldsymbol{b}$, $\boldsymbol{c}$, $\boldsymbol{d}$ decompose vector $\boldsymbol{d}$ according to a basis $\boldsymbol{a}$, $\boldsymbol{b}$, $\boldsymbol{c}$, i.e. express the components of $\boldsymbol{d}$ with respect to a right-handed basis $\boldsymbol{a}$, $\boldsymbol{b}$, $\boldsymbol{c}$ (which is not required to be orthonormal or even orthogonal). From Section 2.7 we know that this is possible. Indeed, given any four non-coplanar and non-zero vectors in $\mathcal{E}^3$ we can express one of the vectors as a linear combination of the remaining three vectors.

By combining Eqs. (2.59) and (2.60) we obtain:

$$(\boldsymbol{a},\boldsymbol{b},\boldsymbol{d})\,\boldsymbol{c}-(\boldsymbol{a},\boldsymbol{b},\boldsymbol{c})\,\boldsymbol{d}=(\boldsymbol{a},\boldsymbol{c},\boldsymbol{d})\,\boldsymbol{b}-(\boldsymbol{b},\boldsymbol{c},\boldsymbol{d})\,\boldsymbol{a}$$

$$\therefore\quad (\boldsymbol{b},\boldsymbol{c},\boldsymbol{d})\,\boldsymbol{a}-(\boldsymbol{a},\boldsymbol{c},\boldsymbol{d})\,\boldsymbol{b}+(\boldsymbol{a},\boldsymbol{b},\boldsymbol{d})\,\boldsymbol{c}-(\boldsymbol{a},\boldsymbol{b},\boldsymbol{c})\,\boldsymbol{d}=\vec{\boldsymbol{0}}$$

Using the non-cyclic permutation rule (R2.15) we obtain:

$$(\boldsymbol{b},\boldsymbol{c},\boldsymbol{d})\,\boldsymbol{a}+(\boldsymbol{a},\boldsymbol{d},\boldsymbol{c})\,\boldsymbol{b}+(\boldsymbol{a},\boldsymbol{b},\boldsymbol{d})\,\boldsymbol{c}-(\boldsymbol{a},\boldsymbol{b},\boldsymbol{c})\,\boldsymbol{d}=\vec{\boldsymbol{0}}$$

But given that any four vectors in $\mathcal{E}^3$ are always linearly dependent (i.e. we can decompose one in terms of the remaining three), we can solve the above equation for $\boldsymbol{d}$:

$$(\boldsymbol{b},\boldsymbol{c},\boldsymbol{d})\,\boldsymbol{a}+(\boldsymbol{a},\boldsymbol{d},\boldsymbol{c})\,\boldsymbol{b}+(\boldsymbol{a},\boldsymbol{b},\boldsymbol{d})\,\boldsymbol{c}=(\boldsymbol{a},\boldsymbol{b},\boldsymbol{c})\,\boldsymbol{d}$$

$$\therefore\quad \boldsymbol{d}=\frac{(\boldsymbol{b},\boldsymbol{c},\boldsymbol{d})\,\boldsymbol{a}+(\boldsymbol{a},\boldsymbol{d},\boldsymbol{c})\,\boldsymbol{b}+(\boldsymbol{a},\boldsymbol{b},\boldsymbol{d})\,\boldsymbol{c}}{(\boldsymbol{a},\boldsymbol{b},\boldsymbol{c})}=\frac{(\boldsymbol{b},\boldsymbol{c},\boldsymbol{d})}{(\boldsymbol{a},\boldsymbol{b},\boldsymbol{c})}\boldsymbol{a}+\frac{(\boldsymbol{a},\boldsymbol{d},\boldsymbol{c})}{(\boldsymbol{a},\boldsymbol{b},\boldsymbol{c})}\boldsymbol{b}+\frac{(\boldsymbol{a},\boldsymbol{b},\boldsymbol{d})}{(\boldsymbol{a},\boldsymbol{b},\boldsymbol{c})}\boldsymbol{c}$$

Expressing $\boldsymbol{d}$ in component form we obtain:

$$\boldsymbol{d}=\left[\frac{(\boldsymbol{b},\boldsymbol{c},\boldsymbol{d})}{(\boldsymbol{a},\boldsymbol{b},\boldsymbol{c})},\frac{(\boldsymbol{a},\boldsymbol{d},\boldsymbol{c})}{(\boldsymbol{a},\boldsymbol{b},\boldsymbol{c})},\frac{(\boldsymbol{a},\boldsymbol{b},\boldsymbol{d})}{(\boldsymbol{a},\boldsymbol{b},\boldsymbol{c})}\right]$$

Using the cyclic permutation rule (R2.14) we obtain:

$$\boldsymbol{d}=\left[\frac{(\boldsymbol{b},\boldsymbol{c},\boldsymbol{d})}{(\boldsymbol{b},\boldsymbol{c},\boldsymbol{a})},\frac{(\boldsymbol{c},\boldsymbol{a},\boldsymbol{d})}{(\boldsymbol{c},\boldsymbol{a},\boldsymbol{b})},\frac{(\boldsymbol{a},\boldsymbol{b},\boldsymbol{d})}{(\boldsymbol{a},\boldsymbol{b},\boldsymbol{c})}\right]$$

Finally, expanding the triple scalar products we obtain:

$$\boldsymbol{d}=\left[\frac{(\boldsymbol{b}\otimes\boldsymbol{c})\odot\boldsymbol{d}}{(\boldsymbol{b}\otimes\boldsymbol{c})\odot\boldsymbol{a}},\frac{(\boldsymbol{c}\otimes\boldsymbol{a})\odot\boldsymbol{d}}{(\boldsymbol{c}\otimes\boldsymbol{a})\odot\boldsymbol{b}},\frac{(\boldsymbol{a}\otimes\boldsymbol{b})\odot\boldsymbol{d}}{(\boldsymbol{a}\otimes\boldsymbol{b})\odot\boldsymbol{c}}\right]$$

The individual components of $\boldsymbol{d}$ are given as:

$$d_a = \frac{(\boldsymbol{b} \otimes \boldsymbol{c}) \odot \boldsymbol{d}}{(\boldsymbol{b} \otimes \boldsymbol{c}) \odot \boldsymbol{a}} = \frac{(\boldsymbol{b}, \boldsymbol{c}, \boldsymbol{d})}{(\boldsymbol{b}, \boldsymbol{c}, \boldsymbol{a})}$$

$$d_b = \frac{(\boldsymbol{c} \otimes \boldsymbol{a}) \odot \boldsymbol{d}}{(\boldsymbol{c} \otimes \boldsymbol{a}) \odot \boldsymbol{b}} = \frac{(\boldsymbol{c}, \boldsymbol{a}, \boldsymbol{d})}{(\boldsymbol{c}, \boldsymbol{a}, \boldsymbol{b})} \tag{2.64}$$

$$d_c = \frac{(\boldsymbol{a} \otimes \boldsymbol{b}) \odot \boldsymbol{d}}{(\boldsymbol{a} \otimes \boldsymbol{b}) \odot \boldsymbol{c}} = \frac{(\boldsymbol{a}, \boldsymbol{b}, \boldsymbol{d})}{(\boldsymbol{a}, \boldsymbol{b}, \boldsymbol{c})}$$

Which accords with the result we obtained in Eq. (2.62). QED

If $\boldsymbol{a}, \boldsymbol{b}, \boldsymbol{c}, \boldsymbol{d}$ have components $[a_x, a_y, a_z]$,$[b_x, b_y, b_z]$,$[c_x, c_y, c_z]$,$[d_x, d_y, d_z]$, respectively, then by expanding the triple scalar products we obtain:

$$\begin{aligned}
d_a &= \frac{(\boldsymbol{b}, \boldsymbol{c}, \boldsymbol{d})}{(\boldsymbol{b}, \boldsymbol{c}, \boldsymbol{a})} = \frac{(b_y c_z - b_z c_y) d_x + (b_z c_x - b_x c_z) d_y + (b_x c_y - b_y c_x) d_z}{(b_y c_z - b_z c_y) a_x + (b_z c_x - b_x c_z) a_y + (b_x c_y - b_y c_x) a_z} \\
d_b &= \frac{(\boldsymbol{c}, \boldsymbol{a}, \boldsymbol{d})}{(\boldsymbol{c}, \boldsymbol{a}, \boldsymbol{b})} = \frac{(c_y a_z - c_z a_y) d_x + (c_z a_x - c_x a_z) d_y + (c_x a_y - c_y a_x) d_z}{(c_y a_z - c_z a_y) b_x + (c_z a_x - c_x a_z) b_y + (c_x a_y - c_y a_x) b_z} \\
d_c &= \frac{(\boldsymbol{a}, \boldsymbol{b}, \boldsymbol{d})}{(\boldsymbol{a}, \boldsymbol{b}, \boldsymbol{c})} = \frac{(a_y b_z - a_z b_y) d_x + (a_z b_x - a_x b_z) d_y + (a_x b_y - a_y b_x) d_z}{(a_y b_z - a_z b_y) c_x + (a_z b_x - a_x b_z) c_y + (a_x b_y - a_y b_x) c_z}
\end{aligned} \tag{2.65}$$

2.22 The Vector Equation of the Line Revisited

2.22.1 *The Line Defined by Two Position Vectors*

In Section 2.5 we have seen that the vector equation of a line $\boldsymbol{\Lambda}$ passing through two points $\boldsymbol{P}_1$ and $\boldsymbol{P}_2$ with position vectors $\boldsymbol{p}_1, \boldsymbol{p}_2$ relative to some origin $\boldsymbol{O}$ is given by:

$$\boldsymbol{p} = (1 - t) \cdot \boldsymbol{p}_1 + t \cdot \boldsymbol{p}_2 \tag{2.66}$$

$$\text{or} \quad \boldsymbol{p} = \boldsymbol{p}_1 + t \cdot (\boldsymbol{p}_2 - \boldsymbol{p}_1) \tag{2.67}$$

where t is a scalar parameter and $\boldsymbol{p}$ is the position vector of the general point $\boldsymbol{P}$ on the line, as shown in Fig. 2.21.

If points $\boldsymbol{P}_1, \boldsymbol{P}_2, \boldsymbol{P}$ have coordinates $[x_1, y_1, z_1]$,$[x_2, y_2, z_2]$,$[x, y, z]$, respectively, then the line equation can be rewritten as:

$$\begin{aligned}
x &= (1 - t) \cdot x_1 + t \cdot x_2 \\
y &= (1 - t) \cdot y_1 + t \cdot y_2 \\
z &= (1 - t) \cdot z_1 + t \cdot z_2
\end{aligned} \tag{2.68}$$

$$\text{or} \quad \begin{aligned}
x &= x_1 + t \cdot (x_2 - x_1) \\
y &= y_1 + t \cdot (y_2 - y_1) \\
z &= z_1 + t \cdot (z_2 - z_1)
\end{aligned} \tag{2.69}$$

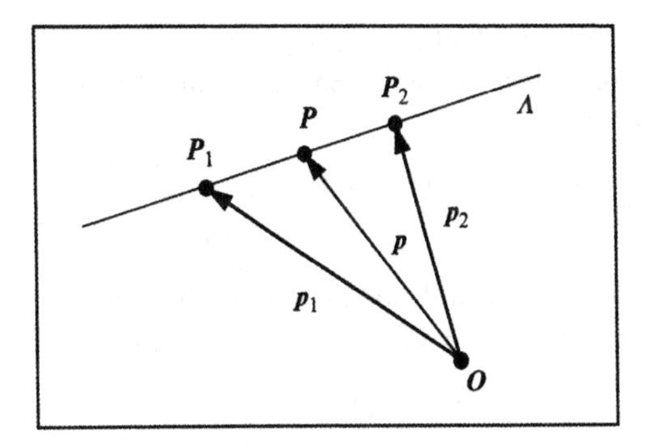

FIGURE 2.21. A line defined by two position vectors.

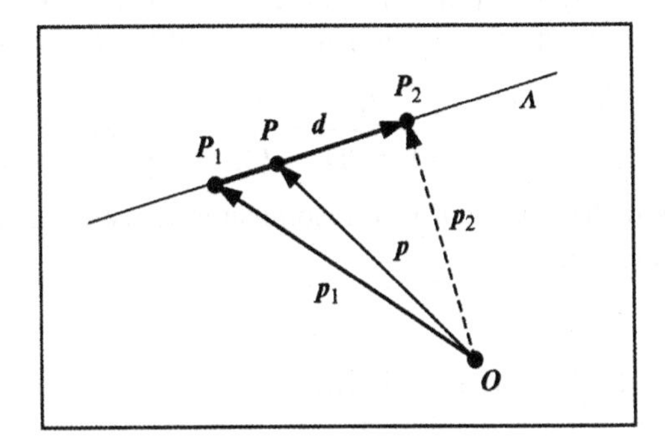

FIGURE 2.22. A line defined by a position vector and a direction vector.

2.22.2 *The Line Defined by a Position Vector and Direction Vector*

An alternative but equivalent representation of the line $\mathbf{\Lambda}$ is derived as follows. Let $\mathbf{\Lambda}$ be a line that passes through a point $\boldsymbol{P}_1$ (having position vector $\boldsymbol{p}_1$) in the direction of a direction unit vector $\boldsymbol{d}$ represented by the directed segment $\overrightarrow{\boldsymbol{P}_1\boldsymbol{P}_2}$ in Fig. 2.22.

Then the line equation can be written as:

$$\boldsymbol{p} = \boldsymbol{p}_1 + t \cdot \boldsymbol{d} \tag{2.70}$$

where, as above, t is a scalar parameter and $\boldsymbol{p}$ is the position vector of the general point $\boldsymbol{P}$ on the line. This equation is reminiscent of the Eq. (2.67). This is not surprising as $\boldsymbol{d} = (\boldsymbol{p}_2 - \boldsymbol{p}_1)$.

If points $\boldsymbol{P}_1, \boldsymbol{P}$ have coordinates $[x_1, y_1, z_1]$, $[x, y, z]$, respectively and vector $\boldsymbol{d}$ has components $[d_x, d_y, d_z]$, then the line equation can be rewritten as:

$$\begin{aligned} x &= x_1 + t \cdot d_x \\ y &= y_1 + t \cdot d_y \\ z &= z_1 + t \cdot d_z \end{aligned} \tag{2.71}$$

Which is reminiscent of Eq. (2.69).

Sometimes it is convenient to rewrite Eq. (2.71) in what is known as the *standard form* of the line equation:

$$\frac{(x - x_1)}{d_x} = \frac{(y - y_1)}{d_y} = \frac{(z - z_1)}{d_z} = t \tag{2.72}$$

Here it is understood that if a denominator vanishes, then so does the corresponding numerator. Thus, if $d_x = 0$, then the above equation becomes:

$$x = x_1, \frac{(y - y_1)}{d_y} = \frac{(z - z_1)}{d_z} = t$$

Since the direction cosines of the direction vector $\boldsymbol{d}$ are $d_x = x_2 - x_1$, $d_y = y_2 - y_1$ and $d_z = z_2 - z_1$, respectively, an alternative form of Eq. (2.72) is:

$$\frac{(x - x_1)}{(x_2 - x_1)} = \frac{(y - y_1)}{(y_2 - y_1)} = \frac{(z - z_1)}{(z_2 - z_1)} = t \tag{2.73}$$

Here again it is understood that if a denominator vanishes, then so does the corresponding numerator.

2.23 The Vector Equation of the Plane

2.23.1 *The Plane Defined by a Position Vector and a Normal Vector*

To derive a vector equation of the plane we use the fact that any line defined by any two points on a plane $\boldsymbol{\Pi}$ is perpendicular to the vector normal to the plane, as seen in Fig. 2.23.

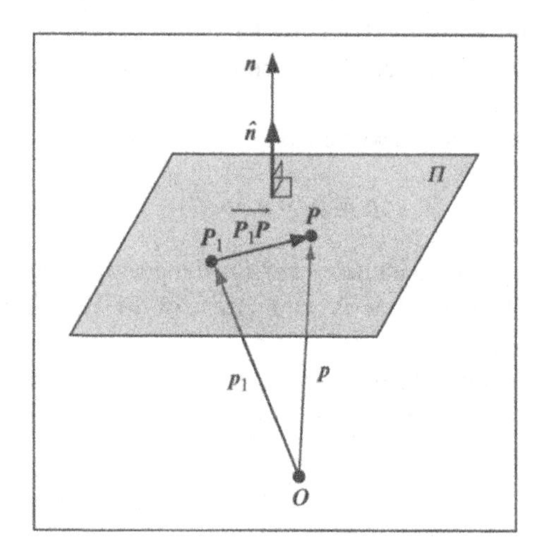

FIGURE 2.23. A plane defined by a position vector and a normal direction vector.

Let $\boldsymbol{P}_1$ be a point on the plane with position vector $\boldsymbol{p}_1$ and let $\boldsymbol{P}$ be the general point on the plane with position vector $\boldsymbol{p}$. Let $\boldsymbol{n}$ be a vector that is normal to the plane and let $\hat{\boldsymbol{n}}$ be the unit normal of the plane. Let vectors $\boldsymbol{p}_1, \boldsymbol{p}, \boldsymbol{n}$ have components $[x_1, y_1, z_1]$,$[x, y, z]$,$[a, b, c]$, respectively. The directed segment $\overrightarrow{\boldsymbol{P}_1\boldsymbol{P}}$ represents vector $(\boldsymbol{p} - \boldsymbol{p}_1)$, which is perpendicular to $\boldsymbol{n}$. Thus:

$$(\boldsymbol{p} - \boldsymbol{p}_1) \odot \boldsymbol{n} = 0 \tag{2.74}$$

$$\therefore \ \boldsymbol{p} \odot \boldsymbol{n} - \boldsymbol{p}_1 \odot \boldsymbol{n} = 0$$

$$\therefore \ \boldsymbol{p} \odot \boldsymbol{n} = \boldsymbol{p}_1 \odot \boldsymbol{n} \tag{2.75}$$

$$\text{or} \quad \boldsymbol{p} \odot \boldsymbol{n} = \boldsymbol{d}_o \tag{2.76}$$

Equations (2.74), (2.75) and (2.76) are all alternative forms of the vector equation of the plane. In the special case where $\boldsymbol{n} = \hat{\boldsymbol{n}}$, then d_o represents the distance of the plane from the origin $\boldsymbol{O}$.

Expressing Eq. (2.74) in component form we obtain:

$$a \cdot (x - x_1) + b \cdot (y - y_1) + c \cdot (z - z_1) = 0 \tag{2.77}$$

This equation provides both *the necessary and sufficient condition* for a point $[x, y, z]$ to lie on the plane containing the point $[x_1, y_1, z_1]$ and being perpendicular to the direction $[a, b, c]$.

Similarly Eq. (2.76) in component form gives:

$$a \cdot x + b \cdot y + c \cdot z + d = 0 \tag{2.78}$$

where $d = -d_o$ is the negative distance of the plane from the origin. Thus the plane is represented by a linear equation of x, y, z.

2.23.2 The Plane Defined by Three Position Vectors

Let $\boldsymbol{P}_1, \boldsymbol{P}_2, \boldsymbol{P}_3$ be three non-collinear points lying on a plane $\boldsymbol{\Pi}$ with position vectors $\boldsymbol{p}_1, \boldsymbol{p}_2, \boldsymbol{p}_3$ and components $[x_1, y_1, z_1]$, $[x_2, y_2, z_2]$, $[x_3, y_3, z_3]$, respectively. Directed segments $\overrightarrow{\boldsymbol{P}_1\boldsymbol{P}_2}, \overrightarrow{\boldsymbol{P}_1\boldsymbol{P}_3}$ represent the vectors $(p_2 - p_1)$, $(p_3 - p_1)$, respectively, as shown in Fig. 2.24.

Both directed segments lie on the plane Π and are therefore perpendicular to the normal of the plane. Thus:

$$\boldsymbol{n} = (\boldsymbol{p}_2 - \boldsymbol{p}_1) \otimes (\boldsymbol{p}_3 - \boldsymbol{p}_1) \tag{2.79}$$

Using the distributive rule (R2.13):

$$\boldsymbol{n} = \boldsymbol{p}_2 \otimes \boldsymbol{p}_3 - \boldsymbol{p}_2 \otimes \boldsymbol{p}_1 - \boldsymbol{p}_1 \otimes \boldsymbol{p}_3 + \boldsymbol{p}_1 \otimes \boldsymbol{p}_1$$

Using axiom (A2.10) and the anti-commutative rule (R2.11):

$$\boldsymbol{n} = \boldsymbol{p}_2 \otimes \boldsymbol{p}_3 + \boldsymbol{p}_1 \otimes \boldsymbol{p}_2 + \boldsymbol{p}_3 \otimes \boldsymbol{p}_1$$

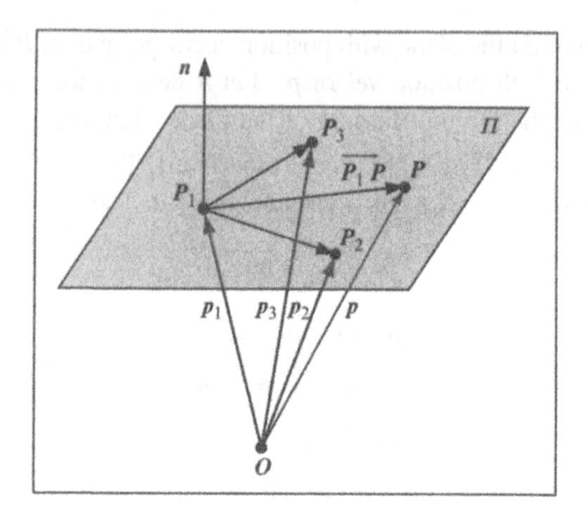

FIGURE 2.24. A plane defined by three position vectors.

Substituting this normal into Eq. (2.74) we obtain:

$$(\boldsymbol{p}-\boldsymbol{p}_1)\odot((\boldsymbol{p}_2-\boldsymbol{p}_1)\otimes(\boldsymbol{p}_3-\boldsymbol{p}_1))=0 \tag{2.80}$$

$$\therefore\ (\boldsymbol{p}-\boldsymbol{p}_1)\odot(\boldsymbol{p}_2\otimes\boldsymbol{p}_3+\boldsymbol{p}_3\otimes\boldsymbol{p}_1+\boldsymbol{p}_1\otimes\boldsymbol{p}_2)=0 \tag{2.81}$$

Rewriting Eq. (2.81) in component form we obtain:

$$\begin{aligned}&(x-x_1)\cdot((y_2z_3-y_3z_2)+(y_3z_1-y_1z_3)+(y_1z_2-y_2z_1))\\&+(y-y_1)\cdot((z_2x_3-z_3x_2)+(z_3x_1-z_1x_3)+(z_1x_2-z_2x_1))\\&+(z-z_1)\cdot((x_2y_3-x_3y_2)+(x_3y_1-x_1y_3)+(x_1y_2-x_2y_1))=0\end{aligned} \tag{2.82}$$

To simplify this expression we label the bracketed vector products as a, b, c. Thus:

$$\begin{aligned}&(x-x_1)\cdot a+(y-y_1)\cdot b+(z-z_1)\cdot c=0\\&\therefore (a\cdot x+b\cdot y+c\cdot z)-(a\cdot x_1+b\cdot y_1+c\cdot z_1)=0\\&\therefore a\cdot x+b\cdot y+c\cdot z+d=0\end{aligned} \tag{2.83}$$

where:

$$\begin{aligned}a&=(y_2z_3-y_3z_2)+(y_3z_1-y_1z_3)+(y_1z_2-y_2z_1)\\b&=(z_2x_3-z_3x_2)+(z_3x_1-z_1x_3)+(z_1x_2-z_2x_1)\\c&=(x_2y_3-x_3y_2)+(x_3y_1-x_1y_3)+(x_1y_2-x_2y_1)\\d&=-(a\cdot x_1+b\cdot y_1+c\cdot z_1)\end{aligned}$$

Here a, b, c are the direction ratios of the plane normal and d is equal to the negative distance of the plane from the origin scaled by the magnitude of the plane normal.

An alternative representation of the plane can be arrived at by expressing the left-hand side of Eq. (2.80) as a triple scalar product:

$$(\boldsymbol{p}-\boldsymbol{p}_1)\odot((\boldsymbol{p}_2-\boldsymbol{p}_1)\otimes(\boldsymbol{p}_3-\boldsymbol{p}_1))=((\boldsymbol{p}-\boldsymbol{p}_1),(\boldsymbol{p}_2-\boldsymbol{p}_1),(\boldsymbol{p}_3-\boldsymbol{p}_1))$$
$$\therefore((\boldsymbol{p}-\boldsymbol{p}_1),(\boldsymbol{p}_2-\boldsymbol{p}_1),(\boldsymbol{p}_3-\boldsymbol{p}_1))=0 \qquad (2.84)$$

Expanding the left-hand side of Eq. (2.81) we obtain:

$$\therefore \boldsymbol{p}\odot(\boldsymbol{p}_2\otimes\boldsymbol{p}_3)+\boldsymbol{p}\odot(\boldsymbol{p}_3\otimes\boldsymbol{p}_1)+\boldsymbol{p}\odot(\boldsymbol{p}_1\otimes\boldsymbol{p}_2)-\boldsymbol{p}_1\odot(\boldsymbol{p}_2\otimes\boldsymbol{p}_3)$$
$$-\boldsymbol{p}_1\odot(\boldsymbol{p}_3\otimes\boldsymbol{p}_1)-\boldsymbol{p}_1\odot(\boldsymbol{p}_1\otimes\boldsymbol{p}_2)=0$$

The last two terms of the above equation cancel out by Axiom (A2.14) giving:

$$\boldsymbol{p}\odot(\boldsymbol{p}_2\otimes\boldsymbol{p}_3)+\boldsymbol{p}\odot(\boldsymbol{p}_3\otimes\boldsymbol{p}_1)+\boldsymbol{p}\odot(\boldsymbol{p}_1\otimes\boldsymbol{p}_2)-\boldsymbol{p}_1\odot(\boldsymbol{p}_2\otimes\boldsymbol{p}_3)=0$$

Which in triple scalar product form is:

$$(\boldsymbol{p},\boldsymbol{p}_2,\boldsymbol{p}_3)+(\boldsymbol{p},\boldsymbol{p}_3,\boldsymbol{p}_1)+(\boldsymbol{p},\boldsymbol{p}_1,\boldsymbol{p}_2)-(\boldsymbol{p}_1,\boldsymbol{p}_2,\boldsymbol{p}_3)=0$$
$$\therefore(\boldsymbol{p},\boldsymbol{p}_2,\boldsymbol{p}_3)+(\boldsymbol{p},\boldsymbol{p}_3,\boldsymbol{p}_1)+(\boldsymbol{p},\boldsymbol{p}_1,\boldsymbol{p}_2)=(\boldsymbol{p}_1,\boldsymbol{p}_2,\boldsymbol{p}_3) \qquad (2.85)$$

2.24 Some Applications of Vector Algebra in Analytical Geometry

2.24.1 The Distance Between Two Points in Space

Given two points $\boldsymbol{P}_1, \boldsymbol{P}_2$ with position vectors $\boldsymbol{p}_1=[x_1,y_1,z_1], \boldsymbol{p}_2=[x_2,y_2,z_2]$, the distance between these points is given by the magnitude of the directed segment $\overrightarrow{\boldsymbol{P}_1\boldsymbol{P}_2}$, which represents the vector $(\boldsymbol{p}_2-\boldsymbol{p}_1)$, as shown in Fig. 2.25. Thus,

$$|\overrightarrow{\boldsymbol{P}_1\boldsymbol{P}_2}|=|\boldsymbol{p}_2-\boldsymbol{p}_1|=\sqrt{(x_2-x_1)^2+(y_2-y_1)^2+(z_2-z_1)^2} \qquad (2.86)$$

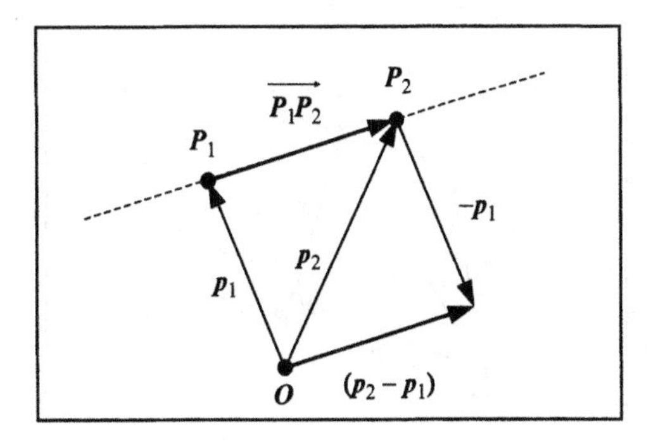

FIGURE 2.25. The distance between two points.

2.24.2 The Perpendicular Distance from a Point to a Line

A line $\boldsymbol{\Lambda}$ is defined by a position vector $\boldsymbol{a} = [a_x, a_y, a_z]$ and a direction unit vector $\boldsymbol{b} = [b_x, b_y, b_z]$. Let $\boldsymbol{P}$ be a point not on the line $\boldsymbol{\Lambda}$ with position vector $\boldsymbol{p} = [p_x, p_y, p_z]$. Suppose we wish to find the foot $\boldsymbol{Q}$ of the perpendicular from point $\boldsymbol{P}$ to the line as well as the perpendicular distance from this point to the line, as seen in Fig. 2.26.

The general point $\boldsymbol{x}$ on the line is given by its parametric equation:

$$\boldsymbol{x} = \boldsymbol{a} + t \cdot \boldsymbol{b}$$

Hence the position vector of point $\boldsymbol{Q}$ is given by:

$$\boldsymbol{q} = \boldsymbol{a} + t_q \cdot \boldsymbol{b} \tag{2.87}$$

for some scalar value t_q. Therefore the directed segment $\overrightarrow{\boldsymbol{PQ}}$ represents the vector:

$$\boldsymbol{v} = \boldsymbol{q} - \boldsymbol{p} = \boldsymbol{a} + t_q \cdot \boldsymbol{b} - \boldsymbol{p} \tag{2.88}$$

and since $\overrightarrow{\boldsymbol{PQ}}$ is perpendicular to $\boldsymbol{\Lambda}$ we have:

$$(\boldsymbol{a} + t_q \cdot \boldsymbol{b} - \boldsymbol{p}) \odot \boldsymbol{b} = 0$$

Expanding this scalar product and recalling that $\boldsymbol{b}$ is a unit vector, we get:

$$\boldsymbol{a} \odot \boldsymbol{b} + t_q \cdot \boldsymbol{b} \odot \boldsymbol{b} - \boldsymbol{p} \odot \boldsymbol{b} = 0$$

$$\therefore\ \boldsymbol{a} \odot \boldsymbol{b} + t_q \cdot 1 - \boldsymbol{p} \odot \boldsymbol{b} = 0$$

Solving for t_q we get:

$$t_q = (\boldsymbol{p} - \boldsymbol{a}) \odot \boldsymbol{b}$$

which in component form is:

$$t_q = (p_x - a_x) \cdot b_x + (p_y - a_y) \cdot b_y + (p_z - a_z) \cdot b_z$$

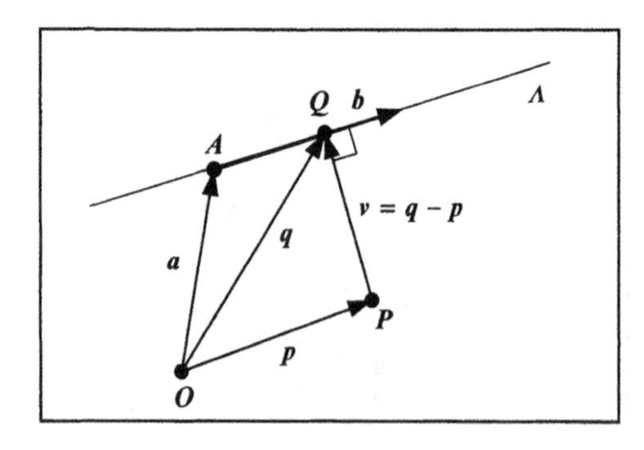

FIGURE 2.26. The foot of the perpendicular from a point to a line.

Substituting this value in Eq. (2.87) we obtain:

$$\boldsymbol{q} = \boldsymbol{a} + ((\boldsymbol{p} - \boldsymbol{a}) \odot \boldsymbol{b}) \cdot \boldsymbol{b} \tag{2.89}$$

which is the required foot of the perpendicular. Rewriting the above equation in component form we get:

$$\begin{aligned} q_x &= a_x + t_q \cdot b_x \\ q_y &= a_y + t_q \cdot b_y \text{ where } t_q = (p_x - a_x) \cdot b_x + \left(p_y - a_y\right) \cdot b_y + (p_z - a_z) \cdot b_z \\ q_z &= a_z + t_q \cdot b_z \end{aligned} \tag{2.90}$$

Now we can calculate the length of the perpendicular distance quite simply as:

$$|\boldsymbol{v}| = |\boldsymbol{q} - \boldsymbol{p}| = \sqrt{(q_x - p_x)^2 + \left(q_y - p_y\right)^2 + (q_z - p_z)^2} \tag{2.91}$$

2.24.3 *The Distance of a Point from a Line*

This is essentially the same problem as in the previous section, but here we do not calculate the foot of the perpendicular. A line $\boldsymbol{\Lambda}$ is defined by a position vector $\boldsymbol{a} = \left[a_x, a_y, a_z\right]$ and a direction unit vector $\boldsymbol{b} = \left[b_x, b_y, b_z\right]$, as shown in Fig. 2.27.

As before, the general point $\boldsymbol{x}$ on the line is given by its parametric equation:

$$\boldsymbol{x} = \boldsymbol{a} + t \cdot \boldsymbol{b}$$

Let the general point $\boldsymbol{P}$ have position vector $\boldsymbol{p} = \left[p_x, p_y, p_z\right]$. The directed segment $\overrightarrow{\boldsymbol{AP}}$ represents the vector $\boldsymbol{c} = (\boldsymbol{p} - \boldsymbol{a})$. The perpendicular distance of the point from the line is the magnitude of the directed segment $\overrightarrow{\boldsymbol{PQ}}$:

$$\left|\overrightarrow{\boldsymbol{PQ}}\right| = |\boldsymbol{p} - \boldsymbol{a}| \cdot \sin\theta = |\boldsymbol{b} \otimes (\boldsymbol{p} - \boldsymbol{a})| \tag{2.92}$$

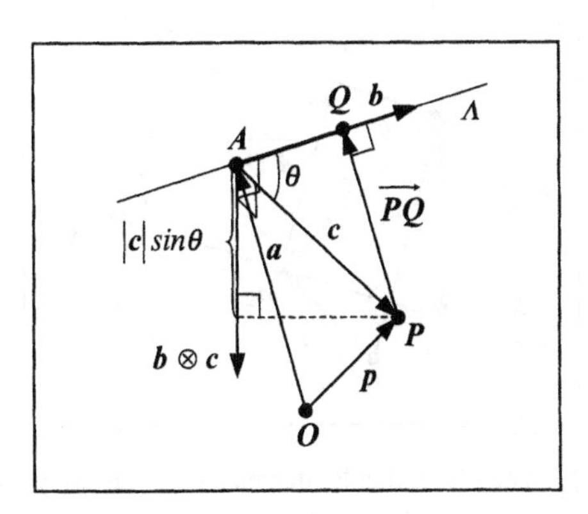

FIGURE 2.27. The perpendicular distance of a point from a line.

The components of the vector product are given by:

$$\boldsymbol{b} \otimes (\boldsymbol{p} - \boldsymbol{a}) = \left[\left(b_y\left(p_z - a_z\right) - b_z\left(p_y - a_y\right)\right),\right.\\ \left.\left(b_z\left(p_x - a_x\right) - b_x\left(p_z - a_z\right)\right), \left(b_x\left(p_y - a_y\right) - b_y\left(p_x - a_x\right)\right)\right]$$

and its magnitude is given by:

$$|\boldsymbol{b} \otimes (\boldsymbol{p} - \boldsymbol{a})| = \sqrt{\begin{array}{l}\left(b_y\left(p_z - a_z\right) - b_z\left(p_y - a_y\right)\right)^2 + \\ \left(b_z\left(p_x - a_x\right) - b_x\left(p_z - a_z\right)\right)^2 + \\ \left(b_x\left(p_y - a_y\right) - b_y\left(p_x - a_x\right)\right)^2\end{array}} \tag{2.93}$$

2.24.4 *The Distance Between Two Parallel Lines*

Let $\boldsymbol{\Lambda}_1, \boldsymbol{\Lambda}_2$ be two parallel lines. Where $\boldsymbol{a}_1 = \left[a_{1x}, a_{1y}, a_{1z}\right], \boldsymbol{a}_2 = \left[a_{2x}, a_{2y}, a_{2z}\right]$ are position vectors of lines $\boldsymbol{\Lambda}_1$, $\boldsymbol{\Lambda}_2$, respectively and $\boldsymbol{b} = \left[b_x, b_y, b_z\right]$ is the direction unit vector of both lines, as shown in Fig. 2.28.

The perpendicular distance between the two parallel lines is equal to the distance of point $\boldsymbol{A}_2$ from line $\boldsymbol{\Lambda}_1$ (or analogously the distance of point $\boldsymbol{A}_1$ from line $\boldsymbol{\Lambda}_2$). As in the previous section, the directed segment $\overrightarrow{\boldsymbol{A}_1\boldsymbol{A}_2}$ represents the vector $\boldsymbol{c} = (\boldsymbol{a}_2 - \boldsymbol{a}_1)$. The perpendicular distance of point $\boldsymbol{A}_2$ from line $\boldsymbol{\Lambda}_1$ is the magnitude of the directed segment $\overrightarrow{\boldsymbol{A}_2\boldsymbol{Q}}$:

$$\left|\overrightarrow{\boldsymbol{A}_2\boldsymbol{Q}}\right| = |\boldsymbol{a}_2 - \boldsymbol{a}_1| \cdot \sin\theta = |\boldsymbol{b} \otimes (\boldsymbol{a}_2 - \boldsymbol{a}_1)| \tag{2.94}$$

The components of the vector product are given by:

$$\boldsymbol{b} \otimes (\boldsymbol{a}_2 - \boldsymbol{a}_1) = \left[\left(b_y\left(a_{2z} - a_{1z}\right) - b_z\left(a_{2y} - a_{1y}\right)\right),\right.\\ \left(b_z\left(a_{2x} - a_{1x}\right) - b_x\left(a_{2z} - a_{1z}\right)\right),\\ \left.\left(b_x\left(a_{2y} - a_{1y}\right) - b_y\left(a_{2x} - a_{1x}\right)\right)\right]$$

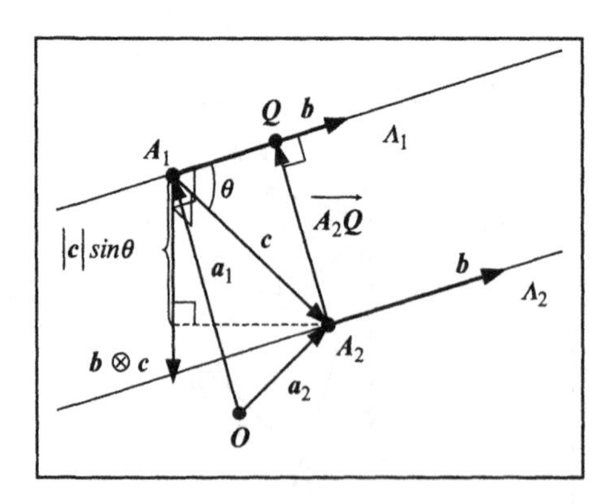

FIGURE 2.28. The perpendicular distance between two parallel lines.

and its magnitude is given by:

$$|\boldsymbol{b} \otimes (\boldsymbol{a}_2 - \boldsymbol{a}_1)| = \sqrt{\begin{array}{l} \left(b_y \left(a_{2z} - a_{1z}\right) - b_z \left(a_{2y} - a_{1y}\right)\right)^2 + \\ \left(b_z \left(a_{2x} - a_{1x}\right) - b_x \left(a_{2z} - a_{1z}\right)\right)^2 + \\ \left(b_x \left(a_{2y} - a_{1y}\right) - b_y \left(a_{2x} - a_{1x}\right)\right)^2 \end{array}} \tag{2.95}$$

2.24.5 The Distance Between Two Non-Parallel Lines

Let $\boldsymbol{\Lambda}_1$,$\boldsymbol{\Lambda}_2$ be two non-parallel lines. Where line $\boldsymbol{\Lambda}_1$ is defined by a position vector $\boldsymbol{a}_1 = \left[a_{1x}, a_{1y}, a_{1z}\right]$ and a direction unit vector $\boldsymbol{b}_1 = \left[b_{1x}, b_{1y}, b_{1z}\right]$ and line is defined by a position vector $\boldsymbol{a}_2 = \left[a_{2x}, a_{2y}, a_{2z}\right]$ and a direction unit vector $\boldsymbol{b}_2 = \left[b_{2x}, b_{2y}, b_{2z}\right]$, as shown in Fig. 2.29.

The directed segment $\overrightarrow{A_1 A_2}$ represents the vector $\boldsymbol{c} = (\boldsymbol{a}_2 - \boldsymbol{a}_1)$. The feet of the line that is mutually perpendicular to the lines $\boldsymbol{\Lambda}_1$,$\boldsymbol{\Lambda}_2$ are $\boldsymbol{Q}_1$, $\boldsymbol{Q}_2$, respectively. The perpendicular distance between the two lines is the magnitude of the directed segment $\overrightarrow{Q_1 Q_2}$. The magnitude of this segment is equal to the ratio of the volume of the parallelepiped having vectors $\boldsymbol{b}_1$, $\boldsymbol{b}_2$, $\boldsymbol{c}$ as concurrent sides over the area of its base:

$$\begin{aligned} \left|\overrightarrow{Q_1 Q_2}\right| &= \frac{volume\ of\ the\ parallelepiped\ \left((\boldsymbol{a}_2 - \boldsymbol{a}_1), \boldsymbol{b}_1, \boldsymbol{b}_2\right)}{area\ of\ the\ base\ of\ the\ parallelepiped\ \left((\boldsymbol{a}_2 - \boldsymbol{a}_1), \boldsymbol{b}_1, \boldsymbol{b}_2\right)} \\ &= \frac{|(\boldsymbol{a}_2 - \boldsymbol{a}_1) \odot (\boldsymbol{b}_1 \otimes \boldsymbol{b}_2)|}{|\boldsymbol{b}_1 \otimes \boldsymbol{b}_2|} \end{aligned} \tag{2.96}$$

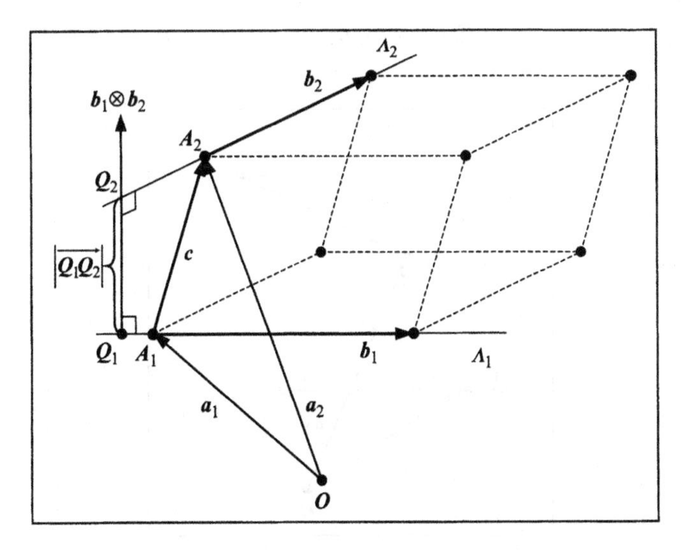

FIGURE 2.29. The distance between two non-parallel lines.

The perpendicular distance between two non-parallel lines in component form is given by:

$$\left|\overrightarrow{Q_1Q_2}\right| = \frac{|\boldsymbol{a} \odot \boldsymbol{b}|}{|\boldsymbol{b}|} = \frac{|a_x b_x + a_y b_y + a_z b_z|}{\sqrt{b_x^2 + b_y^2 + b_z^2}}$$

$$for \begin{cases} \boldsymbol{a} = \boldsymbol{a}_2 - \boldsymbol{a}_1 = [(a_{2x} - a_{1x}), (a_{2y} - a_{1y}), (a_{2z} - a_{1z})] \\ \boldsymbol{b} = \boldsymbol{b}_1 \otimes \boldsymbol{b}_2 = [(b_{1y}b_{2z} - b_{1z}b_{2y}), (b_{1z}b_{2x} - b_{1x}b_{2z}), (b_{1x}b_{2y} - b_{1y}b_{2x})] \end{cases} \tag{2.97}$$

2.24.6 The Cosine of the Angle between Two Lines

Let $\boldsymbol{\Lambda}_1,\boldsymbol{\Lambda}_2$ be two non-parallel lines. Where line $\boldsymbol{\Lambda}_1$ is defined by a position vector $\boldsymbol{a}_1 = [a_{1x}, a_{1y}, a_{1z}]$ and a direction unit vector $\boldsymbol{b}_1 = [b_{1x}, b_{1y}, b_{1z}]$ and line is defined by a position vector $\boldsymbol{a}_2 = [a_{2x}, a_{2y}, a_{2z}]$ and a direction unit vector $\boldsymbol{b}_2 = [b_{2x}, b_{2y}, b_{2z}]$, as shown in Fig. 2.30.

The cosine of the angle between the two lines is given by:

$$\cos\theta = \boldsymbol{b}_1 \odot \boldsymbol{b}_2 = b_{1x}b_{2x} + b_{1y}b_{2y} + b_{1z}b_{2z} \tag{2.98}$$

2.24.7 The Cosine of the Angle between Two Planes

Let $\boldsymbol{\Pi}_1, \boldsymbol{\Pi}_2$ be two non-parallel planes with unit normal vectors $\hat{\boldsymbol{n}}_1 = [\hat{n}_{1x}, \hat{n}_{1y}, \hat{n}_{1z}]$, $\hat{\boldsymbol{n}}_2 = [\hat{n}_{2x}, \hat{n}_{2y}, \hat{n}_{2z}]$, respectively, as seen in Fig. 2.31.

The angle between the planes is by definition the angle between their unit normals, thus:

$$\cos\theta = \hat{\boldsymbol{n}}_1 \odot \hat{\boldsymbol{n}}_2 = \hat{n}_{1x}\hat{n}_{2x} + \hat{n}_{1y}\hat{n}_{2y} + \hat{n}_{1z}\hat{n}_{2z} \tag{2.99}$$

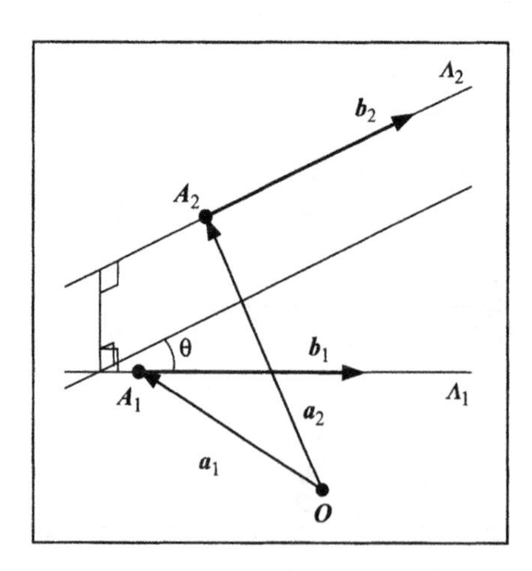

FIGURE 2.30. The cosine of the angle between two non-parallel lines.

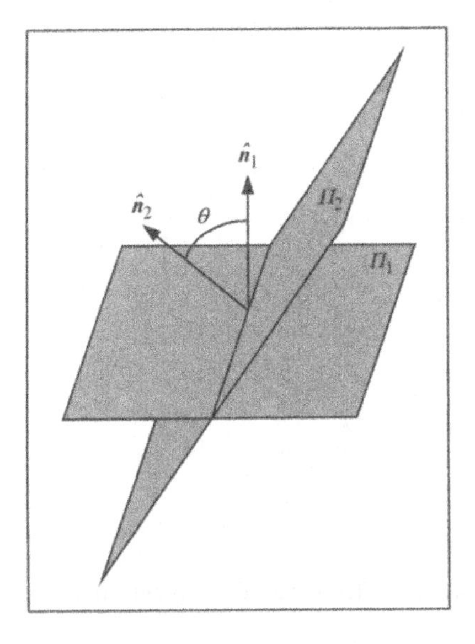

FIGURE 2.31. The cosine of the angle between two planes.

2.24.8 The Distance of a Point from a Plane

From Eq. (2.76) we know that the plane $\boldsymbol{\Pi}$ is defined by:

$$\boldsymbol{x} \odot \hat{\boldsymbol{n}} = d_o$$

where $\boldsymbol{x}$ is the position vector of the general point $\boldsymbol{X}$ on the plane, $\hat{\boldsymbol{n}}$ is the unit normal of the plane and d_o is the distance of the plane from the origin $\boldsymbol{O}$. A point $\boldsymbol{P}$ off the plane has position vector $\boldsymbol{p}$, as shown in Fig. 2.32.

The magnitude of the projection of $\boldsymbol{p}$ onto $\hat{n}$ is given by:

$$\boldsymbol{p} \odot \hat{\boldsymbol{n}} = m_p$$

The signed distance of point $\boldsymbol{P}$ from the plane is given by:

$$d_p = m_p - d_o = \boldsymbol{p} \odot \hat{\boldsymbol{n}} - \boldsymbol{x} \odot \hat{\boldsymbol{n}} = (\boldsymbol{p} - \boldsymbol{x}) \odot \hat{\boldsymbol{n}} \tag{2.100}$$

$$\therefore \; d_p = (p_x - x_x) \cdot \hat{n}_x + \left(p_y - x_y\right) \cdot \hat{n}_y + (p_z - x_z) \cdot \hat{n}_z \tag{2.101}$$

The observant reader would have noticed that the distance of point $\boldsymbol{P}$ from plane $\boldsymbol{\Pi}$ should be $d_p = d_o - m_p$ and not $d_p = m_p - d_o$. This change of sign was done deliberately to satisfy the following convention. Plane $\boldsymbol{\Pi}$ divides three-dimensional space into two *half-spaces*. A *positive half-space* that lies *in front of the plane*, i.e. in the direction that the plane normal points, and a *negative half-space* that lies *behind the plane*. By convention the origin lies in the negative half-space and any point lying on the same half-space as the origin is assumed to have a negative distance from the plane. For example, in CG a cube defined

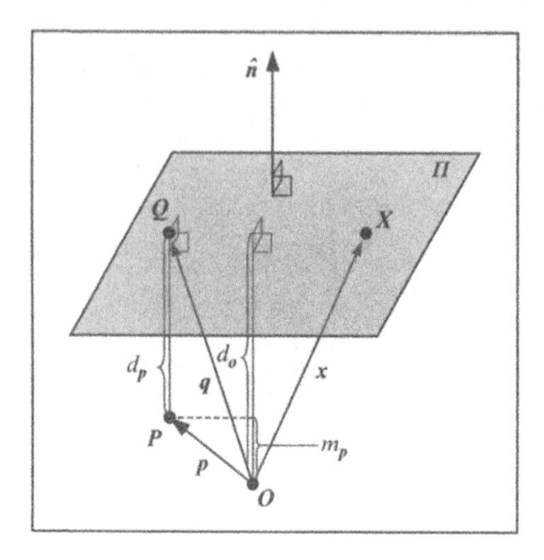

FIGURE 2.32. The distance of a point from a plane.

around the origin has the normals of its faces pointing outwards and the origin is behind the planes that form its faces. Thus:

If $d_p > 0$, then the point $\boldsymbol{P}$ lies on the opposite side of the plane to the origin (i.e. $\boldsymbol{P}$ is in front of the plane).

If $d_p < 0$, then the point $\boldsymbol{P}$ lies on the same side of the plane as the origin (i.e. $\boldsymbol{P}$ is behind of the plane).

If $d_p = 0$, then the point $\boldsymbol{P}$ lies on the plane (i.e. $\boldsymbol{P}$ is on the plane).

Let $\boldsymbol{Q}$ be the foot of the perpendicular from point $\boldsymbol{P}$ to plane $\boldsymbol{\Pi}$. Point $\boldsymbol{Q}$ lies on the plane and thus satisfies the equation of the plane, i.e.

$$\boldsymbol{q} \odot \hat{\boldsymbol{n}} = d_o$$

Point $\boldsymbol{Q}$ also lies on the perpendicular line whose equation is given by:

$$\boldsymbol{x}_l = \boldsymbol{p} + t \cdot \hat{\boldsymbol{n}}$$

Thus, point $\boldsymbol{Q}$ satisfies the perpendicular line equation:

$$\boldsymbol{q} = \boldsymbol{p} + t_q \cdot \hat{\boldsymbol{n}}$$

The parameter t_q represents the distance travelled along $\hat{\boldsymbol{n}}$, starting from $\boldsymbol{P}$ and ending at $\boldsymbol{Q}$, i.e.

$$t_q = d_o - m_p$$

Thus, the position vector of the foot $\boldsymbol{Q}$ of the perpendicular from point $\boldsymbol{P}$ to plane $\boldsymbol{\Pi}$ is given by:

$$\boldsymbol{q} = \boldsymbol{p} + (d_o - m_p) \cdot \hat{\boldsymbol{n}}$$

2.24.9 The Point of Intersection of a Line and a Plane

A line $\boldsymbol{\Lambda}$ is defined by a position vector $\boldsymbol{a} = [a_x, a_y, a_z]$ and a direction unit vector $\boldsymbol{b} = [b_x, b_y, b_z]$. The general point $\boldsymbol{x}$ on the line is given by its parametric equation:

$$\boldsymbol{x} = \boldsymbol{a} + t \cdot \boldsymbol{b}$$

A plane $\boldsymbol{\Pi}$ is defined by:

$$\hat{\boldsymbol{n}} \odot \boldsymbol{x} = d$$

where $\boldsymbol{x}$ is the position vector of the general point $\boldsymbol{X}$ on the plane, $\hat{\boldsymbol{n}} = [\hat{n}_x, \hat{n}_y, \hat{n}_z]$ is the unit normal of the plane and d is the distance of the plane from the origin $\boldsymbol{O}$.

If the line $\boldsymbol{\Lambda}$ intersects the plane $\boldsymbol{\Pi}$, then let $\boldsymbol{X}_0$ be their point of intersection, as shown in Fig. 2.33.

The point of intersection $\boldsymbol{X}_0$ has a position vector given by:

$$\boldsymbol{x}_0 = \boldsymbol{a} + t_0 \cdot \boldsymbol{b}$$

where t_0 is the value of the line parameter at point $\boldsymbol{X}_0$. To determine this value we proceed as follows. First, we determine the magnitude m_a of the projection of position vector $\boldsymbol{a}$ onto the plane normal:

$$\boldsymbol{m}_a = \hat{\boldsymbol{n}} \odot \boldsymbol{a}$$

Then, we find the length of the projection of the directed segment $\overrightarrow{\boldsymbol{AX}_0}$ onto the plane normal (i.e. the value of the line parameter at point $\boldsymbol{X}_0$) by subtracting $\boldsymbol{m}_a$ from the distance of the plane from the origin. See the right-hand diagram of the Fig. Thus:

$$t_0 = d - m_a = d - (\hat{\boldsymbol{n}} \odot \boldsymbol{a}) = d - (n_x a_x + n_y a_y + n_z a_z)$$

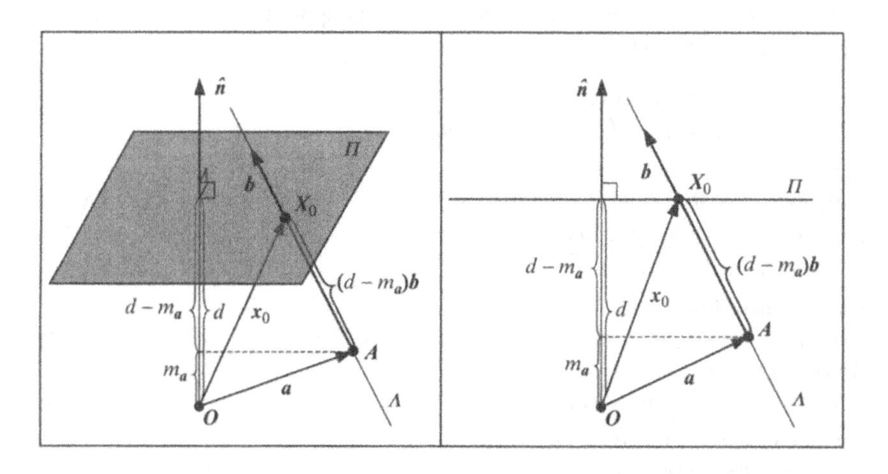

FIGURE 2.33. The distance of a point from a plane.

Now, the position vector of the point of intersection in component form is given by:

$$x_0 = \left[(a_x + t_0 b_x), \left(a_y + t_0 b_y\right), (a_z + t_0 b_z)\right] \quad \text{where}$$
$$t_0 = d - \left(\hat{n}_x a_x + \hat{n}_y a_y + \hat{n}_z a_z\right) \tag{2.102}$$

2.25 Summary of Vector Algebra Axioms and Rules

In this section we collect together, for ease of reference, all the axioms and vector algebra rules that apply to all the vector operations we have examined.

Multiplication of a Vector by a Scalar

The following axioms and rules of vector algebra apply to the vector by scalar product for all vectors $\boldsymbol{v}$, $\boldsymbol{v}_1$, $\boldsymbol{v}_2$ and all scalars α, β:

Existence of the vector by scalar product:

$$\alpha\boldsymbol{v} = \left[\alpha v_x, \alpha v_y, \alpha v_z\right] \tag{A2.1}$$

Existence of the zero element:

$$0\boldsymbol{v} = \boldsymbol{v}0 = \vec{\boldsymbol{0}} \tag{A2.2}$$

Existence of the neutral element:

$$1\boldsymbol{v} = \boldsymbol{v}1 = \boldsymbol{v} \tag{A2.3}$$

Associative law:

$$\alpha(\beta\boldsymbol{v}) = (\alpha\beta)\boldsymbol{v} \tag{R2.1}$$

Distributive laws:

$$(\alpha + \beta)\boldsymbol{v} = \alpha\boldsymbol{v} + \beta\boldsymbol{v} \tag{R2.2}$$
$$\alpha(\boldsymbol{v}_1 + \boldsymbol{v}_2) = \alpha\boldsymbol{v}_1 + \alpha\boldsymbol{v}_2 \tag{R2.3}$$

Vector Addition

The following axioms and rules of vector algebra apply to vector addition hold for all vectors $\boldsymbol{a}, \boldsymbol{b}, \boldsymbol{v}, \boldsymbol{v}_1, \boldsymbol{v}_2, \boldsymbol{v}_3$:

Existence of the vector sum:

$$\boldsymbol{a} + \boldsymbol{b} = \left[(a_x + b_x), \left(a_y + b_y\right), (a_z + b_z)\right] \tag{A2.4}$$

Existence of the neutral element:

$$\vec{\boldsymbol{0}} + \boldsymbol{v} = \boldsymbol{v} + \vec{\boldsymbol{0}} = \boldsymbol{v} \tag{A2.5}$$

Existence of the inverse element:

$$\boldsymbol{v} + (-\boldsymbol{v}) = (-\boldsymbol{v}) + \boldsymbol{v} = \vec{\boldsymbol{0}} \tag{A2.6}$$

Commutative law:

$$\boldsymbol{v}_1 + \boldsymbol{v}_2 = \boldsymbol{v}_2 + \boldsymbol{v}_1 \tag{R2.4}$$

Associative law:

$$(\boldsymbol{v}_1 + \boldsymbol{v}_2) + \boldsymbol{v}_3 = \boldsymbol{v}_1 + (\boldsymbol{v}_2 + \boldsymbol{v}_3) \tag{R2.5}$$

Change of detection rule:

$$-(\boldsymbol{v}_1 - \boldsymbol{v}_2) = (\boldsymbol{v}_2 - \boldsymbol{v}_1) \tag{R2.6}$$

The Scalar Product of Vectors

The following axioms and vector algebra rules apply to the scalar product of vectors for all vectors $\boldsymbol{a}, \boldsymbol{b}, \boldsymbol{v}_1, \boldsymbol{v}_2, \boldsymbol{v}_3, \boldsymbol{v}$ and all scalars α:

Existence of the scalar product:

$$\boldsymbol{a} \odot \boldsymbol{b} = \begin{cases} |\boldsymbol{a}| \cdot |\boldsymbol{b}| \cdot \cos\theta & \\ a_x b_x + a_y b_y + a_z b_z & \text{(Cartesian basis case only)} \end{cases} \quad \text{(A2.7)}$$

Powers of a vector:

$$\boldsymbol{v}^2 = \boldsymbol{v} \odot \boldsymbol{v} = |\boldsymbol{v}|^2 \quad \text{(A2.8)}$$

Commutative law:

$$\boldsymbol{v}_1 \odot \boldsymbol{v}_2 = \boldsymbol{v}_2 \odot \boldsymbol{v}_1 \quad \text{(R2.7)}$$

Distributive laws:

$$(\alpha \boldsymbol{v}_1) \odot \boldsymbol{v}_2 = \boldsymbol{v}_1 \odot (\alpha \boldsymbol{v}_2) = \alpha\, (\boldsymbol{v}_1 \odot \boldsymbol{v}_2) \quad \text{(R2.8)}$$

$$\boldsymbol{v}_1 \odot (\boldsymbol{v}_2 + \boldsymbol{v}_3) = \boldsymbol{v}_1 \odot \boldsymbol{v}_2 + \boldsymbol{v}_1 \odot \boldsymbol{v}_3 \quad \text{(R2.9)}$$

The Vector Product of Two Vectors

The following axioms and vector algebra rules apply to the vector product for all vectors $\boldsymbol{a}, \boldsymbol{b}, \boldsymbol{v}_1, \boldsymbol{v}_2, \boldsymbol{v}_3$ and all scalars α:

Existence of the vector product:

$$\boldsymbol{a} \otimes \boldsymbol{b} = \begin{cases} (|\boldsymbol{a}| \cdot |\boldsymbol{b}| \cdot \sin\theta)\, \hat{\boldsymbol{n}} & \\ \left[\left(a_y b_z - a_z b_y\right), \left(a_z b_x - a_x b_z\right), & \text{(Cartesian basis case only)} \\ \left(a_x b_y - a_y b_x\right)\right] & \end{cases} \quad \text{(A2.9)}$$

Vector product of collinear vectors:

$$\boldsymbol{v}_1 \otimes \boldsymbol{v}_2 = \vec{\boldsymbol{0}} \quad \Leftrightarrow \quad \boldsymbol{v}_1 || \boldsymbol{v}_2 \quad \text{(A2.10)}$$

Associative law:

$$\boldsymbol{v}_1 \otimes (\boldsymbol{v}_2 \otimes \boldsymbol{v}_3) \neq (\boldsymbol{v}_1 \otimes \boldsymbol{v}_2) \otimes \boldsymbol{v}_3 \quad \text{(R2.10)}$$

(does not apply)

Anti-commutative law:

$$\boldsymbol{v}_1 \otimes \boldsymbol{v}_2 = -\, (\boldsymbol{v}_2 \otimes \boldsymbol{v}_1) \quad \text{(R2.11)}$$

Distributive laws:

$$(\alpha \boldsymbol{v}_1) \otimes \boldsymbol{v}_2 = \boldsymbol{v}_1 \otimes (\alpha \boldsymbol{v}_2) = \alpha\, (\boldsymbol{v}_1 \otimes \boldsymbol{v}_2) \quad \text{(R2.12)}$$

$$\boldsymbol{v}_1 \otimes (\boldsymbol{v}_2 + \boldsymbol{v}_3) = \boldsymbol{v}_1 \otimes \boldsymbol{v}_2 + \boldsymbol{v}_1 \otimes \boldsymbol{v}_3 \quad \text{(R2.13)}$$

The Triple Scalar Product

The following axioms and vector algebra rules apply to the triple scalar product for all vectors $\boldsymbol{a}, \boldsymbol{b}, \boldsymbol{v}_1, \boldsymbol{v}_2, \boldsymbol{v}_3$:

Existence of the triple scalar product:

$$(a, b, c) = \begin{cases} (\boldsymbol{a} \otimes \boldsymbol{b}) \odot \boldsymbol{c} = \boldsymbol{a} \odot (\boldsymbol{b} \otimes \boldsymbol{c}) & \\ a_x \left(b_y c_z - b_z c_y\right) + a_y \left(b_z c_x - b_x c_z\right) & \text{(Cartesian basis case only)} \\ + a_z \left(b_x c_y - b_y c_x\right) & \end{cases} \quad \text{(A2.11)}$$

Triple scalar product of coplanar vectors:

$$\text{if all } \boldsymbol{v}_1, \boldsymbol{v}_2, \boldsymbol{v}_3 \text{ are coplanar} \Rightarrow (\boldsymbol{v}_1, \boldsymbol{v}_2, \boldsymbol{v}_3) = 0 \quad \text{(A2.12)}$$

Triple scalar product of zero vectors:

$$\boldsymbol{v}_1 = \vec{\mathbf{0}} \vee \boldsymbol{v}_2 = \vec{\mathbf{0}} \vee \boldsymbol{v}_3 = \vec{\mathbf{0}} \Rightarrow (\boldsymbol{v}_1, \boldsymbol{v}_2, \boldsymbol{v}_3) = 0 \quad \text{(A2.13)}$$

Triple scalar product of collinear vectors:

$$\boldsymbol{v}_1 || \boldsymbol{v}_2 \vee \boldsymbol{v}_2 || \boldsymbol{v}_3 \vee \boldsymbol{v}_3 || \boldsymbol{v}_1 \Rightarrow (\boldsymbol{v}_1, \boldsymbol{v}_2, \boldsymbol{v}_3) = 0 \quad \text{(A2.14)}$$

Cyclic permutation rule:

$$(\boldsymbol{v}_1, \boldsymbol{v}_2, \boldsymbol{v}_3) = (\boldsymbol{v}_2, \boldsymbol{v}_3, \boldsymbol{v}_1) = (\boldsymbol{v}_3, \boldsymbol{v}_1, \boldsymbol{v}_2) \quad \text{(R2.14)}$$

$$\text{i.e., } \boldsymbol{v}_1 \odot (\boldsymbol{v}_2 \otimes \boldsymbol{v}_3) = \boldsymbol{v}_2 \odot (\boldsymbol{v}_3 \otimes \boldsymbol{v}_1) = \boldsymbol{v}_3 \odot (\boldsymbol{v}_1 \otimes \boldsymbol{v}_2)$$

$$\text{and } (\boldsymbol{v}_1 \otimes \boldsymbol{v}_2) \odot \boldsymbol{v}_3 = (\boldsymbol{v}_2 \otimes \boldsymbol{v}_3) \odot \boldsymbol{v}_1 = (\boldsymbol{v}_3 \otimes \boldsymbol{v}_1) \odot \boldsymbol{v}_2$$

Non-cyclic permutation rule:

$$(\boldsymbol{v}_2, \boldsymbol{v}_1, \boldsymbol{v}_3) = (\boldsymbol{v}_3, \boldsymbol{v}_2, \boldsymbol{v}_1) = (\boldsymbol{v}_1, \boldsymbol{v}_3, \boldsymbol{v}_2) = -(\boldsymbol{v}_1, \boldsymbol{v}_2, \boldsymbol{v}_3) \quad \text{(R2.15)}$$

The Triple Vector Product

The following axioms and vector algebra rules apply to the triple vector product for all vectors $\boldsymbol{a}, \boldsymbol{b}, \boldsymbol{v}_1, \boldsymbol{v}_2, \boldsymbol{v}_3$:

Existence of the triple vector product:

$$(\boldsymbol{a} \otimes \boldsymbol{b}) \otimes \boldsymbol{c} = (\boldsymbol{a} \odot \boldsymbol{c})\, \boldsymbol{b} - (\boldsymbol{b} \odot \boldsymbol{c})\, \boldsymbol{a} \quad \text{(A2.15)}$$

Associative law (does not apply):

$$(\boldsymbol{v}_1 \otimes \boldsymbol{v}_2) \otimes \boldsymbol{v}_3 \neq \boldsymbol{v}_1 \otimes (\boldsymbol{v}_2 \otimes \boldsymbol{v}_3) \quad \text{(R2.16)}$$

Permutation rule:

$$(\boldsymbol{v}_1 \otimes \boldsymbol{v}_2) \otimes \boldsymbol{v}_3 = \boldsymbol{v}_3 \otimes (\boldsymbol{v}_2 \otimes \boldsymbol{v}_1) \quad \text{(R2.17)}$$

Expansion rules:

$$(\boldsymbol{v}_1 \otimes \boldsymbol{v}_2) \otimes \boldsymbol{v}_3 = (\boldsymbol{v}_1 \odot \boldsymbol{v}_3)\, \boldsymbol{v}_2 - (\boldsymbol{v}_2 \odot \boldsymbol{v}_3)\, \boldsymbol{v}_1 \quad \text{(R2.18)}$$

$$\boldsymbol{v}_1 \otimes (\boldsymbol{v}_2 \otimes \boldsymbol{v}_3) = (\boldsymbol{v}_1 \odot \boldsymbol{v}_3)\, \boldsymbol{v}_2 - (\boldsymbol{v}_1 \odot \boldsymbol{v}_2)\, \boldsymbol{v}_3 \quad \text{(R2.19)}$$

The Scalar Product of Two Vector Products

The following axiom applies to the scalar product of two vector products for all vectors $\boldsymbol{a}, \boldsymbol{b}, \boldsymbol{c}, \boldsymbol{d}$:

Existence of the scalar product of two vector products:

$$(\boldsymbol{a} \otimes \boldsymbol{b}) \odot (\boldsymbol{c} \otimes \boldsymbol{d}) = (\boldsymbol{c} \odot \boldsymbol{a})\, (\boldsymbol{b} \odot \boldsymbol{d}) - (\boldsymbol{c} \odot \boldsymbol{b})\, (\boldsymbol{d} \odot \boldsymbol{a}) \quad \text{(A2.16)}$$

The Vector Product of Two Vector Products

The following axiom applies to the vector product of two vector products for all vectors $\boldsymbol{a}, \boldsymbol{b}, \boldsymbol{c}, \boldsymbol{d}$:

Existence of the vector product of two vector products:

$$(\boldsymbol{a} \otimes \boldsymbol{b}) \otimes (\boldsymbol{c} \otimes \boldsymbol{d}) = \begin{cases} (\boldsymbol{a}, \boldsymbol{c}, \boldsymbol{d})\, \boldsymbol{b} - (\boldsymbol{b}, \boldsymbol{c}, \boldsymbol{d})\, \boldsymbol{a} \\ (\boldsymbol{a}, \boldsymbol{b}, \boldsymbol{d})\, \boldsymbol{c} - (\boldsymbol{a}, \boldsymbol{b}, \boldsymbol{c})\, \boldsymbol{d} \end{cases} \tag{A2.17}$$

2.26 A Simple Vector Algebra C Library

See *Appendix 1*.

3

Matrix Algebra Survival Kit

The theory of matrices and their determinants is closely related to the problem of solving a system of linear simultaneous equations.

The Babylonians appear to be among the first people to have investigated this problem. In a clay tablet, which has been dated to approximately 400–300 BC, the following problem was formulated.

Two fields have a total area of 1800 *square yards. One field produces grain at the rate of 2/3 of a bushel per square yard, while the other produces grain at the rate of 1/2 a bushel per square yard. If the total yield is* 1100 *bushels, what is the size of each field?*

In today's terminology, this problem can be represented by the following system of linear simultaneous equations:

$$1 \cdot x_1 + 1 \cdot x_2 = 1800$$
$$\tfrac{2}{3} \cdot x_1 + \tfrac{1}{2} \cdot x_2 = 1100$$

which, when solved, yields $x_1 = 1200\ \text{yd}^2$ and $x_2 = 600\ \text{yd}^2$ where x_1, x_2 represent the areas of the first and second fields, respectively.

The earliest example of the use of a determinant appears in the Chinese text entitled *Nine Chapters of the Mathematical Art*, which was written in 200 BC during the Hun Dynasty. In this text the following problem, which is similar to the Babylonian problem, was formulated.

There are three types of corn, of which three bundles of the first, two of the second, and one of the third make 39 measures. Two of the first, three of the second and one of the third make 34 measures. And one of the first, two of the second and three of the third make 26 measures. How many measures of corn are contained in one bundle of each type?

Today we would represent this problem by the following system of linear simultaneous equations:

$$\begin{aligned} 3 \cdot x_1 + 2 \cdot x_2 + 1 \cdot x_3 &= 39 \\ 2 \cdot x_1 + 3 \cdot x_2 + 1 \cdot x_3 &= 34 \\ 1 \cdot x_1 + 2 \cdot x_2 + 3 \cdot x_3 &= 26 \end{aligned}$$

or in matrix form:

$$\begin{bmatrix} 3 & 2 & 1 \\ 2 & 3 & 1 \\ 1 & 2 & 3 \end{bmatrix} \cdot \begin{bmatrix} x_1 \\ x_2 \\ x_3 \end{bmatrix} = \begin{bmatrix} 39 \\ 34 \\ 26 \end{bmatrix}$$

The Chinese mathematician expressed the problem in a tableau of the form:

	c_1	c_2	c_3
x_1	3	2	1
x_2	2	3	2
x_3	1	1	3
	39	34	26

and then proceeded to instruct the reader as to how to solve the problem using a sequence of *elementary column transformations*, the object of which was to introduce two zeros in rows 1 and 2 of column 3, two zeroes in rows 1 and 3 of column 2, and two zeroes in rows 2 and 3 of column 1. Then the value of each unknown could be found by dividing the element at the bottom of a column by the only non-zero element above it. The elementary column transformations allow one to interchange two columns, to multiply a column by a non-zero number and to add to a column or subtract from a column a multiple of another column. This technique was reinvented in Europe, using rows rather than columns, at the beginning of the nineteenth century and is known as the *Gaussian elimination* method.

Here, it is instructive to follow the procedure described in the Chinese text in some detail. In our explanation we will introduce the following shorthand notation. An elementary column transformation of the form

$$c_2 \leftarrow 3 \cdot c_2 - 2 \cdot c_1$$

represents a shorthand notation for the following operation. Multiply the elements of the second column of the tableau by 3 and subtract from them the corresponding elements of the first column multiplied by 2.

According to the Chinese text, the following eight steps must be followed to determine the values of the three unknowns.

Step 1: Construct the initial tableau.

	c_1	c_2	c_3
x_1	3	2	1
x_2	2	3	2
x_3	1	1	3
	39	34	26

Step 2: By setting $c_2 \leftarrow 3 \cdot c_2 - 2 \cdot c_1$ we obtain

	c_1	c_2	c_3
x_1	3	0	1
x_2	2	5	2
x_3	1	1	3
	39	24	26

Step 3: By setting $c_3 \leftarrow 3 \cdot c_3 - 1 \cdot c_1$ we obtain

	c_1	c_2	c_3
x_1	3	0	0
x_2	2	5	4
x_3	1	1	8
	39	24	39

Step 4: By setting $c_3 \leftarrow 5 \cdot c_3 - 4 \cdot c_2$ we obtain

	c_1	c_2	c_3
x_1	3	0	0
x_2	2	5	0
x_3	1	1	36
	39	24	99

Step 5: By setting $c_2 \leftarrow 36 \cdot c_2 - 1 \cdot c_3$ we obtain

	c_1	c_2	c_3
x_1	3	0	0
x_2	2	180	0
x_3	1	0	36
	39	765	99

Step 6: By setting $c_1 \leftarrow 36 \cdot c_1 - 1 \cdot c_3$ we obtain

	c_1	c_2	c_3
x_1	108	0	0
x_2	72	180	0
x_3	0	0	36
	1305	765	99

Step 7: By setting $c_1 \leftarrow 5 \cdot c_1 - 2 \cdot c_2$ we obtain

	c_1	c_2	c_3
x_1	540	0	0
x_2	0	180	0
x_3	0	0	36
	4995	765	99

Step 8: By dividing the bottom element of each column by the non-zero coefficient of the corresponding unknown we obtain the solution of the system of linear simultaneous equations:

$$x_1 = \frac{4995}{540} = 9.25$$

$$x_2 = \frac{765}{180} = 4.25$$

$$x_3 = \frac{99}{36} = 2.75$$

In 1683 Takakazu Seki (1642–1708), a Japanese mathematician, wrote a book entitled *Method of Solving the Dissimulated Problems* in which he introduced matrix and determinant methods, computed the determinants of 2×2, 3×3, 4×4, 5×5 matrices and used them in the solution of systems of linear simultaneous equations. By coincidence, the idea of determinants appeared in Europe in the same year in a letter that the German mathematician Gottfried Leibniz (1646–1716) wrote to the French mathematician Guillaume De l'Hôpital (1661–1704).

Since then, numerous mathematicians worked on the development of matrices and their determinants, until the theory reached its present form in a book entitled *An Introduction to Linear Algebra* published in 1955 by the Russian mathematician Leon Mirsky (1918–1983).

Matrix algebra is of great importance to most fields of study in modern science and it is central to the study of computer graphics. Thus it is important to gain a thorough understanding of matrix algebra.

3.1 The Definition of a Matrix

A matrix is a rectangular array of numbers. An $m \times n$ matrix has *m rows* and *n columns*. The general matrix can be written as

$$\boldsymbol{A}_{(m,n)} = \begin{bmatrix} a_{1,1} & a_{1,2} & \cdots & a_{1,n} \\ a_{2,1} & a_{2,2} & \cdots & a_{2,n} \\ \vdots & \vdots & \ddots & \vdots \\ a_{m,1} & a_{m,2} & \cdots & a_{m,n} \end{bmatrix}$$

All matrices that have the same number of rows and the same number of columns are said to be of the same *type* or *order*.

An individual element of the matrix is written as $a_{i,j}$, where i is the row number and j is the column number of the element. The subscript bracket (m, n) that follows the name of matrix $\boldsymbol{A}_{(m,n)}$ specifies the number of rows and columns of matrix $\boldsymbol{A}$.

Two degenerate cases exist:

1. When $m = 1$, then we have a matrix with a single row $\boldsymbol{B}_{(1,n)} = \begin{bmatrix} b_{1,1} & b_{1,2} & \cdots & b_{1,n} \end{bmatrix}$, which is called a *row matrix*.
2. When $n = 1$, then we have a matrix with a single column $\boldsymbol{C}_{(m,1)} = \begin{bmatrix} c_{1,1} \\ c_{2,1} \\ \vdots \\ c_{m,1} \end{bmatrix}$, which is called a *column matrix*.

Both row and column matrices are called *vectors* and in this case are denoted by a lower case bold letter. Often a simpler notation is adopted for vectors. An order *m row vector* is denoted by $\boldsymbol{v} = [\ v_1 \quad v_2 \quad \cdots \quad v_m\ ]$ and an order *m column vector* is denoted by

$$\boldsymbol{v} = \begin{bmatrix} v_1 \\ v_2 \\ \vdots \\ v_m \end{bmatrix}.$$

3.2 Square Matrices

If a matrix has the same number of rows as columns (i.e. $m = n$), then the matrix is called a *square matrix*. An $m \times m$ square matrix is said to have *order* m. The general form of an order m matrix is

$$\boldsymbol{A}_{(m,m)} = \begin{bmatrix} a_{1,1} & a_{1,2} & \cdots & a_{1,m} \\ a_{2,1} & a_{2,2} & \cdots & a_{2,m} \\ \vdots & \vdots & \ddots & \vdots \\ a_{m,1} & a_{m,2} & \cdots & a_{m,m} \end{bmatrix}$$

In a square matrix of order m, the diagonal from top-left to bottom-right is called the *leading diagonal* or the *main diagonal* or the *principal diagonal* of the matrix:

$$\left\{a_{i,i}\right\}_{i=1}^{m} = \{\ a_{1,1} \quad a_{2,2} \quad \cdots \quad a_{m,m}\ \} \tag{3.1}$$

The diagonal from top-right to bottom-left is called the *trailing diagonal* or the *secondary diagonal* of the matrix:

$$\left\{a_{i,m-(i-1)}\right\}_{i=1}^{m} = \{\ a_{1,m} \quad a_{2,m-1} \quad \cdots \quad a_{m,1}\ \} \tag{3.2}$$

The sum of the elements of the leading diagonal of a square matrix of order m is called the *trace* of the matrix:

$$\text{trace}\ (\boldsymbol{A}) = a_{1,1} + a_{2,2} + \cdots + a_{m,m} = \sum_{i=1}^{m} a_{i,i} \tag{3.3}$$

3.3 Diagonal Matrices

A square matrix that has non-zero elements only in its leading diagonal is called a *diagonal matrix*:

$$D = \begin{bmatrix} a_{1,1} & 0 & \cdots & 0 \\ 0 & a_{2,2} & \vdots & 0 \\ \vdots & \cdots & \ddots & \vdots \\ 0 & 0 & \cdots & a_{m,m} \end{bmatrix} \quad a_{i,j} = 0 \quad \text{for } i \neq j \text{ where , } i, j \in \{1, \ldots, m\} \tag{3.4}$$

3.4 The Identity Matrix

A diagonal matrix that has all its leading diagonal elements set to 1 (i.e. $\{a_{i,i} = 1\}_{i=1}^{m}$) is called the *identity matrix* or the *unit matrix*.

$$I = \begin{bmatrix} 1 & 0 & \cdots & 0 \\ 0 & 1 & \cdots & 0 \\ \vdots & \vdots & \ddots & \vdots \\ 0 & 0 & \cdots & 1 \end{bmatrix} \tag{3.5}$$

The identity matrix of order m is denoted by $\boldsymbol{I}_{(m)}$ or $\boldsymbol{I}$ or $\mathbf{1}$ and, as we shall see later, represents the neutral element in matrix multiplication.

The identity matrix can be written in a condensed form using Kronecker's delta (in a similar way that we have represented orthogonality in vectors in Section 2.12). Thus, the identity matrix has elements, which are defined by

$$\delta_{i,j} = \begin{cases} 1, & i = j \\ 0, & \text{otherwise} \end{cases} \quad \text{with} \quad i, j \in \{1, \ldots, m\} \tag{3.6}$$

3.5 The Zero or Null Matrix

A matrix with all its elements set to zero is called the *zero matrix* or *null matrix*.

$$\mathbf{0} = \begin{bmatrix} 0 & 0 & \cdots & 0 \\ 0 & 0 & \cdots & 0 \\ \vdots & \vdots & \ddots & \vdots \\ 0 & 0 & \cdots & 0 \end{bmatrix} \tag{3.7}$$

The zero matrix of order m is denoted by $\mathbf{0}_{(m)}$ or $\mathbf{0}$ and, as we shall see later, represents the neutral element in matrix addition.

3.6 The Transpose of a Matrix

The transpose of an $m \times n$ matrix $\boldsymbol{A}_{(m,n)}$ is an $n \times m$ matrix $A^T_{(n,m)}$, which is defined by interchanging the rows and the columns of matrix $\boldsymbol{A}$. The elements of the transpose matrix are given by

$$a^T_{j,i} = a_{i,j} \quad \text{where } i \in \{1, \ldots, m\} \wedge j \in \{1, \ldots, n\} \tag{3.8}$$

The transpose of a matrix $\boldsymbol{A}_{(m,n)} = \begin{bmatrix} a_{1,1} & \cdots & a_{1,n} \\ \vdots & \ddots & \vdots \\ a_{m,1} & \cdots & a_{m,n} \end{bmatrix}$ is given by

$$\boldsymbol{A}^T_{(n,m)} = \begin{bmatrix} a_{1,1} & \cdots & a_{m,1} \\ \vdots & \ddots & \vdots \\ a_{1,n} & \cdots & a_{m,n} \end{bmatrix} \tag{3.9}$$

The transpose of a row vector $\boldsymbol{v} = \begin{bmatrix} v_1 & \cdots & v_m \end{bmatrix}$ is given by the column vector

$$\boldsymbol{v}^T = \begin{bmatrix} v_1 \\ \vdots \\ v_m \end{bmatrix} \tag{3.10}$$

The transpose of a column vector $\boldsymbol{v} = \begin{bmatrix} v_1 \\ \vdots \\ v_m \end{bmatrix}$ is given by the row vector

$$\boldsymbol{v}^T = \begin{bmatrix} v_1 & \cdots & v_m \end{bmatrix} \tag{3.11}$$

If matrix $\boldsymbol{A}$ is square, then its transpose matrix is derived from the original by reflecting the matrix about its leading diagonal. Thus, the elements of the transpose are given by

$$a^{\mathrm{T}}_{j,i} = a_{i,j} \quad \text{where } i, j \in \{1, \ldots, m\} \tag{3.12}$$

For example $\begin{bmatrix} 1 & 4 & 7 \\ 2 & 5 & 8 \\ 3 & 6 & 9 \end{bmatrix}^{\mathrm{T}} = \begin{bmatrix} 1 & 2 & 3 \\ 4 & 5 & 6 \\ 7 & 8 & 9 \end{bmatrix}$.

The transpose of the transpose of a matrix is the matrix itself, i.e.

$$\left(A^{\mathrm{T}}\right)^{\mathrm{T}} = A \tag{3.13}$$

3.7 Symmetric and Antisymmetric Matrices

A square matrix that is equal to its transpose (i.e. $\boldsymbol{A}^{\mathrm{T}} = \boldsymbol{A}$) is called a *symmetric matrix*. Symmetric matrices are mirror-symmetric about their leading diagonal. The relationship between the elements of a symmetric matrix is given by

$$a_{i,j} = a_{j,i} \quad \text{where } i, j \in \{1, \ldots, m\} \tag{3.14}$$

For example the matrix $\begin{bmatrix} 1 & 2 & 3 \\ 2 & 4 & 5 \\ 3 & 5 & 6 \end{bmatrix}$ is a symmetric matrix.

A square matrix that is equal to the negative of its transpose (i.e. $\boldsymbol{A}^{\mathrm{T}} = -\boldsymbol{A}$) is called an *antisymmetric matrix* or a *skew-symmetric matrix*. The relationship between the elements of an antisymmetric matrix is given by

$$a_{i,j} = -a_{j,i} \quad \text{where } i, j \in \{1, \ldots, m\} \quad \text{thus, } \left\{a_{i,i} = 0\right\}_{i=1}^{m} \tag{3.15}$$

For example the matrix $\begin{bmatrix} 0 & -2 & -3 \\ 2 & 0 & -5 \\ 3 & 5 & 0 \end{bmatrix}$ is an antisymmetric matrix.

Every square matrix $\boldsymbol{A}$ can be written as the sum of a symmetric matrix $\boldsymbol{A}_S$ and an antisymmetric matrix $\boldsymbol{A}_{\mathrm{A}}$, thus

$$\boldsymbol{A} = \boldsymbol{A}_{\mathrm{S}} + \boldsymbol{A}_{\mathrm{A}} \quad \text{where } \boldsymbol{A}_{\mathrm{S}} = \tfrac{1}{2}\left(\boldsymbol{A} + \boldsymbol{A}^{T}\right) \text{ and } \boldsymbol{A}_{\mathrm{A}} = \tfrac{1}{2}\left(\boldsymbol{A} - \boldsymbol{A}^{\mathrm{T}}\right) \tag{3.16}$$

See Subsection 3.11.1 for a complete explanation.

Example:

Say $\boldsymbol{A} = \begin{bmatrix} 1 & 4 & 7 \\ 2 & 5 & 8 \\ 3 & 6 & 9 \end{bmatrix}$, then its transpose is $\boldsymbol{A}^{\mathrm{T}} = \begin{bmatrix} 1 & 2 & 3 \\ 4 & 5 & 6 \\ 7 & 8 & 9 \end{bmatrix}$.

Thus, $\boldsymbol{A}_{\mathrm{S}} = \begin{bmatrix} \dfrac{1+1}{2} & \dfrac{4+2}{2} & \dfrac{7+3}{2} \\ \dfrac{2+4}{2} & \dfrac{5+5}{2} & \dfrac{8+6}{2} \\ \dfrac{3+7}{2} & \dfrac{6+8}{2} & \dfrac{9+9}{2} \end{bmatrix} = \begin{bmatrix} 1 & 3 & 5 \\ 3 & 5 & 7 \\ 5 & 7 & 9 \end{bmatrix}$ and

$$A_A = \begin{bmatrix} \frac{1-1}{2} & \frac{4-2}{2} & \frac{7-3}{2} \\ \frac{2-4}{2} & \frac{5-5}{2} & \frac{8-6}{2} \\ \frac{3-7}{2} & \frac{6-8}{2} & \frac{9-9}{2} \end{bmatrix} = \begin{bmatrix} 0 & 1 & 2 \\ -1 & 0 & 1 \\ -2 & -1 & 0 \end{bmatrix}.$$

Therefore, $A = A_S + A_A = \begin{bmatrix} 1+0 & 3+1 & 5+2 \\ 3-1 & 5+0 & 7+1 \\ 5-2 & 7-1 & 9+0 \end{bmatrix} = \begin{bmatrix} 1 & 4 & 7 \\ 2 & 5 & 8 \\ 3 & 6 & 9 \end{bmatrix}$.

3.8 Triangular Matrices

An *upper triangular matrix* or *right triangular matrix* is a square matrix in which all the elements below or to the left of its leading diagonal are zero. The elements of an upper triangular matrix are given by

$$a_{i,j} = 0 \quad \text{for } i > j \text{ and where } i, j \in \{1, \ldots, m\} \tag{3.17}$$

Thus

$$U = \begin{bmatrix} a_{1,1} & a_{1,2} & \cdots & a_{1,m} \\ 0 & a_{2,2} & \vdots & a_{2,m} \\ \vdots & \cdots & \ddots & \vdots \\ 0 & 0 & \cdots & a_{m,m} \end{bmatrix} \tag{3.18}$$

For example the matrix $\begin{bmatrix} 1 & 2 & 3 \\ 0 & 4 & 5 \\ 0 & 0 & 6 \end{bmatrix}$ is an upper triangular matrix.

A *lower triangular matrix* or *left triangular matrix* is a square matrix in which all the elements above or to the right of its leading diagonal are zero. The elements of a lower triangular matrix are given by

$$a_{i,j} = 0 \quad \textit{for } i < j \textit{ and where } i, j \in \{1, \ldots, m\} \tag{3.19}$$

$$L = \begin{bmatrix} a_{1,1} & 0 & \cdots & 0 \\ a_{2,1} & a_{2,2} & \vdots & 0 \\ \vdots & \cdots & \ddots & \vdots \\ a_{m,1} & a_{m,2} & \cdots & a_{m,m} \end{bmatrix} \tag{3.20}$$

For example the matrix $\begin{bmatrix} 1 & 0 & 0 \\ 2 & 4 & 0 \\ 3 & 5 & 6 \end{bmatrix}$ is a lower triangular matrix.

The transpose of an upper triangular matrix is a lower triangular matrix and vice versa, thus

$$\boldsymbol{U}^{\mathrm{T}} = \boldsymbol{L} \quad \text{and} \quad \boldsymbol{L}^{\mathrm{T}} = \boldsymbol{U} \tag{3.21}$$

For example, if $\boldsymbol{L} = \begin{bmatrix} 1 & 0 & 0 \\ 2 & 4 & 0 \\ 3 & 5 & 6 \end{bmatrix}$, then its transpose is $\boldsymbol{L}^{\mathrm{T}} = \begin{bmatrix} 1 & 2 & 3 \\ 0 & 4 & 5 \\ 0 & 0 & 6 \end{bmatrix} = \boldsymbol{U}$.

Every diagonal matrix is at the same time both left and right (or upper and lower) triangular. Every diagonal matrix is also a symmetric matrix.

The transpose of a diagonal matrix is equal to the matrix itself, i.e.

$$\boldsymbol{D}^{\mathrm{T}} = \boldsymbol{D} \tag{3.22}$$

For example, if $\boldsymbol{D} = \begin{bmatrix} 1 & 0 & 0 \\ 0 & 2 & 0 \\ 0 & 0 & 3 \end{bmatrix}$, then its transpose is $\boldsymbol{D}^{\mathrm{T}} = \begin{bmatrix} 1 & 0 & 0 \\ 0 & 2 & 0 \\ 0 & 0 & 3 \end{bmatrix} = \boldsymbol{D}$.

3.9 Scalar Matrices

A diagonal matrix that has its leading diagonal elements set to some scalar value s is called a *scalar matrix*, i.e.

$$\boldsymbol{S} = \begin{bmatrix} s & 0 & \cdots & 0 \\ 0 & s & \vdots & 0 \\ \vdots & \cdots & \ddots & \vdots \\ 0 & 0 & \cdots & s \end{bmatrix} = s \cdot \begin{bmatrix} 1 & 0 & \cdots & 0 \\ 0 & 1 & \vdots & 0 \\ \vdots & \cdots & \ddots & \vdots \\ 0 & 0 & \cdots & 1 \end{bmatrix} = s \cdot \boldsymbol{I} \tag{3.23}$$

Note that matrix by scalar multiplication and matrix by matrix multiplication are discussed later.

For example, $\boldsymbol{S} = \begin{bmatrix} 5 & 0 & 0 \\ 0 & 5 & 0 \\ 0 & 0 & 5 \end{bmatrix} = 5 \cdot \begin{bmatrix} 1 & 0 & 0 \\ 0 & 1 & 0 \\ 0 & 0 & 1 \end{bmatrix}$.

3.10 Equality of Matrices

Two matrices $\boldsymbol{A}$ and $\boldsymbol{B}$ are said to be equal, i.e. $\boldsymbol{A} = \boldsymbol{B}$, *if and only if* they are of the same order and their corresponding elements are all equal. Thus

$$\boldsymbol{A}_{(m,n)} = \boldsymbol{B}_{(m,n)} \Leftrightarrow a_{i,j} = b_{i,j} \quad \text{for } i \in \{1, \ldots, m\} \wedge j \in \{1, \cdots, n\} \tag{3.24}$$

3.11 Matrix Operations

3.11.1 Addition and Subtraction of Matrices

Given two matrices $\boldsymbol{A}$ and $\boldsymbol{B}$ of the same order, we may add or subtract these matrices by adding or subtracting their corresponding elements. The result of such an operation is a matrix of the same order. Thus

$$\boldsymbol{C}_{(m,n)} = \boldsymbol{A}_{(m,n)} \pm \boldsymbol{B}_{(m,n)} \Rightarrow c_{i,j} = a_{i,j} \pm b_{i,j} \quad \text{for} \quad i \in \{1, \ldots, m\} \wedge j \in \{1, \cdots, n\} \tag{3.25}$$

Example:

$$\begin{bmatrix} 1 & 3 \\ 2 & 4 \end{bmatrix} + \begin{bmatrix} 5 & 7 \\ 6 & 8 \end{bmatrix} = \begin{bmatrix} 1+5 & 3+7 \\ 2+6 & 4+8 \end{bmatrix} = \begin{bmatrix} 6 & 10 \\ 8 & 12 \end{bmatrix}$$

The following axioms and matrix algebra rules apply to the addition and subtraction of matrices for all matrices $\boldsymbol{A}, \boldsymbol{B}, \boldsymbol{C}$ (which are assumed to be of the same order):

Existence of the matrix sum:

$$\boldsymbol{C}_{(m,n)} = \boldsymbol{A}_{(m,n)} \pm \boldsymbol{B}_{(m,n)} \Rightarrow \left\{ c_{i,j} = a_{i,j} \pm b_{i,j} \right\}_{i,j=1}^{m,n} \tag{A3.1}$$

Existence of the null element:

$$\boldsymbol{A} \pm \boldsymbol{0} = \boldsymbol{A} \tag{A3.2}$$

Existence of the additive inverse:

$$\boldsymbol{A} + (-\boldsymbol{A}) = \boldsymbol{0} \tag{A3.3}$$

Commutative law:

$$\boldsymbol{A} + \boldsymbol{B} = \boldsymbol{B} + \boldsymbol{A} \tag{R3.1}$$

Associative law:

$$(\boldsymbol{A} + \boldsymbol{B}) + \boldsymbol{C} = \boldsymbol{A} + (\boldsymbol{B} + \boldsymbol{C}) \tag{R3.2}$$

Distributive law:

$$s \cdot (\boldsymbol{A} \pm \boldsymbol{B}) = s \cdot \boldsymbol{A} \pm s \cdot \boldsymbol{B} \tag{R3.3}$$

Transposition rule:

$$(\boldsymbol{A} \pm \boldsymbol{B})^T = \boldsymbol{A}^T \pm \boldsymbol{B}^T \tag{R3.4}$$

If $\boldsymbol{A}$ is a square matrix, then

$$\left(\boldsymbol{A}^{\mathrm{T}} + \boldsymbol{A}\right)^{\mathrm{T}} = \left(\boldsymbol{A}^{\mathrm{T}}\right)^{\mathrm{T}} + \boldsymbol{A}^{\mathrm{T}} = \boldsymbol{A} + \boldsymbol{A}^{\mathrm{T}}$$

$$\therefore \quad \left(\boldsymbol{A}^{\mathrm{T}} + \boldsymbol{A}\right)^{\mathrm{T}} = \boldsymbol{A}^{\mathrm{T}} + \boldsymbol{A}$$

Thus $\left(\boldsymbol{A}^{\mathrm{T}} + \boldsymbol{A}\right)^{\mathrm{T}}$ must be a symmetric matrix.

Similarly

$$\left(\boldsymbol{A} - \boldsymbol{A}^{\mathrm{T}}\right)^{\mathrm{T}} = \boldsymbol{A}^{\mathrm{T}} - \left(\boldsymbol{A}^{\mathrm{T}}\right)^{\mathrm{T}} = \boldsymbol{A}^{\mathrm{T}} - \boldsymbol{A}$$

$$\therefore \quad \left(\boldsymbol{A} - \boldsymbol{A}^{\mathrm{T}}\right)^{\mathrm{T}} = -\left(\boldsymbol{A} - \boldsymbol{A}^{\mathrm{T}}\right)$$

Thus $\left(\boldsymbol{A} - \boldsymbol{A}^{\mathrm{T}}\right)^{\mathrm{T}}$ must be an antisymmetric matrix.

As we have seen in Section 3.7, any square matrix can be expressed as the sum of a symmetric and an antisymmetric matrix, thus

$$A = \tfrac{1}{2}\left(A^{\mathrm{T}} + A\right) + \tfrac{1}{2}\left(A - A^{\mathrm{T}}\right) \tag{3.26}$$

3.11.2 *Multiplication of a Matrix by a Scalar*

To multiply a matrix $\boldsymbol{A}$ by a scalar s we multiply every element of the matrix by the scalar. The result is a matrix of the same order as $\boldsymbol{A}$. Thus

$$\boldsymbol{B}_{(m,n)} = \boldsymbol{A}_{(m,n)} \cdot s \quad \Rightarrow \quad b_{i,j} = a_{i,j} \cdot s \quad \text{for } i \in \{1, \ldots, m\} \wedge j \in \{1, \cdots, n\} \tag{3.27}$$

$$\text{or} \quad \boldsymbol{B} = \boldsymbol{A} \cdot s = \begin{bmatrix} a_{1,1} & \cdots & a_{1,n} \\ \vdots & \ddots & \vdots \\ a_{m,1} & \cdots & a_{m,n} \end{bmatrix} \cdot s = \begin{bmatrix} a_{1,1} \cdot s & \cdots & a_{1,n} \cdot s \\ \vdots & \ddots & \vdots \\ a_{m,1} \cdot s & \cdots & a_{m,n} \cdot s \end{bmatrix} \tag{3.28}$$

Matrix by scalar multiplication can also be regarded as matrix by matrix multiplication, where the second matrix is a scalar matrix, i.e.

$$s \cdot \boldsymbol{A} = (s \cdot \boldsymbol{I}) \cdot \boldsymbol{A} = \boldsymbol{S} \cdot \boldsymbol{A} \tag{3.29}$$

Given any non-zero scalar s, matrix division by this scalar is achieved through matrix multiplication by the inverse of the scalar, i.e.

$$\frac{\boldsymbol{A}}{s} = \left(\frac{1}{s}\right) \cdot \boldsymbol{A} \tag{3.30}$$

The following axioms and matrix algebra rules apply to the multiplication of a matrix by a scalar for all matrices $\boldsymbol{A}$, $\boldsymbol{B}$ (which are assumed to be of the same order) and all scalars s, s_1, s_2.

Existence of the matrix by scalar product:

$$\boldsymbol{B}_{(m,n)} = \boldsymbol{A}_{(m,n)} \cdot s \Rightarrow \left\{ b_{i,j} = a_{i,j} \cdot s \right\}_{i,j=1}^{m,n} \tag{A3.4}$$

Existence of the identity map:

$$\boldsymbol{1} \cdot \boldsymbol{A} = \boldsymbol{I} \cdot \boldsymbol{A} = \boldsymbol{A} \tag{A3.5}$$

Commutative law:

$$s \cdot \boldsymbol{A} = \boldsymbol{A} \cdot s \tag{R3.5}$$

Associative law:

$$s_1 \cdot (s_2 \cdot \boldsymbol{A}) = (s_1 \cdot s_2) \cdot \boldsymbol{A} \tag{R3.6}$$

Distributive law:

$$(s_1 \pm s_2) \cdot \boldsymbol{A} = s_1 \cdot \boldsymbol{A} \pm s_2 \cdot \boldsymbol{A} \tag{R3.7}$$

3.11.3 Multiplication of a Vector by a Vector

In vector algebra, we defined two types of vector by vector multiplication, namely the *dot product* $\boldsymbol{a} \odot \boldsymbol{b}$ and the *cross product* $\boldsymbol{a} \otimes \boldsymbol{b}$. In matrix algebra, the definition of the cross product $\boldsymbol{a} \otimes \boldsymbol{b}$ retains the same as in vector algebra and results in a vector. The definition of the dot product $\boldsymbol{a} \odot \boldsymbol{b}$, however, is refined in matrix algebra. By disallowing the commutative property of the dot product we can now define two distinct products, namely the dot product and the *tensor product* or *dyadic product*.

The dot product of an mth order row vector $\boldsymbol{a}$ by an mth order column vector $\boldsymbol{b}$ results in a scalar.

$$\begin{aligned}\boldsymbol{a} \odot \boldsymbol{b} = \boldsymbol{a}^{\mathrm{T}} \cdot \boldsymbol{b} &= \begin{bmatrix} a_1 & a_2 & \cdots & a_m \end{bmatrix} \cdot \begin{bmatrix} b_1 \\ b_2 \\ \vdots \\ b_m \end{bmatrix} \\ &= a_1 b_1 + a_2 b_2 + \cdots + a_m b_m = \sum_{i=1}^{m} a_i \cdot b_i \end{aligned} \tag{3.31}$$

The tensor product of an mth order column vector $\boldsymbol{a}$ and an nth order row vector $\boldsymbol{b}$ results in an $m \times n$ matrix.

$$\boldsymbol{a} \cdot \boldsymbol{b}^{T} = \begin{bmatrix} a_1 \\ \vdots \\ a_m \end{bmatrix} \cdot \begin{bmatrix} b_1 & \cdots & b_n \end{bmatrix} = \begin{bmatrix} a_1 b_1 & \cdots & a_1 b_n \\ \vdots & \ddots & \vdots \\ a_m b_1 & \cdots & a_m b_n \end{bmatrix} \tag{3.32}$$

$$\text{or} \quad \left\{ c_{i,j} = a_i b_j \right\}_{i=1,j=1}^{m,n} \tag{3.33}$$

Example:

$$\begin{bmatrix} 1 \\ 2 \\ 3 \end{bmatrix} \cdot \begin{bmatrix} 4 & 5 & 6 \end{bmatrix} = \begin{bmatrix} 1 \cdot 4 & 1 \cdot 5 & 1 \cdot 6 \\ 2 \cdot 4 & 2 \cdot 5 & 2 \cdot 6 \\ 3 \cdot 4 & 3 \cdot 5 & 3 \cdot 6 \end{bmatrix} = \begin{bmatrix} 4 & 5 & 6 \\ 8 & 10 & 12 \\ 12 & 15 & 18 \end{bmatrix}$$

The following axioms apply to the vector by vector multiplication for all row and column matrices $\boldsymbol{a}, \boldsymbol{b}$ (which are assumed to be conformant).

Existence of the dot product:

$$\boldsymbol{a}^{\mathrm{T}} \cdot \boldsymbol{b} = \sum_{i=1}^{m} a_i \cdot b_i \tag{A3.6}$$

Existence of the tensor product:

$$\boldsymbol{a} \cdot \boldsymbol{b}^{\mathrm{T}} = \left\{ c_{i,j} = a_i b_j \right\}_{i,j=1}^{m,n} \tag{A3.7}$$

3.11.4 Multiplication of a Matrix by a Vector

The post-multiplication of a matrix $\boldsymbol{A}$ of order $m \times n$ by a column vector $\boldsymbol{b}$ of order n results in a column vector $\boldsymbol{c}$ of order m. Thus the product $\boldsymbol{A} \cdot \boldsymbol{b} = \boldsymbol{c}$ is defined as

$$\begin{aligned} \boldsymbol{A} \cdot \boldsymbol{b} &= \begin{bmatrix} a_{1,1} & \cdots & a_{1,n} \\ \vdots & \ddots & \vdots \\ a_{m,1} & \cdots & a_{m,n} \end{bmatrix} \cdot \begin{bmatrix} b_1 \\ \vdots \\ b_n \end{bmatrix} \\ &= \begin{bmatrix} a_{1,1}b_1 + a_{1,2}b_2 + \cdots + a_{1,n}b_n \\ \vdots \\ a_{m,1}b_1 + a_{m,2}b_2 + \cdots + a_{m,n}b_n \end{bmatrix} = \begin{bmatrix} c_1 \\ \vdots \\ c_m \end{bmatrix} = \boldsymbol{c} \end{aligned} \tag{3.34}$$

or

$$\left\{ c_i = \boldsymbol{a}_i \odot \boldsymbol{b} = \sum_{j=1}^{n} a_{i,j} b_j \right\}_{i=1}^{m} \tag{3.35}$$

where $\boldsymbol{a}_i$ is the ith row of matrix $\boldsymbol{A}$. This multiplication is only defined if the matrix $\boldsymbol{A}$ and the column vector $\boldsymbol{b}$ are *conformant*, *conformal* or *conformable*. In this case, *conformity* indicates that the matrix has as many columns as vector has rows.

If matrix $\boldsymbol{A}$ is square, then this operation can be seen as an abbreviated notation for a system of m linear simultaneous equations in m unknowns, i.e.

$$\boldsymbol{A} \cdot \boldsymbol{x} = \boldsymbol{c}$$

where $\boldsymbol{A}$ is the coefficients matrix of the simultaneous equations, $\boldsymbol{x}$ is the vector of unknowns and $\boldsymbol{c}$ is the vector of constants.

For example, the system of three linear simultaneous equations

$$\begin{aligned} a_{1,1}x_1 + a_{1,2}x_2 + a_{1,3}x_3 &= c_1 \\ a_{2,1}x_1 + a_{2,2}x_2 + a_{2,3}x_3 &= c_2 \\ a_{3,1}x_1 + a_{3,2}x_2 + a_{3,3}x_3 &= c_3 \end{aligned}$$

can be represented as

$$\begin{bmatrix} a_{1,1} & a_{1,2} & a_{1,3} \\ a_{2,1} & a_{2,2} & a_{2,3} \\ a_{3,1} & a_{3,2} & a_{3,3} \end{bmatrix} \cdot \begin{bmatrix} x_1 \\ x_2 \\ x_3 \end{bmatrix} = \begin{bmatrix} c_1 \\ c_2 \\ c_3 \end{bmatrix}$$

The multiplication of a matrix by a vector is a special case of the multiplication of two matrices. As we shall see in the next section, the following transposition rule applies to matrix multiplication.

$$(\boldsymbol{A} \cdot \boldsymbol{B})^{\mathrm{T}} = \boldsymbol{B}^{\mathrm{T}} \cdot \boldsymbol{A}^{\mathrm{T}}$$

By applying this transposition rule to Eq. (3.34) the product $\boldsymbol{A} \cdot \boldsymbol{b} = \boldsymbol{c}$ can be rewritten as $\boldsymbol{b}^{\mathrm{T}} \cdot \boldsymbol{A}^{\mathrm{T}} = \boldsymbol{c}^{\mathrm{T}}$. But $\boldsymbol{c}$ and $\boldsymbol{c}^{\mathrm{T}}$ are two different ways of denoting the same vector.

Thus, the pre-multiplication of a matrix $\boldsymbol{A}$ of order $n \times m$ by a row vector $\boldsymbol{b}$ of order n results in a row vector $\boldsymbol{c}$ of order m can be rewritten as

$$\begin{aligned}\boldsymbol{b}^{\mathrm{T}} \cdot \boldsymbol{A}^{\mathrm{T}} &= \begin{bmatrix} b_1 & \cdots & b_n \end{bmatrix} \cdot \begin{bmatrix} a_{1,1} & \cdots & a_{1,m} \\ \vdots & \ddots & \vdots \\ a_{n,1} & \cdots & a_{n,m} \end{bmatrix} \\ &= \left[\left(b_1 a_{1,1} + b_2 a_{1,2} + \cdots + b_n a_{n,1} \right) \cdots \right. \\ &\quad \left. \left(b_1 a_{1,m} + b_2 a_{2,m} + \cdots + b_n a_{n,m} \right) \right] \\ &= \begin{bmatrix} c_1 & \cdots & c_m \end{bmatrix} = \boldsymbol{c}\end{aligned} \tag{3.36}$$

$$\text{or} \quad \left\{ c_j = \sum_{i=1}^{n} b_i a_{i,j} \right\}_{i=1}^{m} \tag{3.37}$$

Here again the row vector and the matrix must be conformant. In this case, conformity indicates that the vector has as many columns as the matrix has rows.

It is easy to verify that Eqs. (3.34)–(3.37) are equivalent and produce identical results.

3.11.5 Multiplication of Two Matrices

Two matrices may be multiplied together if and only if the number of columns of the first matrix is equal to the number rows of the second matrix. Two matrices that meet this constraint are said to be conformant. Multiplying a matrix $\boldsymbol{A}_{(m,l)}$ by a matrix $\boldsymbol{B}_{(l,n)}$ will result in a matrix $\boldsymbol{C}_{(m,n)}$, whose ijth element $\boldsymbol{c}_{i,j}$ is the dot product of the ith row of $\boldsymbol{A}$ and the jth column of $\boldsymbol{B}$ (i.e. the summation of the products of corresponding elements of the ith row of $\boldsymbol{A}$ and the jth column of $\boldsymbol{B}$). Such a product is called a *matrix product* or *scalar product of two matrices*. Thus

$$\begin{aligned}\boldsymbol{A} \cdot \boldsymbol{B} &= \begin{bmatrix} a_{1,1} & \cdots & a_{1,l} \\ \vdots & & \vdots \\ \cdots & \cdots & \cdots \\ a_{i,1} & \cdots & a_{i,l} \\ \cdots & \cdots & \cdots \\ \vdots & & \vdots \\ a_{m,1} & \cdots & a_{m,l} \end{bmatrix} \cdot \begin{bmatrix} b_{1,1} & \cdots & \vdots & b_{1,j} & \vdots & \cdots & b_{1,n} \\ \vdots & & \vdots & \vdots & \vdots & & \vdots \\ b_{l,1} & \cdots & \vdots & b_{l,j} & \vdots & \cdots & b_{l,n} \end{bmatrix} \\ &= \begin{bmatrix} c_{1,1} & \cdots & & \cdots & c_{1,n} \\ \vdots & & & & \vdots \\ & & c_{i,j} & & \\ \vdots & & & & \vdots \\ c_{m,1} & \cdots & & \cdots & c_{m,n} \end{bmatrix} = \boldsymbol{C}\end{aligned} \tag{3.38}$$

If $\boldsymbol{a}_i$ denotes the ith row of $\boldsymbol{A}$ and $\boldsymbol{b}_j$ denotes the jth column of $\boldsymbol{B}$, then element $c_{i,j}$ of the product is defined as

$$\left\{ c_{i,j} = \boldsymbol{a}_i \odot \boldsymbol{b}_j = a_{i,1}b_{1,j} + a_{i,2}b_{2,j} + \cdots + a_{i,l}b_{l,j} = \sum_{k=1}^{l} a_{i,k}b_{k,j} \right\}_{i=1,j=1}^{m,n} \tag{3.39}$$

Matrix by matrix multiplication can be implemented by the following pseudo-code:

for $i = 1..m$ ***do***
for $j = 1..n$ ***do***
begin
$c_{i,j} \leftarrow 0$
for $k = 1..l$ ***do***
$c_{i,j} \leftarrow c_{i,j} + a_{i,k} \cdot b_{k,j}$
end

Division of a matrix $\boldsymbol{A}$ by a matrix $\boldsymbol{B}$ is carried out through the multiplication of $\boldsymbol{A}$ by the *reciprocal* or *inverse* matrix $\boldsymbol{B}^{-1}$. Inversion of matrices will be examined in Section 3.17.

Example:

$$\boldsymbol{C}_{(3,3)} = \boldsymbol{A}_{(3,2)} \cdot \boldsymbol{B}_{(2,3)}$$

$$= \begin{bmatrix} a_{1,1} & a_{1,2} \\ a_{2,1} & a_{2,2} \\ a_{3,1} & a_{3,2} \end{bmatrix} \cdot \begin{bmatrix} b_{1,1} & b_{1,2} & b_{1,3} \\ b_{2,1} & b_{2,2} & b_{2,3} \end{bmatrix}$$

$$= \begin{bmatrix} a_{1,1}b_{1,1} + a_{1,2}b_{2,1} & a_{1,1}b_{1,2} + a_{1,2}b_{2,2} & a_{1,1}b_{1,3} + a_{1,2}b_{2,3} \\ a_{2,1}b_{1,1} + a_{2,2}b_{2,1} & a_{2,1}b_{1,2} + a_{2,2}b_{2,2} & a_{2,1}b_{1,3} + a_{2,2}b_{2,3} \\ a_{3,1}b_{1,1} + a_{3,2}b_{2,1} & a_{3,1}b_{1,2} + a_{3,2}b_{2,2} & a_{3,1}b_{1,3} + a_{3,2}b_{2,3} \end{bmatrix}$$

3.11.6 Powers of Matrices

Given a square matrix $\boldsymbol{A}$, we can define the powers of this matrix through repeated matrix multiplication. The zero power of the matrix is defined to be the identity matrix. And the first power of a matrix is the matrix itself. Thus

$$\boldsymbol{A}^0 = \boldsymbol{I}, \quad \boldsymbol{A}^1 = \boldsymbol{A}, \quad \boldsymbol{A}^2 = \boldsymbol{A} \cdot \boldsymbol{A}, \quad \boldsymbol{A}^3 = \boldsymbol{A}^2 \cdot \boldsymbol{A}, \quad \cdots, \quad \boldsymbol{A}^m = \boldsymbol{A}^{(m-1)} \cdot \boldsymbol{A} \tag{3.40}$$

3.11.7 Axioms and Rules of Matrix Multiplication

The following axioms and matrix algebra rules apply to the matrix product for all matrices $\boldsymbol{A}, \boldsymbol{B}, \boldsymbol{C}$ (which are assumed to be conformant) and all scalars s.

Existence of the matrix product:

$$\boldsymbol{A}_{(m,l)} \cdot \boldsymbol{B}_{(l,n)} = \boldsymbol{C}_{(m,n)} \Rightarrow \left\{ c_{i,j} = \sum_{k=1}^{l} a_{i,k} b_{k,j} \right\}_{i,j=1}^{m,n} \tag{A3.8}$$

Existence of the unit matrix:

$$\boldsymbol{I}_{(m)} \cdot \boldsymbol{A}_{(m,n)} = \boldsymbol{A}_{(m,n)} \cdot \boldsymbol{I}_{(n)} = \boldsymbol{A}_{(m,n)} \tag{A3.9}$$

Existence of the zero matrix

$$\boldsymbol{0}_{(m)} \cdot \boldsymbol{A}_{(m,n)} = \boldsymbol{A}_{(m,n)} \cdot \boldsymbol{0}_{(n)} = \boldsymbol{0}_{(m,n)} \tag{A3.10}$$

Powers of a matrix:

$$\boldsymbol{A}^0 = \boldsymbol{I} \ \wedge \ \boldsymbol{A}^m = \boldsymbol{A}^{(m-1)} \cdot \boldsymbol{A} \tag{A3.11}$$

Commutative law (does not hold):

$$\boldsymbol{A} \cdot \boldsymbol{B} \neq \boldsymbol{B} \cdot \boldsymbol{A} \tag{R3.8}$$

Associative law:

$$\boldsymbol{A} \cdot (\boldsymbol{B} \cdot \boldsymbol{C}) = (\boldsymbol{A} \cdot \boldsymbol{B}) \cdot \boldsymbol{C} \tag{R3.9}$$

Distributive laws:

$$(s \cdot \boldsymbol{A}) \cdot \boldsymbol{B} = \boldsymbol{A} \cdot (s \cdot \boldsymbol{B}) = s \cdot (\boldsymbol{A} \cdot \boldsymbol{B}) \tag{R3.10}$$

$$\boldsymbol{A} \cdot (\boldsymbol{B} \pm \boldsymbol{C}) = \boldsymbol{A} \cdot \boldsymbol{B} \pm \boldsymbol{A} \cdot \boldsymbol{C} \tag{R3.11}$$

$$(\boldsymbol{A} \pm \boldsymbol{B}) \cdot \boldsymbol{C} = \boldsymbol{A} \cdot \boldsymbol{C} \pm \boldsymbol{B} \cdot \boldsymbol{C} \tag{R3.12}$$

Transpose rules:

$$(\boldsymbol{A} \cdot \boldsymbol{B})^{\mathrm{T}} = \boldsymbol{B}^{\mathrm{T}} \cdot \boldsymbol{A}^{\mathrm{T}} \tag{R3.13}$$

$$(\boldsymbol{A} \cdot \boldsymbol{B} \cdot \boldsymbol{C})^{\mathrm{T}} = \boldsymbol{C}^{\mathrm{T}} \cdot \boldsymbol{B}^{\mathrm{T}} \cdot \boldsymbol{A}^{\mathrm{T}} \tag{R3.14}$$

Zero divisor rule:

$$\boldsymbol{A} \cdot \boldsymbol{B} = \boldsymbol{0} \quad \not\Rightarrow /\boldsymbol{A} = \boldsymbol{0} \vee \boldsymbol{B} = \boldsymbol{0} \tag{R3.15}$$

Example:
This example demonstrates the zero divisor rule. Here neither matrix $\boldsymbol{A}$ nor $\boldsymbol{B}$ is a zero matrix but their product is a zero matrix.

$$\boldsymbol{A} \cdot \boldsymbol{B} = \begin{bmatrix} 2 & 0 & 4 \\ 0 & 0 & 0 \\ 1 & 0 & 2 \end{bmatrix} \cdot \begin{bmatrix} 2 & 0 & 4 \\ 0 & 1 & 0 \\ -1 & 0 & -2 \end{bmatrix} = \begin{bmatrix} 0 & 0 & 0 \\ 0 & 0 & 0 \\ 0 & 0 & 0 \end{bmatrix} = \boldsymbol{0}$$

3.12 The Minor of a Matrix

The minor, $\boldsymbol{A}_{i,j}$, of an $m \times n$ matrix $\boldsymbol{A}$ is defined to be an $(m-1) \times (n-1)$ matrix derived by deleting the ith row and the jth column of matrix $\boldsymbol{A}$.

The minor matrix $\boldsymbol{B} = \boldsymbol{A}_{i,j}$ of a matrix $\boldsymbol{A}$ can be computed using the following pseudo-code.

/∗ Copy the Upper Left corner of the Minor∗ /
for $r = 1..(i-1)$ ***do***
for $c = 1..(j-1)$ ***do*** $b_{r,c} \leftarrow a_{r,c}$

/* Copy the Lower Left corner of the Minor* /
for $r = (i+1)..m$ ***do***
for $c = 1..(j-1)$ ***do*** $b_{r-1,c} \leftarrow a_{r,c}$

/* Copy the Upper Right corner of the Minor* /
for $r = 1..(i-1)$ ***do***
for $c = (j+1)..n$ ***do*** $b_{r,c-1} \leftarrow a_{r,c}$

/* Copy the Lower Right corner of the Minor* /
for $r = (i+1)..m$ ***do***
for $c = (j+1)..n$ ***do*** $b_{r-1,c-1} \leftarrow a_{r,c}$

Example:
Given a 4×4 matrix:

$$\boldsymbol{A} = \left[\begin{array}{c:c:cc} a_{1,1} & a_{1,2} & a_{1,3} & a_{1,4} \\ \hdashline a_{2,1} & \boxed{a_{2,2}} & a_{2,3} & a_{2,4} \\ \hdashline a_{3,1} & a_{3,2} & a_{3,3} & a_{3,4} \\ a_{4,1} & a_{4,2} & a_{4,3} & a_{4,4} \end{array}\right]$$

The minor matrix of $\boldsymbol{A}$, with respect to element $a_{2,2}$, is the 3×3 matrix given by

$$\boldsymbol{A}_{2,2} = \left[\begin{array}{c:cc} a_{1,1} & a_{1,3} & a_{1,4} \\ \hdashline a_{3,1} & a_{3,3} & a_{3,4} \\ a_{4,1} & a_{4,3} & a_{4,4} \end{array}\right]$$

Here the indices of the elements of the minor matrix $\boldsymbol{A}_{2,2}$ refer to the elements of the original matrix $\boldsymbol{A}$.

3.13 The Determinant of a Matrix

The *determinant* is a scalar that is calculated from the elements of a square matrix $\boldsymbol{A}_{(m,m)}$ Thus only square matrices have a determinant. The determinant of a matrix $\boldsymbol{A}$ is denoted by $\det(\boldsymbol{A})$ or $|\boldsymbol{A}|$.

In matrix algebra it is frequently important to be able to compute the *reciprocal* or the *inverse* $\boldsymbol{A}^{-1}$ of a square matrix $\boldsymbol{A}$. The inverse of a matrix is required in matrix division, in the solution of systems of linear simultaneous equations and in the reversal of the effect of a transformation in computer graphics. We examine how the inverse of a matrix is computed in a later section.

The determinant of a matrix helps us establish whether the matrix is *invertible* in the first place. If the determinant of a matrix $\boldsymbol{A}$ is non-zero, we say that the matrix is *regular* and thus *invertible*. Alternatively, if the determinant of $\boldsymbol{A}$ is zero, we say that the matrix is *singular* and thus *non-invertible*, i.e.

$$\left.\begin{array}{lcl} \det(\boldsymbol{A}) \neq 0 & \Leftrightarrow & \boldsymbol{A} \text{ is regular} \\ \det(\boldsymbol{A}) = 0 & \Leftrightarrow & \boldsymbol{A} \text{ is singular} \end{array}\right\} \tag{3.41}$$

To avoid confusion between a matrix $\boldsymbol{A}$ and its determinant we denote determinants by enclosing the element of the matrix between vertical parallel bars. Thus the determinant of an $m \times m$ matrix $\boldsymbol{A}$ is given by

$$\det(\boldsymbol{A}) = \begin{vmatrix} a_{1,1} & a_{1,2} & \cdots & a_{1,m} \\ a_{2,1} & a_{2,2} & \cdots & a_{2,m} \\ \vdots & \vdots & \ddots & \vdots \\ a_{m,1} & a_{m,2} & \cdots & a_{m,m} \end{vmatrix} \tag{3.42}$$

The determinant of a 1×1 matrix $\boldsymbol{A}_{(1,1)} = \left[a_{1,1}\right]$ is defined as:

$$\det(\boldsymbol{A}) = \left|a_{1,1}\right| = a_{1,1} \tag{3.43}$$

The determinant of an $m \times m$ matrix $\boldsymbol{A}$ is defined using Laplace's *recursive expansion rule* on a given row or a given column of the matrix.

$$\det(\boldsymbol{A}) = \sum_{i=1}^{m} a_{i,j} \cdot (-1)^{(i+j)} \cdot \det\left(\boldsymbol{A}_{i,j}\right) \text{ for } a \text{ given column } i \in \{1, \ldots, m\} \tag{3.44}$$

$$\text{or} \quad \det(\boldsymbol{A}) = \sum_{i=1}^{m} a_{i,j} \cdot (-1)^{(i+j)} . \det\left(\boldsymbol{A}_{i,j}\right) \text{ for } a \text{ given column } j \in \{1, \ldots, m\} \tag{3.45}$$

Example 1:

$$\begin{aligned} \det(\boldsymbol{A}) &= \det\left(\begin{bmatrix} a_{1,1} & a_{1,2} \\ a_{2,1} & a_{2,2} \end{bmatrix}\right) \\ &= \begin{vmatrix} a_{1,1} & a_{1,2} \\ a_{2,1} & a_{2,2} \end{vmatrix} \\ &= a_{1,1} \cdot (-1)^{(1+1)} \cdot \left|a_{2,2}\right| + a_{2,1} \cdot (-1)^{(2+1)} \cdot \left|a_{1,2}\right| \\ &= a_{1,1}a_{2,2} - a_{2,1}a_{1,2} \end{aligned}$$

Example 2:

$$\begin{aligned} \det(\boldsymbol{A}) &= \det\left(\begin{bmatrix} a_{1,1} & a_{1,2} & a_{1,3} \\ a_{2,1} & a_{2,2} & a_{2,3} \\ a_{3,1} & a_{3,2} & a_{3,3} \end{bmatrix}\right) \\ &= \begin{vmatrix} a_{1,1} & a_{1,2} & a_{1,3} \\ a_{2,1} & a_{2,2} & a_{2,3} \\ a_{3,1} & a_{3,2} & a_{3,3} \end{vmatrix} \end{aligned}$$

$$
\begin{aligned}
&= a_{1,1} \cdot (-1)^{(1+1)} \cdot \begin{vmatrix} a_{2,2} & a_{2,3} \\ a_{3,2} & a_{3,3} \end{vmatrix} \\
&\quad + a_{2,1} \cdot (-1)^{(2+1)} \cdot \begin{vmatrix} a_{1,2} & a_{1,3} \\ a_{3,2} & a_{3,3} \end{vmatrix} \\
&\quad + a_{3,1} \cdot (-1)^{(3+1)} \cdot \begin{vmatrix} a_{1,2} & a_{1,3} \\ a_{2,2} & a_{2,3} \end{vmatrix} \\
&= +a_{1,1} \left(a_{2,2}a_{3,3} - a_{3,2}a_{2,3}\right) \\
&\quad -a_{2,1} \left(a_{1,2}a_{3,3} - a_{3,2}a_{1,3}\right) \\
&\quad +a_{3,1} \left(a_{1,2}a_{2,3} - a_{2,2}a_{1,3}\right)
\end{aligned}
$$

The general method for calculating the determinant of a square matrix can be expressed as follows. We start by selecting a row/column and then we progressively accumulate the product of successive elements (of the selected row/column) by the determinant of the minor of these elements, using the appropriate sign. The sign associated with the determinant of the minor of a particular element is given by

$$
\begin{vmatrix} + & - & + & - & \cdots \\ - & + & - & + & \cdots \\ + & - & + & - & \cdots \\ - & + & - & + & \cdots \\ \vdots & \vdots & \vdots & \vdots & \ddots \end{vmatrix} \tag{3.46}
$$

3.14 The Computational Rules of Determinants

The computational rules for determinants of the mth order are not always obvious and are frequently hard to prove in their full generality. Thus, we will present these rules without proof.

3.14.1 The Transposition Rule

The determinant of a matrix $\boldsymbol{A}$ is equal to the determinant of its transpose $\boldsymbol{A}^{\mathrm{T}}$, i.e.

$$
\det(\boldsymbol{A}) = \det\left(\boldsymbol{A}^{\mathrm{T}}\right) \tag{3.47}
$$

For example:

$$
\det(\boldsymbol{A}) = \begin{vmatrix} a_{1,1} & a_{1,2} \\ a_{2,1} & a_{2,2} \end{vmatrix} = a_{1,1}a_{2,2} - a_{1,2}a_{2,1} \text{ by expanding the first row}
$$

$$
\det\left(\boldsymbol{A}^{\mathrm{T}}\right) = \begin{vmatrix} a_{1,1} & a_{2,1} \\ a_{1,2} & a_{2,2} \end{vmatrix} = a_{1,1}a_{2,2} - a_{1,2}a_{2,1} \text{ by expanding the first column}
$$

$$
\det(\boldsymbol{A}) = \det\left(\boldsymbol{A}^{\mathrm{T}}\right)
$$

3.14.2 The Interchange Rule

Interchanging two rows or two columns of a determinant reverses its sign, i.e.

$$\begin{vmatrix} a_{1,1} & a_{1,2} \\ a_{2,1} & a_{2,2} \end{vmatrix} = -\begin{vmatrix} a_{1,2} & a_{1,1} \\ a_{2,2} & a_{2,1} \end{vmatrix} \wedge \begin{vmatrix} a_{1,1} & a_{1,2} \\ a_{2,1} & a_{2,2} \end{vmatrix} = -\begin{vmatrix} a_{2,1} & a_{2,2} \\ a_{1,1} & a_{1,2} \end{vmatrix} \tag{3.48}$$

For example

$$\det(\boldsymbol{A}) = \begin{vmatrix} a_{1,1} & a_{1,2} \\ a_{2,1} & a_{2,2} \end{vmatrix} = a_{1,1}a_{2,2} - a_{1,2}a_{2,1} \text{ and}$$

$$\det(\boldsymbol{B}) = \begin{vmatrix} a_{1,2} & a_{1,1} \\ a_{2,2} & a_{2,1} \end{vmatrix} = a_{1,2}a_{2,1} - a_{1,1}a_{2,2} = -\left(a_{1,1}a_{2,2} - a_{1,2}a_{2,1}\right)$$

$$\therefore \det(\boldsymbol{A}) = -\det(\boldsymbol{B}) \quad \text{or} \quad \begin{vmatrix} a_{1,1} & a_{1,2} \\ a_{2,1} & a_{2,2} \end{vmatrix} = -\begin{vmatrix} a_{1,2} & a_{1,1} \\ a_{2,2} & a_{2,1} \end{vmatrix}$$

3.14.3 The Factor Rule

Factors common to rows or columns may be removed from a determinant and be multiplied by the resulting determinant, i.e.

$$\begin{vmatrix} \beta a_{1,1} & a_{1,2} \\ \beta a_{2,1} & a_{2,2} \end{vmatrix} = \beta \begin{vmatrix} a_{1,1} & a_{1,2} \\ a_{2,1} & a_{2,2} \end{vmatrix} \wedge \begin{vmatrix} \beta a_{1,1} & \beta a_{1,2} \\ a_{2,1} & a_{2,2} \end{vmatrix} = \beta \begin{vmatrix} a_{1,1} & a_{1,2} \\ a_{2,1} & a_{2,2} \end{vmatrix} \tag{3.49}$$

For example:

$$\begin{aligned} \begin{vmatrix} \beta a_{1,1} & a_{1,2} \\ \beta a_{2,1} & a_{2,2} \end{vmatrix} &= \beta a_{1,1}a_{2,2} - \beta a_{2,1}a_{1,2} \\ &= \beta\left(a_{1,1}a_{2,2} - a_{1,2}a_{2,1}\right) \\ &= \beta \begin{vmatrix} a_{1,1} & a_{1,2} \\ a_{2,1} & a_{2,2} \end{vmatrix} \end{aligned}$$

Conversely, if the determinant is multiplied by a scalar β, then one and only one of its rows or columns is multiplied by this scalar, i.e.

$$\begin{aligned} \beta \begin{vmatrix} a_{1,1} & a_{1,2} \\ a_{2,1} & a_{2,2} \end{vmatrix} &= \begin{vmatrix} \beta a_{1,1} & a_{1,2} \\ \beta a_{2,1} & a_{2,2} \end{vmatrix} \\ &= \begin{vmatrix} a_{1,1} & \beta a_{1,2} \\ a_{2,1} & \beta a_{2,2} \end{vmatrix} = \begin{vmatrix} \beta a_{1,1} & \beta a_{1,2} \\ a_{2,1} & a_{2,2} \end{vmatrix} \\ &= \begin{vmatrix} a_{1,1} & a_{1,2} \\ \beta a_{2,1} & \beta a_{2,2} \end{vmatrix} \end{aligned} \tag{3.50}$$

3.14.4 The Linear Combinations Rule

The value of a determinant is not altered when one of its rows is increased/decreased by equal multiples of another row or one of its columns is increased/decreased by equal multiples of another column, i.e.

$$\begin{vmatrix} a_{1,1} & a_{1,2} \\ a_{2,1} & a_{2,2} \end{vmatrix} = \begin{vmatrix} a_{1,1} \pm \beta a_{2,1} & a_{1,2} \pm \beta a_{2,2} \\ a_{2,1} & a_{2,2} \end{vmatrix} = \begin{vmatrix} a_{1,1} \pm \beta a_{1,2} & a_{1,2} \\ a_{2,1} \pm \beta a_{2,2} & a_{2,2} \end{vmatrix} \tag{3.51}$$

For example:

$$\begin{aligned} \begin{vmatrix} a_{1,1} + \beta a_{1,2} & a_{1,2} \\ a_{2,1} + \beta a_{2,2} & a_{2,2} \end{vmatrix} &= a_{1,1}a_{2,2} + \beta a_{1,2}a_{2,2} - a_{2,1}a_{1,2} - \beta a_{2,2}a_{1,2} \\ &= a_{1,1}a_{2,2} - a_{2,1}a_{1,2} \\ &= \begin{vmatrix} a_{1,1} & a_{1,2} \\ a_{2,1} & a_{2,2} \end{vmatrix} \end{aligned}$$

3.14.5 The Decomposition Rule

The determinant whose row/column consists of the sum or difference of two or more terms can be expanded to the sum or difference of two or more determinants, i.e.

$$\begin{vmatrix} a_{1,1} \pm b_1 & a_{1,2} \\ a_{2,1} \pm b_2 & a_{2,2} \end{vmatrix} = \begin{vmatrix} a_{1,1} & a_{1,2} \\ a_{2,1} & a_{2,2} \end{vmatrix} \pm \begin{vmatrix} b_1 & a_{1,2} \\ b_2 & a_{2,2} \end{vmatrix} \tag{3.52}$$

For example:

$$\begin{aligned} \begin{vmatrix} a_{1,1} + b_1 & a_{1,2} \\ a_{2,1} + b_2 & a_{2,2} \end{vmatrix} &= a_{1,1}a_{2,2} + b_1 a_{2,2} - a_{2,1}a_{1,2} - b_2 a_{1,2} \\ &= \left(a_{1,1}a_{2,2} - a_{2,1}a_{1,2}\right) + \left(b_1 a_{2,2} - b_2 a_{1,2}\right) \\ &= \begin{vmatrix} a_{1,1} & a_{1,2} \\ a_{2,1} & a_{2,2} \end{vmatrix} + \begin{vmatrix} b_1 & a_{1,2} \\ b_2 & a_{2,2} \end{vmatrix} \end{aligned}$$

3.14.6 The Product Rule

The determinant of the product of two matrices is equal to the product of their determinants, i.e.

$$\det(\boldsymbol{A} \cdot \boldsymbol{B}) = \det(\boldsymbol{A}) \cdot \det(\boldsymbol{B}) \tag{3.53}$$

For example:

$$\begin{aligned} \det(\boldsymbol{A} \cdot \boldsymbol{B}) &= \det\left(\begin{bmatrix} a_{1,1} & a_{1,2} \\ a_{2,1} & a_{2,2} \end{bmatrix} \cdot \begin{bmatrix} b_{1,1} & b_{1,2} \\ b_{2,1} & b_{2,2} \end{bmatrix}\right) \\ &= \det\left(\begin{bmatrix} (a_{1,1}b_{1,1} + a_{1,2}b_{2,1}) & (a_{1,1}b_{1,2} + a_{1,2}b_{2,2}) \\ (a_{2,1}b_{1,1} + a_{2,2}b_{2,1}) & (a_{2,1}b_{1,2} + a_{2,2}b_{2,2}) \end{bmatrix}\right) \end{aligned}$$

$$\therefore \quad \det(\boldsymbol{A}\cdot\boldsymbol{B}) = \left(a_{1,1}b_{1,1}+a_{1,2}b_{2,1}\right)\cdot\left(a_{2,1}b_{1,2}+a_{2,2}b_{2,2}\right) \\ -\left(a_{1,1}b_{1,2}+a_{1,2}b_{2,2}\right)\cdot\left(a_{2,1}b_{1,1}+a_{2,2}b_{2,1}\right)$$

$$\begin{aligned}\therefore \quad \det(\boldsymbol{A}\cdot\boldsymbol{B}) = {} & \underline{a_{1,1}a_{2,1}b_{1,1}b_{1,2}} + a_{1,1}a_{2,2}b_{1,1}b_{2,2} + a_{1,2}a_{2,1}b_{2,1}b_{1,2} \\ & + \underline{\underline{a_{1,2}a_{2,2}b_{2,1}b_{2,2}}} \\ & - \underline{a_{1,1}a_{2,1}b_{1,2}b_{1,1}} - a_{1,1}a_{2,2}b_{1,2}b_{2,1} - a_{1,2}a_{2,1}b_{2,2}b_{1,1} \\ & - \underline{\underline{a_{1,2}a_{2,2}b_{2,2}b_{2,1}}}\end{aligned}$$

$$\begin{aligned}\therefore \quad \det(\boldsymbol{A}\cdot\boldsymbol{B}) = {} & a_{1,1}a_{2,2}b_{1,1}b_{2,2} - a_{1,1}a_{2,2}b_{1,2}b_{2,1} \\ & -a_{1,2}a_{2,1}b_{1,1}b_{2,2} + a_{1,2}a_{2,1}b_{1,2}b_{2,1}\end{aligned}$$

$$\begin{aligned}\therefore \quad \det(\boldsymbol{A}\cdot\boldsymbol{B}) &= a_{1,1}a_{2,2}\left(b_{1,1}b_{2,2}-b_{1,2}b_{2,1}\right) - a_{1,2}a_{2,1}\left(b_{1,1}b_{2,2}-b_{1,2}b_{2,1}\right) \\ &= \left(a_{1,1}a_{2,2}-a_{1,2}a_{2,1}\right)\cdot\left(b_{1,1}b_{2,2}-b_{1,2}b_{2,1}\right)\end{aligned}$$

$$\det(\boldsymbol{A}\cdot\boldsymbol{B}) = \det\left(\begin{bmatrix} a_{1,1} & a_{1,2} \\ a_{2,1} & a_{2,2} \end{bmatrix}\right)\det\left(\begin{bmatrix} b_{1,1} & b_{1,2} \\ b_{2,1} & b_{2,2} \end{bmatrix}\right) = \det(\boldsymbol{A})\cdot\det(\boldsymbol{B})$$

3.14.7 The Equality Rule

If the determinants of two matrices $\boldsymbol{A}$ and $\boldsymbol{B}$ are equal this does not guarantee that the matrices are equal, i.e.

$$\det(\boldsymbol{A}) = \det(\boldsymbol{B}) \quad \not\Rightarrow / \boldsymbol{A} = \boldsymbol{B} \tag{3.54}$$

For example, $\det(\boldsymbol{A}) = \begin{vmatrix} 1 & 0 \\ 0 & 1 \end{vmatrix} = 1$ and $\det(\boldsymbol{B}) = \begin{vmatrix} 2 & 0 \\ 0 & 0.5 \end{vmatrix} = 1$, but $\boldsymbol{A} \neq \boldsymbol{B}$.

3.14.8 The Conditions for a Zero Determinant

An mth order determinant is zero if and only if at least one of the following conditions is satisfied.

- An entire row/column of the determinant is zero, i.e.

$$\begin{vmatrix} a_{1,1} & 0 \\ a_{2,1} & 0 \end{vmatrix} = \begin{vmatrix} 0 & 0 \\ a_{2,1} & a_{2,2} \end{vmatrix} = 0 \tag{3.55}$$

- Two rows/columns of the determinant are identical, i.e.

$$\begin{vmatrix} a_{1,1} & a_{1,2} \\ a_{1,1} & a_{1,2} \end{vmatrix} = \begin{vmatrix} a_{1,1} & a_{1,1} \\ a_{2,1} & a_{2,1} \end{vmatrix} = 0 \tag{3.56}$$

- Two rows/columns of the determinant are proportional, i.e.

$$\begin{vmatrix} a_{1,1} & a_{1,2} \\ \beta a_{1,1} & \beta a_{1,2} \end{vmatrix} = \begin{vmatrix} a_{1,1} & \beta a_{1,1} \\ a_{2,1} & \beta a_{2,1} \end{vmatrix} = 0 \tag{3.57}$$

Given the determinant with a second column proportional to its first column $\begin{vmatrix} a_{1,1} & \beta a_{1,1} \\ a_{2,1} & \beta a_{2,1} \end{vmatrix}$, by the linear combination rule we can subtract β times the first column from the second column, which results in a zero second column and a zero determinant:

$$\begin{vmatrix} a_{1,1} & \beta a_{1,1} \\ a_{2,1} & \beta a_{2,1} \end{vmatrix} = \begin{vmatrix} a_{1,1} & \beta a_{1,1} - \beta a_{1,1} \\ a_{2,1} & \beta a_{2,1} - \beta a_{2,1} \end{vmatrix} = \begin{vmatrix} a_{1,1} & 0 \\ a_{2,1} & 0 \end{vmatrix} = 0$$

- One row/column of the determinant is a linear combination of other rows/columns of the determinant, i.e.

$$\begin{vmatrix} a_{1,1} & a_{1,2} & \alpha a_{1,1} + \beta a_{1,2} \\ a_{2,1} & a_{2,2} & \alpha a_{2,1} + \beta a_{2,2} \\ a_{3,1} & a_{3,2} & \alpha a_{3,1} + \beta a_{3,2} \end{vmatrix} = 0 \tag{3.58}$$

By the linear combination rule we may subtract α times the first column plus β times the second column from the third column, thus producing a zero column that gives us a zero determinant.

3.15 The Cofactor of an Element of a Matrix and the Cofactor Matrix

Cofactors are associated with the determinant of a matrix. The *cofactor* of an element of a matrix is the determinant of the minor of that element prefixed by the appropriate sign, as shown in Eq. (3.46). As we have seen in Section 3.13, this sign is given by $(-1)^{i+j}$ where i, j are the row and column numbers of the element, respectively. Thus, the *cofactor*, $c_{i,j}$ of element $a_{i,j}$ of a matrix $\boldsymbol{A}$ is defined as

$$c_{i,j} = (-1)^{(i+j)} \cdot \det\left(\boldsymbol{A}_{i,j}\right) \tag{3.59}$$

The matrix that is formed by the cofactors of all the elements of the matrix $\boldsymbol{A}$ is called the *cofactor matrix*, $\boldsymbol{C}$, of matrix $\boldsymbol{A}$.

Now, the determinant of an $m \times m$ matrix $\boldsymbol{A}$ can be rewritten as

$$\det(\boldsymbol{A}) = \sum_{j=1}^{m} a_{i,j} c_{i,j} \text{ for } a \text{ given row } i, \text{ where } i \in \{1, \ldots, m\} \tag{3.60}$$

or

$$\det(\boldsymbol{A}) = \sum_{i=1}^{m} a_{i,j} c_{i,j} \text{ for } a \text{ given column } j, \text{ where } j \in \{1, \ldots, m\} \tag{3.61}$$

Equations (3.60) and (3.61) are equivalent to Eqs. (3.44) and (3.45).

Example:
Given a 3×3 matrix $\boldsymbol{A}$:

$$\boldsymbol{A} = \begin{bmatrix} a_{1,1} & a_{1,2} & a_{1,3} \\ a_{2,1} & a_{2,2} & a_{2,3} \\ a_{3,1} & a_{3,2} & a_{3,3} \end{bmatrix}$$

The cofactor matrix of $\boldsymbol{A}$ is

$$\boldsymbol{C} = \begin{bmatrix} +\left|\boldsymbol{A}_{1,1}\right| & -\left|\boldsymbol{A}_{1,2}\right| & +\left|\boldsymbol{A}_{1,3}\right| \\ -\left|\boldsymbol{A}_{2,1}\right| & +\left|\boldsymbol{A}_{2,2}\right| & -\left|\boldsymbol{A}_{2,3}\right| \\ +\left|\boldsymbol{A}_{3,1}\right| & -\left|\boldsymbol{A}_{3,2}\right| & +\left|\boldsymbol{A}_{3,3}\right| \end{bmatrix}$$

$$= \begin{bmatrix} +\begin{vmatrix} a_{2,2} & a_{2,3} \\ a_{3,2} & a_{3,3} \end{vmatrix} & -\begin{vmatrix} a_{2,1} & a_{2,3} \\ a_{3,1} & a_{3,3} \end{vmatrix} & +\begin{vmatrix} a_{2,1} & a_{2,2} \\ a_{3,1} & a_{3,2} \end{vmatrix} \\ -\begin{vmatrix} a_{1,2} & a_{1,3} \\ a_{3,2} & a_{3,3} \end{vmatrix} & +\begin{vmatrix} a_{1,1} & a_{1,3} \\ a_{3,1} & a_{3,3} \end{vmatrix} & -\begin{vmatrix} a_{1,1} & a_{1,2} \\ a_{3,1} & a_{3,2} \end{vmatrix} \\ +\begin{vmatrix} a_{1,2} & a_{1,3} \\ a_{2,2} & a_{2,3} \end{vmatrix} & -\begin{vmatrix} a_{1,1} & a_{1,3} \\ a_{2,1} & a_{2,3} \end{vmatrix} & +\begin{vmatrix} a_{1,1} & a_{1,2} \\ a_{2,1} & a_{2,2} \end{vmatrix} \end{bmatrix}$$

$$= \begin{bmatrix} +\left(a_{2,2}a_{3,3} - a_{2,3}a_{3,2}\right) & -\left(a_{2,1}a_{3,3} - a_{2,3}a_{3,1}\right) & +\left(a_{2,1}a_{3,2} - a_{2,2}a_{3,1}\right) \\ -\left(a_{1,2}a_{3,3} - a_{1,3}a_{3,2}\right) & +\left(a_{1,1}a_{3,3} - a_{1,3}a_{3,1}\right) & -\left(a_{1,1}a_{3,2} - a_{1,2}a_{3,1}\right) \\ +\left(a_{1,2}a_{2,3} - a_{1,3}a_{2,2}\right) & -\left(a_{1,1}a_{2,3} - a_{1,3}a_{2,1}\right) & +\left(a_{1,1}a_{2,2} - a_{1,2}a_{2,1}\right) \end{bmatrix}$$

$$\therefore \quad \boldsymbol{C} = \begin{bmatrix} c_{1,1} & c_{1,2} & c_{1,3} \\ c_{2,1} & c_{2,2} & c_{2,3} \\ c_{3,1} & c_{3,2} & c_{3,3} \end{bmatrix}$$

3.16 The Ajoint Matrix or Adjugate Matrix

The transpose of the cofactor matrix, $\boldsymbol{C}^{\mathrm{T}}$, of a matrix $\boldsymbol{A}$ is called the *ajoint matrix* or the *adjugate matrix* of $\boldsymbol{A}$ and it is denoted by adj $\boldsymbol{A}$. Thus

$$\operatorname{adj}(\boldsymbol{A}) = \boldsymbol{C}^{\mathrm{T}} \tag{3.62}$$

Example:
Given 3×3 matrix $\boldsymbol{A}$:

$$\boldsymbol{A} = \begin{bmatrix} a_{1,1} & a_{1,2} & a_{1,3} \\ a_{2,1} & a_{2,2} & a_{2,3} \\ a_{3,1} & a_{3,2} & a_{3,3} \end{bmatrix}$$

The adjoint matrix of $\boldsymbol{A}$ is

$$\operatorname{adj}(\boldsymbol{A}) = \boldsymbol{C}^{\mathrm{T}} = \begin{bmatrix} c_{1,1} & c_{2,1} & c_{3,1} \\ c_{1,2} & c_{2,2} & c_{3,2} \\ c_{1,3} & c_{2,3} & c_{3,3} \end{bmatrix}$$
$$= \begin{bmatrix} +\left(a_{2,2}a_{3,3} - a_{2,3}a_{3,2}\right) & -\left(a_{1,2}a_{3,3} - a_{1,3}a_{3,2}\right) & +\left(a_{1,2}a_{2,3} - a_{1,3}a_{2,2}\right) \\ -\left(a_{2,1}a_{3,3} - a_{2,3}a_{3,1}\right) & +\left(a_{1,1}a_{3,3} - a_{1,3}a_{3,1}\right) & -\left(a_{1,1}a_{2,3} - a_{1,3}a_{2,1}\right) \\ +\left(a_{2,1}a_{3,2} - a_{2,2}a_{3,1}\right) & -\left(a_{1,1}a_{3,2} - a_{1,2}a_{3,1}\right) & +\left(a_{1,1}a_{2,2} - a_{1,2}a_{2,1}\right) \end{bmatrix}$$

3.17 The Reciprocal or Inverse of a Matrix

The inverse $\boldsymbol{A}^{-1}$ of a square matrix $\boldsymbol{A}$ is the matrix which when pre-multiplied or post-multiplied by matrix $\boldsymbol{A}$ results in the identity matrix, i.e.

$$\boldsymbol{A} \cdot \boldsymbol{A}^{-1} = \boldsymbol{A}^{-1} \cdot \boldsymbol{A} = \boldsymbol{I}$$

The determinant of the matrix allows us to establish whether or not the matrix is invertible. If the determinant $\det \boldsymbol{A} \neq 0$, we say that the matrix is *regular* and thus *invertible*, i.e. $\boldsymbol{A}^{-1}$ does exist. Otherwise, if the determinant $\det \boldsymbol{A} = 0$, we say that the matrix is *singular* and thus *non-invertible*, i.e. $\boldsymbol{A}^{-1}$ does not exist.

The elements of $\boldsymbol{A}^{-1}$ are defined as

$$a_{i,j}^{-1} = (-1)^{(i+j)} \cdot \frac{\det\left(\boldsymbol{A}_{j,i}\right)}{\det(A)} = \frac{c_{j,i}}{\det(A)} = \frac{c_{i,j}^{\mathrm{T}}}{\det(A)} \tag{3.63}$$

Thus,

$$\boldsymbol{A}^{-1} = \frac{\boldsymbol{C}^{T}}{\det(A)} = \frac{\operatorname{adj}(A)}{\det(A)} \tag{3.64}$$

The general method for calculating the inverse of a square matrix can be expressed as follows. We start by calculating the determinant of the matrix. If the determinant is zero, then the matrix is singular and has no inverse. Otherwise, the inverse matrix is found by calculating the adjoint matrix and dividing it by the determinant of the original matrix.

The inverse or reciprocal matrix makes matrix division possible. Given a matrix equation:

$$\boldsymbol{A} \cdot X = \boldsymbol{B}$$

To solve for matrix $\boldsymbol{X}$ we pre-multiply both sides of the equation by the reciprocal of $\boldsymbol{A}$ which gives

$$\boldsymbol{A}^{-1} \cdot \boldsymbol{A} \cdot X = \boldsymbol{A}^{-1} \cdot \boldsymbol{B}$$

But since $\boldsymbol{A}^{-1} \cdot \boldsymbol{A} = \boldsymbol{I}$, we get

$$\boldsymbol{X} = \boldsymbol{A}^{-1} \cdot \boldsymbol{B}$$

3.17.1 Justification of the Definition of the Inverse

Consider 3×3 matrix $\boldsymbol{A}$:

$$\boldsymbol{A} = \begin{bmatrix} a_{1,1} & a_{1,2} & a_{1,3} \\ a_{2,1} & a_{2,2} & a_{2,3} \\ a_{3,1} & a_{3,2} & a_{3,3} \end{bmatrix}$$

The minors of the three elements of the top row of A are given by

$$\boldsymbol{A}_{1,1} = \begin{bmatrix} a_{2,2} & a_{2,3} \\ a_{3,2} & a_{3,3} \end{bmatrix}, \quad \boldsymbol{A}_{1,2} = \begin{bmatrix} a_{2,1} & a_{2,3} \\ a_{3,1} & a_{3,3} \end{bmatrix} \text{and } \boldsymbol{A}_{1,3} = \begin{bmatrix} a_{2,1} & a_{2,2} \\ a_{3,1} & a_{3,2} \end{bmatrix}$$

The cofactors of the three elements of the top row of $\boldsymbol{A}$ are

$$c_{1,1} = +\det\left(\boldsymbol{A}_{1,1}\right) = +\left(a_{2,2}a_{3,3} - a_{3,2}a_{2,3}\right)$$

$$c_{1,2} = -\det\left(\boldsymbol{A}_{1,2}\right) = -\left(a_{2,1}a_{3,3} - a_{3,1}a_{2,3}\right)$$

$$c_{1,3} = +\det\left(\boldsymbol{A}_{1,3}\right) = +\left(a_{2,1}a_{3,2} - a_{3,1}a_{2,2}\right)$$

By definition the dot product of the first row of the matrix by the first row of its cofactor matrix yields the determinant of the matrix, thus

$$a_{1,1}c_{1,1} + a_{1,2}c_{1,2} + a_{1,3}c_{1,3} = \det(\boldsymbol{A}) = |A|$$

Next we compute the dot products of the second and third rows of the matrix by the first row of its cofactor matrix.

$$\begin{aligned} a_{2,1}c_{1,1} + a_{2,2}c_{1,2} + a_{2,3}c_{1,3} &= a_{2,1}\left(a_{2,2}a_{3,3} - a_{3,2}a_{2,3}\right) \\ &\quad - a_{2,2}\left(a_{2,1}a_{3,3} - a_{3,1}a_{2,3}\right) \\ &\quad + a_{2,3}\left(a_{2,1}a_{3,2} - a_{3,1}a_{2,2}\right) \\ &= a_{2,1}a_{2,2}a_{3,3} - a_{2,1}a_{3,2}a_{2,3} - a_{2,2}a_{2,1}a_{3,3} \\ &\quad + a_{2,2}a_{3,1}a_{2,3} + a_{2,3}a_{2,1}a_{3,2} - a_{2,3}a_{3,1}a_{2,2} \\ &= 0 \\ a_{3,1}c_{1,1} + a_{3,2}c_{1,2} + a_{3,3}c_{1,3} &= a_{3,1}\left(a_{2,2}a_{3,3} - a_{3,2}a_{2,3}\right) \\ &\quad - a_{3,2}\left(a_{2,1}a_{3,3} - a_{3,1}a_{2,3}\right) \\ &\quad + a_{3,3}\left(a_{2,1}a_{3,2} - a_{3,1}a_{2,2}\right) \\ &= a_{3,1}a_{2,2}a_{3,3} - a_{3,1}a_{3,2}a_{2,3} - a_{3,2}a_{2,1}a_{3,3} \\ &\quad + a_{3,2}a_{3,1}a_{2,3} + a_{3,3}a_{2,1}a_{3,2} - a_{3,3}a_{3,1}a_{2,2} \\ &= 0 \end{aligned}$$

It turns out that this is a general result. That is, if we compute the dot product of any row/column of the matrix by the corresponding row/column of its cofactor matrix it results in the determinant of the matrix. Alternatively, if we compute the dot product of any row/column of the matrix by a different row/column of its cofactor matrix it produces a zero result.

Thus

$$\sum_{k=1}^{m} a_{i,k} c_{j,k} = \begin{cases} |A| & \text{if } i = j \\ 0 & \text{if } i \neq j \end{cases} \tag{3.65}$$

Recalling that $\text{adj}\,\boldsymbol{A} = \boldsymbol{C}^{\mathrm{T}}$, let us calculate the matrix $\boldsymbol{B} = \boldsymbol{A} \cdot \text{adj}\,(\boldsymbol{A})$, concentrating on the *ij*th element of matrix $\boldsymbol{B}$:

$$\begin{aligned} b_{i,j} &= \sum_{k=1}^{m} a_{i,k} c_{k,j}^{T} \\ &= \sum_{k=1}^{m} a_{i,k} c_{j,k} \\ &= \begin{cases} |A| & \text{if } i = j \\ 0 & \text{if } i \neq j \end{cases} \end{aligned}$$

So, for our example

$$\boldsymbol{B} = \boldsymbol{A} \cdot \text{adj}\,(\boldsymbol{A}) = \begin{bmatrix} |\boldsymbol{A}| & 0 & 0 \\ 0 & |\boldsymbol{A}| & 0 \\ 0 & 0 & |\boldsymbol{A}| \end{bmatrix} = |\boldsymbol{A}| \cdot \begin{bmatrix} 1 & 0 & 0 \\ 0 & 1 & 0 \\ 0 & 0 & 1 \end{bmatrix} = |\boldsymbol{A}| \cdot \boldsymbol{I}$$

Thus

$$\begin{aligned} \boldsymbol{A} \cdot \text{adj}\,(\boldsymbol{A}) &= |A| \cdot \boldsymbol{I} \\ \therefore \quad \boldsymbol{A} \cdot \frac{\text{adj}\,(A)}{|A|} &= \boldsymbol{I} \end{aligned}$$

But as $\boldsymbol{A} \cdot \left(\frac{\text{adj}\,(\boldsymbol{A})}{|\boldsymbol{A}|} \right) = \boldsymbol{I}$, the matrix $\left(\frac{\text{adj}\,(\boldsymbol{A})}{|\boldsymbol{A}|} \right)$ must be equal to the inverse of $\boldsymbol{A}$, i.e.

$$\boldsymbol{A}^{-1} = \frac{\text{adj}\,(A)}{|A|}$$

Observe that the inverse of matrix $\boldsymbol{A}$ exists only if $|\boldsymbol{A}| \neq 0$.

3.18 A Theorem on Invertible Matrices and their Determinants

Theorem 3.1 *A square matrix is invertible if and only if it has a non-zero determinant.*

Proof: We will prove this theorem by *elimination*.

Given a 2×2 matrix:

$$\boldsymbol{A} = \begin{bmatrix} a & b \\ c & d \end{bmatrix}$$

The determinant of $\boldsymbol{A}$ is given by

$$\det(\boldsymbol{A}) = |\boldsymbol{A}| = ad - bc$$

Let us, for the moment, assume that $|\boldsymbol{A}| \neq 0$ and let us look for the inverse of matrix $\boldsymbol{A}$. Then

$$\begin{bmatrix} a & b \\ c & d \end{bmatrix} \cdot \begin{bmatrix} x & y \\ z & w \end{bmatrix} = \begin{bmatrix} 1 & 0 \\ 0 & 1 \end{bmatrix}$$

$$\therefore \quad \begin{bmatrix} ax + bz & ay + bw \\ cx + dz & cy + dw \end{bmatrix} = \begin{bmatrix} 1 & 0 \\ 0 & 1 \end{bmatrix}$$

which reduces to solving the following two systems of linear equations in two unknowns:

System 1	*System 2*
$ax + bz = 1$	$cx + dz = 0$
$ay + bw = 0$	$cy + dw = 1$

Now we must solve the two systems for the unknowns x, z and y, w, respectively.

System 1

$$ax + bz = 1 \tag{3.66}$$

$$cx + dz = 0 \tag{3.67}$$

Multiplying Eq. (3.66) by $\frac{d}{b}$ we get

$$a\frac{d}{b}x + dz = \frac{d}{b} \tag{3.68}$$

Now subtracting Eq. (3.68) from Eq. (3.67) we get

$$cx - \frac{ad}{b}x = -\frac{d}{b} \tag{3.69}$$

Multiplying both sides of Eq. (3.69) by b we get

$$bcx - adx = -d$$

$$\therefore \quad adx - bcx = d$$

$$\therefore \quad x\,(ad - bc) = d$$

$$\therefore \quad x = \frac{d}{(ad - bc)}$$

$$\therefore \quad x = \frac{d}{\det(\boldsymbol{A})}$$

Substituting x in Eq. (3.67) we get

$$\frac{dc}{(ad-bc)}+dz=0$$

$$\therefore \quad dz=-\frac{dc}{(ad-bc)}$$

$$\therefore \quad z=-\frac{c}{(ad-bc)}$$

$$\therefore \quad z=\frac{-c}{\det(\mathbf{A})}$$

System 2

$$ay+bw=0 \tag{3.70}$$

$$cy+dw=1 \tag{3.71}$$

Multiplying Eq. (3.70) by $\frac{d}{b}$ we get

$$a\frac{d}{b}y+dw=0 \tag{3.72}$$

Now subtracting Eq. (3.72) from Eq. (3.71) we get

$$cy-\frac{ad}{b}y=1 \tag{3.73}$$

Multiplying both sides of Eq. (3.73) by b we get

$$bcy-ady=b$$

$$\therefore \quad y(bc-ad)=b$$

$$\therefore \quad y=\frac{b}{(bc-ad)}$$

$$\therefore \quad y=\frac{-b}{(ad-bc)}$$

$$\therefore \quad y=\frac{-b}{\det(\mathbf{A})}$$

Substituting y in Eq. (3.70) we get

$$\frac{-ab}{(ad-bc)}+bw=0$$

$$\therefore \quad bw = \frac{ab}{(ad - bc)}$$

$$\therefore \quad w = \frac{a}{(ad - bc)}$$

$$\therefore \quad w = \frac{a}{\det(\boldsymbol{A})}$$

The above result shows that the two systems have a solution (i.e. $\boldsymbol{A}^{-1}$ exists) *if and only if* $|\boldsymbol{A}| \neq 0$. If alternatively $|\boldsymbol{A}| = 0$, then none of the above equations would have a solution and consequently the inverse matrix $\boldsymbol{A}^{-1}$ would not exist. *QED* (*Quod Erat Demonstrandum*, which means "that which is to be proved").

Observation:

Given a matrix $\boldsymbol{A} = \begin{bmatrix} a & b \\ c & d \end{bmatrix}$, then its inverse is $\boldsymbol{A}^{-1} = \begin{bmatrix} a & b \\ c & d \end{bmatrix}^{-1} =$ $\begin{bmatrix} \dfrac{d}{|\boldsymbol{A}|} & \dfrac{-b}{|\boldsymbol{A}|} \\ \dfrac{-c}{|\boldsymbol{A}|} & \dfrac{a}{|\boldsymbol{A}|} \end{bmatrix} = \dfrac{1}{|\boldsymbol{A}|} \cdot \begin{bmatrix} d & -b \\ -c & a \end{bmatrix}$.

The cofactor matrix of matrix $\boldsymbol{A}$ is $\boldsymbol{C} = \begin{bmatrix} d & -c \\ -b & a \end{bmatrix}$. The transpose of its cofactor matrix is $\boldsymbol{C}^T = \begin{bmatrix} d & -b \\ -c & a \end{bmatrix}$.

Thus, we observe that the inverse of matrix $\boldsymbol{A}$ is given by $\boldsymbol{A}^{-1} = \dfrac{\boldsymbol{C}^{\mathrm{T}}}{|\boldsymbol{A}|}$, i.e. by the transpose of its cofactor matrix divided by its determinant.

3.19 Axioms and Rules of Matrix Inversion

The following axioms and matrix algebra rules apply to the operation of matrix inversion for all square matrices $\boldsymbol{A}, \boldsymbol{B}$ (which are assumed to be conformant).

Existence of the inverse of a matrix:

$$\det(\boldsymbol{A}) \neq 0 \quad \Leftrightarrow \quad \boldsymbol{A}^{-1} = \frac{\operatorname{adj}(\boldsymbol{A})}{\det(\boldsymbol{A})} \tag{A3.12}$$

Product of a matrix by its inverse:

$$\boldsymbol{A} \cdot \boldsymbol{A}^{-1} = \boldsymbol{A}^{-1} \cdot \boldsymbol{A} = \boldsymbol{I} \tag{R3.16}$$

Inverse rules:

$$\left(\boldsymbol{A}^{-1}\right)^{-1} = \boldsymbol{A} \tag{R3.17}$$

$$(\boldsymbol{A} \cdot \boldsymbol{B})^{-1} = \boldsymbol{B}^{-1} \cdot \boldsymbol{A}^{-1} \tag{R3.18}$$

Transpose of the inverse rule:

$$\left(\boldsymbol{A}^{-1}\right)^{\mathrm{T}} = \left(\boldsymbol{A}^{\mathrm{T}}\right)^{-1} \tag{R3.19}$$

3.20 Solving a System of Linear Simultaneous Equations

Given a system of m linear simultaneous equations with m unknowns:

$$\begin{aligned} a_{1,1}x_1 + a_{1,2}x_2 + \cdots + a_{1,m}x_m &= c_1 \\ a_{2,1}x_1 + a_{2,2}x_2 + \cdots + a_{2,m}x_m &= c_2 \\ &\vdots \\ a_{m,1}x_1 + a_{m,2}x_2 + \cdots + a_{m,m}x_m &= c_m \end{aligned} \tag{3.74}$$

We may represent this system in matrix form as

$$\begin{bmatrix} a_{1,1} & a_{1,2} & \cdots & a_{1,m} \\ a_{2,1} & a_{2,2} & \cdots & a_{2,m} \\ \vdots & \vdots & \ddots & \vdots \\ a_{m,1} & a_{m,2} & \cdots & a_{m,m} \end{bmatrix} \cdot \begin{bmatrix} x_1 \\ x_2 \\ \vdots \\ x_m \end{bmatrix} = \begin{bmatrix} c_1 \\ c_2 \\ \vdots \\ c_m \end{bmatrix} \tag{3.75}$$

or in a more concise notation as

$$\boldsymbol{A} \cdot \boldsymbol{x} = \boldsymbol{c} \tag{3.76}$$

Where $\boldsymbol{A}$ is the matrix of coefficients, $\boldsymbol{x}$ is the column vector of unknowns and $\boldsymbol{c}$ is the column vector of constants.

To solve this system of equations, we multiply both sides of the above equation by the reciprocal of the coefficients matrix $\boldsymbol{A}^{-1}$, thus

$$\boldsymbol{x} = \boldsymbol{A}^{-1} \cdot \boldsymbol{c} \tag{3.77}$$

3.21 Orthogonal Matrices

A square matrix $\boldsymbol{A}$ for which its inverse $\boldsymbol{A}^{-1}$ is equal to its transpose $\boldsymbol{A}^{\mathrm{T}}$ (i.e. $\boldsymbol{A}^{-1} = \boldsymbol{A}^{\mathrm{T}}$), is called an *orthogonal matrix.* With orthogonal matrices

$$\boldsymbol{A} \cdot \boldsymbol{A}^{\mathrm{T}} = \boldsymbol{A}^{\mathrm{T}} \cdot \boldsymbol{A} = \boldsymbol{I} \tag{3.78}$$

For a more detailed explanation and a proof of this property see Section 3.7.

Example:
Given the orthogonal matrix $\boldsymbol{A}$ defined as:

$$\boldsymbol{A} = \frac{1}{\sqrt{2}} \begin{bmatrix} 1 & 1 & 0 \\ -1 & 1 & 0 \\ 0 & 0 & \sqrt{2} \end{bmatrix}$$

The transpose of this matrix is given by

$$\boldsymbol{A}^{\mathrm{T}} = \frac{1}{\sqrt{2}} \begin{bmatrix} 1 & -1 & 0 \\ 1 & 1 & 0 \\ 0 & 0 & \sqrt{2} \end{bmatrix}$$

Thus

$$A \cdot A^{\mathrm{T}} = \frac{1}{\sqrt{2}} \begin{bmatrix} 1 & 1 & 0 \\ -1 & 1 & 0 \\ 0 & 0 & \sqrt{2} \end{bmatrix} \cdot \frac{1}{\sqrt{2}} \begin{bmatrix} 1 & -1 & 0 \\ 1 & 1 & 0 \\ 0 & 0 & \sqrt{2} \end{bmatrix}$$

$$A \cdot A^{\mathrm{T}} = \frac{1}{2} \begin{bmatrix} 1\cdot 1 + 1\cdot 1 + 0\cdot 0 & -1\cdot 1 + 1\cdot 1 + 0\cdot 0 & 1\cdot 0 + 1\cdot 0 + 0\cdot \sqrt{2} \\ -1\cdot 1 + 1\cdot 1 + 0\cdot 0 & -1\cdot(-1) + 1\cdot 1 + 0\cdot 0 & -1\cdot 0 + 1\cdot 0 + 0\cdot \sqrt{2} \\ 0\cdot 1 + 0\cdot 1 + \sqrt{2}\cdot 0 & 0\cdot(-1) + 0\cdot 1 + \sqrt{2}\cdot 0 & 0\cdot 0 + 0\cdot 0 + \sqrt{2}\cdot\sqrt{2} \end{bmatrix}$$

$$A \cdot A^{\mathrm{T}} = \frac{1}{2} \begin{bmatrix} 2 & 0 & 0 \\ 0 & 2 & 0 \\ 0 & 0 & 2 \end{bmatrix} = I$$

Matrix A represents a 2D counter-clockwise rotation by $-45°$ about the origin and matrix $A^{\mathrm{T}} = A^{-1}$ represents a 2D counter-clockwise rotation by $+45°$ about the origin. Alternately matrix A is the change of basis matrix that rotates the $\boldsymbol{i}, \boldsymbol{j}, \boldsymbol{k}$ Cartesian basis in counter-clockwise fashion about the $\boldsymbol{k}$ basis vector by $+45°$. See Section 3.7.

3.22 Two Theorems on Vector by Matrix Multiplication

Theorem 3.2 *Given the product of a row vector by a conformant matrix, interchanging two elements of the vector and the corresponding rows of the matrix does not alter their product, i.e.*

$$\begin{aligned} &\begin{bmatrix} v_1 & v_2 & v_3 \end{bmatrix} \cdot \begin{bmatrix} m_{1,1} & m_{1,2} & m_{1,3} \\ m_{2,1} & m_{2,2} & m_{2,3} \\ m_{3,1} & m_{3,2} & m_{3,3} \end{bmatrix} \\ &= \begin{bmatrix} v_2 & v_1 & v_3 \end{bmatrix} \cdot \begin{bmatrix} m_{2,1} & m_{2,2} & m_{2,3} \\ m_{1,1} & m_{1,2} & m_{1,3} \\ m_{3,1} & m_{3,2} & m_{3,3} \end{bmatrix} \end{aligned} \tag{3.79}$$

Proof: Performing the matrix multiplication on the left-hand side of Eq. (3.78) we get

$$\begin{bmatrix} v_1 & v_2 & v_3 \end{bmatrix} \cdot \begin{bmatrix} m_{1,1} & m_{1,2} & m_{1,3} \\ m_{2,1} & m_{2,2} & m_{2,3} \\ m_{3,1} & m_{3,2} & m_{3,3} \end{bmatrix} = \big[v_1 m_{1,1} + v_2 m_{2,1} + v_3 m_{3,1}$$

$$v_1 m_{1,2} + v_2 m_{2,2} + v_3 m_{3,2} \quad v_1 m_{1,3} + v_2 m_{2,3} + v_3 m_{3,3} \big]$$

and performing the matrix multiplication on the right-hand side of this equation we get an identical result:

$$\begin{bmatrix} v_2 & v_1 & v_3 \end{bmatrix} \cdot \begin{bmatrix} m_{2,1} & m_{2,2} & m_{2,3} \\ m_{1,1} & m_{1,2} & m_{1,3} \\ m_{3,1} & m_{3,2} & m_{3,3} \end{bmatrix} = \big[v_2 m_{2,1} + v_1 m_{1,1} + v_3 m_{3,1}$$

$$v_2 m_{2,2} + v_1 m_{1,2} + v_3 m_{3,2} \quad v_2 m_{2,3} + v_1 m_{1,3} + v_3 m_{3,3} \big]$$

□

Theorem 3.3 *Similarly, given the product of a matrix by a conformant column vector, interchanging two elements of the vector and the corresponding columns of the matrix does not alter their product, i.e.*

$$\begin{bmatrix} m_{1,1} & m_{1,2} & m_{1,3} \\ m_{2,1} & m_{2,2} & m_{2,3} \\ m_{3,1} & m_{3,2} & m_{3,3} \end{bmatrix} \cdot \begin{bmatrix} v_1 \\ v_2 \\ v_3 \end{bmatrix} = \begin{bmatrix} m_{1,2} & m_{1,1} & m_{1,3} \\ m_{2,2} & m_{2,1} & m_{2,3} \\ m_{3,2} & m_{3,1} & m_{3,3} \end{bmatrix} \cdot \begin{bmatrix} v_2 \\ v_1 \\ v_3 \end{bmatrix} \quad (3.80)$$

Proof: Performing the matrix multiplication on the left-hand side of Eq. (3.80) we get

$$\begin{bmatrix} m_{1,1} & m_{1,2} & m_{1,3} \\ m_{2,1} & m_{2,2} & m_{2,3} \\ m_{3,1} & m_{3,2} & m_{3,3} \end{bmatrix} \cdot \begin{bmatrix} v_1 \\ v_2 \\ v_3 \end{bmatrix} = \begin{bmatrix} m_{1,1}v_1 + m_{1,2}v_2 + m_{1,3}v_3 \\ m_{2,1}v_1 + m_{2,2}v_2 + m_{2,3}v_3 \\ m_{3,1}v_1 + m_{3,2}v_2 + m_{3,3}v_3 \end{bmatrix}$$

and performing the matrix multiplication on the right-hand side of this equation we get an identical result:

$$\begin{bmatrix} m_{1,2} & m_{1,1} & m_{1,3} \\ m_{2,2} & m_{2,1} & m_{2,3} \\ m_{3,2} & m_{3,1} & m_{3,3} \end{bmatrix} \cdot \begin{bmatrix} v_2 \\ v_1 \\ v_3 \end{bmatrix} = \begin{bmatrix} m_{1,2}v_2 + m_{1,1}v_1 + m_{1,3}v_3 \\ m_{2,2}v_2 + m_{2,1}v_1 + m_{2,3}v_3 \\ m_{3,2}v_2 + m_{3,1}v_1 + m_{3,3}v_3 \end{bmatrix}$$

□

3.23 The Row-/Column-Reversal Matrix

There is a special matrix that we shall call the *row-/column-reversal matrix* ***R***, which is a square matrix with all its elements set to 0 apart from the elements of its trailing diagonal which are set to 1. So ***R*** is a *row-reversed* (or a *column-reversed*) version of the identity matrix ***I***. Thus a 3 × 3 row-/column-reversal matrix is given by

$$R = \begin{bmatrix} 0 & 0 & 1 \\ 0 & 1 & 0 \\ 1 & 0 & 0 \end{bmatrix}$$

Pre-multiplying a matrix ***A*** by the row-/column-reversal matrix ***R***, has the effect of row-reversing the matrix, thus

$$\begin{bmatrix} 0 & 0 & 1 \\ 0 & 1 & 0 \\ 1 & 0 & 0 \end{bmatrix} \cdot \begin{bmatrix} a_{1,1} & a_{1,2} & a_{1,3} \\ a_{2,1} & a_{2,2} & a_{2,3} \\ a_{3,1} & a_{3,2} & a_{3,3} \end{bmatrix} = \begin{bmatrix} a_{3,1} & a_{3,2} & a_{3,3} \\ a_{2,1} & a_{2,2} & a_{2,3} \\ a_{1,1} & a_{1,2} & a_{1,3} \end{bmatrix}$$

$$\text{or} \quad \boldsymbol{B} = \boldsymbol{R} \cdot \boldsymbol{A} \quad \text{and} \quad \boldsymbol{A} = \boldsymbol{R} \cdot \boldsymbol{B} \tag{3.81}$$

Similarly, post-multiplying a matrix $\boldsymbol{A}$ by the row-/column-reversal matrix $\boldsymbol{R}$, has the effect of column-reversing the matrix, thus

$$\begin{bmatrix} a_{1,1} & a_{1,2} & a_{1,3} \\ a_{2,1} & a_{2,2} & a_{2,3} \\ a_{3,1} & a_{3,2} & a_{3,3} \end{bmatrix} \cdot \begin{bmatrix} 0 & 0 & 1 \\ 0 & 1 & 0 \\ 1 & 0 & 0 \end{bmatrix} = \begin{bmatrix} a_{1,3} & a_{1,2} & a_{1,1} \\ a_{2,3} & a_{2,2} & a_{2,1} \\ a_{3,3} & a_{3,2} & a_{3,1} \end{bmatrix}$$

$$\text{or} \quad \boldsymbol{C} = \boldsymbol{A} \cdot \boldsymbol{R} \quad \text{and} \quad \boldsymbol{A} = \boldsymbol{C} \cdot \boldsymbol{R} \tag{3.82}$$

Row-reversing the row-reverse matrix yields the identity matrix, i.e.

$$\begin{bmatrix} 0 & 0 & 1 \\ 0 & 1 & 0 \\ 1 & 0 & 0 \end{bmatrix} \cdot \begin{bmatrix} 0 & 0 & 1 \\ 0 & 1 & 0 \\ 1 & 0 & 0 \end{bmatrix} = \begin{bmatrix} 1 & 0 & 0 \\ 0 & 1 & 0 \\ 0 & 0 & 1 \end{bmatrix}$$

Like the identity matrix, the matrix $\boldsymbol{R}$ is equal to its inverse and also to its transpose, i.e.

$$\boldsymbol{R} \cdot \boldsymbol{R} = \boldsymbol{I} \quad \Leftrightarrow \quad \boldsymbol{R} = \boldsymbol{R}^{-1} \tag{3.83}$$

and

$$\boldsymbol{R} = \boldsymbol{R}^{T} \tag{3.84}$$

3.23.1 Summary of Matrix Algebra Axioms and Rules

In this section we collect together, for ease of reference, all the axioms and matrix algebra rules that apply to all the matrix operations we have examined.

Matrix Addition/Subtraction

The following axioms and matrix algebra rules apply to the addition and subtraction of matrices for all matrices $\boldsymbol{A}$, $\boldsymbol{B}$, $\boldsymbol{C}$ (which are assumed to be of the same order).

Existence of the matrix sum:

$$\boldsymbol{C}_{(m,n)} = \boldsymbol{A}_{(m,n)} \pm \boldsymbol{B}_{(m,n)} \Rightarrow \left\{ c_{i,j} = a_{i,j} \pm b_{i,j} \right\}_{i,j=1}^{m,n} \tag{A3.1}$$

Existence of the null element:

$$\boldsymbol{A} \pm \boldsymbol{0} = \boldsymbol{A} \tag{A3.2}$$

Existence of the additive inverse:

$$\boldsymbol{A} + (-\boldsymbol{A}) = \boldsymbol{0} \tag{A3.3}$$

Commutative law:

$$\boldsymbol{A} + \boldsymbol{B} = \boldsymbol{B} + \boldsymbol{A} \tag{R3.1}$$

Associative law:

$$(\boldsymbol{A} + \boldsymbol{B}) + \boldsymbol{C} = \boldsymbol{A} + (\boldsymbol{B} + \boldsymbol{C}) \tag{R3.2}$$

Distributive law:

$$s \cdot (\boldsymbol{A} \pm \boldsymbol{B}) = s \cdot \boldsymbol{A} \pm s \cdot \boldsymbol{B} \tag{R3.3}$$

Transposition rule:

$$(\boldsymbol{A} \pm \boldsymbol{B})^{\mathrm{T}} = \boldsymbol{A}^{\mathrm{T}} \pm \boldsymbol{B}^{\mathrm{T}} \tag{R3.4}$$

Matrix by Scalar Multiplication

The following axioms and matrix algebra rules apply to the multiplication of a matrix by a scalar for all matrices $\boldsymbol{A}$, $\boldsymbol{B}$ (which are assumed to be of the same order) and all scalars s, s_1, s_2.

Existence of the matrix by scalar product:

$$\boldsymbol{B}_{(m,n)} = \boldsymbol{A}_{(m,n)} \cdot s \Rightarrow \left\{ b_{i,j} = s \cdot a_{i,j} \right\}_{i,j=1}^{m,n} \quad \text{(A3.4)}$$

Existence of the identity map:

$$\mathbf{1} \cdot \boldsymbol{A} = \boldsymbol{I} \cdot \boldsymbol{A} = \boldsymbol{A} \quad \text{(A3.5)}$$

Commutative law:

$$s \cdot \boldsymbol{A} = \boldsymbol{A} \cdot s \quad \text{(R3.5)}$$

Associative law:

$$s_1 \cdot (s_2 \cdot \boldsymbol{A}) = (s_1 \cdot s_2) \cdot \boldsymbol{A} \quad \text{(R3.6)}$$

Distributive law:

$$(s_1 \pm s_2) \cdot \boldsymbol{A} = s_1 \cdot \boldsymbol{A} \pm s_2 \cdot \boldsymbol{A} \quad \text{(R3.7)}$$

Vector by Vector Multiplication

The following axioms apply to the vector by vector multiplication for all matrices $\boldsymbol{a}, \boldsymbol{b}$ (which are assumed to be conformant).

Existence of the dot product:

$$\boldsymbol{a}^{\mathrm{T}} \cdot b = \sum_{i=1}^{m} a_i \cdot b_i \quad \text{(A3.6)}$$

Existence of the tensor product:

$$\boldsymbol{a} \cdot \boldsymbol{b}^{\mathrm{T}} = \left\{ c_{i,j} = a_i b_j \right\}_{i,j=1}^{m,n} \quad \text{(A3.7)}$$

Matrix Multiplication

The following axioms and matrix algebra rules apply to the matrix multiplication for all matrices $\boldsymbol{A}$, $\boldsymbol{B}$, $\boldsymbol{C}$ (which are assumed to be conformant) and all scalars s.

Existence of the matrix product:

$$\boldsymbol{A}_{(m,l)} \cdot \boldsymbol{B}_{(l,n)} = \boldsymbol{C}_{(m,n)} \Rightarrow \left\{ c_{i,j} = \sum_{k=1}^{l} a_{i,k} b_{k,j} \right\}_{i,j=1}^{m,n} \quad \text{(A3.8)}$$

Existence of the unit matrix:

$$\boldsymbol{I}_{(m)} \cdot \boldsymbol{A}_{(m,n)} = \boldsymbol{A}_{(m,n)} \cdot \boldsymbol{I}_{(n)} = \boldsymbol{A}_{(m,n)} \quad \text{(A3.9)}$$

Existence of the zero matrix:

$$\mathbf{0}_{(m)} \cdot \boldsymbol{A}_{(m,n)} = \boldsymbol{A}_{(m,n)} \cdot \mathbf{0}_{(n)} = \mathbf{0}_{(m,n)} \quad \text{(A3.10)}$$

Powers of a matrix:

$$\boldsymbol{A}^0 = \boldsymbol{I} \wedge \boldsymbol{A}^m = \boldsymbol{A}^{(m-1)} \cdot \boldsymbol{A} \quad \text{(A3.11)}$$

Commutative law (does not hold):

$$\boldsymbol{A} \cdot \boldsymbol{B} \neq \boldsymbol{B} \cdot \boldsymbol{A} \quad \text{(R3.8)}$$

Associative law:

$$\boldsymbol{A} \cdot (\boldsymbol{B} \cdot \boldsymbol{C}) = (\boldsymbol{A} \cdot \boldsymbol{B}) \cdot \boldsymbol{C} \quad \text{(R3.9)}$$

Distributive laws:

$$(s \cdot \boldsymbol{A}) \cdot \boldsymbol{B} = \boldsymbol{A} \cdot (s \cdot \boldsymbol{B}) = s \cdot (\boldsymbol{A} \cdot \boldsymbol{B}) \qquad \text{(R3.10)}$$

$$\boldsymbol{A} \cdot (\boldsymbol{B} \pm \boldsymbol{C}) = \boldsymbol{A} \cdot \boldsymbol{B} \pm \boldsymbol{A} \cdot \boldsymbol{C} \qquad \text{(R3.11)}$$

$$(\boldsymbol{A} \pm \boldsymbol{B}) \cdot \boldsymbol{C} = \boldsymbol{A} \cdot \boldsymbol{C} \pm \boldsymbol{B} \cdot \boldsymbol{C} \qquad \text{(R3.12)}$$

Transpose rules:

$$(\boldsymbol{A} \cdot \boldsymbol{B})^{\mathrm{T}} = \boldsymbol{B}^{\mathrm{T}} \cdot \boldsymbol{A}^{T} \qquad \text{(R3.13)}$$

$$(\boldsymbol{A} \cdot \boldsymbol{B} \cdot \boldsymbol{C})^{\mathrm{T}} = \boldsymbol{C}^{\mathrm{T}} \cdot \boldsymbol{B}^{\mathrm{T}} \cdot \boldsymbol{A}^{\mathrm{T}} \qquad \text{(R3.14)}$$

Zero divisor rule:

$$\boldsymbol{A} \cdot \boldsymbol{B} = \boldsymbol{0} \quad \not\Rightarrow \boldsymbol{A} = \boldsymbol{0} \vee \boldsymbol{B} = \boldsymbol{0} \qquad \text{(R3.15)}$$

Matrix Inversion

The following axioms and matrix algebra rules apply to the operation of matrix inversion for all matrices $\boldsymbol{A}, \boldsymbol{B}$ (which are assumed to be conformant).

Existence of the inverse of a matrix:

$$\det(\boldsymbol{A}) \neq 0 \quad \Leftrightarrow \quad \boldsymbol{A}^{-1} = \frac{\operatorname{adj}(\boldsymbol{A})}{\det(\boldsymbol{A})} \qquad \text{(A3.12)}$$

Product of a matrix by its inverse:

$$\boldsymbol{A} \cdot \boldsymbol{A}^{-1} = \boldsymbol{A}^{-1} \cdot \boldsymbol{A} = \boldsymbol{I} \qquad \text{(R3.16)}$$

Inverse rules:

$$\left(\boldsymbol{A}^{-1}\right)^{-1} = \boldsymbol{A} \qquad \text{(R3.17)}$$

$$(\boldsymbol{A} \cdot \boldsymbol{B})^{-1} = \boldsymbol{B}^{-1} \cdot \boldsymbol{A}^{-1} \qquad \text{(R3.18)}$$

Transpose of the inverse rule:

$$\left(\boldsymbol{A}^{-1}\right)^{\mathrm{T}} = \left(\boldsymbol{A}^{\mathrm{T}}\right)^{-1} \qquad \text{(R3.19)}$$

3.24 A Simple Matrix Algebra C Library

See *Appendix 2*.

4

Vector Spaces or Linear Spaces

The introduction of Cartesian coordinates in the 1630s by the French mathematicians Pierre de Fermat (1601–1665) and René Descartes (1596–1650) had a profound influence on the development of mathematics. Cartesian coordinates allowed mathematicians to introduce algebraic methods into the study of geometry. By the middle of the 1800s, however, there was a general dissatisfaction with these coordinate methods and mathematicians began to look for more general and abstract ways by which they could study geometry without referring to coordinate systems. This search leads to the development of vectors and matrices, and eventually to the development of *vector spaces* or *linear spaces*.

Among the first to introduce a geometry without coordinates was the Prussian mathematician Hermann Günter Grassman (1809–1877). In his 1844 book *Die Ausdehnungslehre* (*Extension Theory*) he introduced the notion of abstract quantities and he defined the operations of addition and scalar multiplication on these abstract quantities. It was, however, the Italian mathematician Giuseppe Piano (1858–1932) who introduced the first axiomatic definition of a vector space in his 1888 book *Calcolo Geometrico* (*Geometric Calculus*). This book was well in advance of its time and introduced most of the concepts of vector spaces. The complete axiomatic definition of vector spaces appeared in the 1920 doctoral dissertation of the Polish mathematician Stefan Banach (1892–1945).

As the theory of vector spaces underpins much of the work in computer graphics and curve and surface theory, we will briefly introduce some of the basic ideas of vector spaces in this chapter.

Underlying the concept of a vector space is the concept of a *scalar field*. Let us start by defining what a scalar field is.

4.1 Definition of a Scalar Field

A scalar field is defined as a non-empty set of scalars for which the operations of addition and multiplication are defined, such that for any scalars s_1, s_2, s_3 that belong to the scalar field $\mathbf{S}$ the following axioms are satisfied:

Closure of a Scalar Field

Closure under addition:

$s_1, s_2 \in \boldsymbol{S} \;\Rightarrow\; s_1 + s_2 \in \boldsymbol{S}$

Closure under multiplication:

$s_1, s_2 \in \boldsymbol{S} \;\Rightarrow\; s_1 \cdot s_2 \in \boldsymbol{S}$

Addition of Scalars

Comutativity of addition:

$s_1 + s_2 = s_2 + s_1$

Associativity of addition:

$s_1 + (s_2 + s_3) = (s_1 + s_2) + s_3$

Existence of the additive identity element:

$s_1 + 0 = 0 + s_1 = s_1$

Existence of the additive inverse:

$s_1 + (-s_1) = 0,$

i.e. $\forall\, s_1 \in \boldsymbol{S} \,\exists\, (-s_1) \in \boldsymbol{S}$

Multiplication of Scalars

Comutativity of multiplication:

$s_1 \cdot s_2 = s_2 \cdot s_1$

Associativity of multiplication:

$s_1 \cdot (s_2 \cdot s_3) = (s_1 \cdot s_2) \cdot s_3$

Distributivity of multiplication over addition:

$s_1 \cdot (s_2 + s_3) = s_1 \cdot s_2 + s_1 \cdot s_3$

Distributivity of addition over multiplication:

$(s_1 + s_2) \cdot s_3 = s_1 \cdot s_3 + s_2 \cdot s_3$

Existence of the multiplicative identity element:

$1 \cdot s_1 = s_1 \cdot 1 = s_1$

Existence of the multiplicative inverse:

$$s_1 \cdot \frac{1}{s_1} = 1 \;\Leftrightarrow\; s_1 \neq 0$$

Examples of scalar fields are the set of *real numbers* $\mathbb{R}$, the set of *rational numbers* $\mathbb{Q}$ and the set of complex numbers $\mathbb{C}$.

Having defined scalar fields, we are now ready to define a vector space in terms of a scalar field.

4.2 Definition of a Vector Space

A vector space $\boldsymbol{V}$ is a non-empty set of abstract entities called *vectors* taken together with the algebraic rules for *vector addition* and *scalar by vector multiplication.*

Given a scalar field $\boldsymbol{S}$, we can construct a vector as the *ordered n-tuple* $\boldsymbol{v} = [v_1, v_2, \ldots, v_n]$, where the components of the vector $v_1, v_2, \ldots, v_n \in \boldsymbol{S}$ and the vector $\boldsymbol{v} \in \boldsymbol{S}^n$.

Given two vectors $\boldsymbol{v}_1, \boldsymbol{v}_2 \in \boldsymbol{S}^n$, vector addition is defined as

$$\begin{aligned}\boldsymbol{v}_1 + \boldsymbol{v}_2 &= [v_{11}, v_{12}, \ldots, v_{1n}] + [v_{21}, v_{22}, \ldots, v_{2n}] \\ &= [v_{11} + v_{21}, v_{12} + v_{22}, \ldots, v_{1n} + v_{2n}]\end{aligned}$$

where $[v_{11} + v_{21}, v_{12} + v_{22}, \ldots, v_{1n} + v_{2n}] \in \boldsymbol{S}^n$ and $(v_{11} + v_{21})$, $(v_{12} + v_{22})$, $\ldots$, $(v_{1n} + v_{2n}) \in \boldsymbol{S}$.

Given a scalar $s \in \boldsymbol{S}$ and a vector $\boldsymbol{v} \in \boldsymbol{S}^n$, scalar by vector multiplication is defined as

$$s \cdot \boldsymbol{v} = s \cdot [v_1, v_2, \ldots, v_n] = [s \cdot v_1, s \cdot v_2, \ldots, s \cdot v_n]$$

where $[s \cdot v_1, s \cdot v_2, \ldots, s \cdot v_n] \in \boldsymbol{S}^n$ and $s \cdot v_1, s \cdot v_2, \ldots, s \cdot v_n \in \boldsymbol{S}$.

Given two vectors $\boldsymbol{v}_1, \boldsymbol{v}_2 \in \boldsymbol{S}^n$, the equality of two vectors is defined as

$$[v_{11}, v_{12}, \ldots, v_{1n}] = [v_{21}, v_{22}, \ldots, v_{2n}] \quad \Leftrightarrow \quad \{v_{1i} = v_{2i}\}_{i=1}^{n}$$

We say that a non-empty set of vectors $\boldsymbol{V}$ is a vector space over a scalar field $\boldsymbol{S}$, if and only if for any vectors $\boldsymbol{v}_1, \boldsymbol{v}_2, \boldsymbol{v}_3 \in \boldsymbol{V}$ and any scalars $s_1, s_2 \in \boldsymbol{S}$ the following axioms are satisfied.

Closure of a Vector Space

Closure under addition:

$$\boldsymbol{v}_1, \boldsymbol{v}_2 \in \boldsymbol{V} \Rightarrow \boldsymbol{v}_1 + \boldsymbol{v}_2 \in \boldsymbol{V} \tag{A4.1}$$

Closure under multiplication:

$$s_1 \in \boldsymbol{S} \wedge \boldsymbol{v}_1 \in \boldsymbol{V} \Rightarrow s_1 \cdot \boldsymbol{v}_1 \in \boldsymbol{V} \tag{A4.2}$$

Addition of Vectors

Comutativity of addition:

$$\boldsymbol{v}_1 + \boldsymbol{v}_2 = \boldsymbol{v}_2 + \boldsymbol{v}_1 \tag{A4.3}$$

Associativity of addition:

$$\boldsymbol{v}_1 + (\boldsymbol{v}_2 + \boldsymbol{v}_3) = (\boldsymbol{v}_1 + \boldsymbol{v}_2) + \boldsymbol{v}_3 \tag{A4.4}$$

Existence of the additive identity element:

$$\boldsymbol{v}_1 + \vec{\boldsymbol{0}} = \vec{\boldsymbol{0}} + \boldsymbol{v}_1 = \boldsymbol{v}_1 \quad \text{where} \quad \vec{\boldsymbol{0}} \in \boldsymbol{V} \tag{A4.5}$$

Existence of the additive inverse:

$$\boldsymbol{v}_1 + (-\boldsymbol{v}_1) = \vec{\boldsymbol{0}} \quad \text{where} \quad (-\boldsymbol{v}_1), \vec{\boldsymbol{0}} \in \boldsymbol{V} \tag{A4.6}$$

Multiplication of a Vector by a Scalar

Distributivity of scalar multiplication over addition:

$$s_1 \cdot (\boldsymbol{v}_1 + \boldsymbol{v}_2) = s_1 \cdot \boldsymbol{v}_1 + s_1 \cdot \boldsymbol{v}_2 \tag{A4.7}$$

Distributivity of scalar addition over multiplication:

$$(s_1 + s_2) \cdot \boldsymbol{v}_1 = s_1 \cdot \boldsymbol{v}_1 + s_2 \cdot \boldsymbol{v}_1 \tag{A4.8}$$

Associativity of scalar multiplication:

$$s_1 \cdot (s_2 \cdot \boldsymbol{v}_1) = (s_1 \cdot s_2) \cdot \boldsymbol{v}_1 \tag{A4.9}$$

Existence of the multiplicative identity element:

$$1 \cdot \boldsymbol{v}_1 = \boldsymbol{v}_1 \cdot 1 = \boldsymbol{v}_1 \quad \text{where} \quad 1 \in \boldsymbol{S} \tag{A4.10}$$

In the above axioms the zero vector is given by $\vec{\boldsymbol{0}} = [0, 0, \ldots, 0]$, where $\vec{\boldsymbol{0}} \in \boldsymbol{V}$.

A *real vector space* is a vector space V that obeys Axioms (A4.1)–(A4.10) and is defined over the scalar field of real numbers (i.e. $S = \mathbb{R}$). There are many examples of such vector spaces.

Vector Space of Real Numbers
Real numbers are members of the scalar field $S = \mathbb{R}$ and as they satisfy Axioms (A4.1)–(A4.10) they form a vector space. Here $S = \mathbb{R}$, $V = \mathbb{R}^1$, the operation of vector addition reduces to the addition of two real numbers and the operation of scalar by vector multiplication reduces to the multiplication of two real numbers.

Vector Space of 2D Vectors
The vectors of the two-dimensional (2D) Euclidean space $\mathcal{E}^2$ are 2-tuples that are members of the set $S^2 = \mathbb{R}^2$. In this example $S = \mathbb{R}$, $V = \{\boldsymbol{v} \;:\; \boldsymbol{v} = [v_1, v_2] \wedge v_i \in \mathbb{R}\}$, vector addition and scalar by vector multiplication are performed in a component-wise fashion. Thus, it is clear that 2D vectors satisfy Axioms (A4.1)–(A4.10) and form a vector space.

Vector Space of 3D Vectors
The vectors of the three-dimensional Euclidean space $\mathcal{E}^3$ are 3-tuples that are members of the set $S^3 = \mathbb{R}^3$. In this example $S = \mathbb{R}$, $V = \{\boldsymbol{v} \;:\; \boldsymbol{v} = [v_1, v_2, v_3] \wedge v_i \in \mathbb{R}\}$, vector addition and scalar by vector multiplication are performed in a component-wise fashion. Thus, it is clear that 3D vectors satisfy Axioms (A4.1)–(A4.10) and form a vector space.

Vector Space of $m \times n$ Matrices

In this example $S = \mathbb{R}$, $V = \left\{ \boldsymbol{A} \;:\; \boldsymbol{A} = \begin{bmatrix} a_{1,1} & \cdots & a_{1,n} \\ \vdots & \ddots & \vdots \\ a_{m,1} & \cdots & a_{m,n} \end{bmatrix} \wedge a_{i,j} \in \mathbb{R} \right\}$,

matrix addition and scalar by matrix multiplication are performed in a component-wise fashion. Thus, it is clear that set of matrices of order $m \times n$ satisfies Axioms (A4.1)–(A4.10) and forms a vector space.

In the discussion that follows we will concentrate on real vector spaces.

4.3 Linear Combinations of Vectors

Let $\boldsymbol{V}_m = \{\boldsymbol{v}_1, \boldsymbol{v}_2, \ldots, \boldsymbol{v}_m\}$ be any set of vectors in vector space $\boldsymbol{V}$ and let $\{s_1, s_2, \ldots, s_m\}$ be any set of scalars (i.e. $\{s_i\}_{i=1}^m \in \mathbb{R}$). Then any vector $\boldsymbol{v}$ of the form

$$\boldsymbol{v} = s_1 \cdot \boldsymbol{v}_1 + s_2 \cdot \boldsymbol{v}_2 + \cdots + s_m \cdot \boldsymbol{v}_m \tag{4.1}$$

is said to be a *linear combination* of the vectors in $\boldsymbol{V}_m$ and vector $\boldsymbol{v}$ is said to be *linearly dependant* on the vectors in $\boldsymbol{V}_m$. Vector $\boldsymbol{v}$ is also a member of the vector space $\boldsymbol{V}$, as the expression $s_1 \cdot \boldsymbol{v}_1 + s_2 \cdot \boldsymbol{v}_2 + \cdots + s_m \cdot \boldsymbol{v}_m$ obeys Axioms (A4.1)–(A4.10).

4.4 Linear Dependence and Linear Independence of Vectors

Given a set of vectors $\boldsymbol{V}_m = \{\boldsymbol{v}_1, \ \boldsymbol{v}_2, \ldots, \ \boldsymbol{v}_m\}$ from the vector space $\boldsymbol{V}$, we say that the vectors in $\boldsymbol{V}_m$ are *linearly independent* or that $\boldsymbol{V}_m$ is a *free set of vectors* in $\boldsymbol{V}$ if

$$s_1 \cdot \boldsymbol{v}_1 + s_2 \cdot \boldsymbol{v}_2 + \cdots + s_m \cdot \boldsymbol{v}_m = \vec{\boldsymbol{0}} \Leftrightarrow s_1 = s_2 = \cdots = s_m = 0 \tag{4.2}$$

i.e. if the equation $s_1 \cdot \boldsymbol{v}_1 + s_2 \cdot \boldsymbol{v}_2 + \cdots + s_m \cdot \boldsymbol{v}_m = \vec{\boldsymbol{0}}$ only has a solution when $s_1 = s_2 = \cdots = s_m = 0$. We say that the vectors in $\boldsymbol{V}_m$ are *linearly dependent* if the equation $s_1 \cdot \boldsymbol{v}_1 + s_2 \cdot \boldsymbol{v}_2 + \cdots + s_m \cdot \boldsymbol{v}_m = \vec{\boldsymbol{0}}$ has a solution when at least one of the scalars $s_1, s_2, \ldots, s_m$ is non-zero. Thus, a set $\boldsymbol{V}_m$ from a vector space $\boldsymbol{V}$ is said to be linearly dependent if it contains at least one vector $\boldsymbol{v}_i$ that can be written as a linear combination of the remaining $(m - 1)$ vectors in the set $\boldsymbol{V}_m$, i.e.

$$\begin{aligned} -s_i \cdot \boldsymbol{v}_i &= s_1 \cdot \boldsymbol{v}_1 + s_2 \cdot \boldsymbol{v}_2 + \cdots + s_{i-1} \cdot \boldsymbol{v}_{i-1} \\ &\quad + s_{i+1} \cdot \boldsymbol{v}_{i+1} + \cdots + s_m \cdot \boldsymbol{v}_m \end{aligned} \tag{4.3}$$

$$\begin{aligned} \therefore \quad \boldsymbol{v}_i &= \frac{s_1}{-s_i} \cdot \boldsymbol{v}_1 + \frac{s_2}{-s_i} \cdot \boldsymbol{v}_2 + \cdots + \frac{s_{i-1}}{-s_i} \cdot \boldsymbol{v}_{i-1} \\ &\quad + \frac{s_{i+1}}{-s_i} \cdot \boldsymbol{v}_{i+1} + \cdots + \frac{s_m}{-s_i} \cdot \boldsymbol{v}_m \end{aligned} \tag{4.4}$$

4.5 Spans and Bases of a Vector Space

The set that contains all possible linear combinations of the vectors in $\boldsymbol{V}_m$ is called the *span* of $\boldsymbol{V}_m$ and it is denoted by span $(\boldsymbol{V}_m)$ or $\langle \boldsymbol{v}_1, \ \boldsymbol{v}_2, \ldots, \ \boldsymbol{v}_m \rangle$, thus

$$\boldsymbol{V}_S = span\ (\boldsymbol{V}_m) = \left\{ \sum_{i=1}^{m} s_i \cdot \boldsymbol{v}_i | \forall s_i \in \mathbb{R} \right\} \tag{4.5}$$

$\boldsymbol{V}_S$ is always a subset of the vector space $\boldsymbol{V}$. We say that $\boldsymbol{V}_S$ is *spanned* or *generated* by the m vectors in $\boldsymbol{V}_m$. As every vector $\boldsymbol{v}$ in $\boldsymbol{V}_S$ can be expressed as a linear combination of the vectors in $\boldsymbol{V}_m$, we say that $\boldsymbol{V}_m$ is the *spanning set* of $\boldsymbol{V}_S$. The elements of set $\boldsymbol{V}_m$ are called the *generators* of set $\boldsymbol{V}_S$.

Let $\boldsymbol{V}_B = \{\boldsymbol{v}_1, \boldsymbol{v}_2, \ldots, \boldsymbol{v}_m\}$ be a set of vectors from the vector space $\boldsymbol{V}$ and let $\boldsymbol{V}_S$ be the span of $\boldsymbol{V}_B$. If the set of vectors $\boldsymbol{V}_B$ is linearly independent and if $\boldsymbol{V}_B$ is the spanning set of $\boldsymbol{V}$ (i.e. $\boldsymbol{V}_S = \boldsymbol{V}$), then we say that the vectors in $\boldsymbol{V}_B$, taken in the order $\boldsymbol{v}_1, \boldsymbol{v}_2, \ldots, \boldsymbol{v}_m$, form a basis for $\boldsymbol{V}$ and that $\boldsymbol{V}$ has a *dimension* m (i.e. dim $(\boldsymbol{V}) = m$).

Since the vectors in V_B span V, any vector $v \in V$ can be written uniquely as

$$v = \alpha_1 \cdot v_1 + \alpha_2 \cdot v_2 + \cdots + \alpha_m \cdot v_m \tag{4.6}$$

To establish the uniqueness of this representation, let us assume that v can also be written as

$$v = \beta_1 \cdot v_1 + \beta_2 \cdot v_2 + \cdots + \beta_m \cdot v_m \tag{4.7}$$

Then

$$\begin{aligned}
&\alpha_1 \cdot v_1 + \alpha_2 \cdot v_2 + \cdots + \alpha_m \cdot v_m = \beta_1 \cdot v_1 + \beta_2 \cdot v_2 + \cdots + \beta_m \cdot v_m \\
\therefore \quad &\alpha_1 \cdot v_1 + \alpha_2 \cdot v_2 + \cdots + \alpha_m \cdot v_m - \beta_1 \cdot v_1 - \beta_2 \cdot v_2 - \cdots - \beta_m \cdot v_m = \vec{0} \\
\therefore \quad &(\alpha_1 - \beta_1) \cdot v_1 + (\alpha_2 - \beta_2) \cdot v_2 + \cdots + (\alpha_m - \beta_m) \cdot v_m = \vec{0}
\end{aligned} \tag{4.8}$$

But since $v_1, v_2, \ldots, v_m$ are linearly independent, the only possible solution is

$$\alpha_1 = \beta_1, \; \alpha_2 = \beta_2, \; \ldots, \; \alpha_m = \beta_m \tag{4.9}$$

Thus the two expressions on the right hand side of Eqs. (4.6) and (4.7) are the same and there is a unique way of writing v.

As shown above, given a basis for a vector space, any vector in the vector space can be expressed as a linear combination of the elements of this basis. All the bases of a vector space have the same number of elements, which is called the dimension of the vector space.

In 3D Euclidean space the vectors $v_1 = [1, 0, 0]$, $v_2 = [0, 1, 0]$, $v_3 = [0, 0, 1]$ form a basis of the vector space of 3D vectors. Similarly, $u_1 = [1, 0, 0]$, $u_2 = [1, 1, 0]$, $u_3 = [1, 1, 1]$ form an alternative valid basis. While the basis of a vector space may change, the number of vectors in a basis of the vector space does not as it is an intrinsic property of this vector space.

4.6 Transformations Between Bases

Since any valid basis of a particular vector space contains the same number of vectors (equal to the dimension of the vector space) there exists a transformation that allows us to convert from one basis to another. We call such a transformation a *change of basis transformation*.

We shall start by considering a 3D vector space, but our results can be extended to m-dimensional vector spaces.

Given a 3D vector space V, let s_1, s_2, s_3 and d_1, d_2, d_3 be two valid bases in V. Assume that s_1, s_2, s_3 is the source basis and that d_1, d_2, d_3 is the destination basis. From the definition of the basis, we know that we can rewrite vectors d_1, d_2, d_3 as linear combinations of the source basis s_1, s_2, s_3, i.e. we can express the components of d_1, d_2, d_3 with respect to the basis s_1, s_2, s_3.

Thus

$$\begin{aligned}
d_1 &= d_{1,1} \cdot s_1 + d_{1,2} \cdot s_2 + d_{1,3} \cdot s_3 \\
d_2 &= d_{2,1} \cdot s_1 + d_{2,2} \cdot s_2 + d_{2,3} \cdot s_3 \\
d_3 &= d_{3,1} \cdot s_1 + d_{3,2} \cdot s_2 + d_{3,3} \cdot s_3
\end{aligned} \tag{4.10}$$

where $d_{i,j}$ represents the jth component of the ith destination base vector with respect to the source basis $\boldsymbol{s}_1, \boldsymbol{s}_2, \boldsymbol{s}_3$.

Any vector $\boldsymbol{v}$ in $\boldsymbol{V}$ can be expressed either with respect to the source basis:

$$\boldsymbol{v} = v_1 \cdot \boldsymbol{s}_1 + v_2 \cdot \boldsymbol{s}_2 + v_3 \cdot \boldsymbol{s}_3 \tag{4.11}$$

or with respect to the destination basis:

$$\boldsymbol{v} = v'_1 \cdot \boldsymbol{d}_1 + v'_2 \cdot \boldsymbol{d}_2 + v'_3 \cdot \boldsymbol{d}_3 \tag{4.12}$$

Substituting Eq. (4.10) into (4.12) we get:

$$\begin{aligned}
\boldsymbol{v} &= v'_1 \cdot \left(d_{1,1} \cdot \boldsymbol{s}_1 + d_{1,2} \cdot \boldsymbol{s}_2 + d_{1,3} \cdot \boldsymbol{s}_3\right) \\
&\quad + v'_2 \cdot \left(d_{2,1} \cdot \boldsymbol{s}_1 + d_{2,2} \cdot \boldsymbol{s}_2 + d_{2,3} \cdot \boldsymbol{s}_3\right) \\
&\quad + v'_3 \cdot \left(d_{3,1} \cdot \boldsymbol{s}_1 + d_{3,2} \cdot \boldsymbol{s}_2 + d_{3,3} \cdot \boldsymbol{s}_3\right) \\
\therefore \quad \boldsymbol{v} &= v'_1 \cdot d_{1,1} \cdot \boldsymbol{s}_1 + v'_1 \cdot d_{1,2} \cdot \boldsymbol{s}_2 + v'_1 \cdot d_{1,3} \cdot \boldsymbol{s}_3 \\
&\quad + v'_2 \cdot d_{2,1} \cdot \boldsymbol{s}_1 + v'_2 \cdot d_{2,2} \cdot \boldsymbol{s}_2 + v'_2 \cdot d_{2,3} \cdot \boldsymbol{s}_3 \\
&\quad + v'_3 \cdot d_{3,1} \cdot \boldsymbol{s}_1 + v'_3 \cdot d_{3,2} \cdot \boldsymbol{s}_2 + v'_3 \cdot d_{3,3} \cdot \boldsymbol{s}_3 \\
\therefore \quad \boldsymbol{v} &= \left(v'_1 \cdot d_{1,1} + v'_2 \cdot d_{2,1} + v'_3 \cdot d_{3,1}\right) \cdot \boldsymbol{s}_1 \\
&\quad + \left(v'_1 \cdot d_{1,2} + v'_2 \cdot d_{2,2} + v'_3 \cdot d_{3,2}\right) \cdot \boldsymbol{s}_2 \\
&\quad + \left(v'_1 \cdot d_{1,3} + v'_2 \cdot d_{2,3} + v'_3 \cdot d_{3,3}\right) \cdot \boldsymbol{s}_3
\end{aligned} \tag{4.13}$$

Equating the corresponding coefficients in Eqs. (4.11) and (4.13) we obtain

$$\begin{aligned}
v_1 &= v'_1 \cdot d_{1,1} + v'_2 \cdot d_{2,1} + v'_3 \cdot d_{3,1} \\
v_2 &= v'_1 \cdot d_{1,2} + v'_2 \cdot d_{2,2} + v'_3 \cdot d_{3,2} \\
v_3 &= v'_1 \cdot d_{1,3} + v'_2 \cdot d_{2,3} + v'_3 \cdot d_{3,3}
\end{aligned} \tag{4.14}$$

Rewriting Eq. (4.14) in matrix form we get

$$\begin{bmatrix} v_1 \\ v_2 \\ v_3 \end{bmatrix} = \begin{bmatrix} d_{1,1} & d_{2,1} & d_{3,1} \\ d_{1,2} & d_{2,2} & d_{3,2} \\ d_{1,3} & d_{2,3} & d_{3,3} \end{bmatrix} \cdot \begin{bmatrix} v'_1 \\ v'_2 \\ v'_3 \end{bmatrix} \tag{4.15}$$

$$\text{or} \quad \boldsymbol{v} = \boldsymbol{D} \cdot \boldsymbol{v}' \tag{4.16}$$

where matrix $\boldsymbol{D}$ is given by

$$\boldsymbol{D} = \begin{bmatrix} d_{1,1} & d_{2,1} & d_{3,1} \\ d_{1,2} & d_{2,2} & d_{3,2} \\ d_{1,3} & d_{2,3} & d_{3,3} \end{bmatrix} \tag{4.17}$$

Here $\boldsymbol{v}$ is the vector expressed in the source basis, $\boldsymbol{v}'$ is the vector expressed in the destination basis and $\boldsymbol{D}$ is a *change of basis transformation matrix*. Matrix $\boldsymbol{D}$ transforms vectors whose components are expressed relative to the destination basis into vectors whose components are expressed relative to the source basis, i.e. it changes the basis from destination to source.

From Eq. (4.17) we can see that the elements of the columns of matrix $\boldsymbol{D}$ are the components of the destination base vectors $\boldsymbol{d}_1$, $\boldsymbol{d}_2$, $\boldsymbol{d}_3$ with respect to the source basis.

From the above it is apparent that changing from a source basis to a destination basis involves pre-multiplying the original vector by the change of basis transformation matrix.

If we pre-multiply both sides of Eq. (4.16) by the inverse of matrix $\boldsymbol{D}$ we obtain

$$\boldsymbol{D}^{-1} \cdot \boldsymbol{v} = \boldsymbol{D}^{-1} \cdot \boldsymbol{D} \cdot \boldsymbol{v}'$$

$$\therefore \quad \boldsymbol{v}' = \boldsymbol{D}^{-1} \cdot \boldsymbol{v} \tag{4.18}$$

Or if we label $\boldsymbol{D}^{-1}$ as $\boldsymbol{S}$ we have

$$\boldsymbol{v}' = \boldsymbol{S} \cdot \boldsymbol{v} \tag{4.19}$$

where $\boldsymbol{S}$ is the change of basis matrix that changes the basis from source to destination.

The multiplication by $\boldsymbol{D}^{-1}$ in Eq. (4.18) is only possible if $\boldsymbol{D}$ is a regular matrix, i.e. invertible. But since by definition vector $\boldsymbol{v}$ can be expressed in both the source and destination bases, the change of basis transformation matrix *must* be invertible.

For m-dimensional vector spaces Eq. (4.15) can be generalised to

$$\begin{bmatrix} v_1 \\ \vdots \\ v_m \end{bmatrix} = \begin{bmatrix} d_{1,1} & \cdots & d_{m,1} \\ \vdots & \ddots & \vdots \\ d_{1,m} & \cdots & d_{m,m} \end{bmatrix} \cdot \begin{bmatrix} v'_1 \\ \vdots \\ v'_m \end{bmatrix} \tag{4.20}$$

where the elements of the columns of the matrix are the components of the destination base vectors $\boldsymbol{d}_1, \ldots, \boldsymbol{d}_m$ with respect to the source basis $\boldsymbol{s}_1, \ldots, \boldsymbol{s}_m$.

Successive transformations into a second or third basis are given by

$$\boldsymbol{v}' = \boldsymbol{S}_1 \cdot \boldsymbol{v}$$
$$\boldsymbol{v}'' = \boldsymbol{S}_2 \cdot \boldsymbol{v}'$$
$$\boldsymbol{v}''' = \boldsymbol{S}_3 \cdot \boldsymbol{v}''$$

Thus

$$\boldsymbol{v}''' = \boldsymbol{S}_3 \cdot (\boldsymbol{S}_2 \cdot (\boldsymbol{S}_1 \cdot \boldsymbol{v}))$$

$$\therefore \quad \boldsymbol{v}''' = (\boldsymbol{S}_3 \cdot \boldsymbol{S}_2 \cdot \boldsymbol{S}_1) \cdot \boldsymbol{v} \tag{4.21}$$

which follows the normal matrix multiplication rules.

4.7 Transformations Between Orthonormal Bases

A noteworthy case occurs when both the source and destination bases are orthonormal. In this case, using the definition of the dot product from matrix algebra, we have

$$\boldsymbol{s}_i^{\mathrm{T}} \odot \boldsymbol{s}_j = \delta_{i,j} \quad \wedge \quad \boldsymbol{d}_i^{\mathrm{T}} \odot \boldsymbol{d}_j = \delta_{i,j} \quad \text{for} \quad i, j \in \{1, 2, 3\} \tag{4.22}$$

where $\delta_{i,j}$ is Kronecker's delta, which was defined in Eqs. (2.24) and (3.6).

If we expand the dot product $\boldsymbol{d}_i^{\mathrm{T}} \odot \boldsymbol{d}_j$ using Eq. (4.10) we obtain

$$\begin{aligned}
\boldsymbol{d}_i^{\mathrm{T}} \odot \boldsymbol{d}_j &= \left(\sum_{k=1}^{3} d_{i,k} \cdot \boldsymbol{s}_k^{\mathrm{T}}\right) \cdot \left(\sum_{l=1}^{3} d_{j,l} \cdot \boldsymbol{s}_l\right) = \sum_{k=1}^{3}\sum_{l=1}^{3} d_{i,k} d_{j,l} \left(\boldsymbol{s}_k^{\mathrm{T}} \odot \boldsymbol{s}_l\right) \\
&= \sum_{k=1}^{3}\sum_{l=1}^{3} d_{i,k} d_{j,l} \delta_{k,l} \\
&= d_{i,1}d_{j,1}\delta_{1,1} + d_{i,1}d_{j,2}\delta_{1,2} + d_{i,1}d_{j,3}\delta_{1,3} \\
&\quad + d_{i,2}d_{j,1}\delta_{2,1} + d_{i,2}d_{j,2}\delta_{2,2} + d_{i,2}d_{j,3}\delta_{2,3} \\
&\quad + d_{i,3}d_{j,1}\delta_{3,1} + d_{i,3}d_{j,2}\delta_{3,2} + d_{i,3}d_{j,3}\delta_{3,3} \\
&= d_{i,1}d_{j,1}1 + d_{i,1}d_{j,2}0 + d_{i,1}d_{j,3}0 \\
&\quad + d_{i,2}d_{j,1}0 + d_{i,2}d_{j,2}1 + d_{i,2}d_{j,3}0 \\
&\quad + d_{i,3}d_{j,1}0 + d_{i,3}d_{j,2}0 + d_{i,3}d_{j,3}1 \\
&= d_{i,1}d_{j,1} + d_{i,2}d_{j,2} + d_{i,3}d_{j,3}
\end{aligned}$$

$$\therefore \quad \boldsymbol{d}_i^{\mathrm{T}} \odot \boldsymbol{d}_j = \sum_{m=1}^{3} d_{i,m} d_{j,m} \tag{4.23}$$

But since $\boldsymbol{d}_i^{\mathrm{T}} \odot \boldsymbol{d}_j = \delta_{i,j}$ so

$$\sum_{m=1}^{3} d_{i,m} d_{j,m} = \delta_{i,j} \tag{4.24}$$

which expressed in matrix form yields

$$\boldsymbol{D}^{\mathrm{T}} \cdot \boldsymbol{D} = \boldsymbol{I} \tag{4.25}$$

Since

$$\begin{bmatrix} d_{1,1} & d_{1,2} & d_{1,3} \\ d_{2,1} & d_{2,2} & d_{2,3} \\ d_{3,1} & d_{3,2} & d_{3,3} \end{bmatrix} \cdot \begin{bmatrix} d_{1,1} & d_{2,1} & d_{3,1} \\ d_{1,2} & d_{2,2} & d_{3,2} \\ d_{1,3} & d_{2,3} & d_{3,3} \end{bmatrix}$$

$$= \begin{bmatrix} d_{1,1}d_{1,1} + d_{1,2}d_{1,2} + d_{1,3}d_{1,3} & d_{1,1}d_{2,1} + d_{1,2}d_{2,2} + d_{1,3}d_{2,3} & d_{1,1}d_{3,1} + d_{1,2}d_{3,2} + d_{1,3}d_{3,3} \\ d_{2,1}d_{1,1} + d_{2,2}d_{1,2} + d_{2,3}d_{1,3} & d_{2,1}d_{2,1} + d_{2,2}d_{2,2} + d_{2,3}d_{2,3} & d_{2,1}d_{3,1} + d_{2,2}d_{3,2} + d_{2,3}d_{3,3} \\ d_{3,1}d_{1,1} + d_{3,2}d_{1,2} + d_{3,3}d_{1,3} & d_{3,1}d_{2,1} + d_{3,2}d_{2,2} + d_{3,3}d_{2,3} & d_{3,1}d_{3,1} + d_{3,2}d_{3,2} + d_{3,3}d_{3,3} \end{bmatrix}$$

$$= \begin{bmatrix} 1 & 0 & 0 \\ 0 & 1 & 0 \\ 0 & 0 & 1 \end{bmatrix}$$

so $\boldsymbol{D}^{\mathrm{T}}$ must be equal to $\boldsymbol{D}^{-1}$ and $\boldsymbol{D}$ must be an *orthogonal matrix*. Which means that

$$\boldsymbol{S} = \boldsymbol{D}^{\mathrm{T}} \tag{4.26}$$

Thus the elements of the rows of matrix $\boldsymbol{S}$ are the components of the destination base vectors $\boldsymbol{d}_1, \boldsymbol{d}_2, \boldsymbol{d}_3$ with respect to the source basis. Thus

$$\boldsymbol{S} = \begin{bmatrix} d_{1,1} & d_{1,2} & d_{1,3} \\ d_{2,1} & d_{2,2} & d_{2,3} \\ d_{3,1} & d_{3,2} & d_{3,3} \end{bmatrix} \tag{4.27}$$

For m-dimensional vector spaces Eq. (4.19) can now be generalised to

$$\begin{bmatrix} v'_1 \\ \vdots \\ v'_m \end{bmatrix} = \begin{bmatrix} d_{1,1} & \cdots & d_{1,m} \\ \vdots & \ddots & \vdots \\ d_{m,1} & \cdots & d_{m,m} \end{bmatrix} \cdot \begin{bmatrix} v_1 \\ \vdots \\ v_m \end{bmatrix} \tag{4.28}$$

where the elements of the rows of the matrix are the components of the destination base vectors $\boldsymbol{d}_1, \ldots, \boldsymbol{d}_m$ with respect to the source basis $\boldsymbol{s}_1, \ldots, \boldsymbol{s}_m$.

4.8 Alternative Notation for Change of Basis Transformations

An alternative notation, which is found in some books, can be arrived at using the matrix multiplication transposition rules (R3.13) and (R3.14). In this notation vectors are assumed to be row vectors rather than column vectors, matrices are replaced by their transpose and changing from a source basis to a destination basis involves post-multiplying the source basis representation of the vector by the change of basis transformation matrix. Thus Eq. (4.15) yields

$$[v_1, v_2, v_3] = [v'_1, v'_2, v'_3] \cdot \begin{bmatrix} d_{1,1} & d_{1,2} & d_{1,3} \\ d_{2,1} & d_{2,2} & d_{2,3} \\ d_{3,1} & d_{3,2} & d_{3,3} \end{bmatrix} \tag{4.29}$$

Equation (4.16) yields

$$\boldsymbol{v} = \boldsymbol{v}' \cdot \boldsymbol{D} \tag{4.30}$$

where matrix $\boldsymbol{D}$ is given by

$$\boldsymbol{D} = \begin{bmatrix} d_{1,1} & d_{1,2} & d_{1,3} \\ d_{2,1} & d_{2,2} & d_{2,3} \\ d_{3,1} & d_{3,2} & d_{3,3} \end{bmatrix} \tag{4.31}$$

In this notation, the elements of the rows of matrix $\boldsymbol{D}$ are the components of the destination base vectors $\boldsymbol{d}_1, \boldsymbol{d}_2, \boldsymbol{d}_3$ with respect to the source basis $\boldsymbol{s}_1, \boldsymbol{s}_2, \boldsymbol{s}_3$.

Equation (4.19) yields

$$\boldsymbol{v}' = \boldsymbol{v} \cdot \boldsymbol{S} \tag{4.32}$$

where matrix $\boldsymbol{S}$ is given by

$$\boldsymbol{S} = \begin{bmatrix} d_{1,1} & d_{2,1} & d_{3,1} \\ d_{1,2} & d_{2,2} & d_{3,2} \\ d_{1,3} & d_{2,3} & d_{3,3} \end{bmatrix} \tag{4.33}$$

In this notation, the elements of the columns of matrix $\boldsymbol{S}$ are the components of the destination base vectors $\boldsymbol{d}_1$, $\boldsymbol{d}_2$, $\boldsymbol{d}_3$ with respect to the source basis $\boldsymbol{s}_1$, $\boldsymbol{s}_2$, $\boldsymbol{s}_3$. Here the two bases are assumed to be orthonormal.

Equation (4.21) yields

$$\boldsymbol{v}''' = \boldsymbol{v} \cdot (\boldsymbol{S}_1 \cdot \boldsymbol{S}_2 \cdot \boldsymbol{S}_3) \tag{4.34}$$

Finally, for m-dimensional vector spaces Eqs. (4.30) and (4.32) can be generalised to

$$[v_1, \ldots, v_m] = \left[v'_1, \ldots, v'_m\right] \cdot \begin{bmatrix} d_{1,1} & \cdots & d_{1,m} \\ \vdots & \ddots & \vdots \\ d_{m,1} & \cdots & d_{m,m} \end{bmatrix} \tag{4.35}$$

$$\left[v'_1, \ldots, v'_m\right] = [v_1, \ldots, v_m] \cdot \begin{bmatrix} d_{1,1} & \cdots & d_{1,m} \\ \vdots & \ddots & \vdots \\ d_{m,1} & \cdots & d_{m,m} \end{bmatrix}^{\mathrm{T}} \tag{4.36}$$

where the elements of the rows (or after transposition, the elements of the columns) of the matrix are the components of the destination base vectors $\boldsymbol{d}_1, \ldots, \boldsymbol{d}_m$ with respect to the source basis $\boldsymbol{s}_1, \ldots, \boldsymbol{s}_m$. Here the two bases are assumed to be orthonormal.

where matrix A is given by

5

Two-Dimensional Transformations

5.1 Definition of a 2D Transformation

A 2D transformation is a function or a mapping, which when applied to a 2D point $\boldsymbol{P} = \langle x, y \rangle$ will transform it (map it) onto another 2D point $P' = \langle x', y' \rangle$. So a 2D transformation transforms a set of original points, $\boldsymbol{O}$, into a set of transformed points, $\boldsymbol{T}$.

In this chapter we will focus our attention on the type of transformation in which any point $\boldsymbol{P}_i = \langle x_i, y_i \rangle$ of set $\boldsymbol{O}$ can be transformed to *one and only one* point $\boldsymbol{P}'_i = \langle x'_i, y'_i \rangle$ of set $\boldsymbol{T}$. Also no two points of set $\boldsymbol{O}$ can be transformed into the same point of set $\boldsymbol{T}$. Such a transformation is said to be a *one to one mapping* (see Fig. 5.1).

So a transformation $\boldsymbol{t}$ can transform a 2D point $\langle x, y \rangle$ into a 2D point $\langle x', y' \rangle$, i.e. $\boldsymbol{t}(\langle x, y \rangle) = \langle x', y' \rangle$ or $\boldsymbol{P}' = \boldsymbol{t}(\boldsymbol{P})$

Note that the elements of the sets $\boldsymbol{O}$ and $\boldsymbol{T}$ have a *one-to-one correspondence*, i.e. by applying the transformation function $\boldsymbol{t}$ to the point $\langle x, y \rangle$ we transform it to the point $\langle x', y' \rangle$ and if we apply the inverse transformation function $\boldsymbol{t}^{-1}$ to the transformed point $\langle x', y' \rangle$ we will transform it back to the original point $\langle x, y \rangle$.

5.2 Concatenation of Transformations

The *concatenation of transformations* is sometimes called the *product of transformations*. Let $\boldsymbol{t}_1$ be a transformation that maps elements of the set $\boldsymbol{O}$ onto elements of the set $\boldsymbol{T}_1$. Let $\boldsymbol{t}_2$ be a transformation that maps elements of the set $\boldsymbol{T}_1$ onto elements of the set $\boldsymbol{T}_2$ (see Fig. 5.2).

So

$$\begin{aligned} \langle x', y' \rangle &= \boldsymbol{t}_1(\langle x, y \rangle) \\ \text{and } \langle x'', y'' \rangle &= \boldsymbol{t}_2(\langle x', y' \rangle) \end{aligned}$$

By substitution we get

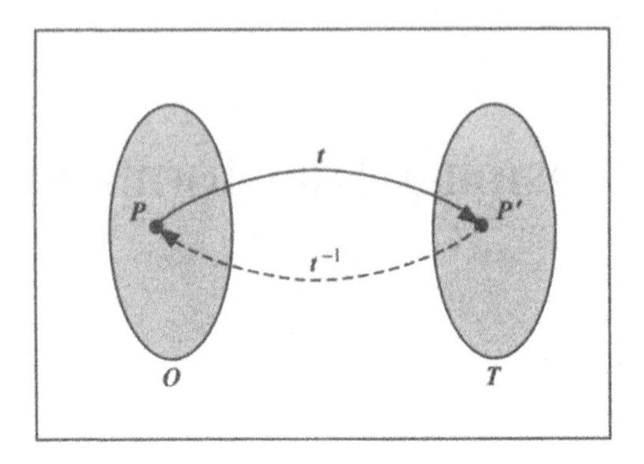

FIGURE 5.1. A transformation mapping.

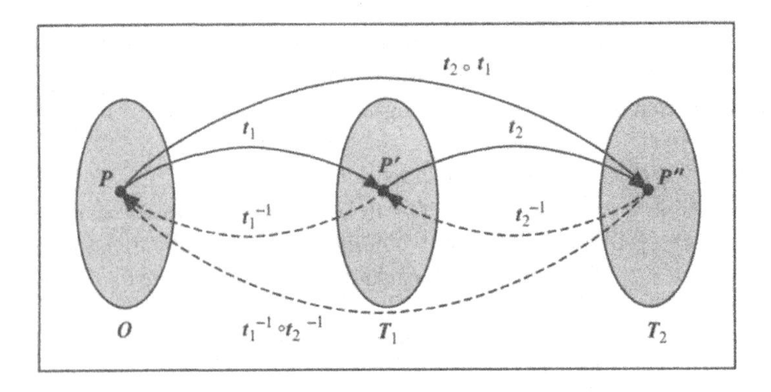

FIGURE 5.2. The concatenation of transformations.

$$\langle x'', y'' \rangle = t_2 (t_1 (\langle x, y \rangle))$$

which can be written as

$$\langle x'', y'' \rangle = t_2 \circ t_1 (\langle x, y \rangle)$$

where $t_2 \circ t_1$ is the *product* or *concatenation* of the transformations t_1 and t_2 in this exact order. So applying the concatenation of a series of transformations is equivalent to applying the individual transformations one after the other. A transformation composed of the concatenation of a series of *primitive* (simple) transformations is called a *composite* transformation.

Observe that if we wish to undo a sequence of transformations t_1 and t_2, applied in this order, we must undo transformation t_2 before we undo transformation t_1, i.e.

$$\langle x, y \rangle = t_1^{-1} \circ t_2^{-1} (\langle x'', y'' \rangle)$$

Thus

$$(t_2 \circ t_1)^{-1} = t_1^{-1} \circ t_2^{-1} \tag{5.1}$$

This result is also true for the concatenation of any series of n transformations:

$$(t_n \circ \cdots \circ t_2 \circ t_1)^{-1} = t_1^{-1} \circ t_2^{-1} \circ \cdots \circ t_n^{-1} \tag{5.2}$$

5.3 2D Graphics Transformations

The principles of 2D transformations, mentioned above, can be directly applied to the transformation of points of the 2D Euclidean space $\mathcal{E}^2$ into new points of this space. Lines defined in $\mathcal{E}^2$ can also be transformed in a similar way. All we have to do is transform the endpoints of the line and then join these together to obtain the transformed line.

Some examples of 2D transformations:

- scaling
- translation
- rotation
- reflection
- shearing

Two-dimensional transformations can be categorised either as *primitive transformations* or as *composite transformations*. Composite transformations can be constructed by *composing* (concatenating together) a sequence of primitive transformations.

5.4 2D Primitive Transformations

5.4.1 Scaling Transformation Relative to the Origin

The scaling transformation relative to the origin of $\mathcal{E}^2$ is defined as

$$\langle x', y' \rangle = t(\langle x, y \rangle) = \langle x \cdot s_x, y \cdot s_y \rangle$$

or in component form as

$$\begin{aligned} x' &= x \cdot s_x \\ y' &= y \cdot s_y \end{aligned} \tag{5.3}$$

where s_x and s_y are the *scale factors* along the x and y axes, respectively. These scale factors can assume any non-zero value, as zero scale factors lead to a *non-invertible* transformation that collapses every transformed point to the origin. If the two scaling factors are equal (i.e. $s_x = s_y$), then this transformation is known as *uniform scaling*, otherwise it is known as *non-uniform scaling*. Uniform scaling is a special case of non-uniform scaling.

$$\begin{aligned} x' &= x \cdot s \\ y' &= y \cdot s \end{aligned} \tag{5.4}$$

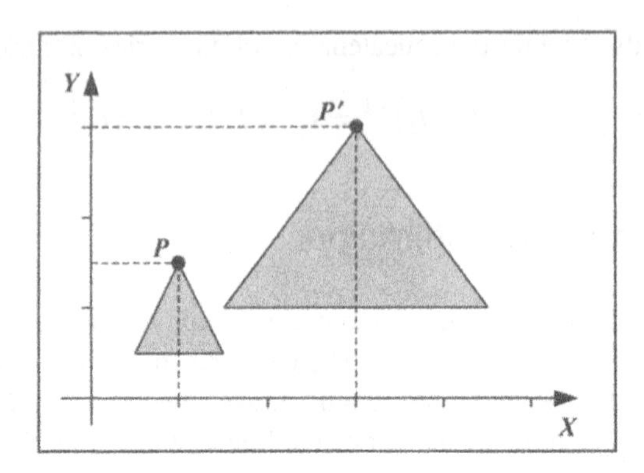

FIGURE 5.3. A scaling transformation relative to the origin.

Uniform scaling can also be expressed in vector form as follows.

$$\boldsymbol{P}' = \boldsymbol{t}(\boldsymbol{P}) = s \cdot \boldsymbol{P} \tag{5.5}$$

Figure 5.3 illustrates a scaling transformation where the scaling factors along the x and y axes are $s_x = 3$ and $s_y = 2$, respectively.

5.4.2 *Translation Transformation*

The translation transformation is defined as

$$\langle x', y' \rangle = \boldsymbol{t}(\langle x, y \rangle) = \langle x + d_x, y + d_y \rangle$$

or in component form as

$$\begin{aligned} x' &= x + d_x \\ y' &= y + d_y \end{aligned} \tag{5.6}$$

where d_x and d_y are the *displacements* along the x and y axes, respectively.

The translation transformation can also be expressed in vector form as follows.

$$\boldsymbol{P}' = \boldsymbol{t}(\boldsymbol{P}) = \boldsymbol{P} + \boldsymbol{d} = \boldsymbol{P} + [d_x, d_y] \tag{5.7}$$

Figure 5.4 illustrates a translation transformation where the displacements along the x and y axes are $d_x = 2$ and $d_y = 1$, respectively.

5.4.3 *Rotation Transformation about the Origin*

The rotation transformation relative to the origin of $\mathcal{E}^2$ is defined as

$$\langle x', y' \rangle = \boldsymbol{t}(\langle x, y \rangle) = \langle x \cdot \cos\theta - y \cdot \sin\theta, x \cdot \sin\theta + y \cdot \cos\theta \rangle$$

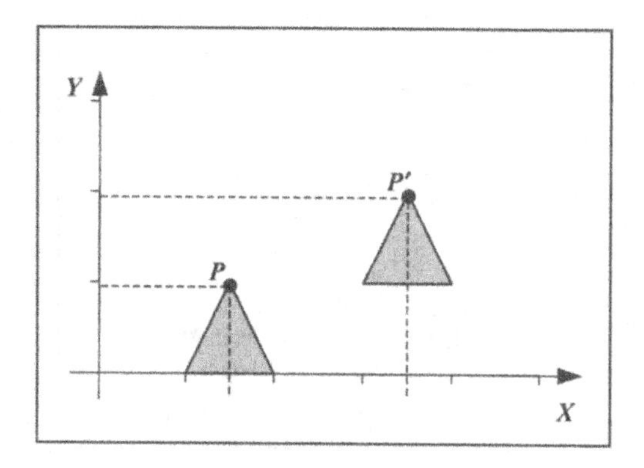

FIGURE 5.4. A translation transformation.

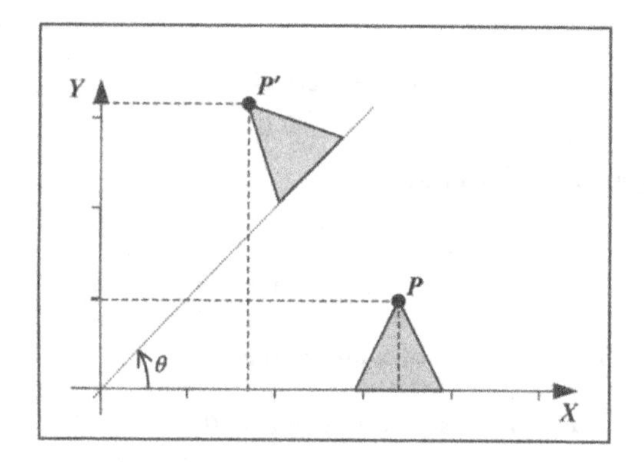

FIGURE 5.5. A rotation transformation about the origin.

or in component form as

$$x' = x \cdot \cos\theta - y \cdot \sin\theta$$
$$y' = x \cdot \sin\theta + y \cdot \cos\theta \tag{5.8}$$

where θ is the angle of rotation about the origin measured in a counter-clockwise positive fashion.

To derive the rotation transformation we reason as follows. Rotating the point $\boldsymbol{P}$ through a counter-clockwise angle θ about the origin of the coordinate system $\boldsymbol{XOY}$ is equivalent to rotating these axes through a clockwise angle θ and leaving the point $\boldsymbol{P}$ stationary. As shown in Fig. 5.6, the system $\boldsymbol{XOY}$ now becomes $\boldsymbol{X'OY'}$.

The polar coordinates of point $\boldsymbol{P}$ in the $\boldsymbol{XOY}$ system are $\langle r, \alpha\rangle$ and in the $\boldsymbol{X'OY'}$ system are $\langle r, \beta\rangle$. The Cartesian coordinates of point $\boldsymbol{P}$ in the two systems are

$$x = r \cdot \cos\alpha$$
$$y = r \cdot \sin\alpha$$

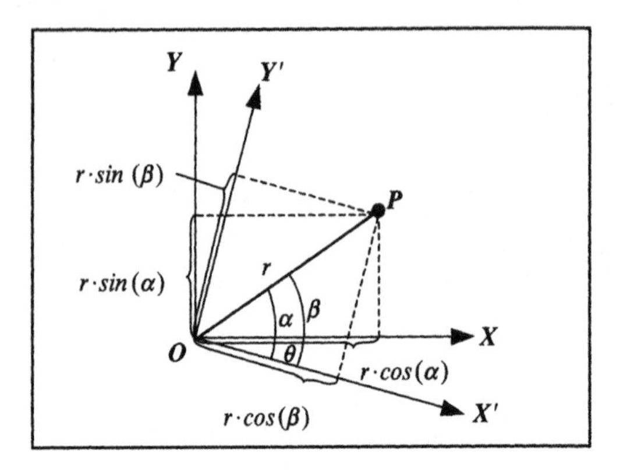

FIGURE 5.6. The coordinate system rotation.

and

$$x' = r \cdot \cos \beta$$
$$y' = r \cdot \sin \beta$$

Given the trigonometric identities:

$$\cos(a + b) = \cos a \cdot \cos b - \sin a \cdot \sin b$$
$$\sin(a + b) = \sin a \cdot \cos b + \cos a \cdot \sin b$$

we can express $x'$and y' in terms of x and y to get

$$\begin{aligned} x' &= r \cdot \cos \beta = r \cdot \cos(\alpha + \theta) \\ &= r \cdot (\cos \alpha \cdot \cos \theta - \sin \alpha \cdot \sin \theta) \\ &= r \cdot \cos \alpha \cdot \cos \theta - r \cdot \sin \alpha \cdot \sin \theta \\ &= x \cdot \cos \theta - y \cdot \sin \theta \end{aligned}$$

and

$$\begin{aligned} y' &= r \cdot \sin \beta = r \cdot \sin(\alpha + \theta) \\ &= r \cdot (\sin \alpha \cdot \cos \theta + \cos \alpha \cdot \sin \theta) \\ &= r \cdot \sin \alpha \cdot \cos \theta + r \cdot \cos \alpha \cdot \sin \theta \\ &= x \cdot \sin \theta + y \cdot \cos \theta \end{aligned}$$

Similarly, the rotation transformation relative to the origin through a clockwise positive angle θ is given by

$$\begin{aligned} x' &= x \cdot \cos \theta + y \cdot \sin \theta \\ y' &= -x \cdot \sin \theta + y \cdot \cos \theta \end{aligned} \tag{5.9}$$

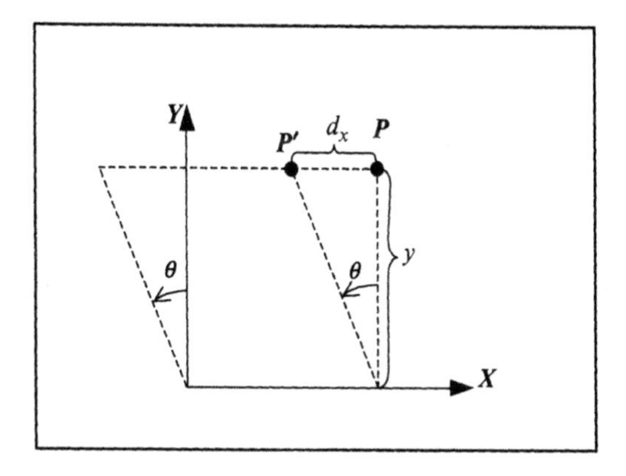

FIGURE 5.7. A shearing transformation along the x-axis.

5.4.4 *Shearing Transformation Along the x-Axis*

This transformation involves shearing a point $\boldsymbol{P}$ along the x-axis thus modifying its x coordinate while leaving its y coordinate unchanged. The shearing transformation is defined in terms of the *angle of shear* θ, which is measured in a counter-clockwise positive fashion. The effect of this transformation is to *shear the y-axis parallel to the x-axis*. This transformation is defined as follows.

From Fig. 5.7 we have

$$\tan\theta = \frac{d_x}{y}$$
$$\therefore \quad d_x = y \cdot \tan\theta$$

Thus

$$\begin{aligned} x' &= x - d_x = x - y \cdot \tan\theta \\ y' &= y \end{aligned} \tag{5.10}$$

5.4.5 *Shearing Transformation Along the y-Axis*

This transformation involves shearing a point $\boldsymbol{P}$ along the y-axis thus modifying its y coordinate while leaving its x coordinate unchanged. The shearing transformation is defined in terms of the *angle of shear* θ, which is measured in a counter-clockwise positive fashion. The effect of this transformation is to *shear the x-axis parallel to the y-axis*. This transformation is defined as follows.

From Fig. 5.8 we have

$$\tan\theta = \frac{d_y}{x}$$
$$\therefore \quad d_y = x \cdot \tan\theta$$

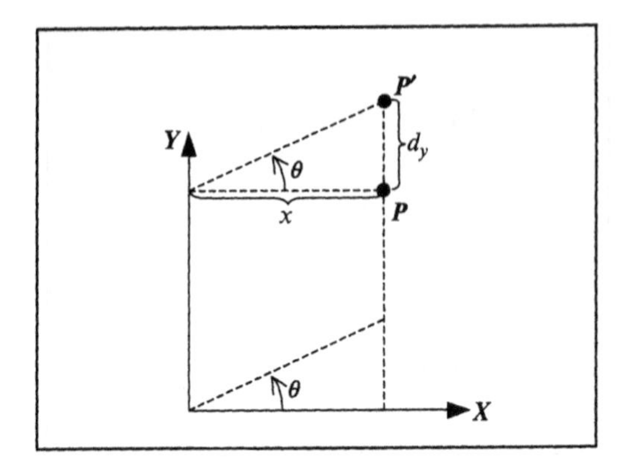

FIGURE 5.8. A shearing transformation along the y-axis.

Thus

$$x' = x$$
$$y' = y + d_y = y + x \cdot \tan\theta \qquad (5.11)$$

5.5 2D Composite Transformations

5.5.1 *Reflection Transformations About One- or Two-Coordinate Axes*

Reflections about one or more of the axes of the coordinate system can be achieved by applying a scaling transformation with either or both scaling factors having a negative value (see Fig. 5.9).

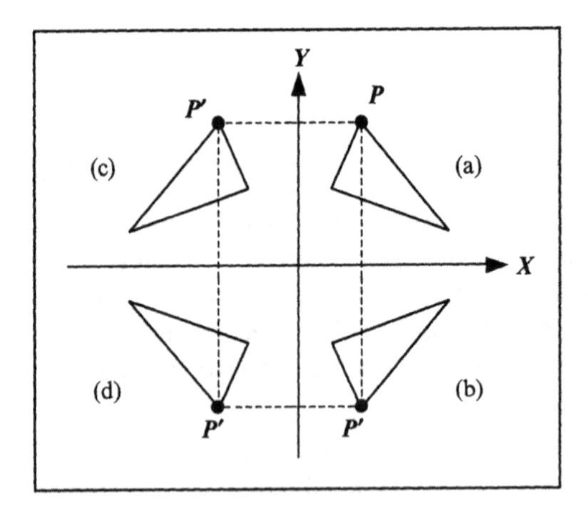

FIGURE 5.9. Three reflection transformations.

In Fig. 5.9, triangle (a) represents the original image, triangle (b) is the result of a reflection about the x-axis, triangle (c) is the result of a reflection about the y-axis and triangle (d) is the result of a reflection about the origin.

The reflection about the x-axis [case (b)] is defined as

$$\begin{aligned} x' &= x \cdot (+1) \\ y' &= y \cdot (-1) \end{aligned} \tag{5.12}$$

The reflection about the y-axis [case (c)] is defined as

$$\begin{aligned} x' &= x \cdot (-1) \\ y' &= y \cdot (+1) \end{aligned} \tag{5.13}$$

The reflection about the origin [case (d)] is defined as

$$\begin{aligned} x' &= x \cdot (-1) \\ y' &= y \cdot (-1) \end{aligned} \tag{5.14}$$

If the scaling factors are other than (+1) or (–1) then scaling as well as reflection takes place.

As we have seen above some non-primitive transformations can be achieved by modifying the parameters of a primitive transformation. In general however, composite transformations can only be achieved by applying a sequence of primitive transformations one after another, i.e. by concatenating this sequence of transformations.

Usually for a transformation problem that is too complex to solve in its current domain we adopt the following procedure. First, through a series of primitive transformation steps, we transform the geometry of the problem into a much simpler domain defined around the origin of the coordinate space. Here the problem can usually be solved using a primitive transformation. Then, having solved the problem in this simpler domain we return the geometry to its original domain, through a series of inverse transformation steps, i.e. by undoing the steps that transformed the geometry from its original domain to the simpler domain.

5.5.2 Scaling Transformation About an Arbitrary Point

Scaling a point $\boldsymbol{P} = \langle x, y \rangle$ by scale factors s_x and s_y about an arbitrary point $\boldsymbol{P}_c = \langle x_c, y_c \rangle$ is achieved through the following steps.

Step 1: Translate the arbitrary point to the origin

$$\begin{aligned} x_1 &= x - x_c \\ y_1 &= y - y_c \end{aligned}$$

Step 2: Scale the point $\boldsymbol{P}_1$ about the origin

$$\begin{aligned} x_2 &= x_1 \cdot s_x \\ y_2 &= y_1 \cdot s_y \end{aligned}$$

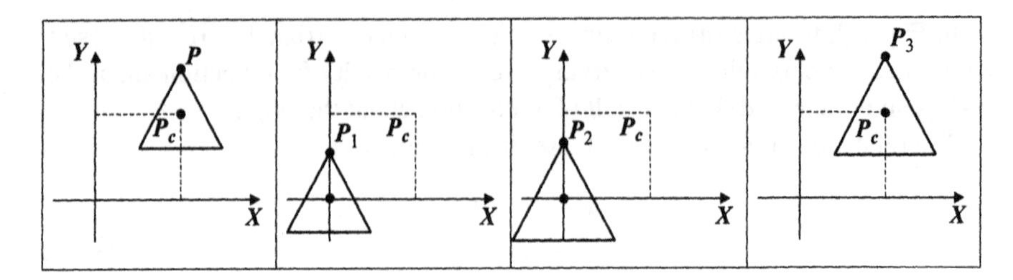

FIGURE 5.10. A scaling transformation about an arbitrary point.

Step 3: Translate the arbitrary point back to its original position

$$x_3 = x_2 + x_c$$
$$y_3 = y_2 + y_c$$

Observe that the third transformation is the inverse of the first transformation, i.e. it undoes what the first transformation did.

These three transformations can be concatenated into a composite transformation:

$$x_3 = x_2 + x_c = x_1 \cdot s_x + x_c$$
$$y_3 = y_2 + y_c = y_1 \cdot s_y + y_c$$

$\therefore$

$$x_3 = (x - x_c) \cdot s_x + x_c$$
$$y_3 = (y - y_c) \cdot s_y + y_c \tag{5.15}$$

5.5.3 *Rotation Transformation About an Arbitrary Point*

Rotating a point $\boldsymbol{P} = \langle x, y \rangle$ through a counter-clockwise angle θ about an arbitrary point $\boldsymbol{P}_c = \langle x_c, y_c \rangle$ is achieved through the following steps.

Step 1: Translate the arbitrary point to the origin

$$x_1 = x - x_c$$
$$y_1 = y - y_c$$

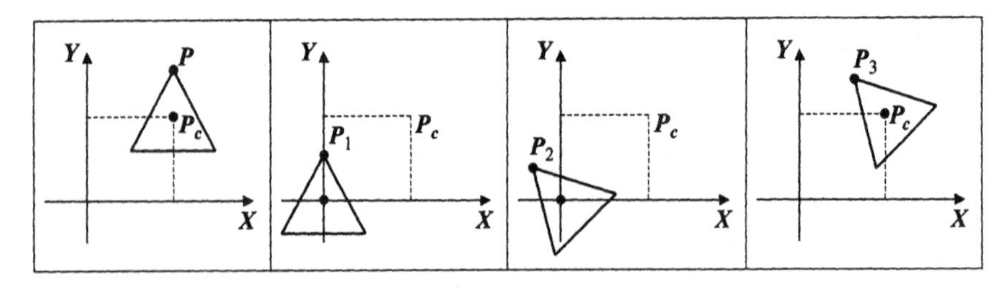

FIGURE 5.11. A rotation transformation about an arbitrary point.

Step 2: Rotate the point $\boldsymbol{P}_1$ about the origin

$$x_2 = x_1 \cdot \cos\theta - y_1 \cdot \sin\theta$$
$$y_2 = x_1 \cdot \sin\theta + y_1 \cdot \cos\theta$$

Step 3: Translate the arbitrary point back to its original position

$$x_3 = x_2 + x_c$$
$$y_3 = y_2 + y_c$$

Observe that the third transformation is the inverse of the first transformation, i.e. it undoes what the first transformation did.

These three transformations can be concatenated into a composite transformation:

$$x_3 = x_2 + x_c = x_1 \cdot \cos\theta - y_1 \cdot \sin\theta + x_c$$
$$y_3 = y_2 + y_c = x_1 \cdot \sin\theta + y_1 \cdot \cos\theta + y_c$$

$\therefore$

$$x_3 = (x - x_c) \cdot \cos\theta - (y - y_c) \cdot \sin\theta + x_c$$
$$y_3 = (x - x_c) \cdot \sin\theta + (y - y_c) \cdot \cos\theta + y_c \qquad (5.16)$$

5.5.4 Reflection Transformation About an Arbitrary Axis

Reflecting a point $\boldsymbol{P} = \langle x, y \rangle$about an arbitrary axis passing through points $\boldsymbol{P}_a = \langle x_a, y_a \rangle$ and $\boldsymbol{P}_b = \langle x_b, y_b \rangle$ is achieved through the following steps, shown in Fig. 5.12.

Step 1: Translate the arbitrary point $\boldsymbol{P}_a$ to the origin

$$x_1 = x - x_a$$
$$y_1 = y - y_a$$

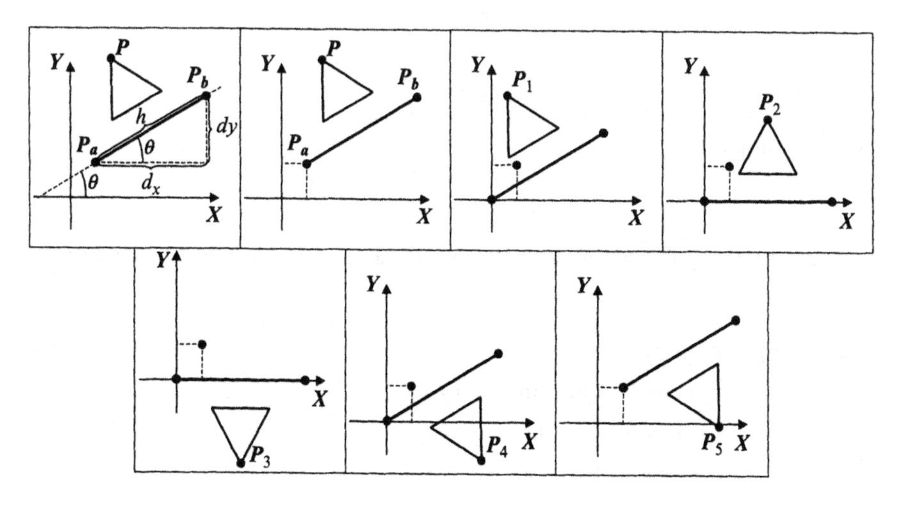

FIGURE 5.12. A reflection transformation about an arbitrary axis.

Step 2: Rotate about the origin through a clockwise angle θ, so that the arbitrary axis coincides with the x-axis

$$x_2 = x_1 \cdot \cos(-\theta) - y_1 \cdot \sin(-\theta) = x_1 \cdot \cos\theta + y_1 \cdot \sin\theta$$
$$y_2 = x_1 \cdot \sin(-\theta) + y_1 \cdot \cos(-\theta) = -x_1 \cdot \sin\theta + y_1 \cdot \cos\theta$$

Step 3: Reflect the point $\boldsymbol{P}_2$ about the x-axis

$$x_3 = x_2 \cdot (+1)$$
$$y_3 = y_2 \cdot (-1)$$

Step 4: Rotate about the origin through a counter-clockwise angle θ, so that the arbitrary axis assumes its original angle with the x-axis

$$x_4 = x_3 \cdot \cos\theta - y_3 \cdot \sin\theta$$
$$y_4 = x_3 \cdot \sin\theta + y_3 \cdot \cos\theta$$

Step 5: Translate the arbitrary point $\boldsymbol{P}_a$ back to its original position

$$x_5 = x_4 + x_a$$
$$y_5 = y_4 + y_a$$

Observe that the fourth step is the inverse of the second step and that the fifth step is the inverse of the first step.

From Fig. 5.12 we see that

$$d_x = x_b - x_a$$
$$d_y = y_b - y_a$$

and $h = \sqrt{d_x^2 + d_y^2}$

Thus

$$\cos\theta = \frac{d_x}{h}$$
$$\sin\theta = \frac{d_y}{h}$$

Also recall that

$$\cos(-\theta) = \cos\theta$$
$$\sin(-\theta) = -\sin\theta$$

These five transformations can be concatenated to the following transformation:

$$x_5 = x_4 + x_a$$
$$\therefore \quad x_5 = x_3 \cdot \cos\theta - y_3 \cdot \sin\theta + x_a$$
$$\therefore \quad x_5 = x_2 \cdot \cos\theta + y_2 \cdot \sin\theta + x_a$$
$$\therefore \quad x_5 = (x_1 \cdot \cos\theta + y_1 \cdot \sin\theta) \cdot \cos\theta$$
$$+ (-x_1 \cdot \sin\theta + y_1 \cdot \cos\theta) \cdot \sin\theta + x_a$$

$$\therefore \quad x_5 = [(x - x_a) \cdot \cos\theta + (y - y_a) \cdot \sin\theta] \cdot \cos\theta$$
$$+ [-(x - x_a) \cdot \sin\theta + (y - y_a) \cdot \cos\theta] \cdot \sin\theta + x_a$$
$$\therefore \quad x_5 = (x - x_a) \cdot \cos^2\theta + (y - y_a) \cdot \sin\theta \cdot \cos\theta - (x - x_a) \cdot \sin^2\theta$$
$$+ (y - y_a) \cdot \cos\theta \cdot \sin\theta + x_a$$
$$\therefore \quad x_5 = \left[(x - x_a) \cdot \cos^2\theta - (x - x_a) \cdot \sin^2\theta\right]$$
$$+ [(y - y_a) \cdot \sin\theta \cdot \cos\theta + (y - y_a) \cdot \cos\theta \cdot \sin\theta] + x_a$$
$$\therefore \quad x_5 = (x - x_a) \cdot \cos(2\theta) + (y - y_a) \cdot \sin(2\theta) + x_a$$

and

$$y_5 = y_4 + y_a$$
$$\therefore \quad y_5 = x_3 \cdot \sin\theta + y_3 \cdot \cos\theta + y_a$$
$$\therefore \quad y_5 = x_2 \cdot \sin\theta - y_2 \cdot \cos\theta + y_a$$
$$\therefore \quad y_5 = (x_1 \cdot \cos\theta + y_1 \cdot \sin\theta) \cdot \sin\theta$$
$$- (-x_1 \cdot \sin\theta + y_1 \cdot \cos\theta) \cdot \cos\theta + y_a$$
$$\therefore \quad y_5 = [(x - x_a) \cdot \cos\theta + (y - y_a) \cdot \sin\theta] \cdot \sin\theta$$
$$- [-(x - x_a) \cdot \sin\theta + (y - y_a) \cdot \cos\theta] \cdot \cos\theta + y_a$$
$$\therefore \quad y_5 = (x - x_a) \cdot \cos\theta \cdot \sin\theta + (y - y_a) \cdot \sin^2\theta$$
$$+ (x - x_a) \cdot \sin\theta \cdot \cos\theta - (y - y_a) \cdot \cos^2\theta + y_a$$
$$\therefore \quad y_5 = [(x - x_a) \cdot \cos\theta \cdot \sin\theta + (x - x_a) \cdot \sin\theta \cdot \cos\theta]$$
$$+ \left[(y - y_a) \cdot \sin^2\theta - (y - y_a) \cdot \cos^2\theta\right] + y_a$$
$$\therefore \quad y_5 = (x - x_a) \cdot \sin(2\theta) - (y - y_a) \cdot \cos(2\theta) + y_a$$

Thus the concatenated transformation is given by

$$\begin{aligned} x_5 &= (x - x_a) \cdot \cos(2\theta) + (y - y_a) \cdot \sin(2\theta) + x_a \\ y_5 &= (x - x_a) \cdot \sin(2\theta) - (y - y_a) \cdot \cos(2\theta) + y_a \end{aligned} \tag{5.17}$$

5.6 Sign of the Angles in Transformations

In our discussion so far all the angles used in transformations were assumed to be measured in a counter-clockwise positive fashion. This was a purely arbitrary choice. Some writers choose to define their angles as being counter-clockwise positive and others as being clockwise positive. This is a matter of taste and different graphics libraries and systems do it differently. The important thing to remember is that, in order to convert from one system to the other we just need to negate the angles in our transformations recalling that

$$\sin(-\theta) = -\sin\theta$$
$$\cos(-\theta) = \cos\theta$$
$$\tan(-\theta) = -\tan\theta$$

5.7 Some Important Observations

It is worth observing the following with regard to the analytical form of the transformations that we have used so far.

1. A sequence of transformations can be concatenated into a single analytical equation.
2. The ordering of the sequence of individual transformations must be preserved in the concatenated transformation.
3. Concatenated transformations have a number of advantages over a sequence of individual transformations:
   i. they can be represented more compactly,
   ii. they require fewer arithmetic operations than if we were to apply each individual transformation in sequence one after the other.
4. Concatenated transformations have one main disadvantage. The rules of concatenating transformations are quite complex.

In order to overcome this disadvantage of the concatenated transformations we can represent transformations as matrices, then their concatenation reduces to a matrix multiplication.

5.8 Matrix Representation of 2D Transformations

Two-dimensional transformations can be represented by a 3×3 matrix and 2D points can be represented in *homogeneous form* by a three-element row or column vector. (See Section 5.15 for a detailed discussion of the homogeneous form.)

There are two distinct and equivalent notations that can be used to represent the transformation of a 2D point. The first notation is used most widely by mathematicians, engineers and computer scientists around the world.

$$\boldsymbol{P}' = \boldsymbol{P} \cdot \boldsymbol{T}$$

where $\boldsymbol{P}$ and $\boldsymbol{P}'$ are the original and the transformed points, expressed in homogeneous form, and $\boldsymbol{T}$ is the square matrix representing the transformation. For a 2D transformation in this notation we would write

$$\left[x', y', 1\right] = [x, y, 1] \cdot \begin{bmatrix} t_{1,1} & t_{1,2} & t_{1,3} \\ t_{2,1} & t_{2,2} & t_{2,3} \\ t_{3,1} & t_{3,2} & t_{3,3} \end{bmatrix}$$

American computer scientists use an alternative notation

$$\boldsymbol{P}'^{\mathrm{T}} = \boldsymbol{T}^{\mathrm{T}} \cdot \boldsymbol{P}^{\mathrm{T}}$$

where $\boldsymbol{P}^{\mathrm{T}}$ and $\boldsymbol{P}'^{\mathrm{T}}$ are the transposed original and transformed points, respectively and $\boldsymbol{T}^{\mathrm{T}}$ is the transpose of the square matrix representing the transformation. For a 2D transformation in this notation we would write

$$\begin{bmatrix} x' \\ y' \\ 1 \end{bmatrix} = \begin{bmatrix} t_{1,1} & t_{2,1} & t_{3,1} \\ t_{1,2} & t_{2,2} & t_{3,2} \\ t_{1,3} & t_{2,3} & t_{3,3} \end{bmatrix} \cdot \begin{bmatrix} x \\ y \\ 1 \end{bmatrix}$$

Observe that as $\boldsymbol{P}' = \boldsymbol{P}'^{\mathrm{T}}$ so is $\boldsymbol{P} \cdot \boldsymbol{T} = \boldsymbol{T}^{\mathrm{T}} \cdot \boldsymbol{P}^{\mathrm{T}}$.

More precisely, the transformation of a 2D point $\langle x, y \rangle$ to a new point $\langle x', y' \rangle$ can then be represented by

$$[x', y', 1] = [x, y, 1] \cdot \begin{bmatrix} a & d & 0 \\ b & e & 0 \\ c & f & 1 \end{bmatrix} \tag{5.18a}$$

or alternatively by

$$\begin{bmatrix} x' \\ y' \\ 1 \end{bmatrix} = \begin{bmatrix} a & b & c \\ d & e & f \\ 0 & 0 & 1 \end{bmatrix} \cdot \begin{bmatrix} x \\ y \\ 1 \end{bmatrix} \tag{5.18b}$$

where the 3×3 matrix completely specifies the transformation. As we have seen in Section 3.11.4, those two representations are equivalent and may be used interchangeably. The third element in the vector $[x, y, 1]$ is added so that the matrix and the vector are *conformant* for matrix multiplication. This product yields a row/column vector that contains the transformed point $\langle x', y' \rangle$. From Eq. (5.18a) we get

$$[x', y', 1] = [ax + by + c,\ dx + ey + f,\ 1] \tag{5.19a}$$

Alternatively, from Eq. (5.18b) we get

$$\begin{bmatrix} x' \\ y' \\ 1 \end{bmatrix} = \begin{bmatrix} ax + by + c \\ dx + ey + f \\ 1 \end{bmatrix} \tag{5.19b}$$

In both the cases this reduces to

$$\begin{aligned} x' &= ax + by + c \\ y' &= dx + ey + f \end{aligned} \tag{5.20}$$

which requires only four multiplication and four addition operations. It would therefore appear that we only need the first two columns/rows of the transformation matrix. However, the third column/row is necessary if we wish to concatenate a sequence transformations by multiplying them together or to invert a transformation represented by a matrix. Recall that matrices have to be conformant in order to be multiplied together and that we can only compute the inverse of a square matrix.

5.9 Matrix Representation of Primitive Transformations

The 2D primitive transformations are represented in matrix notation as follows.

Scaling Relative to the Origin

$$[x', y', 1] = [x, y, 1] \cdot \begin{bmatrix} s_x & 0 & 0 \\ 0 & s_y & 0 \\ 0 & 0 & 1 \end{bmatrix} \tag{5.21a}$$

or alternatively by

$$\begin{bmatrix} x' \\ y' \\ 1 \end{bmatrix} = \begin{bmatrix} s_x & 0 & 0 \\ 0 & s_y & 0 \\ 0 & 0 & 1 \end{bmatrix} \cdot \begin{bmatrix} x \\ y \\ 1 \end{bmatrix} \tag{5.21b}$$

Translation

$$[x', y', 1] = [x, y, 1] \cdot \begin{bmatrix} 1 & 0 & 0 \\ 0 & 1 & 0 \\ d_x & d_y & 1 \end{bmatrix} \tag{5.22a}$$

or alternatively by

$$\begin{bmatrix} x' \\ y' \\ 1 \end{bmatrix} = \begin{bmatrix} 1 & 0 & d_x \\ 0 & 1 & d_y \\ 0 & 0 & 1 \end{bmatrix} \cdot \begin{bmatrix} x \\ y \\ 1 \end{bmatrix} \tag{5.22b}$$

Rotation About the Origin

$$[x', y', 1] = [x, y, 1] \cdot \begin{bmatrix} \cos\theta & \sin\theta & 0 \\ -\sin\theta & \cos\theta & 0 \\ 0 & 0 & 1 \end{bmatrix} \tag{5.23a}$$

or alternatively by

$$\begin{bmatrix} x' \\ y' \\ 1 \end{bmatrix} = \begin{bmatrix} \cos\theta & -\sin\theta & 0 \\ \sin\theta & \cos\theta & 0 \\ 0 & 0 & 1 \end{bmatrix} \cdot \begin{bmatrix} x \\ y \\ 1 \end{bmatrix} \tag{5.23b}$$

Shearing the y-Axis Parallel to the x-Axis

$$[x', y', 1] = [x, y, 1] \cdot \begin{bmatrix} 1 & 0 & 0 \\ -\tan\theta & 1 & 0 \\ 0 & 0 & 1 \end{bmatrix} \tag{5.24a}$$

or alternatively by

$$\begin{bmatrix} x' \\ y' \\ 1 \end{bmatrix} = \begin{bmatrix} 1 & -\tan\theta & 0 \\ 0 & 1 & 0 \\ 0 & 0 & 1 \end{bmatrix} \cdot \begin{bmatrix} x \\ y \\ 1 \end{bmatrix} \tag{5.24b}$$

Shearing the x-Axis Parallel to the y-Axis:

$$[x', y', 1] = [x, y, 1] \cdot \begin{bmatrix} 1 & \tan\theta & 0 \\ 0 & 1 & 0 \\ 0 & 0 & 1 \end{bmatrix} \tag{5.25a}$$

or alternatively by

$$\begin{bmatrix} x' \\ y' \\ 1 \end{bmatrix} = \begin{bmatrix} 1 & 0 & 0 \\ \tan\theta & 1 & 0 \\ 0 & 0 & 1 \end{bmatrix} \cdot \begin{bmatrix} x \\ y \\ 1 \end{bmatrix} \tag{5.25b}$$

5.10 Some Transformation Matrix Properties

A matrix representing a sequence of rotation and translation transformations is called an *orthogonal matrix* and is of the form

$$\boldsymbol{T} = \begin{bmatrix} r_{1,1} & r_{1,2} & 0 \\ r_{2,1} & r_{2,2} & 0 \\ t_x & t_y & 1 \end{bmatrix} \quad \text{or} \quad \boldsymbol{T}^{\mathrm{T}} = \begin{bmatrix} r_{1,1} & r_{2,1} & t_x \\ r_{1,2} & r_{2,2} & t_y \\ 0 & 0 & 1 \end{bmatrix}$$

A unit square transformed by such matrices retains its original shape and size. Such transformations are called *rigid body transformations*. In such matrices the elements labelled $r_{i,j}$ are involved in the rotation transformations and the elements labelled t_x and t_y are involved in the translation transformations.

A matrix representing a sequence of rotation, translation, scaling, and shearing transformations is called an *affine transformation*. A unit square transformed by such a matrix may become a parallelogram. Affine transformations preserve parallelism of lines but do not preserve the lengths of edges or the angles between edges.

In such matrices the elements labelled $r_{i,j}$ are involved in the rotation transformations, the elements labelled t_x and t_y are involved in the translation transformations, the diagonal elements labelled $r_{i,j}$ are involved in the scale transformations and the off-diagonal elements labelled $r_{2,1}$ and $r_{1,2}$ are involved in the shearing transformations.

5.11 Concatenation of Transformation Matrices

If we wish to apply a sequence of transformations $\boldsymbol{t}_1, \boldsymbol{t}_2, \ldots, \boldsymbol{t}_n$ to a point $\langle x, y \rangle$, the transformed point $\langle x', y' \rangle$ is given by

$$\begin{aligned} \langle x', y' \rangle &= \boldsymbol{t}_n (\cdots \boldsymbol{t}_2 (\boldsymbol{t}_1 (\langle x, y \rangle))) \\ &= \boldsymbol{t}_n \circ \cdots \circ \boldsymbol{t}_2 \circ \boldsymbol{t}_1 (\langle x, y \rangle) \end{aligned} \tag{5.26}$$

To represent this concatenated transformation in matrix form we must represent each transformation by an individual matrix and then concatenate these transformations by multiplying the corresponding matrices together in the right order.

The order in which we multiply these matrices is related to the way we choose to represent a point as a row or a column vector. If we choose to represent a point by a row vector, then Eq. (5.26) can be represented in matrix form as follows.

$$[x', y', 1] = [x, y, 1] \cdot \boldsymbol{T}_1 \cdot \boldsymbol{T}_2 \cdot \cdots \cdot \boldsymbol{T}_n = [x, y, 1] \cdot \boldsymbol{T} \tag{5.27a}$$

Alternatively, if we choose to represent a point by a column vector, then Eq. (5.26) can be represented in matrix form as follows.

$$\begin{bmatrix} x' \\ y' \\ 1 \end{bmatrix} = \boldsymbol{T}_n^{\mathrm{T}} \cdot \cdots \cdot \boldsymbol{T}_2^{\mathrm{T}} \cdot \boldsymbol{T}_1^{\mathrm{T}} \cdot \begin{bmatrix} x \\ y \\ 1 \end{bmatrix} = \boldsymbol{T}^{\mathrm{T}} \cdot \begin{bmatrix} x \\ y \\ 1 \end{bmatrix} \tag{5.27b}$$

So, if we choose the row representation of a point we must post-multiply the matrices in sequence and if we choose the column representation of a point we must pre-multiply the transpose matrices in sequence.

As an example let us represent in matrix form the transformation of rotation about an arbitrary point which we have already examined in analytical form in Section 5.5.3.

First let us represent the point $\langle x, y \rangle$ as a row vector. Recall that this transformation was achieved through the following three steps.

Step 1: Translate the arbitrary point to the origin

$$[x_1, y_1, 1] = [x, y, 1] \cdot \begin{bmatrix} 1 & 0 & 0 \\ 0 & 1 & 0 \\ -x_c & -y_c & 1 \end{bmatrix}$$

Step 2: Rotate the point $\boldsymbol{P}_1$ about the origin

$$[x_2, y_2, 1] = [x_1, y_1, 1] \cdot \begin{bmatrix} \cos\theta & \sin\theta & 0 \\ -\sin\theta & \cos\theta & 0 \\ 0 & 0 & 1 \end{bmatrix}$$

Step 3: Translate the arbitrary point back to its original position

$$[x_3, y_3, 1] = [x_2, y_2, 1] \cdot \begin{bmatrix} 1 & 0 & 0 \\ 0 & 1 & 0 \\ x_c & y_c & 1 \end{bmatrix}$$

These three transformations can be concatenated to

$$\begin{aligned} [x_3, y_3, 1] \\ &= [x, y, 1] \cdot \begin{bmatrix} 1 & 0 & 0 \\ 0 & 1 & 0 \\ -x_c & -y_c & 1 \end{bmatrix} \cdot \begin{bmatrix} \cos\theta & \sin\theta & 0 \\ -\sin\theta & \cos\theta & 0 \\ 0 & 0 & 1 \end{bmatrix} \cdot \begin{bmatrix} 1 & 0 & 0 \\ 0 & 1 & 0 \\ x_c & y_c & 1 \end{bmatrix} \\ &= [x, y, 1] \cdot \begin{bmatrix} \cos\theta & \sin\theta & 0 \\ -\sin\theta & \cos\theta & 0 \\ -x_c\cos\theta + y_c\sin\theta & -x_c\sin\theta - y_c\cos\theta & 1 \end{bmatrix} \cdot \begin{bmatrix} 1 & 0 & 0 \\ 0 & 1 & 0 \\ x_c & y_c & 1 \end{bmatrix}, \end{aligned}$$

$$= [x, y, 1] \cdot \begin{bmatrix} \cos\theta & \sin\theta & 0 \\ -\sin\theta & \cos\theta & 0 \\ -x_c\cos\theta + y_c\sin\theta + x_c & -x_c\sin\theta - y_c\cos\theta + y_c & 1 \end{bmatrix}$$

$$= [x\cos\theta - y\sin\theta - x_c\cos\theta + y_c\sin\theta + x_c,\quad x\sin\theta + y\cos\theta - x_c\sin\theta - y_c\cos\theta + y_c,\ \ 1]$$

$$= [(x - x_c)\cdot\cos\theta - (y - y_c)\cdot\sin\theta + x_c,\quad (x - x_c)\cdot\sin\theta + (y - y_c)\cdot\cos\theta + y_c,\ \ 1]$$

Thus,

$$x_3 = (x - x_c)\cdot\cos\theta - (y - y_c)\cdot\sin\theta + x_c$$
$$y_3 = (x - x_c)\cdot\sin\theta + (y - y_c)\cdot\cos\theta + y_c$$

Alternatively, representing the point $\langle x, y\rangle$ as a column vector leads to the following three transformations.

Step 1: Translate the arbitrary point to the origin

$$\begin{bmatrix} x_1 \\ y_1 \\ 1 \end{bmatrix} = \begin{bmatrix} 1 & 0 & -x_c \\ 0 & 1 & -y_c \\ 0 & 0 & 1 \end{bmatrix} \cdot \begin{bmatrix} x \\ y \\ 1 \end{bmatrix}$$

Step 2: Rotate the point $\boldsymbol{P}_1$ about the origin

$$\begin{bmatrix} x_2 \\ y_2 \\ 1 \end{bmatrix} = \begin{bmatrix} \cos\theta & -\sin\theta & 0 \\ \sin\theta & \cos\theta & 0 \\ 0 & 0 & 1 \end{bmatrix} \cdot \begin{bmatrix} x_1 \\ y_1 \\ 1 \end{bmatrix}$$

Step 3: Translate the arbitrary point back to its original position

$$\begin{bmatrix} x_3 \\ y_3 \\ 1 \end{bmatrix} = \begin{bmatrix} 1 & 0 & x_c \\ 0 & 1 & y_c \\ 0 & 0 & 1 \end{bmatrix} \cdot \begin{bmatrix} x_2 \\ y_2 \\ 1 \end{bmatrix}$$

These three transformations can be concatenated to

$$\begin{bmatrix} x_3 \\ y_3 \\ 1 \end{bmatrix} = \begin{bmatrix} 1 & 0 & x_c \\ 0 & 1 & y_c \\ 0 & 0 & 1 \end{bmatrix} \cdot \begin{bmatrix} \cos\theta & -\sin\theta & 0 \\ \sin\theta & \cos\theta & 0 \\ 0 & 0 & 1 \end{bmatrix} \cdot \begin{bmatrix} 1 & 0 & -x_c \\ 0 & 1 & -y_c \\ 0 & 0 & 1 \end{bmatrix} \cdot \begin{bmatrix} x \\ y \\ 1 \end{bmatrix}$$

$$= \begin{bmatrix} 1 & 0 & x_c \\ 0 & 1 & y_c \\ 0 & 0 & 1 \end{bmatrix} \cdot \begin{bmatrix} \cos\theta & -\sin\theta & -\cos\theta\, x_c + \sin\theta\, y_c \\ \sin\theta & \cos\theta & -\sin\theta\, x_c - \cos\theta\, y_c \\ 0 & 0 & 1 \end{bmatrix} \cdot \begin{bmatrix} x \\ y \\ 1 \end{bmatrix}$$

$$= \begin{bmatrix} \cos\theta & -\sin\theta & -\cos\theta\, x_c + \sin\theta\, y_c + x_c \\ \sin\theta & \cos\theta & -\sin\theta\, x_c - \cos\theta\, y_c + y_c \\ 0 & 0 & 1 \end{bmatrix} \cdot \begin{bmatrix} x \\ y \\ 1 \end{bmatrix}$$

$$= \begin{bmatrix} \cos\theta\, x - \sin\theta\, y - \cos\theta\, x_c + \sin\theta y_c + x_c \\ \sin\theta\, x + \cos\theta\, y - \sin\theta\, x_c - \cos\theta\, y_c + y_c \\ 1 \end{bmatrix}$$

$$\therefore \begin{aligned} x_3 &= (x - x_c) \cdot \cos\theta - (y - y_c) \cdot \sin\theta + x_c \\ y_3 &= (x - x_c) \cdot \sin\theta + (y - y_c) \cdot \cos\theta + y_c \end{aligned}$$

Observe that the results are identical in both the cases.

5.12 Local Frame and Global Frame Transformations

We can perform two different types of transformation: *global transformations* and *local transformations*. Global transformations are performed and expressed with respect to the *global frame of reference* or *world space origin and axes*, while local transformations are performed and expressed with respect to the *local frame of reference* or *local space origin and axes*.

When a global transformation is applied to an entity the origin and axes of the global space remain unchanged. In contrast, when a local transformation is applied to an entity the origin and axes of its local frame can be thought as reflecting the transformations applied to the entity itself. As an entity is translated with a *local translation*, its *local frame* origin and axes will also be translated together with the entity and when an entity is rotated with a *local rotation*, its local frame origin and axes will also be rotated.

To better understand the difference between global and local transformations let us consider the concatenation of the same two transformations applied in the global and local frames, respectively. Figure 5.13 shows the concatenation of a global translation by 3 units along the x-axis followed by a global rotation by 45° about the world space origin, while Fig. 5.14 shows the concatenation of a local

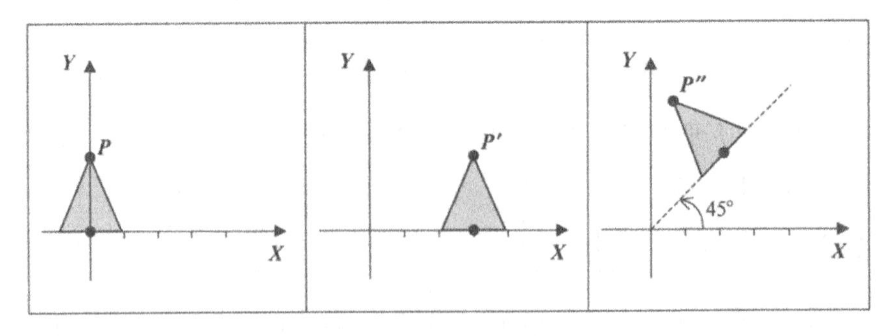

FIGURE 5.13. The concatenation of a global translation by 3 units along the x-axis followed by a global rotation of 45°.

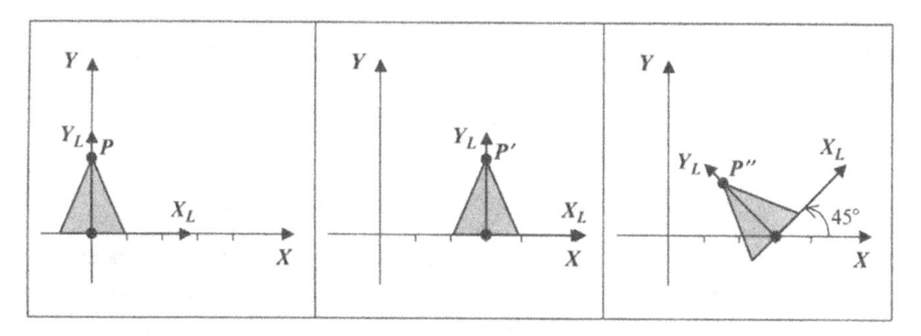

FIGURE 5.14. The concatenation of a local translation by 3 units along the x-axis followed by a local rotation of 45 °.

translation by 3 units along the x-axis followed by a local rotation by 45° about the local space origin. Observe that after the local translation the local frame of the triangle has also been translated and that after the local rotation its local frame has also been rotated.

Below we show how local and global composite transformations can be constructed by pre- or post-concatenating a transformation into the current transformation matrix. We will discover that the results are symmetrical for the two alternative matrix notations introduced in Section 5.8.

5.12.1 Concatenation of Global Transformations

Here we denote the global transformation matrices by $\boldsymbol{G_i}$ and the concatenated transformation matrix by $\boldsymbol{T}$. As we have already seen in Section 5.11, global transformations are concatenated in ascending order of application (i.e. from the least recently applied to the most recently applied) as follows.

When using the $\boldsymbol{P}' = \boldsymbol{P} \cdot \boldsymbol{T}$ *notation*

$$\begin{aligned} \boldsymbol{P}' &= (((\boldsymbol{P} \cdot \boldsymbol{G}_1) \cdot \boldsymbol{G}_2) \cdot \boldsymbol{G}_3) \\ &= \boldsymbol{P} \cdot (\boldsymbol{G}_1 \cdot \boldsymbol{G}_2 \cdot \boldsymbol{G}_3) \\ &= \boldsymbol{P} \cdot \boldsymbol{T} \end{aligned}$$

Thus, the concatenated global transformation is constructed by post-multiplying the individual transformations from left to right in ascending order of application, i.e.

$$\boldsymbol{T} = (\boldsymbol{G}_1 \cdot \boldsymbol{G}_2) \cdot \boldsymbol{G}_3$$

When using the $\boldsymbol{P}'^{\mathrm{T}} = \boldsymbol{T}^{\mathrm{T}} \cdot \boldsymbol{P}^{\mathrm{T}}$ *notation*

$$\begin{aligned} \boldsymbol{P}'^{\mathrm{T}} &= \left(\boldsymbol{G}_3^{\mathrm{T}} \cdot \left(\boldsymbol{G}_2^{\mathrm{T}} \cdot \left(\boldsymbol{G}_1^{\mathrm{T}} \cdot \boldsymbol{P}^{\mathrm{T}}\right)\right)\right) \\ &= \left(\boldsymbol{G}_3^{\mathrm{T}} \cdot \boldsymbol{G}_2^{\mathrm{T}} \cdot \boldsymbol{G}_1^{\mathrm{T}}\right) \cdot \boldsymbol{P}^{\mathrm{T}} \\ &= \boldsymbol{T}^{\mathrm{T}} \cdot \boldsymbol{P}^{\mathrm{T}} \end{aligned}$$

Thus, the concatenated global transformation is constructed by pre-multiplying the individual transformations from right to left in ascending order of application, i.e.

$$T^{\mathrm{T}} = G_3^{\mathrm{T}} \cdot \left(G_2^{\mathrm{T}} \cdot G_1^{\mathrm{T}}\right)$$

5.12.2 *Concatenation of Local Transformations*

Here we denote the local transformation matrices by $\boldsymbol{L}_i$ and the concatenated transformation matrix by $\boldsymbol{T}$. Local transformations are concatenated in descending order of application (i.e. from the most recently applied to the least recently applied) as follows.

When using the $\boldsymbol{P}' = \boldsymbol{P} \cdot \boldsymbol{T}$ *notation*

$$\begin{aligned} P' &= (((P \cdot L_3) \cdot L_2) \cdot L_1) \\ &= P \cdot (L_3 \cdot L_2 \cdot L_1) \\ &= P \cdot T \end{aligned}$$

Thus, the concatenated local transformation is constructed by pre-multiplying the individual transformations from right to left in ascending order of application, i.e.

$$T = L_3 \cdot (L_2 \cdot L_1)$$

When using the $\boldsymbol{P}'^{\mathrm{T}} = \boldsymbol{T}^{\mathrm{T}} \cdot \boldsymbol{P}^{\mathrm{T}}$ *notation:*

$$\begin{aligned} P'^{\mathrm{T}} &= (L_1^{\mathrm{T}} \cdot (L_2^{\mathrm{T}} \cdot (L_3^{\mathrm{T}} \cdot P^{\mathrm{T}}))) \\ &= (L_1^{\mathrm{T}} \cdot L_2^{\mathrm{T}} \cdot L_3^{\mathrm{T}}) \cdot P^{\mathrm{T}} \\ &= T^{\mathrm{T}} \cdot P^{\mathrm{T}} \end{aligned}$$

Thus, the concatenated local transformation is constructed by post-multiplying the individual transformations from left to right in ascending order of application, i.e.

$$T^{\mathrm{T}} = \left(L_1^{\mathrm{T}} \cdot L_2^{\mathrm{T}}\right) \cdot L_3^{\mathrm{T}}$$

5.13 Transformations of the Frame of Reference or Coordinate System

All the transformations we have examined so far are transformations of a point relative to a given frame of reference. An alternative way of viewing such transformations is to think of the point in question remaining stationary while its frame of reference is transformed by the inverse transformation. Thus transforming a point $\boldsymbol{P}$ by a transformation matrix $\boldsymbol{T}$ will produce the same result as transforming its frame of reference by the transformation matrix $\boldsymbol{T}^{-1}$. In Fig. 5.15, point $\boldsymbol{P}$ in diagram (a) is transformed to point P' in diagram (b) by translating it by 2 units

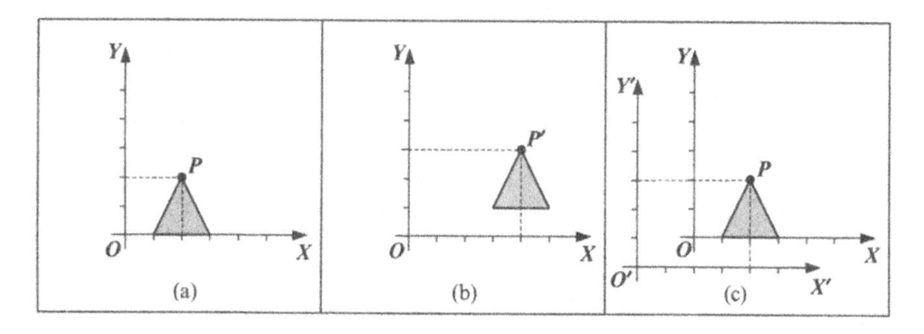

FIGURE 5.15. The transformation of a point vs. the transformation of the frame of reference.

along the x-axis and by 1 unit along the y-axis. Point $\boldsymbol{P}$ in diagram (c) remains stationary while its frame of reference is translated by –2 units along the x-axis and by –1 unit along the y-axis. Observe that after these translations, point $\boldsymbol{P}'$ in the $\{\boldsymbol{X}, \boldsymbol{Y}, \boldsymbol{O}\}$ frame of reference and point $\boldsymbol{P}$ in the $\{\boldsymbol{X}', \boldsymbol{Y}', \boldsymbol{O}'\}$ frame of reference have identical coordinates. Thus, it is apparent that these two transformations are equivalent.

5.14 Viewing Transformation

In computer graphics we normally define our picture in the *world coordinate system*. This system can have units to fit the particular application. When we view our picture we may choose to project the entire image or a small rectangular area of the image onto the entire screen or a small rectangular area of the screen.

The process of mapping a world coordinate system point onto the *screen coordinate system* point is called the *viewing transformation.* The world coordinate system is chosen to suit the application, while the screen coordinate system is fixed once the output device has been selected.

The viewing transformation is composed of two simpler transformations:

- *the windowing transformation,*
- *the viewporting transformation.*

The viewing transformation can be represented by a 3×3 matrix.

5.14.1 Windowing Transformation

The windowing transformation is named so because it involves specifying a *window* which surrounds a rectangular area of the picture. Only the portion of the picture which falls inside the window will be displayed on the screen. The window defines its own coordinate system called the *window coordinate space.* This coordinate space has its origin at the lower left corner of the window. The window coordinate system is sometimes called the *normalised coordinate system* as

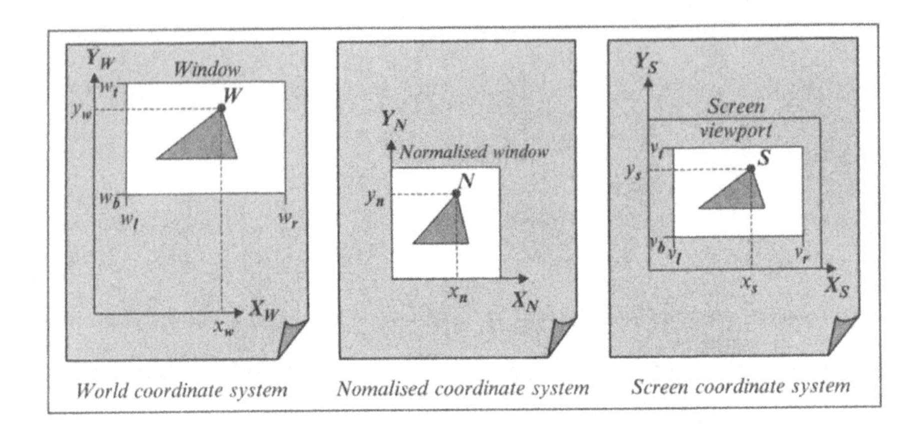

FIGURE 5.16. The world, the normalised and the screen coordinate systems.

it transforms the rectangular window defined in the world coordinate system into a square window defined in the window coordinate system with dimensions 1 unit by 1 unit.

Let w_l, w_r, w_b and w_t be the left, right, bottom and top limits of the window, defined in the world coordinate space. Also, let $\boldsymbol{W} = [x_w, y_w]$ be a point defined in the world coordinate system (see Fig. 5.16).

The world coordinate system point $\boldsymbol{W}$ can be mapped into the window coordinate system (normalised coordinate system) point $\boldsymbol{N} = [x_n, y_n]$ as follows.

$$x_n = \frac{x_w - w_l}{w_r - w_l}$$

$$y_n = \frac{y_w - w_b}{w_t - w_b} \tag{5.28}$$

Thus points inside the window have normalised coordinates, i.e.

$$0 \leq x_n \leq 1 \quad \text{and} \quad 0 \leq y_n \leq 1 \tag{5.29}$$

5.14.2 Viewporting Transformation

In addition to the window we can also define a *viewport*. The viewport is a rectangular area of the screen where we wish to display the contents of the window. It is often useful to specify a viewport which covers only a small portion of the screen, so that we can leave room for command menus and system messages.

The mapping from a window coordinate system point $\boldsymbol{N}$ to a screen coordinate system point $\boldsymbol{S} = [x_s, y_s]$ is called the *viewporting transformation*.

Let v_l, v_r, v_b and v_t be the left, right, bottom and top limits of the viewport, defined in the screen coordinate space (see Fig. 5.16).

The viewporting transformation is defined as follows.

$$x_s = x_n \cdot (v_r - v_l) + v_l$$
$$y_s = y_n \cdot (v_t - v_b) + v_b \tag{5.30}$$

So the points inside the viewport have coordinates in the range

$$v_l \le x_s \le v_r \quad \text{and} \quad v_b \le y_s \le v_t \tag{5.31}$$

The windowing and the viewporting transformations can be concatenated into a composite viewing transformation as follows.

$$x_s = \frac{x_w - w_l}{w_r - w_l} \cdot (v_r - v_l) + v_l$$
$$y_s = \frac{y_w - w_b}{w_t - w_b} \cdot (v_t - v_b) + v_b$$

which can be rewritten as

$$x_s = (x_w - w_l) \cdot \frac{v_r - v_l}{w_r - w_l} + v_l$$
$$y_s = (y_w - w_b) \cdot \frac{v_t - v_b}{w_t - w_b} + v_b \tag{5.32}$$

which expands to

$$x_s = x_w \cdot \frac{v_r - v_l}{w_r - w_l} - w_l \cdot \frac{v_r - v_l}{w_r - w_l} + v_l$$
$$y_s = y_w \cdot \frac{v_t - v_b}{w_t - w_b} - w_b \cdot \frac{v_t - v_b}{w_t - w_b} + v_b$$

which can be rewritten as

$$x_s = x_w \cdot s_x + d_x$$
$$y_s = y_w \cdot s_y + d_y \tag{5.33}$$

where

$$s_x = \frac{v_r - v_l}{w_r - w_l}$$
$$d_x = -w_l \cdot \frac{v_r - v_l}{w_r - w_l} + v_l$$
$$s_y = \frac{v_t - v_b}{w_t - w_b} \tag{5.34}$$
$$d_y = -w_b \cdot \frac{v_t - v_b}{w_t - w_b} + v_b$$

We only need to compute the values of s_x, d_x, s_y and d_y once, when the window and the viewport are defined. And so we can transform every point in our picture by a computation involving only two multiplication and two addition operations.

Clipping can be applied either to the screen space points using the viewport perimeter as the clipping boundary or to the world space points using the window perimeter as the clipping boundary.

The analytical equations, developed above, can be rewritten in matrix form as follows.

$$[x_s, y_s, 1] = [x_w, y_w, 1] \cdot \begin{bmatrix} s_x & 0 & 0 \\ 0 & s_y & 0 \\ d_x & d_y & 1 \end{bmatrix} \tag{5.34a}$$

or alternatively as

$$\begin{bmatrix} x_s \\ y_s \\ 1 \end{bmatrix} = \begin{bmatrix} s_x & 0 & d_x \\ 0 & s_y & d_y \\ 0 & 0 & 1 \end{bmatrix} \cdot \begin{bmatrix} x_w \\ y_w \\ 1 \end{bmatrix} \tag{5.34b}$$

The viewing transformation matrix can be concatenated with the matrix containing all other transformations, thus allowing us to apply all transformations with one matrix multiplication operation. In this case, clipping must be applied to the screen space points using the viewport perimeter as the clipping boundary.

5.15 Homogeneous Coordinates

The homogeneous representation of an entity defined in n-dimensional space is another entity defined in $(n+1)$-dimensional space. The mapping from n-space to $(n+1)$-space is a *one-to-many* mapping. Thus, the n-space entity has an infinity of *images* (representations) in $(n+1)$-space. The inverse mapping, which is a *many-to-one* mapping, is a *central projection* that uses the origin as its *centre of projection*.

A point/vector in n-dimensional Euclidean space can be represented as an $(n+1)$-dimensional homogeneous space point/vector by the introduction of an additional coordinate or component that also acts as a scale factor.

The homogeneous representation of a 2D point $\boldsymbol{P} = \begin{bmatrix} x & y \end{bmatrix}$ is the homogeneous point $\boldsymbol{P}^{\mathrm{h}} = \begin{bmatrix} w \cdot x & w \cdot y & w \end{bmatrix}$, where w is the homogeneous coordinate which is a non-zero scalar frequently known as the *weight* of the point. By selecting different values for w we produce different homogeneous mappings of the original point. Thus, the homogeneous points

$$\boldsymbol{P}_1^{\mathrm{h}} = \begin{bmatrix} w_1 \cdot x & w_1 \cdot y & w_1 \end{bmatrix}, \boldsymbol{P}_2^{\mathrm{h}} = \begin{bmatrix} w_2 \cdot x & w_2 \cdot y & w_2 \end{bmatrix},$$
$$\boldsymbol{P}_3^{\mathrm{h}} = \begin{bmatrix} w_3 \cdot x & w_3 \cdot y & w_3 \end{bmatrix}, \ldots$$

are all valid homogeneous mappings of the point $\boldsymbol{P} = \begin{bmatrix} x & y \end{bmatrix}$ and there is an infinity of such mappings.

Two homogeneous points $\boldsymbol{P}_1^{\mathrm{h}} = [\, X_1 \quad Y_1 \quad W_1 \,]$ and $\boldsymbol{P}_2^{\mathrm{h}} = [\, X_2 \quad Y_2 \quad W_2 \,]$ are said to be equal if and only if $\boldsymbol{P}_2^{\mathrm{h}}$ is a multiple of $\boldsymbol{P}_1^{\mathrm{h}}$, i.e. if and only if

$$X_2 = \alpha \cdot X_1$$
$$Y_2 = \alpha \cdot Y_1$$
$$W_2 = \alpha \cdot W_1$$

For example, $\boldsymbol{P}_1^{\mathrm{h}} = [\, 1 \quad 3 \quad 4 \,]$ and $\boldsymbol{P}_2^{\mathrm{h}} = [\, 2 \quad 6 \quad 8 \,]$ are equal as $\boldsymbol{P}_2^{\mathrm{h}} = 2\boldsymbol{P}_1^{\mathrm{h}}$.

If we take all the triplets representing the same homogeneous point, i.e., $[\, (t \cdot w) \cdot x \quad (t \cdot w) \cdot y \quad (t \cdot w) \,]$ where $t \neq 0$, they represent a line in 3D Euclidean space $\mathcal{E}^3$. From Fig. 5.17 we can see that the $\mathcal{E}^2$ point P maps onto an infinity of homogeneous points forming a line in $\mathcal{E}^3$. Thus, we say that a point in $\mathcal{E}^2$ maps onto a line in homogeneous space (in this case $\mathcal{E}^3$).

A homogeneous point is said to be *homogenised* if we divide all its components by its $\boldsymbol{W}$ component, thus obtaining a homogeneous point of the form $[\, \frac{X}{W} \quad \frac{Y}{W} \quad 1 \,]$. Geometrically this represents a central projection of the point $[\, X \quad Y \quad W \,]$ onto the hyperplane $W = 1$, using the origin as the centre of projection. Thus, the set of all homogenised points forms the $W = 1$ hyperplane.

Given a homogeneous point $\boldsymbol{P}^{\mathrm{h}} = [\, X \quad Y \quad W \,]$ we can map it back to the original 2D Euclidean point $\boldsymbol{P} = [\, x \quad y \,]$ (from which it has originated) by first homogenising $\boldsymbol{P}^h$ to obtain $[\, \frac{X}{W} \quad \frac{Y}{W} \quad 1 \,]$ and then ignoring its third component to obtain $\boldsymbol{P} = [\, \frac{X}{W} \quad \frac{Y}{W} \,]$ (as seen in Fig. 5.17), where

$$x = \frac{X}{W}$$
$$y = \frac{Y}{W} \tag{5.35}$$

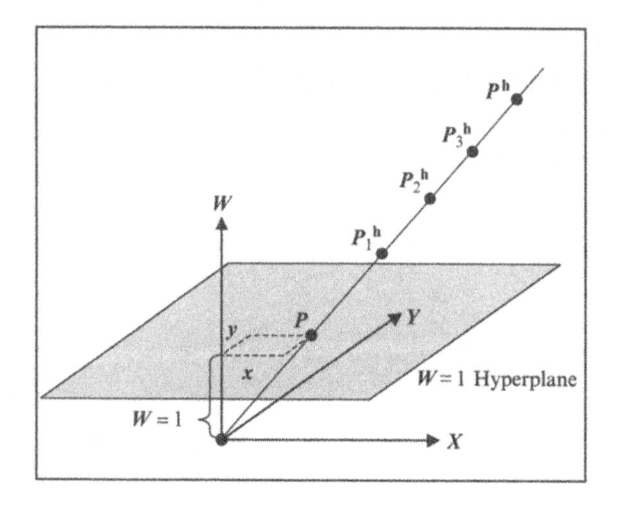

FIGURE 5.17. The 3D homogeneous space for E^2.

This process is called a *projective map* $\boldsymbol{H}$ and is the inverse of the mapping that originally mapped point $\boldsymbol{P}$ onto the homogeneous point $\boldsymbol{P}^h$.

In 2D Euclidean space both points and vectors are represented by two components and can only be distinguished by their context. They are however quite different. A vector has a magnitude and a direction, but no fixed position, while a point has a position but no magnitude or direction.

By examining Eq. (5.35) we can see that as W approaches 0 both x and y approach infinity from a particular direction. Thus, we may think of the homogeneous point $\begin{bmatrix} w \cdot x & w \cdot y & w \end{bmatrix}$ as representing the point $\begin{bmatrix} x & y \end{bmatrix}$ and the homogeneous point $\begin{bmatrix} x & y & 0 \end{bmatrix}$ as representing the vector $\begin{bmatrix} x & y \end{bmatrix}$. So, in homogeneous space we have a way of distinguishing between points and vectors.

Three-dimensional Euclidean points/vectors are represented analogously although their homogeneous representation is more difficult to visualise. More formally, a point/vector $\boldsymbol{P} = \begin{bmatrix} x & y & z \end{bmatrix}$ in 3D Euclidean space may be represented by the 4D homogeneous point

$$\boldsymbol{P}^{\mathrm{h}} = \begin{cases} \begin{bmatrix} w \cdot x & w \cdot y & w \cdot z & w \end{bmatrix}, & w \neq 0 \\ \begin{bmatrix} x & y & z & 0 \end{bmatrix}, & w = 0 \end{cases} \tag{5.36}$$

The original point/vector $\boldsymbol{P}$ may be retrieved from its homogeneous representation $\boldsymbol{P}^h$ by using the projective map $\boldsymbol{H}$, which is defined as

$$H\left(\begin{bmatrix} X & Y & Z & W \end{bmatrix}\right) = \begin{cases} \begin{bmatrix} \frac{X}{W} & \frac{Y}{W} & \frac{Z}{W} \end{bmatrix}, & W \neq 0 \\ \text{direction}\left(\begin{bmatrix} X & Y & Z \end{bmatrix}\right), & W = 0 \end{cases} \tag{5.37}$$

Thus, the projective map $\boldsymbol{H}$ uses the origin as the centre of projection to project the homogeneous point/vector $\boldsymbol{P}^h$ onto the hyperplane $W = 1$ in order to retrieve the 3D point/vector $\boldsymbol{P}$.

$$\boldsymbol{P} = \boldsymbol{H}\left(\boldsymbol{P}^{\mathrm{h}}\right) \tag{5.38}$$

5.16 A Simple C Library for 2D Transformations

See *Appendix 3*.

6

Two-Dimensional Clipping

Clipping is a process that subdivides each element of a picture to be displayed into its visible and invisible parts, thus allowing us to discard the invisible parts of the picture. In 2D, the clipping process can be applied to a variety of graphics primitives such as points, lines, polygons and curves. Clipping is performed with respect to a *clipping boundary*, which may be a *convex* or *concave polygonal boundary*. Clipping to a convex polygonal boundary is much simpler, thus we will start our discussion by looking at various algorithms that clip graphics primitives against a convex polygonal boundary. In its simplest form, a convex polygonal boundary consists of the four edges of the world window or the screen viewport. In this case, x_l (x-left), x_r (x-right), y_b (y-bottom) and y_t (y-top) represent the four extremities of this rectangular clipping boundary.

6.1 Clipping a 2D Point to a Rectangular Clipping Boundary

Clipping 2D points to a rectangular clipping boundary is very simple. All we have to do is to determine whether the point $\langle x, y\rangle$ satisfies the following four inequalities.

$$x_l \leq x \leq x_r \quad \text{and} \quad y_b \leq y \leq y_t \tag{6.1}$$

If the point satisfies all four inequalities, then it is visible otherwise it is invisible (see Fig. 6.1).

The following **C** Boolean function clips a point to a rectangular clipping boundary.

```
/* Define the Boolean type and constants */

typedef unsigned char boolean_t;

boolean_t clip_point(double x, double y,
                     double xl, double xr, double yb, double yt)
{
```

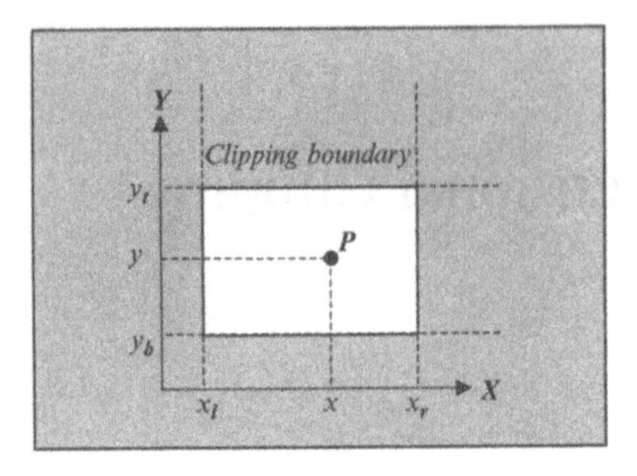

FIGURE 6.1. Clipping a point to a rectangular clipping boundary.

```
return((boolean_t) ((xl <= x) && (x <= xr) &&
                    (yb <= y) && (y <= yt));

} /* clip_point */
```

This function provides us with a very simple method of clipping primitives on a point by point basis. It would however be quite inappropriate to clip primitives by decomposing them into points and using this function to determine their visibility. The clipping process would take far too long and leave the primitives in a form no longer suitable for a line-drawing display. We must instead attempt to clip different primitives by developing more powerful clipping algorithms that can determine the visible and invisible portions of line segments and polygons.

6.2 Clipping a 2D Line Segment to a Rectangular Clipping Boundary

Figure 6.2 shows a number of different attitudes that a straight line segment may assume with respect to a rectangular clipping boundary. Notice that the lines that

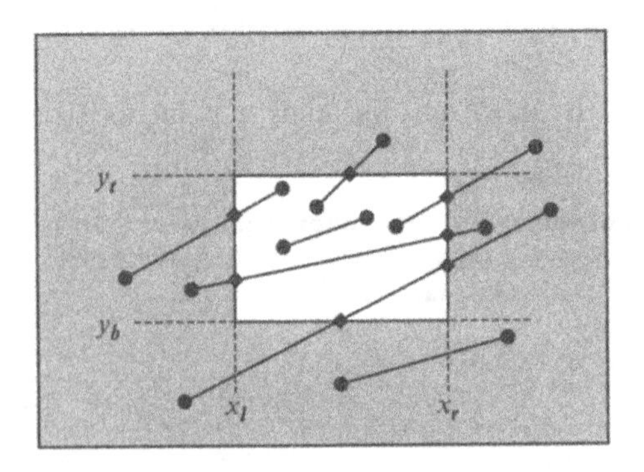

FIGURE 6.2. Examples of clipped line segments.

are partially visible are divided by the clipping boundary into one or two invisible segments but only one visible segment. It is an extremely useful fact that clipping to a convex clipping boundary never generates more than one visible segment of a straight line. This means that for a partially visible line segment we have to determine its new end points as it enters and/or leaves the clipping region.

An efficient line-clipping algorithm has to determine whether a line segment is entirely within the clipping boundary or entirely outside the clipping boundary and dispense with such cases quickly. If however a part of the line segment falls within the clipping boundary and one or two parts lie outside it, then the algorithm has to determine the point or points of intersection of the line segment with one or more edges of the clipping boundary. Given that the line segment to be clipped has endpoints $\boldsymbol{P}_1 = \langle x_1, y_1 \rangle$ and $\boldsymbol{P}_2 = \langle x_2, y_2 \rangle$, the coordinates of the point of intersection $\boldsymbol{P}_i = \langle x_i, y_i \rangle$ with each of the rectangular clipping boundary extremities are computed as follows.

The intersection of the line segment $\overrightarrow{\boldsymbol{P}_1\boldsymbol{P}_2}$ with the *left clipping boundary* is shown in Fig. 6.3a. Here the x coordinate of the point of intersection is

$$x_i = x_l$$

From the similar triangles in this figure we have

$$\frac{y_i - y_1}{y_2 - y_1} = \frac{x_l - x_1}{x_2 - x_1}$$

$$\therefore \quad y_i - y_1 = \frac{x_l - x_1}{x_2 - x_1}(y_2 - y_1)$$

Thus

$$\begin{aligned} x_i &= x_l \\ y_i &= \frac{x_l - x_1}{x_2 - x_1}(y_2 - y_1) + y_1 \end{aligned} \tag{6.2}$$

The intersection of the line segment $\overrightarrow{\boldsymbol{P}_1\boldsymbol{P}_2}$ with the *right clipping boundary* is shown in Fig. 6.3b. Here the x coordinate of the point of intersection is

$$x_i = x_r$$

From the similar triangles in this figure we have

$$\frac{y_i - y_1}{y_2 - y_1} = \frac{x_r - x_1}{x_2 - x_1}$$

$$\therefore \quad y_i - y_1 = \frac{x_r - x_1}{x_2 - x_1}(y_2 - y_1)$$

Thus

$$\begin{aligned} x_i &= x_r \\ y_i &= \frac{x_r - x_1}{x_2 - x_1}(y_2 - y_1) + y_1 \end{aligned} \tag{6.3}$$

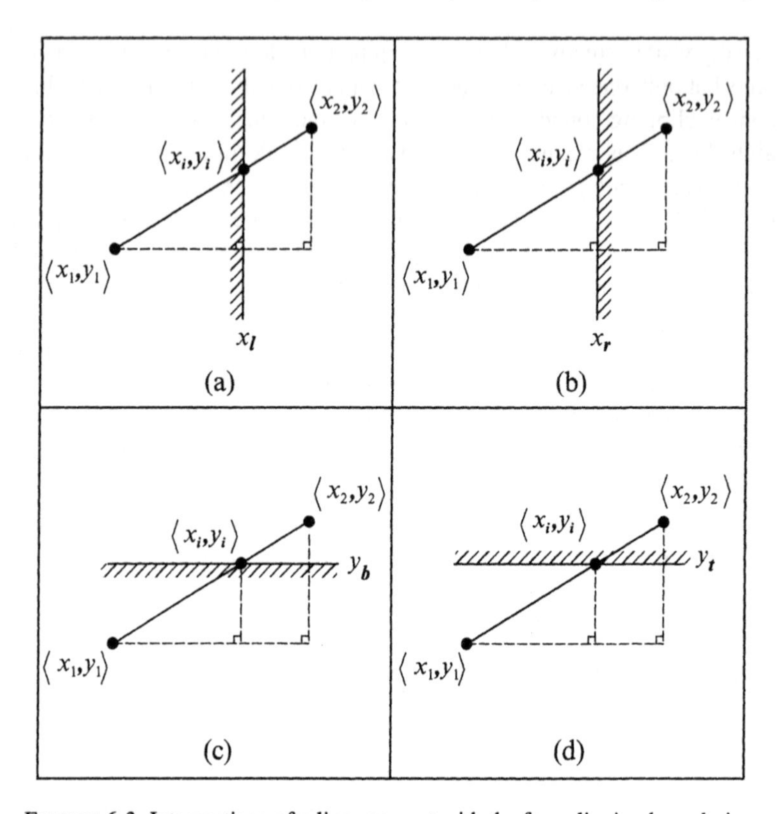

FIGURE 6.3. Intersections of a line segment with the four clipping boundaries.

The intersection of the line segment $\overrightarrow{P_1P_2}$ with the *bottom clipping boundary* is shown in Fig. 6.3c. Here the y coordinate of the point of intersection is

$$y_i = y_b$$

From the similar triangles in this figure we have

$$\frac{x_i - x_1}{x_2 - x_1} = \frac{y_b - y_1}{y_2 - y_1}$$

$$\therefore \quad x_i - x_1 = \frac{y_b - y_1}{y_2 - y_1}(x_2 - x_1)$$

Thus

$$\begin{aligned} y_i &= y_b \\ x_i &= \frac{y_b - y_1}{y_2 - y_1}(x_2 - x_1) + x_1 \end{aligned} \tag{6.4}$$

Finally, the intersection of the line segment $\overrightarrow{P_1P_2}$ with the *top clipping boundary* is shown in Fig. 6.3d. Here the y coordinate of the point of intersection is

$$y_i = y_t$$

From the similar triangles in this figure we have

$$\frac{x_i - x_1}{x_2 - x_1} = \frac{y_t - y_1}{y_2 - y_1}$$

$$\therefore \quad x_i - x_1 = \frac{y_t - y_1}{y_2 - y_1}(x_2 - x_1)$$

Thus

$$y_i = y_t$$

$$x_i = \frac{y_t - y_1}{y_2 - y_1}(x_2 - x_1) + x_1 \tag{6.5}$$

6.3 The Cohen and Sutherland 2D Line-Clipping Algorithm

The Cohen and Sutherland 2D line-clipping algorithm clips a line segment $\overrightarrow{\boldsymbol{P}_1\boldsymbol{P}_2}$ to a rectangular clipping boundary defined by its four extremities: x_l, x_r, y_b and y_t. This algorithm speeds up the decision that allows us to reject a line segment as being entirely invisible or accept it as being entirely visible. It achieves this by the following procedure.

The edges of the clipping region are extended so that they divide the plane into nine regions, as shown in Fig. 6.4. Each of these regions is given a 4-bit code and the two endpoints of the line segment $\overrightarrow{\boldsymbol{P}_1\boldsymbol{P}_2}$ are assigned a code according to the region they are in.

The 4-bit codes have the following meaning when set:

1st bit (least significant bit) being set indicates that the point is to the *left of the left* clipping edge x_l

2nd bit indicates that the point is to the *right of the right* clipping edge x_r

3rd bit indicates that the point is to the *bottom of the bottom* clipping edge y_b

4th bit (most significant bit) indicates that the point is to the *top of the top* clipping edge y_t

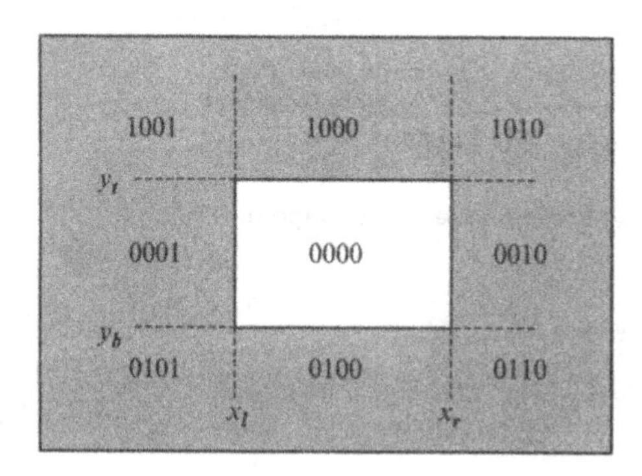

FIGURE 6.4 The 4-bit region codes of the Cohen and Sutherland 2D line-clipping algorithm.

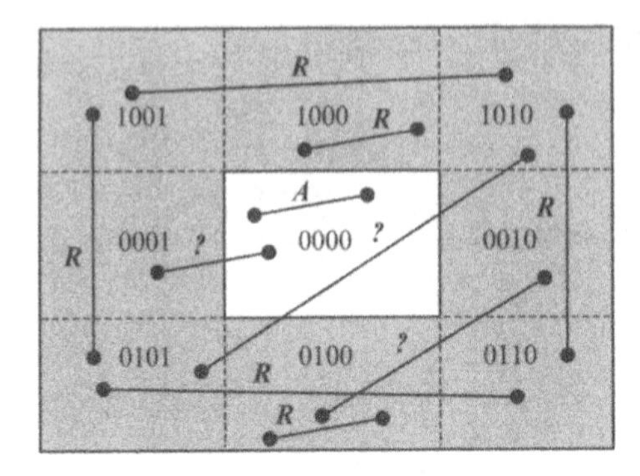

FIGURE 6.5. Examples of trivial acceptance and trivial rejection.

If the 4-bit codes of both endpoints are zero, then the line segment is entirely inside the clipping boundary. This is known as the *trivial acceptance* case. In Fig. 6.5, the line segment labelled ***A*** can be trivially accepted as being visible in its entirety.

If the *bit-wise and* of the two 4-bit codes is non-zero, then the line segment is entirely outside the clipping region. This is known as the *trivial rejection* case. In Fig. 6.5, the line segments labelled ***R*** can be trivially rejected as being entirely invisible.

Finally, if the line segment can not be eliminated by either of these tests, then it must be subdivided by intersecting it with one of the clipping boundary edges and the two tests repeated. This process may have to be repeated a number of times. In Fig. 6.5, the line segments labelled ***?*** can neither be trivially accepted nor trivially rejected even though they might be totally invisible.

Algorithm 6.1 outlines the steps of the Cohen and Sutherland 2D line-clipping algorithm.

```
1.  c1 ←compute the 4-bit code for P1;
    c2 ←compute the 4-bit code for P2;
2.  if (c1 = 0) and (c2 = 0) then
    {
      trivially accept the line segment;
      done;
    }
3.  if (c1 bit − wise AND c2) then
    {
      trivially reject the line segment;
      done;
    }
4.  if (c1 = 0) then c = c2;
                else c = c1;
```

```
5.if the first bit of c is set then
  {
    compute the point of intersection ⟨xi, yi⟩ of the line
    segment with the left clipping boundary;
    goto step 9;
  }
6.if the second bit of c is set then
  {
    compute the point of intersection ⟨xi, yi⟩ of the line
    segment with the right clipping boundary;
    goto step 9;
  }
7.if the third bit of c is set then
  {
    compute the point of intersection ⟨xi, yi⟩ of the line
    segment with the bottom clipping boundary;
    goto step 9;
  }
8.if the fouth bit of c is set then
  {
    compute the point of intersection ⟨xi, yi⟩ of the line
    segment with the top clipping boundary;
    goto step 9;
  }
9.if (c = c1) then
  {
    x1 ← xi;
    y1 ← yi;
    c1 ←recompute the 4-bit code for this new point;
    goto step 2;
  }
  else
  {
    x2 ← xi;
    y2 ← yi;
    c2 ←recompute the 4-bit code for this new point;
    goto step 2;
  }
```

Algorithm 6.1 The Cohen and Sutherland 2D line-clipping algorithm.

The following **C** functions implement the Cohen and Sutherland 2D line-clipping algorithm.

```
/*
 * Common constants and typedefs.
 */

typedef unsigned char boolean_t;

#define False (boolean_t) 0
#define True  (boolean_t) 1
```

```
/*
 * Constants and typedef for Cohen and Sutherland 2D line-clipping
 * routine.
 */

#define small_t 0.000000005
typedef unsigned char region_code_t;

#define left_region   ((region_code_t) 1)
#define right_region  ((region_code_t) 2)
#define bottom_region ((region_code_t) 4)
#define top_region    ((region_code_t) 8)

/*-------------------------------------------------------------*/

void cs_get_region_code
(
 double        xl,  /* clipping Boundary Limits (In) */
 double        yb,
 double        xr,
 double        yt,
 double        x,   /* Test Point (In) */
 double        y,
 region_code_t *c   /* Region code (Out) */
)
{

*c = 0;

 if (x < xl) *c = left_region; else
 if (x > xr) *c = right_region;

 if (y < yb) *c |= bottom_region; else
 if (y > yt) *c |= top_region;

} /* cs_get_region_code */

/*-------------------------------------------------------------*/

boolean_t cs_clip_line
(
 double *x1, /* Test Line (In/Out) */
 double *y1,
 double *x2,
 double *y2,
 double xl,  /* clipping Boundary Limits (In) */
 double yb,
 double xr,
 double yt
)
{

region_code_t c, c1, c2;
double        x, y, t;
```

```
cs_get_region_code(xl, yb, xr, yt, *x1, *y1, &c1);
cs_get_region_code(xl, yb, xr, yt, *x2, *y2, &c2);

while ((c1 != 0) || (c2!= 0))
 {
  if ((c1 & c2) != 0) return(False); /* Trivial Rejection */
  /*
   * The line segment may be partially inside the clipping boundary.
   */

  if (c1 == 0) c = c2;
  else         c = c1;

   if (c & left_region)
   {
    /*
     * Compute the intersection with the xl boundary edge.
     */

    t = *x2 - *x1;

    if (fabs(t) < small_t) t = small_t;

    y = (xl - *x1) / t * (*y2 - *y1) + *y1;
    x = xl;
   }
  else

  if (c & right_region)
   {
    /*
     * Compute the intersection with the xr boundary edge.
     */

    t = *x2 - *x1;

    if (fabs(t) < small_t) t = small_t;

    y = (xr - *x1) / t * (*y2 - *y1) + *y1;
    x = xr;
   }
  else

  if (c & bottom_region)
   {
    /*
     * Compute the intersection with the yb boundary edge.
     */

    t = *y2 - *y1;

    if (fabs(t) < small_t) t = small_t;

    x = (yb - *y1) / t * (*x2 - *x1) + *x1;
```

```
      y = yb;
    }
  else

  if (c & top_region)
    {
     /*
      * Compute the intersection with the yt boundary edge.
      */

     t = *y2 - *y1;

     if (fabs(t) < small_t) t = small_t;

     x = (yt - *y1) / t * (*x2 - *x1) + *x1;
     y = yt;
    }

  if (c == c1)
    {
     *x1 = x;
     *y1 = y;
     cs_get_region_code(xl, yb, xr, yt, x, y, &c1);
    }
  else
    {
     *x2 = x;
     *y2 = y;
     cs_get_region_code(xl, yb, xr, yt, x, y, &c2);
    }
  } /* while */

 return(True);

} /* cs_clip_line */
```

6.4 2D Polygon Clipping

If we wish to display a polygon in line drawn form, we can clip each of its edges in sequence and display their visible portions. For a polygon that is partially visible this process will result in a series of disconnected edges, as seen in Fig. 6.6b. If, on the other hand, we wish to display a polygon as an enclosed area filled with colour, the polygon-clipping process must produce a closed polygon. To do so it must piece together the visible segments of the original polygon edges and sections of the clipping boundary edges, as seen in Fig. 6.6c.

An elegant solution to the polygon-clipping problem has been suggested in a paper by Ivan Sutherland and Gary Hodgman [Sutherland 74].

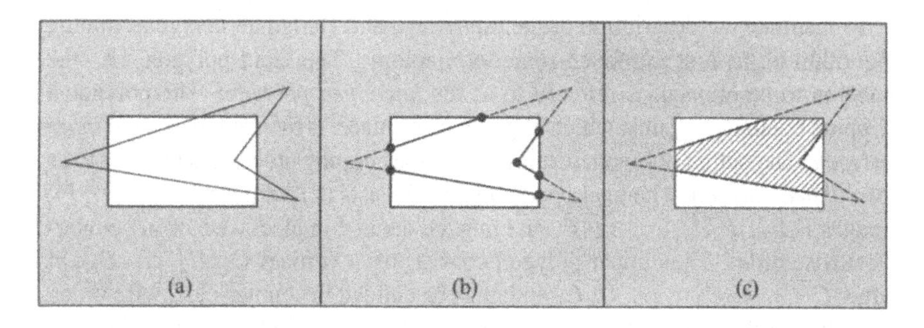

FIGURE 6.6 Edge and polygon clipping: (a) unclipped polygon, (b) edge clipping, (c) polygon clipping.

6.4.1 The Sutherland and Hodgman Polygon-Clipping Algorithm

The Sutherland and Hodgman algorithm allows us to clip a convex or concave polygon to a convex polygonal clipping boundary (i.e. we are not restricted to a rectangular clipping boundary). The algorithm abandons the idea of clipping each separate edge of the polygon against all clipping boundaries in sequence in favour of clipping the entire polygon against each clipping boundary in sequence, as shown in Fig. 6.7.

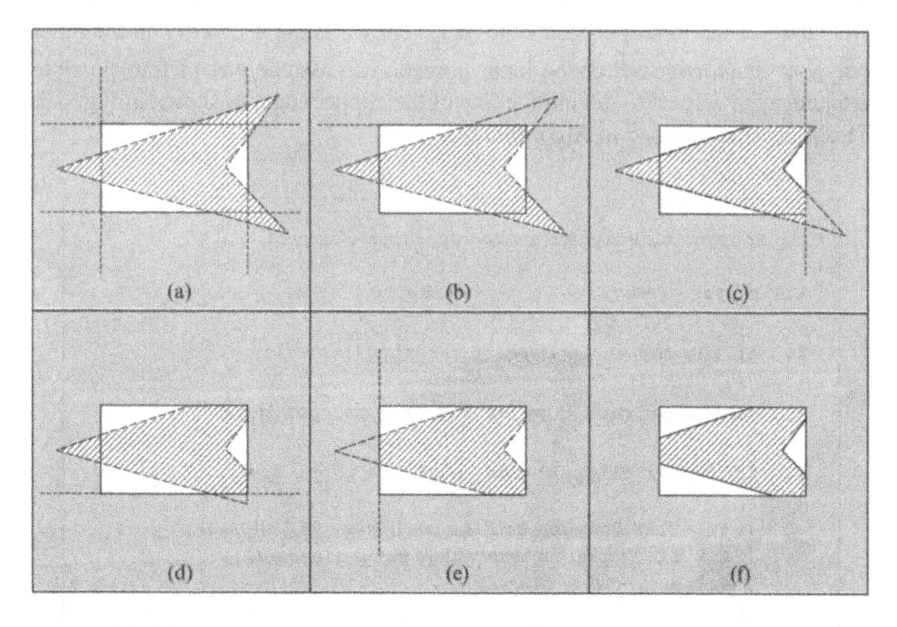

FIGURE 6.7 Clipping against each clipping boundary in turn: (a) the unclipped polygon, (b) clipping against the top boundary, (c) clipping against the right boundary, (d) clipping against the bottom boundary, (e) clipping against the left boundary, (f) the clipped polygon.

To facilitate the description of the Sutherland and Hodgman polygon-clipping algorithm let us first introduce some terminology. The input polygon, i.e. the polygon to be clipped, is referred to as the *unclipped polygon*. The polygonal clipping boundary, against which clipping takes place, is referred to as the *clipper polygon*. The output polygon, resulting from the clipping process, is referred to as the *clipped polygon*. The unclipped polygon consists of l vertices $\boldsymbol{P}_1, \boldsymbol{P}_2, \ldots, \boldsymbol{P}_l$, l edges $\overline{\boldsymbol{P}_1\boldsymbol{P}_2}$, $\overline{\boldsymbol{P}_2\boldsymbol{P}_3}$, ..., $\overline{\boldsymbol{P}_l\boldsymbol{P}_1}$ and may be defined in clockwise or in counter-clockwise order. The clipper polygon consists of m vertices $\boldsymbol{C}_1$, $\boldsymbol{C}_2, \ldots, \boldsymbol{C}_m$, m edges $\overline{\boldsymbol{C}_1\boldsymbol{C}_2}$, $\overline{\boldsymbol{C}_2\boldsymbol{C}_3}$, ..., $\overline{\boldsymbol{C}_m\boldsymbol{C}_1}$ and must be defined in counter-clockwise order. The clipped polygon consists of n vertices $\boldsymbol{O}_1$, $\boldsymbol{O}_2$, ..., $\boldsymbol{O}_n$, and is defined in the same order as the unclipped polygon. The numbers of vertices of these three polygons obey the relationship $n \leq (l + m)$.

The Sutherland and Hodgman polygon-clipping algorithm can be seen as consisting of two sub-algorithms. The first sub-algorithm constructs *clipper edges* and clips the entire polygon against each such edge in turn, by calling the second sub-algorithm. In the first clipping stage, the input polygon is clipped against the first clipper edge. The output polygon resulting from the first clipping stages then becomes the input polygon for the second clipping stage and so forth. This process is outlined in Algorithm 6.2. The second sub-algorithm clips the entire input polygon against one clipper edge. It considers the list of vertices of the input polygon one at a time. For each such vertex $\boldsymbol{P}_i$ zero, one or two output vertices $\boldsymbol{O}_j$ will be generated. Every vertex $\boldsymbol{P}_i$ is considered to be the terminal vertex of an edge $\overrightarrow{\boldsymbol{S}\boldsymbol{P}_i}$, where $\boldsymbol{S}$ is the previous (saved) vertex in the polygon vertex loop. Note that for the input polygon edge $\overrightarrow{\boldsymbol{S}\boldsymbol{P}_1}$, vertex $\boldsymbol{S}$ is the last vertex in the input polygon. Each edge $\overrightarrow{\boldsymbol{S}\boldsymbol{P}}$ of the input polygon can assume one of four possible attitudes with respect to the infinite line of the clipper edge, as shown in Fig. 6.8. This process is outlined in Algorithm 6.3.

```
Clip_Polygon_to_Polygon(unclipped, clipper, clipped)
{
  1.  clipped ← empty;

  2.  if the clipper is empty then return(clipped);

  3.  make a private copy of the unclipped polygon;

  4.  for every clipper edge ce of the clipper do
       {
        4.1 Clip_Polygon_to_Edge(unclipped, ce, clipped);
        4.2 if clipped is empty then return(clipped);
        4.3 unclipped ← clipped;
       }
  5.  return(clipped);
}
```

Algorithm 6.2 The *Clip_Polygon_to_Polygon* algorithm.

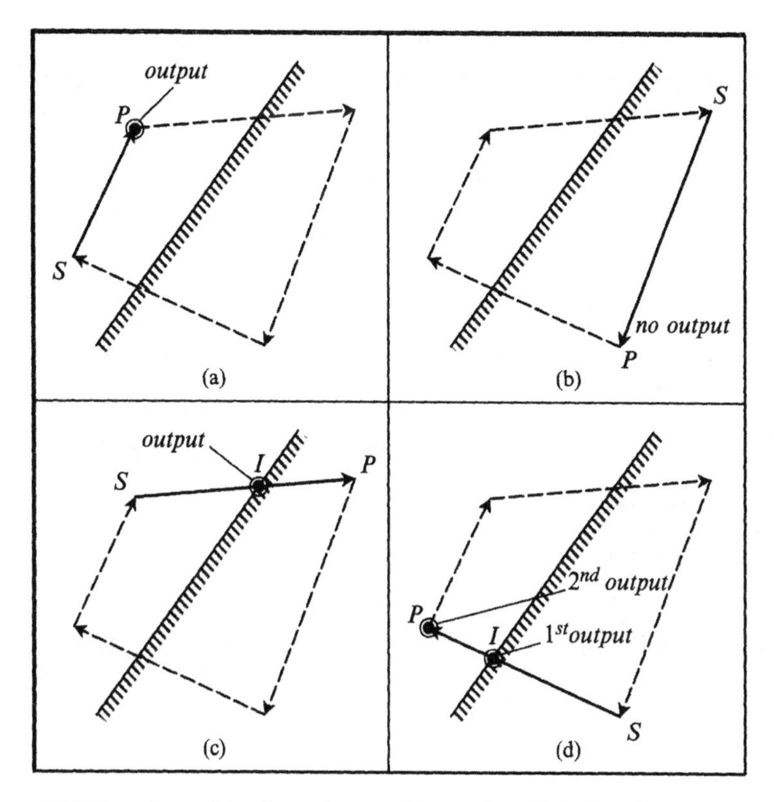

FIGURE 6.8 The relationship of an edge to a clipper edge: (a) both endpoints are visible, (b) both endpoints are invisible, (c) the edge is leaving the visible region, (d) the edge is entering the visible region.

If both endpoints of the input polygon edge lie on the visible half-space of the clipper edge, then only its terminal vertex ***P*** need be output since its initial vertex ***S*** would have already been output. This case is depicted in Fig. 6.8a. If both endpoints of the edge lie on the invisible half-space of the clipper edge, then nothing need be output. This case is depicted in Fig. 6.8b. If the edge is leaving the visible half-space of the clipper edge, with ***S*** visible and ***P*** invisible, then only the point of intersection ***I*** of the input polygon edge and the clipper edge need be output, as the initial vertex ***S*** would have already been output. This case is depicted in Fig. 6.8c. If the edge is entering the visible half-space of the clipper edge, with ***S*** invisible and ***P*** visible, then two outputs must be generated, the point of intersection ***I*** of the edge and the clipper edge, followed by the terminal vertex ***P***. This case is depicted in Fig. 6.8d.

Figure 6.9 illustrates the clipping process steps when clipping a concave polygon against a convex clipping boundary. In this case two portions of the unclipped polygon fall within the clipping boundary. The Sutherland and Hodgman polygon-clipping algorithm deals with such polygons correctly by inserting so-called *bridging edges* that connect the separate parts of the polygon. These bridging edges are indicated by arrows in the figure. In this example the unclipped polygon is defined in counter-clockwise order.

In order to implement this algorithm we must be able to determine if a point of the input polygon lies on the visible half-plane defined by a clipper edge and we must be able to calculate the coordinates of the point of intersection of an input polygon edge and a clipper edge. Let us develop the mathematical formulae that will allow us to make those two determinations.

```
Clip_Polygon_to_Edge(unclipped, ce, clipped)
{
 1.   Determine the visibility of every vertex of the
      unclipped polygon;

 2.   if the unclipped polygon can be trivially rejected then
       {
 2.1     dispose of the unclipped polygon;
 2.2     clipped ← empty;
 2.3     return(clipped);
       }

 3.   if the unclipped polygon can be trivially accepted then
       {
 3.1     clipped ← unclipped;
 3.2     return(clipped);
       }

 4.   /* get the current and previous input vertex pointers */
      civ ← unclipped;
      piv ←last vertex of the unclipped polygon;

 5.   while(civ ≠ empty) do
       {
 5.1     if civ is visible then
         {
 5.1.1     if piv is visible then Output_Vertex(civ); else
           {
 5.1.2.1     Output_Intersection(piv, civ, ce);
 5.1.2.2     Output_Vertex(civ);
           }
         }else
 5.2  if civis invisible then
         {
 5.2.1     if piv is visible then Output_Intersection(piv,civ,ce);
         }
 5.3  piv ← civ;
 5.4  civ ←next vertex of the unclipped polygon;
       }
 6.   return(clipped);
}
```

Algorithm 6.3 The *Clip_Polygon_to_Edge* algorithm.

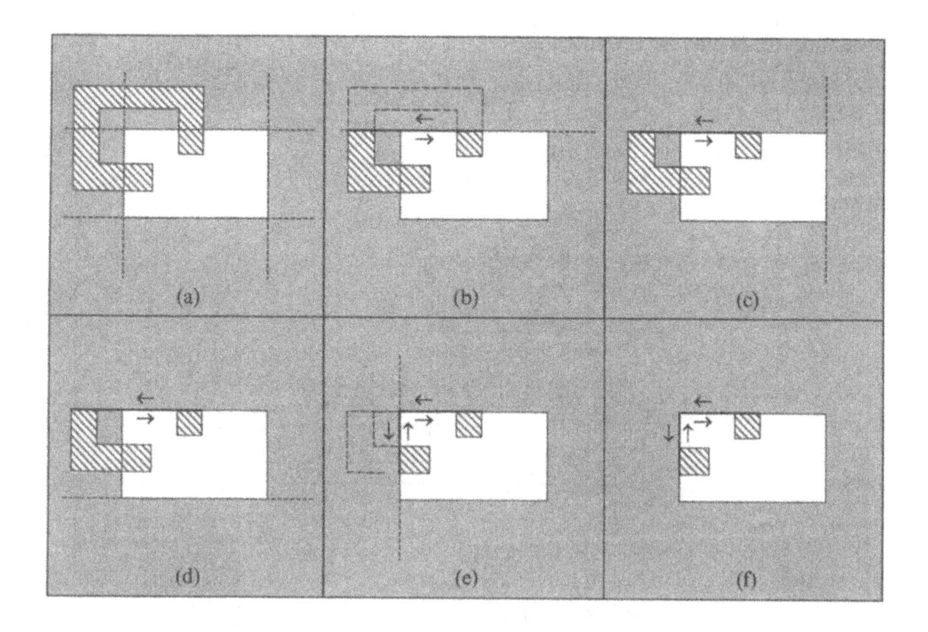

FIGURE 6.9 Clipping a polygon, with two visible parts, against each clipping boundary in turn: (a) the unclipped polygon, (b) clipping against the top boundary, (c) clipping against the right boundary, (d) clipping against the bottom boundary, (e) clipping against the left boundary, (f) the clipped polygon.

Given a line segment defined by two points $\boldsymbol{P}_1 = \langle x_1, y_1 \rangle$, $\boldsymbol{P}_2 = \langle x_2, y_2 \rangle$ and the given general point on this line is $P = \langle x, y \rangle$, then the slope of the line is given by

$$m = \frac{d_y}{d_x} = \frac{y_2 - y_1}{x_2 - x_1} = \frac{y - y_1}{x - x_1} \tag{6.6}$$

$$\begin{aligned}
\therefore \quad & (y - y_1) \cdot (x_2 - x_1) = (y_2 - y_1) \cdot (x - x_1) \\
\therefore \quad & y \cdot (x_2 - x_1) - y_1 \cdot (x_2 - x_1) = x \cdot (y_2 - y_1) - x_1 \cdot (y_2 - y_1) \\
\therefore \quad & -x \cdot (y_2 - y_1) + y \cdot (x_2 - x_1) + x_1 \cdot (y_2 - y_1) - y_1 \cdot (x_2 - x_1) = 0 \\
\therefore \quad & -(y_2 - y_1) \cdot x + (x_2 - x_1) \cdot y + x_1 \cdot y_2 - x_1 \cdot y_1 - y_1 \cdot x_2 + y_1 \cdot x_1 = 0
\end{aligned}$$

which can be written as

$$\therefore \quad a \cdot x + b \cdot y + c = 0 \tag{6.7}$$

where

$$\begin{aligned}
a &= -(y_2 - y_1) = (y_1 - y_2) = -d_y \\
b &= (x_2 - x_1) = d_x \\
c &= (x_1 \cdot y_2 - y_1 \cdot x_2)
\end{aligned} \tag{6.8}$$

Equation (6.7) may be rewritten as

$$x = -\frac{b}{a}y - \frac{c}{a}$$
$$\text{or} \quad x = d \cdot y + e \tag{6.9}$$

where

$$d = -\frac{b}{a} = -\frac{d_x}{-d_y} = \frac{d_x}{d_y}$$
$$\begin{aligned} e = -\frac{c}{a} &= \frac{x_1 \cdot y_2 - y_1 \cdot x_2}{d_y} \\ &= \frac{x_1 \cdot y_2 - x_1 \cdot y_1 - y_1 \cdot x_2 + x_1 \cdot y_1}{d_y} \\ &= \frac{(y_2 - y_1) \cdot x_1 - (x_2 - x_1) \cdot y_1}{d_y} \\ &= -\frac{d_x}{d_y} \cdot y_1 + x_1 \\ &= -d \cdot y_1 + x_1 \end{aligned}$$

Thus

$$d = \frac{d_x}{d_y}$$
$$e = -d \cdot y_1 + x_1 \tag{6.10}$$

Finding the Point of Intersection of Two Lines

Assuming that the two lines intersect at a point $\boldsymbol{P}_i = \langle x_i, y_i \rangle$, then the two line equations are given by

$$a_1 \cdot x_i + b_1 \cdot y_i + c_1 = 0$$
$$a_2 \cdot x_i + b_2 \cdot y_i + c_2 = 0$$

or alternatively by

$$\begin{aligned} & x_i = d_1 \cdot y_i + e_1 \\ & x_i = d_2 \cdot y_i + e_2 \\ \therefore \quad & d_1 \cdot y_i + e_1 = d_2 \cdot y_i + e_2 \\ \therefore \quad & (d_1 - d_2) \cdot y_i = e_2 - e_1 \\ \therefore \quad & y_i = \frac{e_2 - e_1}{d_1 - d_2} \end{aligned} \tag{6.11}$$

Once we have calculated y_i we can substitute it in Eq. (6.9) to compute x_i.

$$\begin{aligned} x_i &= d_1 \cdot y_i + e_1 \\ &= d_1 \cdot y_i - d_1 \cdot y_1 + x_1 \\ &= d_1 \cdot (y_i - y_1) + x_1 \end{aligned} \tag{6.12}$$

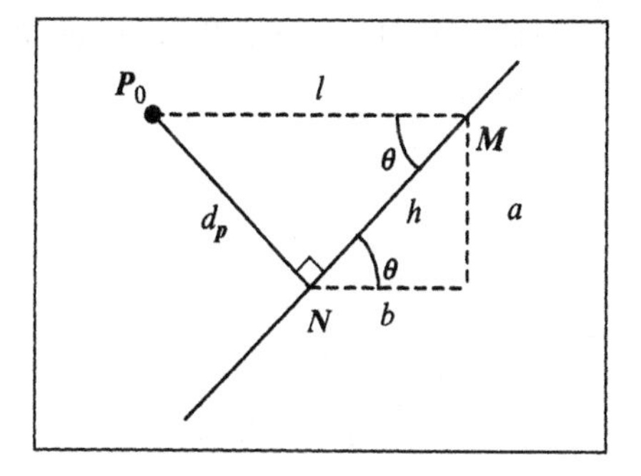

FIGURE 6.10. The Perpendicular distance from a point to a line.

Finding the Perpendicular Distance from a Point to a Line

Let $\boldsymbol{P}_0 = \langle x_0, y_0 \rangle$ be a given point and $\overrightarrow{\boldsymbol{P}_0\boldsymbol{N}}$ be the perpendicular from point $\boldsymbol{P}_0$ to the line with equation

$$a \cdot x + b \cdot y + c = 0$$

Let $\overrightarrow{\boldsymbol{P}_0\boldsymbol{M}}$ be a horizontal line through point $\boldsymbol{P}_0$.

Recall from Eq. (6.8) that $a = -d_y$ and $b = d_x$. Now from Fig. 6.10 we have

$$h = \sqrt{(-a)^2 + b^2} = \sqrt{a^2 + b^2}$$

Point $\boldsymbol{M}$ has $y = y_0$, its x coordinate can be found from Eq. (6.7) as

$$x = \frac{-(b \cdot y_0 + c)}{a}$$

Hence line segment $\overrightarrow{\boldsymbol{P}_0\boldsymbol{M}}$ has length

$$l = x - x_0 = \frac{-(b \cdot y_0 + c)}{a} - x_0$$

$$\therefore \quad l = \frac{-(a \cdot x_0 + b \cdot y_0 + c)}{a}$$

From Fig. 6.10 we have

$$\sin\theta = \frac{-a}{h}$$

Also from this figure we have

$$\sin\theta = \frac{d_p}{l}$$

Thus the perpendicular distance d_p of the point $\boldsymbol{P}_0$ from the line is given by

$$\begin{aligned} d_p &= l \cdot \sin\theta \\ &= -\frac{(a \cdot x_0 + b \cdot y_0 + c)}{a} \cdot \frac{-a}{h} \\ &= \frac{a \cdot x_0 + b \cdot y_0 + c}{h} \\ \therefore \quad d_p &= \frac{a \cdot x_0 + b \cdot y_0 + c}{\sqrt{a^2 + b^2}} \end{aligned} \tag{6.13}$$

The sign of this expression is sometimes positive and sometimes negative, as the line divides the plane in two half-planes.

If the line segment $\overrightarrow{\mathbf{NM}}$ is directed and forms part of a counter-clockwise polygonal outline, then d_p is positive when the point $\boldsymbol{P}_0$ lies inside the polygon and negative when $\boldsymbol{P}_0$ lies outside the polygon. Observe that the denominator of the fraction in Eq. (6.13) will always be positive. Thus, if we are only interested in the sign of d_p we can simplify this expression to

$$\text{sign}\left(d_p\right) = \text{sign}\left(a \cdot x_0 + b \cdot y_0 + c\right) \tag{6.14}$$

The following **C** functions implement the Sutherland and Hodgman polygon-clipping algorithm.

```
/*
 * Common constants and typedefs.
 */

#include <stdio.h>
#include <math.h>

#define small_t 0.000000005

typedef unsigned char boolean_t;

#define False (boolean_t) 0
#define True  (boolean_t) 1

/*
 * Constants and typedefs for the Sutherland and Hodgman
 * polygon-clipping routine.
 */

typedef struct poly_vtx_t *poly_vtx_ptr_t;

typedef struct poly_vtx_t
{
 double          x, y;      /* Point coordinates       */
 poly_vtx_ptr_t next_vtx;  /* Next vertex pointer     */
 boolean_t       inside;    /* Inside clipper indicator */
} poly_vtx_t;
```

```
typedef struct clipper_edge_t
{
 double x1, y1;    /* Clipper Edge 1st Endpoint Coordinates    */
 double x2, y2;    /* Clipper Edge 2nd Endpoint Coordinates    */
 double a, b, c;   /* Clipper Edge Line Equation Coefficients */
} clipper_edge_t;

/*---------------------------------------------------------------------*/

void sh_output_vertex
(
 poly_vtx_ptr_t civp,    /* Current Input Clip Vertex Pointer  */
 poly_vtx_ptr_t *fovp,   /* First Output Clip Vertex Pointer   */
 poly_vtx_ptr_t *covp    /* Current Output Clip Vertex Pointer */
)
{
 /*
  * Output the current input vertex.
  */

 poly_vtx_ptr_t nvp;


 nvp = (poly_vtx_ptr_t) malloc(sizeof(poly_vtx_t));

 /*
  * Copy the structure.
  */
 *nvp = *civp;

 nvp->next_vtx = NULL;

 if (*fovp == NULL) *fovp = nvp;
 else               (*covp)->next_vtx = nvp;

 *covp = nvp;

} /* sh_output_vertex */

/*---------------------------------------------------------------------*/

void sh_output_intersection
(
 poly_vtx_ptr_t civp,    /* Current Input Clip Vtx ptr  */
 poly_vtx_ptr_t pivp,    /* Previous Input Clip Vtx ptr */
 poly_vtx_ptr_t *fovp,   /* First Output Clip Vtx ptr   */
 poly_vtx_ptr_t *covp,   /* Current Output Clip Vtx ptr */
 clipper_edge_t ce       /* Clipper Edge Data           */
)
{
 poly_vtx_ptr_t nvp;
 double         t,
                ux1, uy1,
                ux2, uy2,
```

```
                ua, ub,
                d1, d2,
                xi, yi;

  /*
   * Compute and output the point of intersection of the current
   * unclipped polygon edge with the current clipper edge.
   */

  /*
   * Create a new vertex.
   */

  nvp = (poly_vtx_ptr_t) malloc(sizeof(poly_vtx_t));

  if (*fovp == NULL) *fovp = nvp;
  else               (*covp)->next_vtx = nvp;

  *covp = nvp;

  /*
   * Compute the intersection of the unclipped edge with the
   * clipper edge.
   */

  ux1 = pivp->x;
  uy1 = pivp->y;

  ux2 = civp->x;
  uy2 = civp->y;

  ua = uy1 - uy2;
  ub = ux2 - ux1;

  if (fabs(ua) < small_t)
   {
    /*
     * Unclipped edge is vertical.
     */

    t = ce.y2 - ce.y1;

    if (fabs(t) < small_t) t = small_t;

    xi = (uy1 - ce.y1) / t * ce.b + ce.x1;
    yi = uy1;
   }
  else

  if (fabs (ce.a) < small_t)
   {
    /*
     * Clipper edge is horizontal.
     */
```

```
   t = uy2 - uy1;

   if (fabs(t) < small_t) t = small_t;

   xi = (ce.y1 - uy1) / t * ub + ux1;
   yi = ce.y1;
   }
  else
   {
    /*
     * General case.
     */

   d1 = ub / -ua;
   d2 = ce.b / -ce.a;

   /*
    * From: d1 * yi + e1 = d2 * yi + e2
    *
    * we get: yi = (e2 - e1) / (d1 - d2)
    *
    * where: e1 = (ux1 - uy1 * d1) and e2 = (ce.x1 - ce.y1 * d2).
    */

   yi = (ce.x1 - ce.y1 * d2 - ux1 + uy1 * d1) / (d1 - d2);

   /*
    * From: (xi - ux1) / (yi - uy1) = d1
    *
    * we get: xi = (yi - uy1) * d1 + ux1.
    */

   xi = (yi - uy1) * d1 + ux1;
  }

  (*covp)->x = xi;
  (*covp)->y = yi;
  (*covp)->next_vtx = NULL;

} /* sh_output_intersection */

/*------------------------------------------------------------------*/

void sh_clip_polygon_to_edge
(
 poly_vtx_ptr_t ivp,    /* Input Polygon (In)      */
 poly_vtx_ptr_t *ovp,   /* Output Polygon (Out)    */
 clipper_edge_t ce      /* Clipper Edge Data (In) */
)
{
 poly_vtx_ptr_t civp; /* Current Input Clip Vertex Pointer  */
 poly_vtx_ptr_t pivp; /* Previous Input Clip Vertex Pointer */
 poly_vtx_ptr_t livp; /* Last Input Clip Vertex Pointer     */
```

```
  poly_vtx_ptr_t fovp; /* First Output Clip Vertex Pointer   */
  poly_vtx_ptr_t covp; /* Current Output Clip Vertex Pointer */
  boolean_t      all_inside,
                 all_outside;
  double         d;

/*
 * Determine the visibility of each vertex of the input polygon
 * relative to the current clipping boundary.
 *
 * Determine if the input polygon can be trivially accepted or
 * rejected.
 *
 * Also find the last vertex of the input polygon and use it as the
 * start vertex.
 */

  all_inside  = True;
  all_outside = True;

  pivp = NULL;
  civp = ivp;

  while (civp != NULL)
   {
    pivp = civp;

    /*
     * Compute the signed distance of the point from the clipper edge.
     *
     * The signed distance of a point <x, y> from the line with
     * equation:
     *
     * a*x+b*y+c=0
     *
     * is given by:
     *
     * d = (a * x + b * y + c) / sqrt(a^2 + b^2).
     *
     * In this case we only need the sign of d, so we only need to
     * compute numerator of the above fraction.
     */

    d = ce.a * civp->x + ce.b * civp->y + ce.c;

    /*
     * Do the inside test.
     */

    civp->inside = (d >= 0);
    all_inside   = (all_inside & civp->inside);

    /*
     * Do the outside test.
```

```
    */

   all_outside = (all_outside & (d <= 0));
   civp        = civp->next_vtx;
  }


/*
 * Save the pointer to the last vertex in the input polygon.
 */

livp = pivp;

if (all_inside)
 {
  /*
   * Trivially accepted.
   */

  *ovp = ivp;
  return;
 }

if (all_outside)
 {
  /*
   * Trivially rejected.
   */

  *ovp = NULL;

  /*
   * Dispose of the input polygon.
   */

  while (ivp != NULL)
   {
    civp = ivp;
    ivp = ivp->next_vtx;
    free(civp);
   }

  return;
 }

/*
 * Must process further.
 */

fovp = NULL;
covp = NULL;
civp = ivp;

while (civp != NULL)
```

```
{
 /*
  * For every vertex in the input polygon.
  */


 if (civp->inside)
  {
   /*
    * Current input vertex is inside.
    */

   if (pivp->inside)
    {
     /*
      * Previous input vertex is inside.
      */

     sh_output_vertex(civp, &fovp, &covp);
    }
   else
    {
     /*
      * Previous Input Vertex is Outside.
      */

     sh_output_intersection(civp, pivp, &fovp, &covp, ce);
     sh_output_vertex(civp, &fovp, &covp);
    }
  }
 else
  {
   /*
    * Current input vertex is outside.
    */

   if (pivp->inside)
    {
     /*
      * Previous input vertex is inside.
      */

     sh_output_intersection(civp, pivp, &fovp, &covp, ce);
    }
  }

 pivp = civp;
 civp = civp->next_vtx;
 }


*ovp = fovp;

/*
 * Dispose of the input polygon.
```

```
  */

 while (ivp != NULL)
  {
   civp = ivp;
   ivp = ivp->next_vtx;
   free(civp);
  }
} /* sh_clip_polygon_to_edge */

/*------------------------------------------------------------*/

poly_vtx_ptr_t sh_copy_polygon(poly_vtx_ptr_t ivp)
{
 poly_vtx_ptr_t fvp, nvp, lvp;


 fvp = NULL;
 nvp = NULL;

 while (ivp != NULL)
  {
   lvp = nvp;

   nvp = (poly_vtx_ptr_t) malloc(sizeof(poly_vtx_t));

   if (fvp == NULL) fvp = nvp;

   nvp->x        = ivp->x;
   nvp->y        = ivp->y;
   nvp->next_vtx = NULL;

   if (lvp != NULL) lvp->next_vtx = nvp;

   ivp = ivp->next_vtx;
  }

 return(fvp);

} /* sh_copy_polygon */

/*------------------------------------------------------------*/

void sh_clip_polygon_to_polygon
(
 poly_vtx_ptr_t unclipped,  /* Input Poly (In)   */
 poly_vtx_ptr_t clipper,    /* Clipper Poly (In) */
 poly_vtx_ptr_t *clipped    /* Output Poly (Out) */
)
{
 /*
 * This is an implementation of the Sutherland and Hodgman
 * polygon-clipping algorithm. This version clips the unclipped
 * polygon to the clipping boundary specified by the
```

```
* clipper polygon.
*
* The unclipped polygon pointer points to a linked list of polygon
* vertices provided by the caller.
*
* The clipped polygon pointer points to a linked list of polygon
* vertices returned by this function. If the clipped polygon is
* empty a NULL pointer is returned.
*
* The clipper polygon pointer points to a linked list of polygon
* vertices provided by the caller. The clipper must be specified
* in counter-clockwise order and must be a convex polygon.
*
* This function works with empty input polygons as well!
*/

poly_vtx_ptr_t fcvp;  /* First Clipper Vertex Pointer   */
poly_vtx_ptr_t ccvp;  /* Current Clipper Vertex Pointer */
poly_vtx_ptr_t ncvp;  /* Next Clipper Vertex Pointer    */
clipper_edge_t ce;    /* Clipper Edge Data              */
double         dx, dy;

/*
 * First check if the clipper is empty.
 */

*clipped = NULL;

if (!clipper) return;

/*
 * Make a private copy of the unclipped polygon. This is required
 * as this function disposes of the unclipped polygon and the
 * caller might not like this!
 */

unclipped = sh_copy_polygon(unclipped);

/*
 * Clip the unclipped polygon against all the clipper edges.
 */

fcvp = NULL;
ccvp = clipper;

while (ccvp != fcvp)
 {
  if (!fcvp) fcvp = ccvp;

  ncvp = ccvp->next_vtx;

  if (!ncvp) ncvp = fcvp;

  /*
   * Compute the line equation of the clipper edge.
```

```
    */

    ce.x1 = ccvp->x;
    ce.y1 = ccvp->y;
    ce.x2 = ncvp->x;
    ce.y2 = ncvp->y;

    dx = ce.x2 - ce.x1;
    dy = ce.y2 - ce.y1;

    ce.a = -dy;
    ce.b = dx;
    ce.c = ce.x1 * dy - ce.y1 * dx;

    sh_clip_polygon_to_edge(unclipped, clipped, ce);

    if (!(*clipped)) return;

    unclipped = *clipped;

    ccvp = ncvp;
  }

} /* sh_clip_polygon_to_polygon */
```

6.4.2 The Weiler and Atherton Polygon-Clipping Algorithm

The Sutherland–Hodgman algorithm allows us to clip a convex or concave polygon to a convex clipping boundary. If we wish to clip a to a convex or concave clipping boundary we have to use the Weiler–Atherton algorithm [Weiler 77]. This is a generalised-polygon clipper that is capable of clipping a concave polygon with holes to the borders of a concave polygon with holes.

The polygon to be clipped is called the *subject polygon* and clipping is performed to the borders of the *clip polygon*.

The algorithm represents a polygon as a set of contours. These contours can be *outline contours* or *hole contours*. Each contour is represented as a circular list of vertices. Unlike the Sutherland–Hodgman algorithm, the vertices of an outline contour are linked in a clockwise order and those of a hole contour are linked in counter-clockwise order. Using this order, as one follows along the chain of vertices of the polygon, the interior of the polygon is always to the right of the border (see Fig. 6.11).

The clipping process may fragment the subject polygon in more than one visible polygons (see Fig. 6.12).

The clipping process has the following steps:

1. The borders of the two polygons are compared for intersections. At each intersection a new *false vertex* is added into the contour chain of each of the two intersecting polygons and it is marked as an *intersection vertex*. A link is established between each pair of intersection vertices, thus allowing us to switch (jump) between the two polygons whenever they intersect.

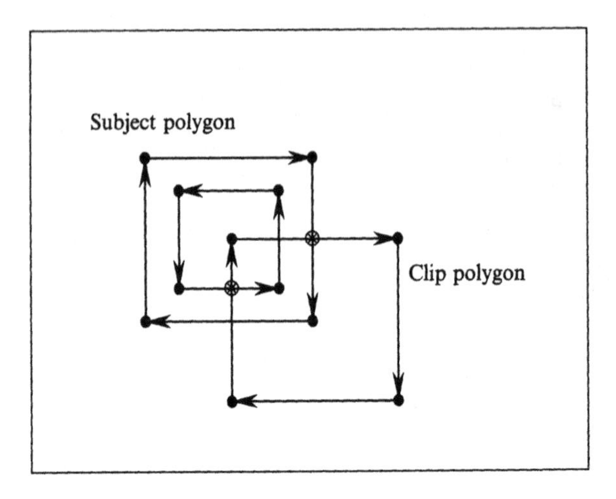

FIGURE 6.11. A subject polygon with a hole and the clip polygon.

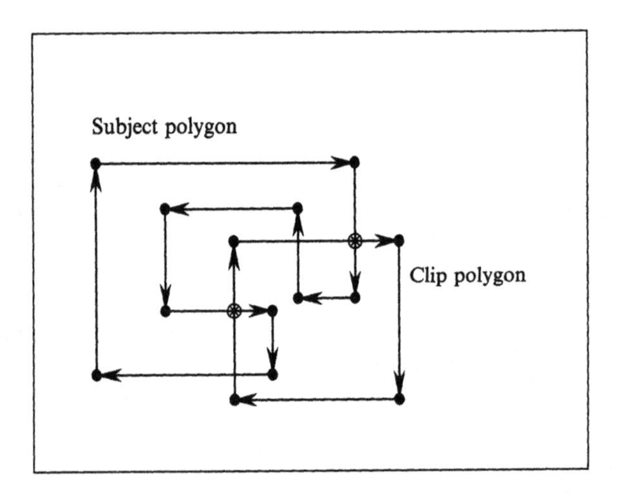

FIGURE 6.12. Subject polygon fragmentation.

2. Next we process all contours that have no intersections. Each contour of the subject polygon that has no intersections and lies inside the clip polygon is placed in a *holding list*. Clip polygon contours that lie outside the subject polygon are ignored. Clip polygon contours that lie inside the subject polygon are put in the holding list, as such contours in effect cut holes in the subject polygon (see Fig. 6.13).
3. A list of the intersection vertices found on all the subject polygon contours is formed. This *intersection list* contains only those intersections where *the clip polygon border passes to the outside of the subject polygon* (i.e. the intersection points indicated by the symbol ⊗ in Figs. 6.11–6.13).

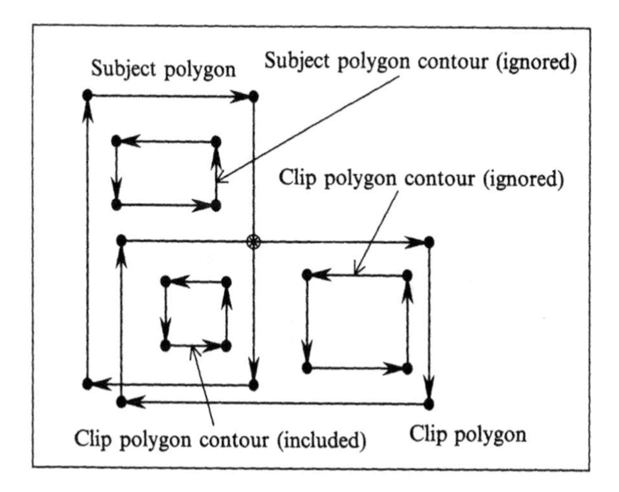

FIGURE 6.13. New hole cut in the subject polygon.

4. The actual clipping can now be done. It consists of six steps:
   a. An intersection vertex is removed from the intersection list and used as the starting point of a new output contour. If this list is exhausted, then the clipping is complete; so go to step 5.
   b. Follow along the subject polygon vertex chain until the next intersection is reached.
   c. Jump to the clip polygon (using the link created in step 1).
   d. Follow along the clip polygon vertex chain until the next intersection is reached.
   e. Jump to the subject polygon (using the link created in step 1). If this intersection vertex is not the start vertex, then remove it from the intersection list.
   f. Repeat steps (b)–(e) until the start intersection point has been reached. When this point is reached a new contour has just been closed and can be put in the holding list. Go to step (a).
5. All the holes in the holding list are attached to their corresponding outlines.

Example

Let us consider the subject and clip polygons shown in Fig. 6.14.

The subject polygon is described by the chain of vertices $\{s_1, s_2, s_3, s_4, s_5, s_6, s_7, s_8, s_9, s_{10}, s_1\}$ and the clip polygon is described by the chain of vertices $\{c_1, c_2, c_3, c_4, c_1\}$. The two polygons intersect at the points $\{i_1^{\otimes}, i_2, i_3^{\otimes}, i_4\}$ but only vertices $i_1^{\otimes}$ and $i_3^{\otimes}$ are included in the intersection list.

After steps (1)–(3) of the algorithm our lists look like this:

subject polygon: $\{s_1, ({}^s i_1 \rightarrow {}^c i_1), s_2, s_3, ({}^s i_2 \rightarrow {}^c i_2), s_4, s_5, s_6, ({}^s i_3 \rightarrow {}^c i_3), s_7, s_8, ({}^s i_4 \rightarrow {}^c i_4), s_9, s_{10}, s_1\}$

clip polygon: $\{c_1, c_2, ({}^c i_4 \rightarrow {}^s i_4), ({}^c i_3 \rightarrow {}^s i_3), c_3, ({}^c i_2 \rightarrow {}^s i_2), ({}^c i_1 \rightarrow {}^s i_1), c_4, c_1\}$

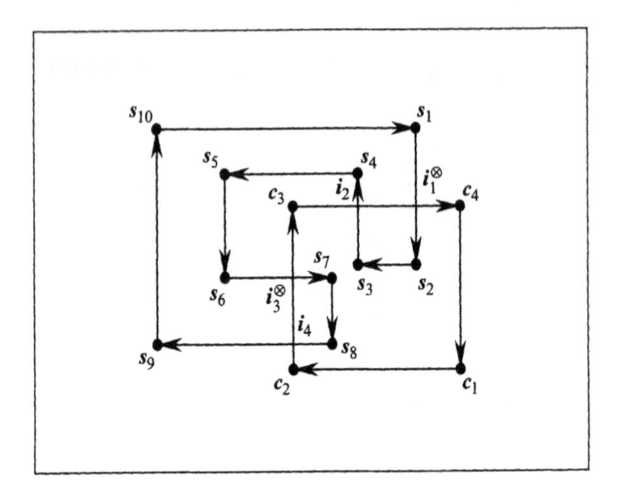

FIGURE 6.14. A clipping example.

intersection list: $\{{}^s i_1, {}^s i_3\}$
output polygon: { }
holding list: { }

where $({}^s\mathbf{i}_j \rightarrow {}^c\mathbf{i}_j)$ means that *the intersection vertex* $\mathbf{i}_j$ *in the subject polygon points to the intersection point* $\mathbf{i}_j$ *in the clip polygon* and analogously $({}^c\mathbf{i}_j \rightarrow {}^s\mathbf{i}_j)$ means that *the intersection vertex* $\mathbf{i}_j$ *in the clip polygon points to the intersection point* $\mathbf{i}_j$ *in the subject polygon.*

At step (4a) we pickup $i_1^{\otimes}$ from the intersection list, we remove it from this list and we output i_1. At this stage our output polygon list looks like this: $\{i_1\}$.

At step (4b), we jump to the subject polygon and we output s_2, s_3 and i_2. At this stage our output polygon list looks like this: $\{i_1, s_2, s_3, i_2\}$.

At step (4c), we jump to the clip polygon and at step (4d), we output i_1. After steps (4e) and (4f), we have closed the output polygon $\{i_1, s_2, s_3, i_2, i_1\}$ and thus we can put it in the holding list and continue with any remaining output. At this stages our lists look like this:

intersection list: $\{{}^s i_3\}$
output polygon: { }
holding list: $\{\{i_1, s_2, s_3, i_2, i_1\}\}$

The intersection list is not empty and so we go back to step (4a). We pickup $i_3^{\otimes}$ from the intersection list, we remove it from this list and we output i_3. At this stage our output polygon list looks like this: $\{i_3\}$.

At step (4b), we jump to the subject polygon and we output s_7, s_8 and i_4. At this stage our output polygon list looks like this: $\{i_3, s_7, s_8, i_4\}$.

At step (4c), we jump to the clip polygon and at step (4d), we output i_3. After steps (4e) and (4f), we have closed the output polygon $\{i_3, s_7, s_8, i_4, i_3\}$ and thus we can put it in the holding list. As the intersection list is empty we can go to step (5). At this stage our lists look like this:

intersection list: {}
output polygon: {}
holding list: $\{\{i_1, s_2, s_3, i_2, i_1\}, \{i_3, s_7, s_8, i_4, i_3\}\}$
The holding list contains two outline contours and we are done.

References

[Sutherland 74] Sutherland, I. E. and Hodgman, G. W. Reentrant polygon clipping. *CACM*, Vol. 17, No. 1, p.p. 32–42, Jan 1974.

[Weiler 77] Weiler, K. and Atherton, P. Hidden surface removal using polygon area sorting. *Computer Graphics*, Vol. 11, No. 2, p.p. 214–222 (1977).

7

Three-Dimensional Transformations

7.1 Introduction

Geometric transformations play an important part in the visualisation of three-dimensional scenes. The ability to rotate, translate and scale an object is fundamental to the understanding of its shape. This can easily be demonstrated by picking up a relatively complex and unfamiliar object. In an effort to understand its shape one rotates the object and looks at it from close or at arms length. In the generation of different views of a given scene with the computer, transformations are used to achieve the effect of different viewing positions and directions.

The techniques we shall develop in this chapter for expressing 3D transformations will be an extension of the 2D techniques that we have developed in the chapter concerning 2D transformations.

Two important points to remember are

1. A transformation can be represented by a transformation matrix.
2. Complex transformations can be expressed as a sequence of primitive transformations and can be concatenated to yield a single transformation matrix, which has the same effect as the sequence of the primitive transformations.

Three-dimensional transformations can be represented by a 4×4 matrix and 3D points can be represented in *homogeneous form* by a four-element row or column vector. As with 2D transformations there are two distinct and equivalent notations that can be used to represent the transformation of a 3D point. The first notation is

$$\boldsymbol{P}' = \boldsymbol{P} \cdot \boldsymbol{T}$$

where $\boldsymbol{P}$ and $\boldsymbol{P}'$ are the original and the transformed points, expressed in homogeneous form, and $\boldsymbol{T}$ is the square matrix representing the transformation. For a 3D transformation in this notation we would write

$$[x', y', z', 1] = [x, y, z, 1] \cdot \begin{bmatrix} r_{1,1} & r_{1,2} & r_{1,3} & 0 \\ r_{2,1} & r_{2,2} & r_{2,3} & 0 \\ r_{3,1} & r_{3,2} & r_{3,3} & 0 \\ t_x & t_y & t_z & 1 \end{bmatrix}$$

Or using the alternative notation

$$\boldsymbol{P}'^{\mathrm{T}} = \boldsymbol{T}^{\mathrm{T}} \cdot \boldsymbol{P}^{\mathrm{T}}$$

where $\boldsymbol{P}^{\mathrm{T}}$ and $\boldsymbol{P}'^{\mathrm{T}}$ are the transposed original and transformed points, respectively and $\boldsymbol{T}^{\mathrm{T}}$ is the transpose of the square matrix representing the transformation. For a 3D transformation in this notation we would write

$$\begin{bmatrix} x' \\ y' \\ z' \\ 1 \end{bmatrix} = \begin{bmatrix} r_{1,1} & r_{2,1} & r_{3,1} & t_x \\ r_{1,2} & r_{2,2} & r_{3,2} & t_y \\ r_{1,3} & r_{2,3} & r_{3,3} & t_z \\ 0 & 0 & 0 & 1 \end{bmatrix} \cdot \begin{bmatrix} x \\ y \\ z \\ 1 \end{bmatrix}$$

Observe that as in the 2D case $\boldsymbol{P}' = \boldsymbol{P}'^{\mathrm{T}}$ and so is $\boldsymbol{P} \cdot \boldsymbol{T} = \boldsymbol{T}^{\mathrm{T}} \cdot \boldsymbol{P}^{\mathrm{T}}$.

The generalised transformation matrix

$$\boldsymbol{T} = \left[\begin{array}{ccc:c} r_{1,1} & r_{1,2} & r_{1,3} & 0 \\ r_{2,1} & r_{2,2} & r_{2,3} & 0 \\ r_{3,1} & r_{3,2} & r_{3,3} & 0 \\ \hdashline t_x & t_y & t_z & 1 \end{array}\right] \quad or \quad \boldsymbol{T}^{\mathrm{T}} = \left[\begin{array}{ccc:c} r_{1,1} & r_{2,1} & r_{3,1} & t_x \\ r_{1,2} & r_{2,2} & r_{3,2} & t_y \\ r_{1,3} & r_{2,3} & r_{3,3} & t_z \\ \hdashline 0 & 0 & 0 & 1 \end{array}\right] \tag{7.1}$$

can be partitioned into three separate sub-matrices. The 3×3 sub-matrix of elements labelled $r_{i,j}$ is used to represent *linear transformations* such as scaling, rotation, shearing and reflection transformations. The diagonal elements of this sub-matrix are used to represent scaling and reflection transformations, and the off-diagonal elements are used to represent shearing transformations. A *linear transformation* is one that transforms a linear combination of vectors into some linear combination of transformed vectors. The 1×3 row sub-matrix or the 3×1 column sub-matrix of elements labelled t_x, t_y, t_z is used to represent translations. Finally, the 4×1 column sub-matrix or the 1×4 row sub-matrix must always be set to

$$\begin{bmatrix} 0 \\ 0 \\ 0 \\ 1 \end{bmatrix} \quad \text{or} \quad \begin{bmatrix} 0 & 0 & 0 & 1 \end{bmatrix}$$

so that we have an *affine transformation*. An affine transformation is a combination of linear transformations followed by a translation. Under an affine transformation every straight line maps onto a straight line, parallel lines map onto parallel lines, and if a point divides a segment into a given ratio, its image divides the image of this segment into the same ratio [Ahuja 68].

7.2 Primitive 3D Transformations

The following primitive 3D transformations will be discussed.

1. Translation transformations along the ***X***-, ***Y***- and ***Z***-axes.
2. Scaling transformations relative to the origin and along the ***X***-, ***Y***- and ***Z***-axes.

3. Rotation transformations about the **X**-, **Y**- and **Z**-axes.
4. Shearing transformations parallel to the **X**-, **Y**- and **Z**-axes.

7.2.1 Scaling Transformation Relative to the Origin

The scaling transformation relative to the origin of E^3 is defined in analytical form as

$$\begin{aligned} x' &= x \cdot s_x \\ y' &= y \cdot s_y \\ z' &= z \cdot s_z \end{aligned} \tag{7.2}$$

where s_x, s_y and s_z are the *scale factors* along the **X**-, **Y**- and **Z**-axes, respectively. These scale factors can assume any value except from the value zero, which leads to a *non-invertible* transformation that collapses every transformed point to the origin. If all three scaling factors are equal (i.e. $s_x = s_y = s_z$), then this transformation is known as *uniform scaling*, otherwise it is known as *non-uniform scaling*. Uniform scaling is a special case of non-uniform scaling.

$$\begin{aligned} x' &= x \cdot s \\ y' &= y \cdot s \\ z' &= z \cdot s \end{aligned} \tag{7.3}$$

Uniform scaling can also be expressed in vector form as follows.

$$\boldsymbol{P}' = t(\boldsymbol{P}') = s \cdot \boldsymbol{P}' \tag{7.4}$$

In matrix form the scaling transformation is given by

$$[x', y', z', 1] = [x, y, z, 1] \cdot \begin{bmatrix} s_x & 0 & 0 & 0 \\ 0 & s_y & 0 & 0 \\ 0 & 0 & s_z & 0 \\ 0 & 0 & 0 & 1 \end{bmatrix} \tag{7.5a}$$

or alternatively by

$$\begin{bmatrix} x' \\ y' \\ z' \\ 1 \end{bmatrix} = \begin{bmatrix} s_x & 0 & 0 & 0 \\ 0 & s_y & 0 & 0 \\ 0 & 0 & s_z & 0 \\ 0 & 0 & 0 & 1 \end{bmatrix} \cdot \begin{bmatrix} x \\ y \\ z \\ 1 \end{bmatrix} \tag{7.5b}$$

7.2.2 Translation Transformation

The translation transformation in analytical form is defined as

$$\begin{aligned} x' &= x + d_x \\ y' &= y + d_y \\ z' &= z + d_z \end{aligned} \tag{7.6}$$

where d_x, d_y and d_z are the *displacements* along the **X**-, **Y**- and **Z**-axes, respectively.

The translation transformation can also be expressed in vector form as follows.

$$\boldsymbol{P}' = \boldsymbol{P} + \boldsymbol{d} = \boldsymbol{P} + \left[d_x, d_y, d_z\right] \tag{7.7}$$

In matrix form the transformation is given by

$$\left[x', y', z', 1\right] = [x, y, z, 1] \cdot \begin{bmatrix} 1 & 0 & 0 & 0 \\ 0 & 1 & 0 & 0 \\ 0 & 0 & 1 & 0 \\ d_x & d_y & d_z & 1 \end{bmatrix} \tag{7.8a}$$

or alternatively by

$$\begin{bmatrix} x' \\ y' \\ z' \\ 1 \end{bmatrix} = \begin{bmatrix} 1 & 0 & 0 & d_x \\ 0 & 1 & 0 & d_y \\ 0 & 0 & 1 & d_z \\ 0 & 0 & 0 & 1 \end{bmatrix} \cdot \begin{bmatrix} x \\ y \\ z \\ 1 \end{bmatrix} \tag{7.8b}$$

7.2.3 Rotation About a Coordinate Axis

In three dimensions it is necessary to devise transformations for rotations about all three coordinate axes. The *rotation angle* θ is measured in a *counter-clockwise positive fashion* about a given axis, when looking at the origin from a point on the positive half of this axis. An important point to note is that the transformation leaves the coordinate values associated with the axis of rotation unchanged and only affects the coordinate values associated with the other two axes.

7.2.3.1 Rotation About the Z-Axis

We start with the rotation about the **Z**-axis because it can be seen as a simple extension of the 2D rotation about the origin. The rotation transformation about the **Z**-axis in analytical form is defined as

$$\begin{aligned} x' &= x \cdot \cos\theta - y \cdot \sin\theta \\ y' &= x \cdot \sin\theta + y \cdot \cos\theta \\ z' &= z \end{aligned} \tag{7.9}$$

where θ is the *angle of rotation* about the **Z**-axis measured in a *counter-clockwise positive fashion*, as seen in Fig. 7.1.

In matrix form the transformation is given by

$$\left[x', y', z', 1\right] = [x, y, z, 1] \cdot \begin{bmatrix} \cos\theta & \sin\theta & 0 & 0 \\ -\sin\theta & \cos\theta & 0 & 0 \\ 0 & 0 & 1 & 0 \\ 0 & 0 & 0 & 1 \end{bmatrix} \tag{7.10a}$$

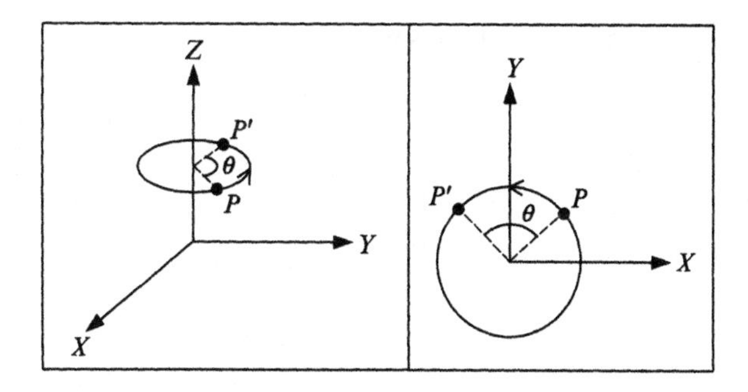

FIGURE 7.1. Rotation about the **Z**-axis.

or alternatively by

$$\begin{bmatrix} x' \\ y' \\ z' \\ 1 \end{bmatrix} = \begin{bmatrix} \cos\theta & -\sin\theta & 0 & 0 \\ \sin\theta & \cos\theta & 0 & 0 \\ 0 & 0 & 1 & 0 \\ 0 & 0 & 0 & 1 \end{bmatrix} \cdot \begin{bmatrix} x \\ y \\ z \\ 1 \end{bmatrix} \tag{7.10b}$$

Observe the similarity of this matrix with that for 2D rotation about the origin.

Transformations of rotation about the ***X***-axis and the ***Y***-axis can be derived from the rotation about the **Z**-axis by permuting the axes in a cyclic fashion, i.e. $x \mapsto y,\ y \mapsto z,\ z \mapsto x$.

7.2.3.2 Rotation About the X-Axis

The rotation transformation about the ***X***-axis in analytical form is defined as

$$\begin{aligned} x' &= x \\ y' &= y \cdot \cos\theta - z \cdot \sin\theta \\ z' &= y \cdot \sin\theta + z \cdot \cos\theta \end{aligned} \tag{7.11}$$

where θ is the *angle of rotation* about the ***X***-axis measured in a *counter-clockwise positive fashion*, as seen in Fig. 7.2.

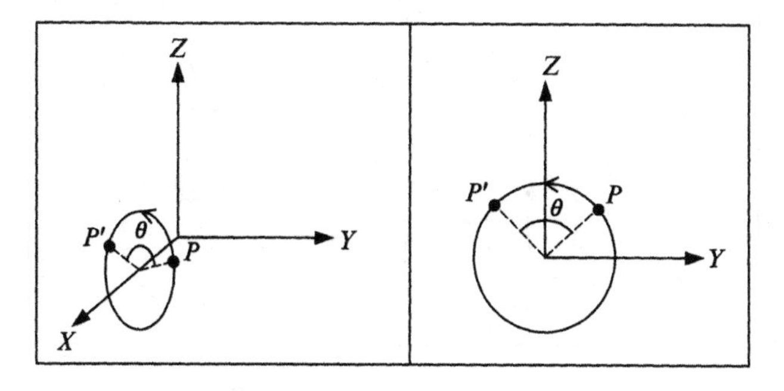

FIGURE 7.2. Rotation about the ***X***-axis.

In matrix form the transformation is given by

$$[x', y', z', 1] = [x, y, z, 1] \cdot \begin{bmatrix} 1 & 0 & 0 & 0 \\ 0 & \cos\theta & \sin\theta & 0 \\ 0 & -\sin\theta & \cos\theta & 0 \\ 0 & 0 & 0 & 1 \end{bmatrix} \tag{7.12a}$$

or alternatively by

$$\begin{bmatrix} x' \\ y' \\ z' \\ 1 \end{bmatrix} = \begin{bmatrix} 1 & 0 & 0 & 0 \\ 0 & \cos\theta & -\sin\theta & 0 \\ 0 & \sin\theta & \cos\theta & 0 \\ 0 & 0 & 0 & 1 \end{bmatrix} \cdot \begin{bmatrix} x \\ y \\ z \\ 1 \end{bmatrix} \tag{7.12b}$$

7.2.3.3 Rotation About the Y-Axis

The rotation transformation about the ***Y***-axis in analytical form is defined as

$$\begin{aligned} x' &= x \cdot \cos\theta + z \cdot \sin\theta \\ y' &= y \\ z' &= -x \cdot \sin\theta + z \cdot \cos\theta \end{aligned} \tag{7.13}$$

where θ is the *angle of rotation* about the ***Y***-axis measured in a *counter-clockwise positive fashion*, as seen in Fig. 7.3.

In matrix form the transformation is given by

$$[x', y', z', 1] = [x, y, z, 1] \cdot \begin{bmatrix} \cos\theta & 0 & -\sin\theta & 0 \\ 0 & 1 & 0 & 0 \\ \sin\theta & 0 & \cos\theta & 0 \\ 0 & 0 & 0 & 1 \end{bmatrix} \tag{7.14a}$$

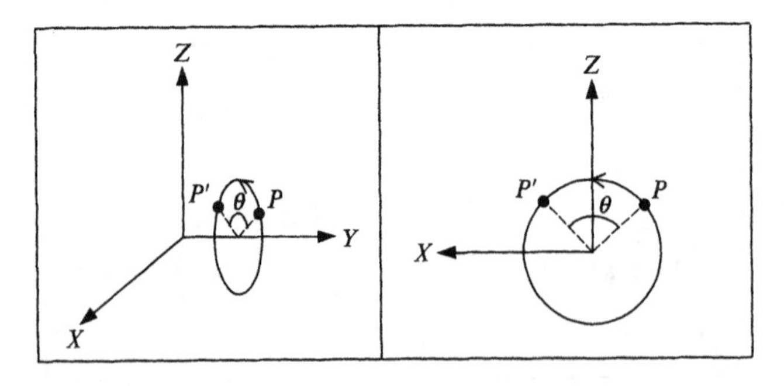

FIGURE 7.3. Rotation about the ***Y***-axis.

or alternatively by

$$\begin{bmatrix} x' \\ y' \\ z' \\ 1 \end{bmatrix} = \begin{bmatrix} \cos\theta & 0 & \sin\theta & 0 \\ 0 & 1 & 0 & 0 \\ -\sin\theta & 0 & \cos\theta & 0 \\ 0 & 0 & 0 & 1 \end{bmatrix} \cdot \begin{bmatrix} x \\ y \\ z \\ 1 \end{bmatrix} \tag{7.14b}$$

7.2.4 Shearing Transformations

In the generalised 4 × 4 transformation matrix of Eq. (7.1), the off-diagonal elements of the upper left 3 × 3 sub-matrix can be used to represent the six primitive 3D shearing transformations, as shown below.

$$[x', y', z', 1] = [x, y, z, 1] \cdot \begin{bmatrix} 1 & \mathbf{sx}\,\|\mathbf{y} & \mathbf{sx}\,\|\mathbf{z} & 0 \\ \mathbf{sy}\,\|\mathbf{x} & 1 & \mathbf{sy}\,\|\mathbf{z} & 0 \\ \mathbf{sz}\,\|\mathbf{x} & \mathbf{sz}\,\|\mathbf{y} & 1 & 0 \\ 0 & 0 & 0 & 1 \end{bmatrix}$$

$$= [x, y, z, 1] \cdot \begin{bmatrix} 1 & \tan(\theta_{xy}) & -\tan(\theta_{xz}) & 0 \\ -\tan(\theta_{yx}) & 1 & \tan(\theta_{yz}) & 0 \\ \tan(\theta_{zx}) & -\tan(\theta_{zy}) & 1 & 0 \\ 0 & 0 & 0 & 1 \end{bmatrix} \tag{7.15a}$$

or alternatively as

$$\begin{bmatrix} x' \\ y' \\ z' \\ 1 \end{bmatrix} = \begin{bmatrix} 1 & \mathbf{sy}\,\|\mathbf{x} & \mathbf{sz}\,\|\mathbf{x} & 0 \\ \mathbf{sx}\,\|\mathbf{y} & 1 & \mathbf{sz}\,\|\mathbf{y} & 0 \\ \mathbf{sx}\,\|\mathbf{z} & \mathbf{sy}\,\|\mathbf{z} & 1 & 0 \\ 0 & 0 & 0 & 1 \end{bmatrix} \cdot \begin{bmatrix} x \\ y \\ z \\ 1 \end{bmatrix}$$

$$= \begin{bmatrix} 1 & -\tan(\theta_{yx}) & \tan(\theta_{zx}) & 0 \\ \tan(\theta_{xy}) & 1 & -\tan(\theta_{zy}) & 0 \\ -\tan(\theta_{xz}) & \tan(\theta_{yz}) & 1 & 0 \\ 0 & 0 & 0 & 1 \end{bmatrix} \cdot \begin{bmatrix} x \\ y \\ z \\ 1 \end{bmatrix} \tag{7.15b}$$

In the transformation matrices of Eqs. (7.15a) and (7.15b), the element labelled $\mathbf{sx}\,\|\mathbf{y}$ represents the transformation of shearing the $\boldsymbol{X}$-axis parallel to the $\boldsymbol{Y}$-axis, the element labelled $\mathbf{sx}\,\|\mathbf{z}$ represents the transformation of shearing the $\boldsymbol{X}$-axis parallel to the $\boldsymbol{Z}$-axis and so on for all other shearing transformations. All shearing transformations can be defined in terms of the six primitive shearing transformations that we introduce below.

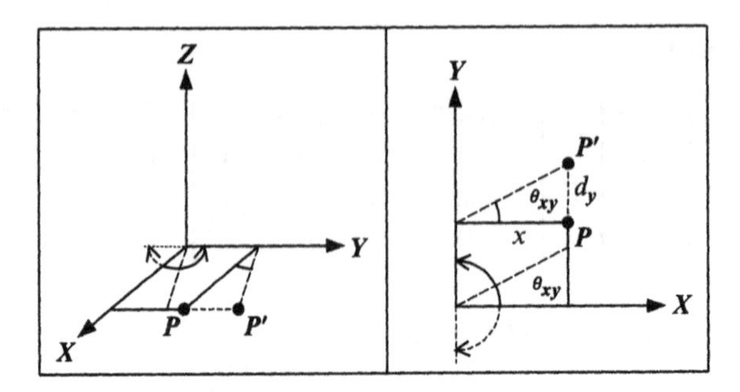

FIGURE 7.4. Shearing the **X**-axis parallel to the **Y**-axis.

7.2.4.1 Shearing the X-Axis Parallel to the Y-Axis ($sx \| y$)

Shearing the **X**-axis parallel to the **Y**-axis results in a transformation which modifies the y coordinate of the transformed point without altering its x and z coordinates. Thus, in the transformed point, only the coordinate measured along the axis parallel to which we are shearing is modified, as seen in Fig. 7.4.

The analytical form of this shearing transformation can be derived as follows. From the above figure we have

$$\frac{d_y}{x} = \tan\left(\theta_{xy}\right)$$
$$\therefore \quad d_y = x \cdot \tan\left(\theta_{xy}\right)$$

Thus

$$\begin{aligned} x' &= x \\ y' &= y + d_y = y + x \cdot \tan\left(\theta_{xy}\right) \\ z' &= z \end{aligned} \tag{7.16}$$

where θ_{xy} is the *angle of shear* of the **X**-axis parallel to the **Y**-axis measured in counter-clockwise positive direction on the **XY** plane.

In matrix form this transformation is given by

$$\left[x', y', z', 1\right] = [x, y, z, 1] \cdot \begin{bmatrix} 1 & \tan\left(\theta_{xy}\right) & 0 & 0 \\ 0 & 1 & 0 & 0 \\ 0 & 0 & 1 & 0 \\ 0 & 0 & 0 & 1 \end{bmatrix} \tag{7.17a}$$

or alternatively by

$$\begin{bmatrix} x' \\ y' \\ z' \\ 1 \end{bmatrix} = \begin{bmatrix} 1 & 0 & 0 & 0 \\ \tan\left(\theta_{xy}\right) & 1 & 0 & 0 \\ 0 & 0 & 1 & 0 \\ 0 & 0 & 0 & 1 \end{bmatrix} \cdot \begin{bmatrix} x \\ y \\ z \\ 1 \end{bmatrix} \tag{7.17b}$$

Note that the shearing angle θ_{xy} must lie in the range $-\frac{\pi}{2} < \theta_{xy} < \frac{\pi}{2}$ as $\tan\left(-\frac{\pi}{2}\right) = -\infty$ and $\tan\left(\frac{\pi}{2}\right) = +\infty$.

7.2.4.2 Shearing the X-Axis Parallel to the Z-Axis ($sx \| z$)

Shearing the **X**-axis parallel to the **Z**-axis results in a transformation which modifies the z coordinate of the transformed point without altering its x and y coordinates. Thus, in the transformed point, only the coordinate measured along the axis parallel to which we are shearing is modified, as seen in Fig. 7.5.

The analytical form of this shearing transformation can be derived as follows. From the figure below we have

$$\frac{d_z}{x} = \tan(\theta_{xz})$$
$$\therefore\ d_z = x \cdot \tan(\theta_{xz})$$

Thus

$$\begin{aligned} x' &= x \\ y' &= y \\ z' &= z - d_z = z - x \cdot \tan(\theta_{xz}) \end{aligned} \tag{7.18}$$

where θ_{xz} is the *angle of shear* of the **X**-axis parallel to the **Z**-axis measured in counter-clockwise positive direction on the **XZ** plane.

In matrix form this transformation is given by

$$[x', y', z', 1] = [x, y, z, 1] \cdot \begin{bmatrix} 1 & 0 & -\tan(\theta_{xz}) & 0 \\ 0 & 1 & 0 & 0 \\ 0 & 0 & 1 & 0 \\ 0 & 0 & 0 & 1 \end{bmatrix} \tag{7.19a}$$

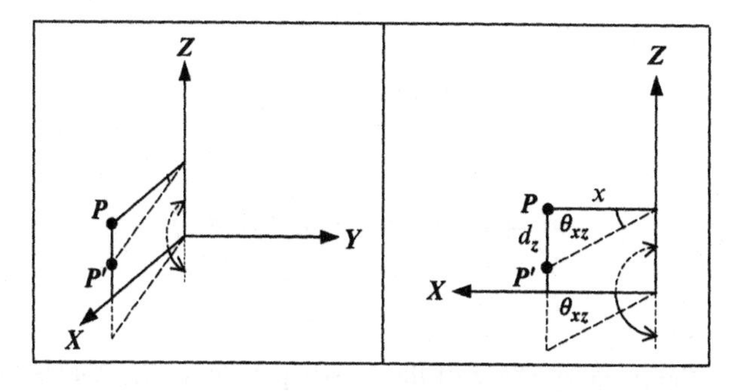

FIGURE 7.5. Shearing the **X**-axis parallel to the **Z**-axis.

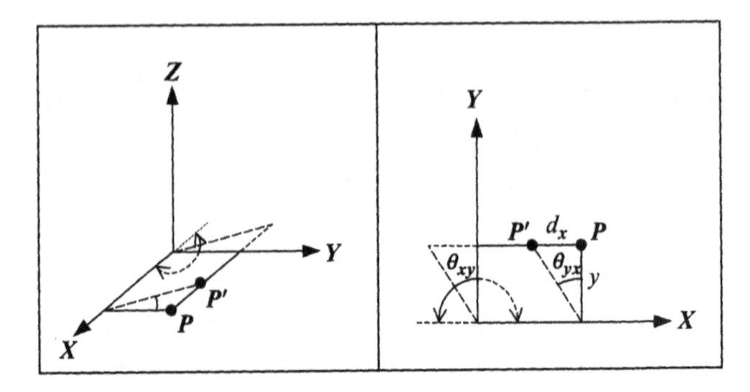

FIGURE 7.6. Shearing the $\boldsymbol{Y}$-axis parallel to the $\boldsymbol{X}$-axis.

or alternatively by

$$\begin{bmatrix} x' \\ y' \\ z' \\ 1 \end{bmatrix} = \begin{bmatrix} 1 & 0 & 0 & 0 \\ 0 & 1 & 0 & 0 \\ -\tan(\theta_{xz}) & 0 & 1 & 0 \\ 0 & 0 & 0 & 1 \end{bmatrix} \cdot \begin{bmatrix} x \\ y \\ z \\ 1 \end{bmatrix} \tag{7.19b}$$

Note that the shearing angle θ_{xz} must lie in the range $-\frac{\pi}{2} < \theta_{xz} < \frac{\pi}{2}$ as $\tan\left(-\frac{\pi}{2}\right) = -\infty$ and $\tan\left(\frac{\pi}{2}\right) = +\infty$.

7.2.4.3 Shearing the Y-Axis Parallel to the X-Axis ($sy \parallel x$)

Shearing the $\boldsymbol{Y}$-axis parallel to the $\boldsymbol{X}$-axis results in a transformation which modifies the x coordinate of the transformed point without altering its y and z coordinates. Thus, in the transformed point, only the coordinate measured along the axis parallel to which we are shearing is modified, as seen in Fig. 7.6.

The analytical form of this shearing transformation can be derived as follows. From the above figure we have

$$\frac{d_x}{y} = \tan(\theta_{yx})$$
$$\therefore\ d_x = y \cdot \tan(\theta_{yx})$$

Thus

$$\begin{aligned} x' &= x - d_x = x - y \cdot \tan(\theta_{yx}) \\ y' &= y \\ z' &= z \end{aligned} \tag{7.20}$$

where θ_{yx} is the *angle of shear* of the $\boldsymbol{Y}$-axis parallel to the $\boldsymbol{X}$-axis measured in counter-clockwise positive direction on the $\boldsymbol{XY}$ plane.

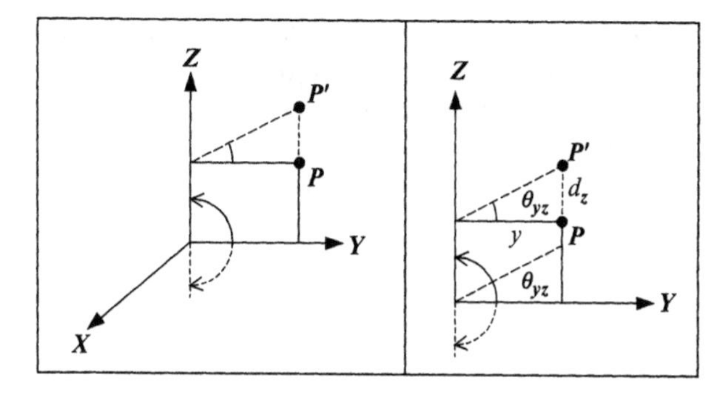

FIGURE 7.7. Shearing the **Y**-axis parallel to the **Z**-axis.

In matrix form this transformation is given by

$$[x', y', z', 1] = [x, y, z, 1] \cdot \begin{bmatrix} 1 & 0 & 0 & 0 \\ -\tan(\theta_{yx}) & 1 & 0 & 0 \\ 0 & 0 & 1 & 0 \\ 0 & 0 & 0 & 1 \end{bmatrix} \tag{7.21a}$$

or alternatively by

$$\begin{bmatrix} x' \\ y' \\ z' \\ 1 \end{bmatrix} = \begin{bmatrix} 1 & -\tan(\theta_{yx}) & 0 & 0 \\ 0 & 1 & 0 & 0 \\ 0 & 0 & 1 & 0 \\ 0 & 0 & 0 & 1 \end{bmatrix} \cdot \begin{bmatrix} x \\ y \\ z \\ 1 \end{bmatrix} \tag{7.21b}$$

Note that the shearing angle θ_{yx} must lie in the range $-\frac{\pi}{2} < \theta_{yx} < \frac{\pi}{2}$ as $\tan\left(-\frac{\pi}{2}\right) = -\infty$ and $\tan\left(\frac{\pi}{2}\right) = +\infty$.

7.2.4.4 Shearing the Y-Axis Parallel to the Z-Axis ($sy \| z$)

Shearing the **Y**-axis parallel to the **Z**-axis results in a transformation which modifies the z coordinate of the transformed point without altering its x and y coordinates. Thus, in the transformed point, only the coordinate measured along the axis parallel to which we are shearing is modified, as seen in Fig. 7.7.

The analytical form of this shearing transformation can be derived as follows. From the above figure we have

$$\frac{d_z}{y} = \tan(\theta_{yz})$$
$$\therefore\ d_z = y \cdot \tan(\theta_{yz})$$

Thus

$$\begin{aligned} x' &= x \\ y' &= y \\ z' &= z + d_z = z + y \cdot \tan(\theta_{yz}) \end{aligned} \tag{7.22}$$

where θ_{yz} is the *angle of shear* of the **Y**-axis parallel to the **Z**-axis measured in counter-clockwise positive direction on the **YZ** plane.

In matrix form this transformation is given by

$$[x', y', z', 1] = [x, y, z, 1] \cdot \begin{bmatrix} 1 & 0 & 0 & 0 \\ 0 & 1 & \tan(\theta_{yz}) & 0 \\ 0 & 0 & 1 & 0 \\ 0 & 0 & 0 & 1 \end{bmatrix} \tag{7.23a}$$

or alternatively by

$$\begin{bmatrix} x' \\ y' \\ z' \\ 1 \end{bmatrix} = \begin{bmatrix} 1 & 0 & 0 & 0 \\ 0 & 1 & 0 & 0 \\ 0 & \tan(\theta_{yz}) & 1 & 0 \\ 0 & 0 & 0 & 1 \end{bmatrix} \cdot \begin{bmatrix} x \\ y \\ z \\ 1 \end{bmatrix} \tag{7.23b}$$

Note that the shearing angle θ_{yz} must lie in the range $-\frac{\pi}{2} < \theta_{yz} < \frac{\pi}{2}$ as $\tan\left(-\frac{\pi}{2}\right) = -\infty$ and $\tan\left(\frac{\pi}{2}\right) = +\infty$.

7.2.4.5 Shearing the Z-Axis Parallel to the X-Axis ($sz \,\|x$)

Shearing the **Z**-axis parallel to the **X**-axis results in a transformation which modifies the x coordinate of the transformed point without altering its y and z coordinates. Thus, in the transformed point, only the coordinate measured along the axis parallel to which we are shearing is modified, as seen in Fig. 7.8.

The analytical form of this shearing transformation can be derived as follows. From the figure below we have

$$\frac{d_x}{z} = \tan(\theta_{zx})$$
$$\therefore\ d_x = z \cdot \tan(\theta_{zx})$$

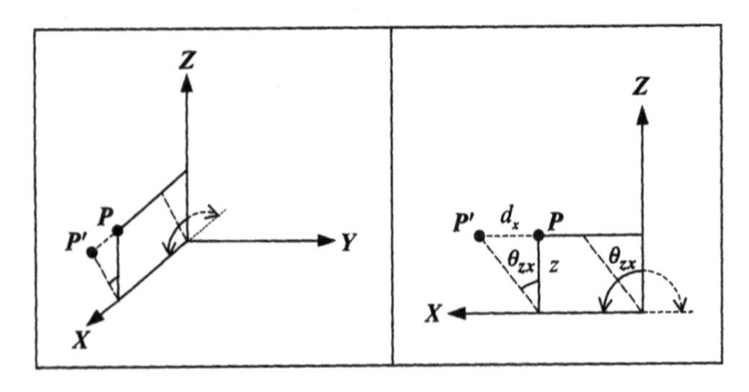

FIGURE 7.8. Shearing the **Z**-axis parallel to the **X**-axis.

Thus

$$x' = x + d_x = x + z \cdot \tan(\theta_{zx})$$
$$y' = y \qquad (7.24)$$
$$z' = z$$

where θ_{zx} is the *angle of shear* of the **Z**-axis parallel to the **X**-axis measured in counter-clockwise positive direction on the **XZ** plane.

In matrix form this transformation is given by

$$[x', y', z', 1] = [x, y, z, 1] \cdot \begin{bmatrix} 1 & 0 & 0 & 0 \\ 0 & 1 & 0 & 0 \\ \tan(\theta_{zx}) & 0 & 1 & 0 \\ 0 & 0 & 0 & 1 \end{bmatrix} \qquad (7.25a)$$

or alternatively by

$$\begin{bmatrix} x' \\ y' \\ z' \\ 1 \end{bmatrix} = \begin{bmatrix} 1 & 0 & \tan(\theta_{zx}) & 0 \\ 0 & 1 & 0 & 0 \\ 0 & 0 & 1 & 0 \\ 0 & 0 & 0 & 1 \end{bmatrix} \cdot \begin{bmatrix} x \\ y \\ z \\ 1 \end{bmatrix} \qquad (7.25b)$$

Note that the shearing angle θ_{zx} must lie in the range $-\frac{\pi}{2} < \theta_{zx} < \frac{\pi}{2}$ as $\tan\left(-\frac{\pi}{2}\right) = -\infty$ and $\tan\left(\frac{\pi}{2}\right) = +\infty$.

7.2.4.6 Shearing the Z-Axis Parallel to the Y-Axis ($sz \parallel y$)

Shearing the **Z**-axis parallel to the **Y**-axis results in a transformation which modifies the y coordinate of the transformed point without altering its x and z coordinates. Thus, in the transformed point, only the coordinate measured along the axis parallel to which we are shearing is modified, as seen in Fig. 7.9.

The analytical form of this shearing transformation can be derived as follows. From the figure below we have

$$\frac{d_y}{z} = \tan\left(\theta_{zy}\right)$$
$$\therefore\ d_y = z \cdot \tan\left(\theta_{zy}\right)$$

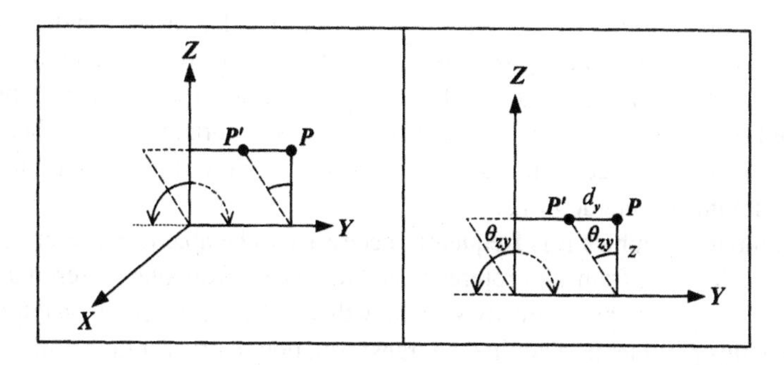

FIGURE 7.9. Shearing the **Z**-axis parallel to the **Y**-axis.

Thus

$$\begin{aligned} x' &= x \\ y' &= y - d_y = y - z \cdot \tan\left(\theta_{zy}\right) \\ z' &= z \end{aligned} \tag{7.26}$$

where θ_{zy} is the *angle of shear* of the **Z**-axis parallel to the **Y**-axis measured in counter-clockwise positive direction on the ***YZ*** plane.

In matrix form this transformation is given by

$$\left[x', y', z', 1\right] = [x, y, z, 1] \cdot \begin{bmatrix} 1 & 0 & 0 & 0 \\ 0 & 1 & 0 & 0 \\ 0 & -\tan\left(\theta_{zy}\right) & 1 & 0 \\ 0 & 0 & 0 & 1 \end{bmatrix} \tag{7.27a}$$

or alternatively by

$$\begin{bmatrix} x' \\ y' \\ z' \\ 1 \end{bmatrix} = \begin{bmatrix} 1 & 0 & 0 & 0 \\ 0 & 1 & -\tan\left(\theta_{zy}\right) & 0 \\ 0 & 0 & 1 & 0 \\ 0 & 0 & 0 & 1 \end{bmatrix} \cdot \begin{bmatrix} x \\ y \\ z \\ 1 \end{bmatrix} \tag{7.27b}$$

Note that the shearing angle θ_{zy} must lie in the range $-\frac{\pi}{2} < \theta_{zy} < \frac{\pi}{2}$ as $\tan\left(-\frac{\pi}{2}\right) = -\infty$ and $\tan\left(\frac{\pi}{2}\right) = +\infty$.

7.3 Global and Local Frames of Reference

In computer graphics a *frame of reference* is normally defined by selecting an orthonormal basis and a point representing the origin of the frame. The base vectors of the frame are often referred to as the *axes* of the frame.

Commonly the orthonormal basis $\boldsymbol{X} = [1, 0, 0]$, $\boldsymbol{Y} = [0, 1, 0]$, $\boldsymbol{Z} = [0, 0, 1]$ and the origin $\boldsymbol{O} = [0, 0, 0]$ are taken to represent the *global frame of reference*, which is the *Cartesian coordinate system.* This frame of reference is unchanging and is referred to as the *world coordinate system.*

Other frames of reference may be defined with respect to the global frame of reference. Such frames of reference may be right-handed or left-handed and are referred to as *local frames of reference.* Local frames of reference are normally associated with objects, lights or the camera. The local frame of an object is assumed to be attached to it. Thus, whenever a transformation is applied to an object it is also applied to its local frame, which follows the orientation of the object as it moves and deforms.

In computer graphics it is frequently necessary to be able to represent a point relative to different frames of reference and to convert from one representation to the other. Indeed we have already seen how this can be achieved in Chapter 4. To illustrate this point let us examine the transformation matrix for the rotation about the **Z**-axis.

First we will use the notation that represents points as row vectors. From Eq. (7.10a) the transformation matrix is given by

$$
\boldsymbol{R}_Z = \begin{bmatrix} \cos\theta & \sin\theta & 0 & \vdots & 0 \\ -\sin\theta & \cos\theta & 0 & \vdots & 0 \\ 0 & 0 & 1 & \vdots & 0 \\ \cdots & \cdots & \cdots & \cdots & \cdots \\ 0 & 0 & 0 & \vdots & 1 \end{bmatrix}
$$

This matrix transforms point $\boldsymbol{P}$ into point $\boldsymbol{P}'$. As the point is rotated about the $\boldsymbol{Z}$-axis by a counter-clockwise angle θ, so is its local frame of reference, as seen in Fig. 7.10.

As can be readily seen from Fig. 7.10, the three rows of the upper-left 3 × 3 sub-matrix of the $\boldsymbol{R}_Z$ transformation matrix represent the transformed local frame base vectors (*local axes*):

$$
\begin{aligned}
\boldsymbol{X}'_L &= [\ \cos\theta \quad \sin\theta \quad 0\ ] \\
\boldsymbol{Y}'_L &= [\ -\sin\theta \quad \cos\theta \quad 0\ ] \\
\boldsymbol{Z}'_L &= [\ 0 \quad 0 \quad 1\ ]
\end{aligned} \tag{7.28}
$$

The transformed local frame remains a Cartesian basis, i.e. it continues to have base vectors which are unit length, perpendicular to each other and forming a right-handed system. Thus the rotation of a 3D point can be expressed as

$$
[x', y', z'] = [x, y, z] \cdot \begin{bmatrix} \boldsymbol{X}'_{Lx} & \boldsymbol{X}'_{Ly} & \boldsymbol{X}'_{Lz} \\ \boldsymbol{Y}'_{Lx} & \boldsymbol{Y}'_{Ly} & \boldsymbol{Y}'_{Lz} \\ \boldsymbol{Z}'_{Lx} & \boldsymbol{Z}'_{Ly} & \boldsymbol{Z}'_{Lz} \end{bmatrix} \tag{7.29}
$$

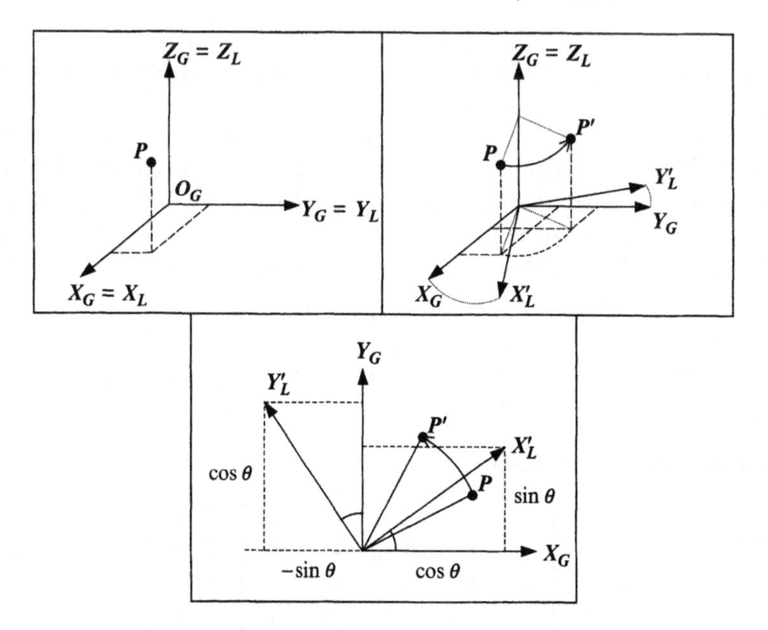

FIGURE 7.10. The rotation of a point and its local frame of reference.

The matrix in Eq. (7.29) is the transpose (inverse) of matrix $\boldsymbol{S}$ in Eq. (4.33). The reason for this is that here we are transforming the point and we are not simply changing its basis. As we have seen in Section 4.7, this is an orthogonal matrix and its inverse is equal to its transpose. Thus

$$[x, y, z] = [x', y', z'] \cdot \begin{bmatrix} X'_{Lx} & Y'_{Lx} & Z'_{Lx} \\ X'_{Ly} & Y'_{Ly} & Z'_{Ly} \\ X'_{Lz} & Y'_{Lz} & Z'_{Lz} \end{bmatrix} \tag{7.30}$$

For the same reason, the matrix in Eq. (7.30) is the transpose (inverse) of matrix $\boldsymbol{D}$ in Eq. (4.31).

Using the above result we can make the following generalisation. Given an object the local frame of which coincides with the global frame and an arbitrary orthonormal basis $\boldsymbol{V}_1$, $\boldsymbol{V}_2$, $\boldsymbol{V}_3$ we can construct a rotation transformation matrix that aligns the local frame of reference of the object with the basis $\boldsymbol{V}_1$, $\boldsymbol{V}_2$, $\boldsymbol{V}_3$. This transformation matrix is given by

$$\boldsymbol{R_A} = \begin{bmatrix} V_{1x} & V_{1y} & V_{1z} & 0 \\ V_{2x} & V_{2y} & V_{2z} & 0 \\ V_{3x} & V_{3y} & V_{3z} & 0 \\ 0 & 0 & 0 & 1 \end{bmatrix} \tag{7.31}$$

Conversely, an object the local frame of which is aligned with an arbitrary orthonormal basis $\boldsymbol{V}_1$, $\boldsymbol{V}_2$, $\boldsymbol{V}_3$ can be realigned to coincide with the global frame using the transformation matrix

$$\boldsymbol{R_G} = \boldsymbol{R}_{\boldsymbol{A}}^{-1} = \begin{bmatrix} V_{1x} & V_{2x} & V_{3x} & 0 \\ V_{1y} & V_{2y} & V_{3y} & 0 \\ V_{1z} & V_{2z} & V_{3z} & 0 \\ 0 & 0 & 0 & 1 \end{bmatrix} \tag{7.32}$$

When using the notation that represents points as column vectors the rotation matrices $\boldsymbol{R_A}$ and $\boldsymbol{R_G}$ are transposed.

Let us recapitulate. In this section we have developed a very powerful mechanism that allows us to align the local frame of reference of an object and thus the object itself with an arbitrary orthonormal basis $\boldsymbol{V}_1$, $\boldsymbol{V}_2$, $\boldsymbol{V}_3$. If the local frame of reference is originally aligned with the global frame of reference (i.e. $X_L = X_G$, $Y_L = Y_G$ and $Z_L = Z_G$), then this transformation is effected by

$$[x', y', z', 1] = [x, y, z, 1] \cdot \begin{bmatrix} V_{1x} & V_{1y} & V_{1z} & 0 \\ V_{2x} & V_{2y} & V_{2z} & 0 \\ V_{3x} & V_{3y} & V_{3z} & 0 \\ 0 & 0 & 0 & 1 \end{bmatrix} \tag{7.33a}$$

Alternatively, it is effected by

$$\begin{bmatrix} x' \\ y' \\ z' \\ 1 \end{bmatrix} = \begin{bmatrix} V_{1x} & V_{2x} & V_{3x} & 0 \\ V_{1y} & V_{2y} & V_{3y} & 0 \\ V_{1z} & V_{2z} & V_{3z} & 0 \\ 0 & 0 & 0 & 1 \end{bmatrix} \cdot \begin{bmatrix} x \\ y \\ z \\ 1 \end{bmatrix} \tag{7.33b}$$

If the local frame of reference is not originally aligned with the global frame of reference, then this transformation has two steps. First, we must align it with the global frame of reference and then align it with the arbitrary orthonormal basis, by applying these two transformations in sequence.

$$[x', y', z', 1] = [x, y, z, 1] \cdot \begin{bmatrix} X_{Lx} & Y_{Lx} & Z_{Lx} & 0 \\ X_{Ly} & Y_{Ly} & Z_{Ly} & 0 \\ X_{Lz} & Y_{Lz} & Z_{Lz} & 0 \\ 0 & 0 & 0 & 1 \end{bmatrix} \cdot \begin{bmatrix} V_{1x} & V_{1y} & V_{1z} & 0 \\ V_{2x} & V_{2y} & V_{2z} & 0 \\ V_{3x} & V_{3y} & V_{3z} & 0 \\ 0 & 0 & 0 & 1 \end{bmatrix} \tag{7.34a}$$

Alternatively, by applying these two transformations in sequence

$$\begin{bmatrix} x' \\ y' \\ z' \\ 1 \end{bmatrix} = \begin{bmatrix} V_{1x} & V_{2x} & V_{3x} & 0 \\ V_{1y} & V_{2y} & V_{3y} & 0 \\ V_{1z} & V_{2z} & V_{3z} & 0 \\ 0 & 0 & 0 & 1 \end{bmatrix} \cdot \begin{bmatrix} X_{Lx} & X_{Ly} & X_{Lz} & 0 \\ Y_{Lx} & Y_{Ly} & Y_{Lz} & 0 \\ Z_{Lx} & Z_{Ly} & Z_{Lz} & 0 \\ 0 & 0 & 0 & 1 \end{bmatrix} \cdot \begin{bmatrix} x \\ y \\ z \\ 1 \end{bmatrix} \tag{7.34b}$$

7.4 Aiming Transformations

In computer graphics we often need to be able to align one of the local axes of an object with an arbitrary vector. This problem is under-specified, i.e. we do not have enough information to determine a unique solution to the problem. Indeed there is no unique solution to this problem, as the other two local axes of the object are free to assume an infinity of orientations around the fixed axis, as shown in Fig. 7.11.

This is a classic problem in 3D geometry and it is known as a *gimbal lock* or, in plain English, being *lost in space*. A problem dreaded by spacecraft designers

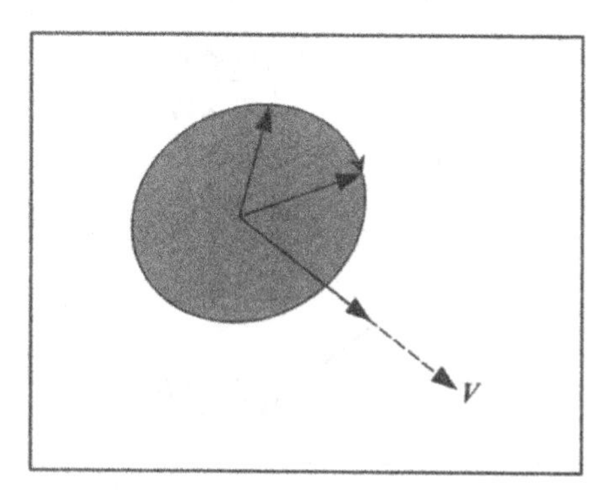

FIGURE 7.11. Aiming a local axis of an object along an arbitrary vector.

and navigators, which is due to the fact that we have lost track of the *up direction* (i.e. which way is up). This, of course, is a problem that is easily solved on the surface of a planet as gravity tells us which way is down and thus up is in the opposite direction. In space however to avoid the gimbal lock problem we require two vectors: one indicating the *heading direction* or *aim direction* and another indicating the *relative up direction*. This is cumbersome and so to avoid having to specify the relative up direction we often use the *absolute up direction* (i.e. the $\boldsymbol{Z_G}$ axis) as an *approximate up direction*. This simplifying assumption, however, does not work when our heading direction is parallel or antiparallel to the $\boldsymbol{Z_G}$ axis. Another option we have is to use the relative up direction of the object (i.e. its $\boldsymbol{Z_L}$ axis) before the transformation as the approximate up direction. This may work if the aiming operation is part of an animation sequence and develops gradually over a number of frames. Failing that, we are *gimbal locked* and we must resort to asking the user to specify the relative up direction otherwise our animated object will spin unpredictably around its aim axis.

7.4.1 Aiming the Local X*-Axis in the Direction of an Arbitrary Unit Vector* V

Let us assume that the local frame of an object is aligned with the global frame and that we wish to align the local $\boldsymbol{X}$-axis of this object with an arbitrary unit vector $\boldsymbol{V}$. As we have seen above, this is an under-specified problem. So we will need to determine the new relative up direction (i.e. the new orientation of the local $\boldsymbol{Z}$-axis) of the transformed object before we can compute the aiming transformation matrix. To help us do this we reason as follows. If the arbitrary unit vector $\boldsymbol{V}$ is parallel or antiparallel to the global up direction (i.e. if $\boldsymbol{V} = \boldsymbol{Z_G}$ or $\boldsymbol{V} = -\boldsymbol{Z_G}$), then we have no option but to require the user to specify the approximate up direction $\boldsymbol{U}$, otherwise we set the approximate up direction to the global up direction, i.e. $\boldsymbol{U} = \boldsymbol{Z_G}$, as shown in Fig. 7.12.

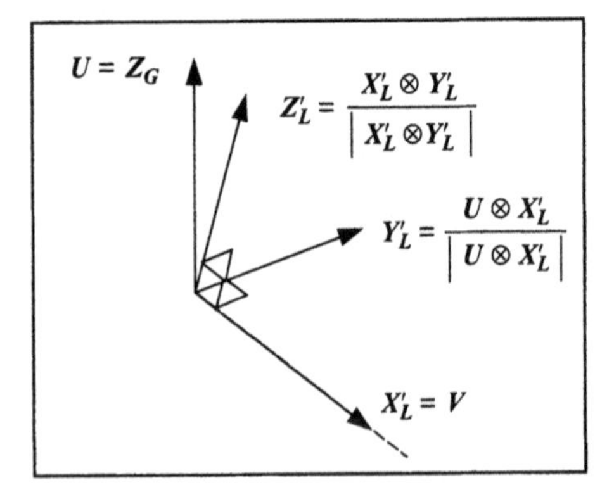

FIGURE 7.12. Aiming the local $\boldsymbol{X}$-axis along an arbitrary unit vector $\boldsymbol{V}$.

Next we calculate the new orientation of the transformed local axes.

$$\begin{aligned} \boldsymbol{X}'_L &= \boldsymbol{V} \\ \boldsymbol{Y}'_L &= \frac{\boldsymbol{U} \otimes \boldsymbol{X}'_L}{|\boldsymbol{U} \otimes \boldsymbol{X}'_L|} \\ \boldsymbol{Z}'_L &= \frac{\boldsymbol{X}'_L \otimes \boldsymbol{Y}'_L}{|\boldsymbol{X}'_L \otimes \boldsymbol{Y}'_L|} \end{aligned} \tag{7.35}$$

Now we can construct the rotation transformation matrix that will reorient the local frame of the object in a manner similar to that of Eq. (7.31). Thus, the transformation for aiming the local $\boldsymbol{X}$-axis, of an object, in the direction of an arbitrary unit vector $\boldsymbol{V}$ is given by

$$[x', y', z', 1] = [x, y, z, 1] \cdot \begin{bmatrix} X'_{Lx} & X'_{Ly} & X'_{Lz} & 0 \\ Y'_{Lx} & Y'_{Ly} & Y'_{Lz} & 0 \\ Z'_{Lx} & Z'_{Ly} & Z'_{Lz} & 0 \\ 0 & 0 & 0 & 1 \end{bmatrix} \tag{7.36a}$$

or alternatively by

$$\begin{bmatrix} x' \\ y' \\ z' \\ 1 \end{bmatrix} = \begin{bmatrix} X'_{Lx} & Y'_{Lx} & Z'_{Lx} & 0 \\ X'_{Ly} & Y'_{Ly} & Z'_{Ly} & 0 \\ X'_{Lz} & Y'_{Lz} & Z'_{Lz} & 0 \\ 0 & 0 & 0 & 1 \end{bmatrix} \cdot \begin{bmatrix} x \\ y \\ z \\ 1 \end{bmatrix} \tag{7.36b}$$

7.4.2 *Aiming the Local* Y*-Axis in the Direction of an Arbitrary Unit Vector* V

The transformation for aiming the local $\boldsymbol{Y}$-axis, of an object, in the direction of an arbitrary unit vector $\boldsymbol{V}$ is similar to that for aiming its $\boldsymbol{X}$-axis. As above, if the arbitrary unit vector $\boldsymbol{V}$ is parallel or antiparallel to the global up direction (i.e. if $\boldsymbol{V} = \boldsymbol{Z_G}$ or $\boldsymbol{V} = -\boldsymbol{Z_G}$), then we require the user to specify the approximate up direction $\boldsymbol{U}$, otherwise we set it to the global up direction, i.e. $\boldsymbol{U} = \boldsymbol{Z_G}$, as shown in Fig. 7.13.

The new orientation of the transformed local axes is given as

$$\begin{aligned} \boldsymbol{Y}'_L &= \boldsymbol{V} \\ \boldsymbol{X}'_L &= \frac{\boldsymbol{Y}'_L \otimes \boldsymbol{U}}{|\boldsymbol{Y}'_L \otimes \boldsymbol{U}|} \\ \boldsymbol{Z}'_L &= \frac{\boldsymbol{X}'_L \otimes \boldsymbol{Y}'_L}{|\boldsymbol{X}'_L \otimes \boldsymbol{Y}'_L|} \end{aligned} \tag{7.37}$$

Thus, the transformation for aiming the local $\boldsymbol{Y}$-axis, of an object, in the direction of an arbitrary unit vector $\boldsymbol{V}$ is given by Eq. (7.36a) or (7.36b).

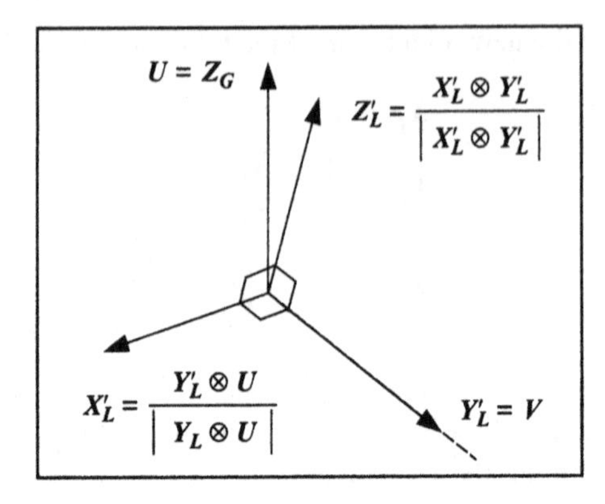

FIGURE 7.13. Aiming the local Y-axis along an arbitrary unit vector V.

7.4.3 *Aiming the Local* Z*-Axis in the Direction of an Arbitrary Unit Vector* V

The transformation for aiming the local Z-axis, of an object, in the direction of an arbitrary unit vector V is similar to that for aiming its X-axis. As above, if the arbitrary unit vector V is parallel or antiparallel to the global up direction (i.e. if $V = Z_G$ or $V = -Z_G$), then we require the user to specify the approximate up direction U, otherwise we set it to the global up direction, i.e. $U = Z_G$, as shown in Fig. 7.14.

The new orientation of the transformed local axes is given as

$$\begin{aligned} Z'_L &= V \\ Y'_L &= \frac{U \otimes Z'_L}{|U \otimes Z'_L|} \\ X'_L &= \frac{Y'_L \otimes Z'_L}{|Y'_L \otimes Z'_L|} \end{aligned} \tag{7.38}$$

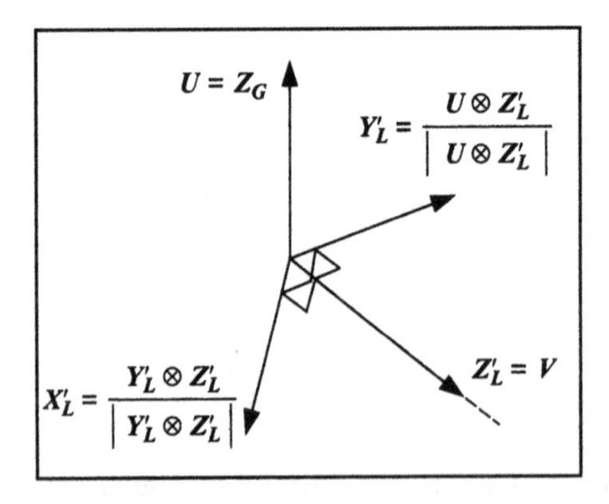

FIGURE 7.14. Aiming the local Z-axis along an arbitrary unit vector V.

Thus, the transformation for aiming the local **Z**-axis, of an object, in the direction of an arbitrary unit vector ***V*** is given by equation (7.36a) or (7.36b).

7.5 Composite Transformations

As we have already seen in the case of 2D transformations, composite transformations can be constructed by composing (concatenating together) a sequence of primitive transformations. As in the 2D case, some non-primitive transformations can be constructed by modifying the parameters of a primitive transformation. In general however, composite transformations can only be constructed by applying a sequence of primitive transformations one after another, i.e. by concatenating this sequence of transformations.

Usually for a transformation problem that is too complex to solve in its current domain, we adopt the following procedure. First, through a series of primitive transformation steps, we transform the geometry of the problem into a much simpler domain defined around the origin of the frame of reference. Here the problem can usually be solved using a primitive transformation. Then, having solved the problem in this simpler domain we return the geometry to its original domain, through a series of inverse transformation steps, i.e. by undoing the steps that transformed the geometry from its original domain to the simpler domain.

In 3D there is a proliferation of composite transformations. In an effort to simplify the examination of these transformations we will categorise them as being performed relative to a point, an axis or a plane.

7.5.1 Composite Transformations Relative to a Point

Composite transformations relative to a point can be further subdivided into *composite transformations relative to the origin of the frame of reference* and *composite transformations relative to an arbitrary point.*

7.5.1.1 Composite Transformations Relative to the Origin of the Frame

This is only one composite transformation that falls into this category, namely the *reflection transformation about the origin of the frame of reference.* When reflecting a point about the origin of the frame of reference, we scale all three of its coordinates by -1. Thus

$$\begin{aligned} x' &= x \cdot (-1) \\ y' &= y \cdot (-1) \\ z' &= z \cdot (-1) \end{aligned} \tag{7.39}$$

which in matrix form is given by

$$\left[x', y', z', 1\right] = [x, y, z, 1] \cdot \begin{bmatrix} -1 & 0 & 0 & 0 \\ 0 & -1 & 0 & 0 \\ 0 & 0 & -1 & 0 \\ 0 & 0 & 0 & 1 \end{bmatrix} \tag{7.40a}$$

or alternatively by

$$\begin{bmatrix} x' \\ y' \\ z' \\ 1 \end{bmatrix} = \begin{bmatrix} -1 & 0 & 0 & 0 \\ 0 & -1 & 0 & 0 \\ 0 & 0 & -1 & 0 \\ 0 & 0 & 0 & 1 \end{bmatrix} \cdot \begin{bmatrix} x \\ y \\ z \\ 1 \end{bmatrix} \tag{7.40b}$$

7.5.1.2 Composite Transformations Relative to an Arbitrary Point

There are two composite transformations that fall in this category, namely the *scaling* and *reflection transformations about an arbitrary point*. The transformation of a point $\boldsymbol{P}$ about an arbitrary point $\boldsymbol{P_c}$ can be constructed by concatenating the transformations representing the following three steps.

Step 1: Translate the arbitrary point $\boldsymbol{P_c}$ to the origin of the frame. Call this transformation $\boldsymbol{T_P}$.

Step 2: Perform the required transformation about the origin of the frame. Call this transformation $\boldsymbol{T_O}$.

Step 3: Translate the arbitrary point $\boldsymbol{P_c}$ back to its original position, by applying the inverse of the transformation from step 1. Call this transformation $\boldsymbol{T_P^{-1}}$.

Thus the composite transformation about an arbitrary point is given by

$$[x', y', z', 1] = [x, y, z, 1] \cdot \boldsymbol{T_P} \cdot \boldsymbol{T_O} \cdot \boldsymbol{T_P^{-1}} \tag{7.41a}$$

or alternatively by

$$\begin{bmatrix} x' \\ y' \\ z' \\ 1 \end{bmatrix} = \boldsymbol{T_P^{-1}} \cdot \boldsymbol{T_O} \cdot \boldsymbol{T_P} \cdot \begin{bmatrix} x \\ y \\ z \\ 1 \end{bmatrix} \tag{7.41b}$$

In the above two equations it is assumed that we have used the appropriate rules in the construction of the transformation matrices that correspond to the row/column representation of points.

7.5.2 Composite Transformations Relative to an Axis

Composite transformations relative to an axis can be further subdivided into *composite transformations relative to a major axis* (i.e. the $\boldsymbol{X}$-, $\boldsymbol{Y}$-, $\boldsymbol{Z}$-axes of the frame), *composite transformations relative to an axis parallel to a major axis* and *composite transformations relative to an arbitrary axis.*

7.5.2.1 Composite Transformations Relative to a Major Axis

There are three composite transformations that fall in this category, namely the *composite transformations of reflection about the* $\boldsymbol{X}$-, $\boldsymbol{Y}$- *and* $\boldsymbol{Z}$-*axes.*

When reflecting a point about the $\boldsymbol{X}$-axis, its y and z coordinates are scaled by –1. Thus

$$\begin{aligned} s_x &= +1 \\ s_y &= -1 \\ s_z &= -1 \end{aligned} \tag{7.42a}$$

When reflecting a point about the $\boldsymbol{Y}$-axis, its x and z coordinates are scaled by –1. Thus

$$\begin{aligned} s_x &= -1 \\ s_y &= +1 \\ s_z &= -1 \end{aligned} \tag{7.42b}$$

When reflecting a point about the $\boldsymbol{Z}$-axis, its x and y coordinates are scaled by –1. Thus

$$\begin{aligned} s_x &= -1 \\ s_y &= -1 \\ s_z &= +1 \end{aligned} \tag{7.42c}$$

Using one of the Eqs. (7.42a)–(7.42c), we can construct the reflection about one of the major axes as

$$\begin{aligned} x' &= x \cdot s_x \\ y' &= y \cdot s_y \\ z' &= z \cdot s_z \end{aligned} \tag{7.43}$$

which in matrix form is given by

$$\left[x', y', z', 1\right] = [x, y, z, 1] \cdot \begin{bmatrix} s_x & 0 & 0 & 0 \\ 0 & s_y & 0 & 0 \\ 0 & 0 & s_z & 0 \\ 0 & 0 & 0 & 1 \end{bmatrix} \tag{7.44a}$$

or alternatively by

$$\begin{bmatrix} x' \\ y' \\ z' \\ 1 \end{bmatrix} = \begin{bmatrix} s_x & 0 & 0 & 0 \\ 0 & s_y & 0 & 0 \\ 0 & 0 & s_z & 0 \\ 0 & 0 & 0 & 1 \end{bmatrix} \cdot \begin{bmatrix} x \\ y \\ z \\ 1 \end{bmatrix} \tag{7.44b}$$

7.5.2.2 Composite Transformations Relative to an Axis Parallel to a Major Axis

There are three composite transformations that fall in this category, namely the *composite transformations of reflection, rotation and scaling about/along an axis parallel to a major axis.* Given that such an axis is defined by a point $\boldsymbol{P_a}$ and a direction unit vector $\boldsymbol{V}$, the transformation of point $\boldsymbol{P}$ about/along this axis can be constructed by concatenating the transformations representing the following three steps.

Step 1: Translate the point $\boldsymbol{P_a}$ to the origin of the frame. This has the effect of aligning the parallel axis with the corresponding major axis. Call this transformation $\boldsymbol{T_P}$.

Step 2: Perform the required transformation about/along this major axis. Call this transformation $\boldsymbol{T_A}$.

Step 3: Translate the point $\boldsymbol{P_a}$ back to its original position, by applying the inverse of the transformation from step 1. Call this transformation $\boldsymbol{T_P^{-1}}$.

Thus the composite transformation about/along an axis parallel to a major axis is given by

$$[x', y', z', 1] = [x, y, z, 1] \cdot \boldsymbol{T_P} \cdot \boldsymbol{T_A} \cdot \boldsymbol{T_P^{-1}} \tag{7.45a}$$

or alternatively by

$$\begin{bmatrix} x' \\ y' \\ z' \\ 1 \end{bmatrix} = \boldsymbol{T_P^{-1}} \cdot \boldsymbol{T_A} \cdot \boldsymbol{T_P} \cdot \begin{bmatrix} x \\ y \\ z \\ 1 \end{bmatrix} \tag{7.45b}$$

In the above two equations it is assumed that we have used the appropriate rules in the construction of the transformation matrices that correspond to the row/column representation of points.

7.5.2.3 Composite Transformations Relative to an Arbitrary Axis

There are four composite transformations that fall in this category, namely the *composite transformations of reflection, rotation, scaling and translation about/along an arbitrary axis*. Given that such an axis is defined by a point $\boldsymbol{P_a}$ and a direction unit vector $\boldsymbol{V}$, the transformation of point $\boldsymbol{P}$ about/along this axis can be constructed by concatenating the transformations representing the following five steps.

Step 1: Translate the point $\boldsymbol{P_a}$ to the origin of the frame. Call this transformation $\boldsymbol{T_P}$.

Step 2: Align the unit vector V with one of the major axes, using the inverse of the aiming transformation that would align the selected axis with this vector. Call this transformation $\boldsymbol{A_V}$.

Step 3: Perform the required transformation about/along this major axis. Call this transformation $\boldsymbol{T_A}$.

Step 4: Return the $\boldsymbol{V}$ vector to its original orientation, by applying the inverse of the transformation from step 2. Call this transformation $\boldsymbol{A_V^{-1}}$.

Step 5: Translate the point $\boldsymbol{P_a}$ back to its original position, by applying the inverse of the transformation from step 1. Call this transformation $\boldsymbol{T_P^{-1}}$.

Thus the composite transformation about/along an arbitrary axis is given by

$$[x', y', z', 1] = [x, y, z, 1] \cdot \boldsymbol{T_P} \cdot \boldsymbol{A_V} \cdot \boldsymbol{T_A} \cdot \boldsymbol{A_V^{-1}} \cdot \boldsymbol{T_P^{-1}} \tag{7.46a}$$

or alternatively by

$$\begin{bmatrix} x' \\ y' \\ z' \\ 1 \end{bmatrix} = \boldsymbol{T_P^{-1}} \cdot \boldsymbol{A_V^{-1}} \cdot \boldsymbol{T_A} \cdot \boldsymbol{A_V} \cdot \boldsymbol{T_P} \cdot \begin{bmatrix} x \\ y \\ z \\ 1 \end{bmatrix} \tag{7.46b}$$

In the above two equations it is assumed that we have used the appropriate rules in the construction of the transformation matrices that correspond to the row/column representation of points. Also recall that the inverse of the aim transformation is equal to its transpose, i.e. $\boldsymbol{A_V^{-1}} = \boldsymbol{A_V^{\mathrm{T}}}$.

7.5.3 Composite Transformations Relative to a Plane

Composite transformations relative to a plane can be further subdivided into *composite transformations relative to a major plane* (i.e. the ***XY***, ***XZ*** or ***YZ*** plane) and *composite transformations relative to an arbitrary plane.*

7.5.3.1 Composite Transformations Relative to a Major Plane

There are three composite transformations that fall in this category, namely the *composite transformations of reflection about the **XY**, **XZ** or **YZ** planes.*

When reflecting a point about the ***XY*** plane, its z coordinate is scaled by –1. Thus

$$\begin{aligned} s_x &= +1 \\ s_y &= +1 \\ s_z &= -1 \end{aligned} \tag{7.47a}$$

When reflecting a point about the ***XZ*** plane, its y coordinate is scaled by -1. Thus

$$\begin{aligned} s_x &= +1 \\ s_y &= -1 \\ s_z &= +1 \end{aligned} \tag{7.47b}$$

When reflecting a point about the ***YZ*** plane, its x coordinate is scaled by -1. Thus

$$\begin{aligned} s_x &= -1 \\ s_y &= +1 \\ s_z &= +1 \end{aligned} \tag{7.47c}$$

Using one of Eqs. (7.47a)–(7.47c), we can construct the reflection transformation about one of the major planes as

$$\begin{aligned} x' &= x \cdot s_x \\ y' &= y \cdot s_y \\ z' &= z \cdot s_z \end{aligned} \tag{7.48}$$

which in matrix form is given by

$$[x', y', z', 1] = [x, y, z, 1] \cdot \begin{bmatrix} s_x & 0 & 0 & 0 \\ 0 & s_y & 0 & 0 \\ 0 & 0 & s_z & 0 \\ 0 & 0 & 0 & 1 \end{bmatrix} \tag{7.49a}$$

or alternatively by

$$\begin{bmatrix} x' \\ y' \\ z' \\ 1 \end{bmatrix} = \begin{bmatrix} s_x & 0 & 0 & 0 \\ 0 & s_y & 0 & 0 \\ 0 & 0 & s_z & 0 \\ 0 & 0 & 0 & 1 \end{bmatrix} \cdot \begin{bmatrix} x \\ y \\ z \\ 1 \end{bmatrix} \tag{7.49b}$$

7.5.3.2 Composite Transformations Relative to an Arbitrary Plane

There is only one composite transformation that falls into this category, namely the *reflection transformation about an arbitrary plane*.

Given an arbitrary plane Π defined by a point $\boldsymbol{P_a} = [x_a, y_a, z_a]$ and a unit normal vector $\boldsymbol{N} = [x_n, y_n, z_n]$, the transformation that reflects a point $\boldsymbol{P}$ about this plane can be constructed by concatenating transformations representing the following five steps.

Step 1: Translate the arbitrary point $\boldsymbol{P_a}$ to the origin of the frame. Call this transformation $\boldsymbol{T_P}$.

Step 2: Align the unit normal vector $\boldsymbol{N}$ of the arbitrary plane with one of the major axes, using the inverse of the aiming transformation that would align the selected axis with the unit normal vector. This has the effect of aligning the arbitrary plane Π with one of the major planes. Call this transformation $\boldsymbol{A_N}$.

Step 3: Perform the reflection about this major plane. Call this transformation $\boldsymbol{R_P}$.

Step 4: Return the unit normal vector $\boldsymbol{N}$ to its original orientation, by applying the inverse of the transformation from step 2. Call this transformation $\boldsymbol{A_N^{-1}}$.

Step 5: Translate the arbitrary point $\boldsymbol{P_a}$ back to its original position, by applying the inverse of the transformation from step 1. Call this transformation $\boldsymbol{T_P^{-1}}$.

Thus the composite transformation for the reflection about an arbitrary plane is given by

$$[x', y', z', 1] = [x, y, z, 1] \cdot \boldsymbol{T_P} \cdot \boldsymbol{A_N} \cdot \boldsymbol{R_P} \cdot \boldsymbol{A_N^{-1}} \cdot \boldsymbol{T_P^{-1}} \tag{7.50a}$$

or alternatively by

$$\begin{bmatrix} x' \\ y' \\ z' \\ 1 \end{bmatrix} = \boldsymbol{T_P^{-1}} \cdot \boldsymbol{A_N^{-1}} \cdot \boldsymbol{R_P} \cdot \boldsymbol{A_N} \cdot \boldsymbol{T_P} \cdot \begin{bmatrix} x \\ y \\ z \\ 1 \end{bmatrix} \tag{7.50b}$$

In the above two equations it is assumed that we have used the appropriate rules in the construction of the transformation matrices that correspond to the row/column representation of points. Also recall that the inverse of the aim transformation is equal to its transpose, i.e. $\boldsymbol{A}_{\boldsymbol{N}}^{-1} = \boldsymbol{A}_{\boldsymbol{N}}^{\mathrm{T}}$.

7.6 Local Frame and Global Frame Transformations

As in the 2D case, in 3D we can perform *global transformations* and *local transformations*. In Section 5.12, we have seen that global transformations are performed and are expressed with respect to the *global frame of reference* or *world space origin and axes*, while local transformations are performed and are expressed with respect to the *local frame of reference* or *local space origin and axes* of an object. In Section 5.12.1, we have observed that the rules for concatenating a series of global/local transformations are as follows.

When using the row representation of points, a series of n global transformations is concatenated by post-multiplying the individual transformations from left to right in ascending order of application, i.e.

$$\boldsymbol{P}' = \left(\left(\left(\boldsymbol{P}\cdot\boldsymbol{G}_1\right)\cdot\boldsymbol{G}_2\right)\cdot\ldots\right)\cdot\boldsymbol{G}_n = \boldsymbol{P}\cdot\left(\boldsymbol{G}_1\cdot\boldsymbol{G}_2\cdot\ldots\cdot\boldsymbol{G}_n\right) = \boldsymbol{P}\cdot\boldsymbol{T_G} \quad (7.51a)$$

While, a series of n local transformations is concatenated by pre-multiplying the individual transformations from right to left in ascending order of application, i.e.

$$\boldsymbol{P}' = \left(\left(\left(\left(\boldsymbol{P}\cdot\boldsymbol{L}_n\right)\cdot\ldots\right)\cdot\boldsymbol{L}_2\right)\cdot\boldsymbol{L}_1\right) = \boldsymbol{P}\cdot\left(\boldsymbol{L}_n\cdot\ldots\cdot\boldsymbol{L}_2\cdot\boldsymbol{L}_1\right) = \boldsymbol{P}\cdot\boldsymbol{T_L} \quad (7.52a)$$

When using the column representation of points, a series of n global transformations is concatenated by pre-multiplying the individual transformations from right to left in ascending order of application, i.e.

$$\boldsymbol{P}'^{\mathrm{T}} = \left(\boldsymbol{G}_n^{\mathrm{T}}\cdot\left(\ldots\cdot\left(\boldsymbol{G}_2^{\mathrm{T}}\cdot\left(\boldsymbol{G}_1^{\mathrm{T}}\cdot\boldsymbol{P}^{\mathrm{T}}\right)\right)\right)\right) = \left(\boldsymbol{G}_n^{\mathrm{T}}\cdot\ldots\cdot\boldsymbol{G}_2^{\mathrm{T}}\cdot\boldsymbol{G}_1^{\mathrm{T}}\right)\cdot\boldsymbol{P}^{\mathrm{T}} = \boldsymbol{T}_{\boldsymbol{G}}^{\mathrm{T}}\cdot\boldsymbol{P}^{\mathrm{T}} \quad (7.51b)$$

While, a series of n local transformations is concatenated by post-multiplying the individual transformations from left to right in ascending order of application, i.e.

$$\boldsymbol{P}'^{\mathrm{T}} = \left(\boldsymbol{L}_1^{\mathrm{T}}\cdot\left(\boldsymbol{L}_2^{\mathrm{T}}\cdot\left(\ldots\cdot\left(\boldsymbol{L}_n^{\mathrm{T}}\cdot\boldsymbol{P}^{\mathrm{T}}\right)\right)\right)\right) = \left(\boldsymbol{L}_1^{\mathrm{T}}\cdot\boldsymbol{L}_2^{\mathrm{T}}\cdot\ldots\cdot\boldsymbol{L}_n^{\mathrm{T}}\right)\cdot\boldsymbol{P}^{\mathrm{T}} = \boldsymbol{T}_{\boldsymbol{L}}^{\mathrm{T}}\cdot\boldsymbol{P}^{\mathrm{T}} \quad (7.52b)$$

7.7 Transformations of the Frame of Reference or Coordinate System

Analogously to the 2D case, which we have examined in Section 5.13, transforming a point $\boldsymbol{P}$ by a transformation matrix $\boldsymbol{T}$ relative to a frame of reference $\boldsymbol{F}$ is equivalent to leaving the point $\boldsymbol{P}$ stationary while transforming the frame of reference $\boldsymbol{F}$ by the inverse transformation matrix $\boldsymbol{T}^{-1}$.

References

[Ahuja 68] Ahuja, D. V. and Coons, S. A. Geometry for construction and display. *IBM System Journal*, vol. 3 & 4, p.p. 188–205 1968.

8

Viewing and Projection Transformations

In computer graphics we are often concerned with representing three-dimensional scenes on a two-dimensional surface. In order to generate an image of a 3D object on a 2D display, we must first project this object onto a *projection plane* and then display this projected image.

A point of a 3D object is expressed in terms of the *object-space coordinate system* (sometimes referred to as the *world-space coordinate system*), the projection of such a point onto the projection plane is expressed in terms of the *projection-space coordinate system* and the final image of this point on the display device is expressed in terms of the *screen-space coordinate system* (sometimes referred to as the *image-space coordinate system*). The complete transformation of a point from the object-space coordinate system to the screen-space system is achieved through the concatenation of three distinct transformations.

- The *viewing transformation* (or the *object-space to eye-space coordinate transformation*) expresses the location of object-space points relative to the observer's eye, accounting for the observer's position and direction of view.
- The *projection transformation* (or the *eye-space to projection-space coordinate transformation*) expresses the location of eye-space points relative to the projection plane, accounting for the effects of the type of projection used.
- The *viewporting transformation* (or the *projection-space to screen-space coordinate transformation*) expresses the location of projection-space points relative to a viewport of the screen, accounting for the viewport size and displacement.

In order to best understand the viewing transformation we will attempt to explain it in terms of a conceptual camera model.

8.1 Conceptual Camera Model

When we shoot a picture with a camera, we are looking at a 3D scene, namely the world, every point of which may be expressed in terms of three coordinates measured from a given origin. In our conceptual camera model, this coordinate system is called the object-space or world-space coordinate system. By convention, this

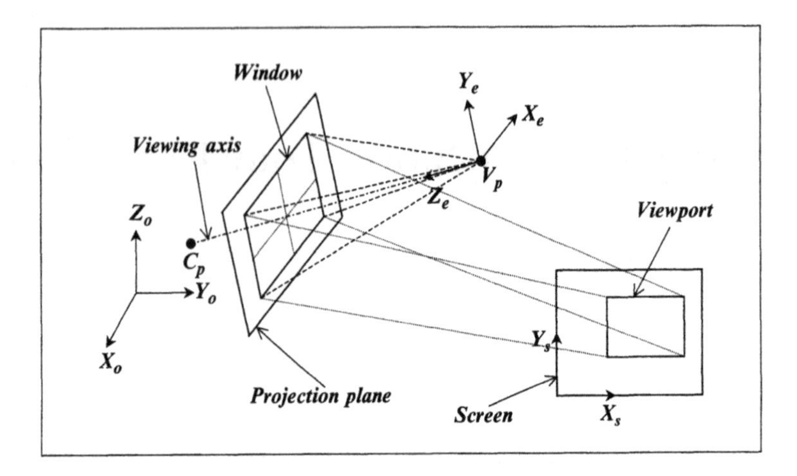

FIGURE 8.1. The object-space, eye-space and screen-space coordinate systems.

coordinate system is *right-handed*, as shown in Fig. 8.1. In this 3D space the camera (i.e. the observers eye) lies at a point called the *viewing point* $\boldsymbol{V_p}$. The point towards which we focus our attention is called the *centre point* $\boldsymbol{C_p}$. The viewing point and the centre point may be placed anywhere in the object space but they may not coincide. These two points define the *viewing axis*, i.e. the *direction of view*.

A new coordinate system can now be defined with its origin at the viewing point, its $\boldsymbol{Z_e}$-axis pointing towards the centre point, its $\boldsymbol{X_e}$-axis pointing to the right and its $\boldsymbol{Y_e}$-axis pointing up as the observer sees them. This coordinate system is called the *eye-space coordinate system* and it is a *left-handed* coordinate system, as shown in Figs. 8.1 and 8.2. The left-handedness of this system comes about from our desire to make the z_e coordinate a direct measure of the depth of the point

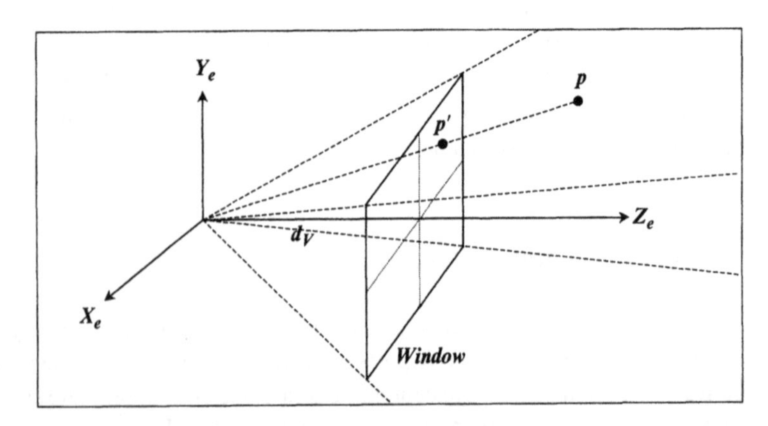

FIGURE 8.2. The eye-space coordinate system and the viewing pyramid.

(with respect to the viewing point) and to keep the $\boldsymbol{X}_e$ and $\boldsymbol{Y}_e$ axes in their most familiar positions. Later we will explain how this transformation is achieved.

In order to transform the 3D eye-space coordinates into 2D screen-space coordinates we must first project the eye-space point onto a projection plane. This is achieved through the projection transformation that we will examine in some detail later. In our camera model the projection plane is placed between the 3D scene and the viewing point and is perpendicular to the viewing axis, as seen in Fig. 8.1. We are generally only interested in a small portion of the projection plane, a rectangular area centred about the viewing axis known as the *window*. The perimeter of the window and the viewing point form the *viewing pyramid*. Any part of the scene that falls outside this pyramid will not be visible to the eye, so we say that the scene is clipped to the viewing pyramid. The distance between the viewing point and the centre of the window is known as the *viewing distance* d_V. Finally, the portion of the projection plane that falls within the window will be mapped onto a rectangular portion of the screen (or other graphic device) known as the *viewport*. This is achieved through the viewporting transformation that we will examine later.

Let us start by looking at the viewing transformation in more detail.

8.2 Viewing Transformation

The object-space to eye-space coordinate transformation is often referred to as the viewing transformation. The viewing transformation, represented by the matrix $\boldsymbol{T_V}$, converts an object-space point $\boldsymbol{p_o} = \langle x_o, y_o, z_o \rangle$ into an eye-space point $\boldsymbol{p_e} = \langle x_e, y_e, z_e \rangle$. Thus

$$[x_e, y_e, z_e, 1] = [x_o, y_o, z_o, 1] \cdot \boldsymbol{T_V} \tag{8.1a}$$

or alternatively

$$[x_e, y_e, z_e, 1]^{\mathrm{T}} = \boldsymbol{T}_{\boldsymbol{V}}^{\mathrm{T}} \cdot [x_o, y_o, z_o, 1]^{\mathrm{T}} \tag{8.1b}$$

The viewing transformation can be constructed by concatenating a number of primitive transformations that are determined by the viewing parameters. In constructing the viewing transformation we have to recall that "a transformation that transforms a point expressed relative to a frame of reference is the inverse of the transformation that transforms the frame of reference relative to this point". For example, given a point $\boldsymbol{p}$ expressed in the frame $\{\boldsymbol{X}, \boldsymbol{Y}, \boldsymbol{Z}, \boldsymbol{O}\}$, translating this point along to the $\boldsymbol{X}$-axis by 5 units is equivalent to translating the frame along to the $\boldsymbol{X}$-axis by -5 units. Similarly, rotating this point about the $\boldsymbol{Z}$-axis by 30° is equivalent to rotating the frame about the $\boldsymbol{Z}$-axis by $-30°$.

In the discussion that follows the scene will be observed from the viewing point $\boldsymbol{V_p} = \langle V_{px}, V_{py}, V_{pz} \rangle$, looking towards the centre point $\boldsymbol{C_p} = \langle C_{px}, C_{py}, C_{pz} \rangle$. Those two points define the viewing axis $\boldsymbol{v}$ of the camera.

$$\boldsymbol{v} = \left[(C_{px} - V_{px})(C_{py} - V_{py})(C_{pz} - V_{pz}) \right]$$

which can be normalised to

$$\hat{\boldsymbol{v}} = \frac{\boldsymbol{v}}{|\boldsymbol{v}|}$$

To unambiguously define the orientation of the camera and to avoid any potential gimbal-lock problems we require an additional vector $\boldsymbol{u}$ that defines the approximate up direction of the camera. If the viewing axis is not parallel or antiparallel to the $\boldsymbol{Z_o}$-axis, then the up-direction vector can be automatically set to $\boldsymbol{u} = [0 \quad 0 \quad 1]$, otherwise it will have to be specified by the user (or determined otherwise from the previous orientation of the camera). See Fig. 8.3. This vector must also be normalised to

$$\hat{\boldsymbol{u}} = \frac{\boldsymbol{u}}{|\boldsymbol{u}|}$$

We assume that the object-space coordinate system is defined by the frame $\{\boldsymbol{i_o}, \boldsymbol{j_o}, \boldsymbol{k_o}, \boldsymbol{O}\}$, where $\boldsymbol{i_o} = \begin{bmatrix}1 & 0 & 0\end{bmatrix}, \boldsymbol{j_o} = \begin{bmatrix}0 & 1 & 0\end{bmatrix}$ and $\boldsymbol{k_o} = \begin{bmatrix}0 & 0 & 1\end{bmatrix}$. Similarly, we assume that the eye-space coordinate system is defined by the frame $\{\boldsymbol{i_e}, \boldsymbol{j_e}, \boldsymbol{k_e}, \boldsymbol{V_p}\}$, where $\boldsymbol{i_e} = \begin{bmatrix}x_{ie} & y_{ie} & z_{ie}\end{bmatrix}, \boldsymbol{j_e} = \begin{bmatrix}x_{je} & y_{je} & z_{je}\end{bmatrix}$ and $\boldsymbol{k_e} = \begin{bmatrix}x_{ke} & y_{ke} & z_{ke}\end{bmatrix}$, which need to be determined. See Fig. 8.4.

To determine the viewing transformation we proceed as follows. First, we translate the object-space frame to the to the viewing point, as shown in Fig. 8.5. Using the point row representation, this is achieved with the following transformation.

$$T_T = \begin{bmatrix} 1 & 0 & 0 & 0 \\ 0 & 1 & 0 & 0 \\ 0 & 0 & 1 & 0 \\ V_{px} & V_{py} & V_{pz} & 1 \end{bmatrix} \tag{8.2}$$

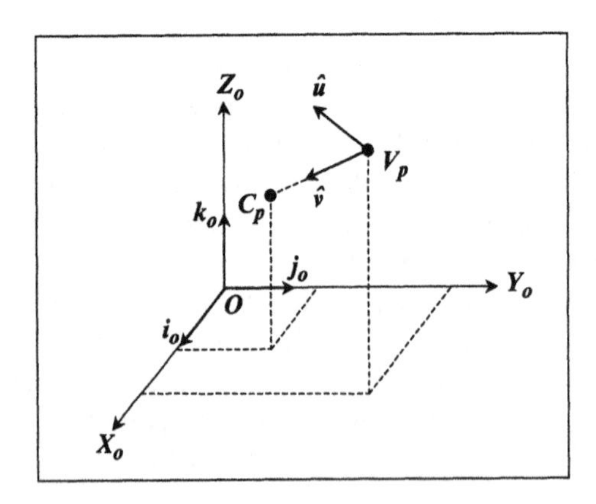

FIGURE 8.3. The $\boldsymbol{v}$ and $\boldsymbol{u}$ vectors.

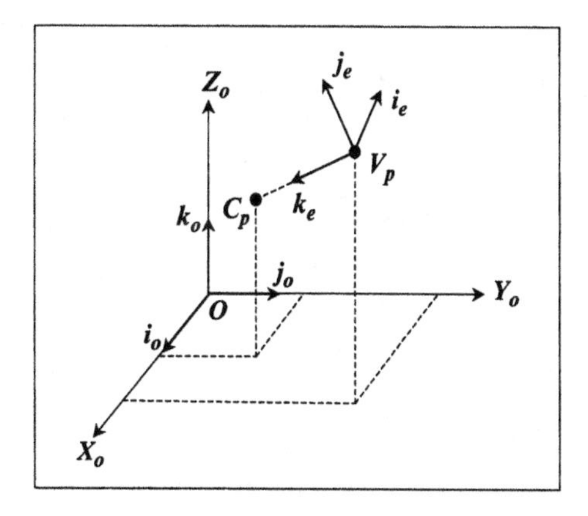

FIGURE 8.4. The eye-space coordinate system frame.

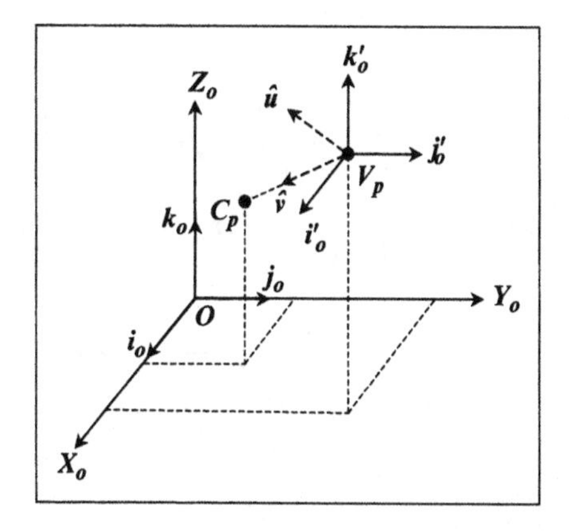

FIGURE 8.5. Moving the object-space frame to the viewing point.

Next, we aim the $\boldsymbol{k}'_o$ vector along the normalised viewing axis $\hat{\boldsymbol{v}}$, as shown in Fig. 8.6. This is achieved by re-computing the three frame vectors as follows.

$$\begin{aligned} \boldsymbol{k}''_o &\leftarrow \hat{\boldsymbol{v}} \\ \boldsymbol{i}''_o &\leftarrow \frac{\hat{\boldsymbol{u}} \otimes \boldsymbol{k}''_o}{\left|\hat{\boldsymbol{u}} \otimes \boldsymbol{k}''_o\right|} \\ \boldsymbol{j}''_o &\leftarrow \frac{\boldsymbol{k}''_o \otimes \boldsymbol{i}''_o}{\left|\boldsymbol{k}''_o \otimes \boldsymbol{i}''_o\right|} \end{aligned} \tag{8.3}$$

Finally, we reverse the direction of the $\boldsymbol{i}''_o$ vector, as the eye-space system is a left-handed coordinate system. See Fig. 8.7. Thus, we can compute the eye-space frame $\{\boldsymbol{i}_e, \boldsymbol{j}_e, \boldsymbol{k}_e, \boldsymbol{V}_p\}$ as follows.

$$\begin{aligned} \boldsymbol{i}_e &\leftarrow -\boldsymbol{i}''_o \\ \boldsymbol{j}_e &\leftarrow \boldsymbol{j}''_o \\ \boldsymbol{k}_e &\leftarrow \boldsymbol{k}''_o \end{aligned} \tag{8.4}$$

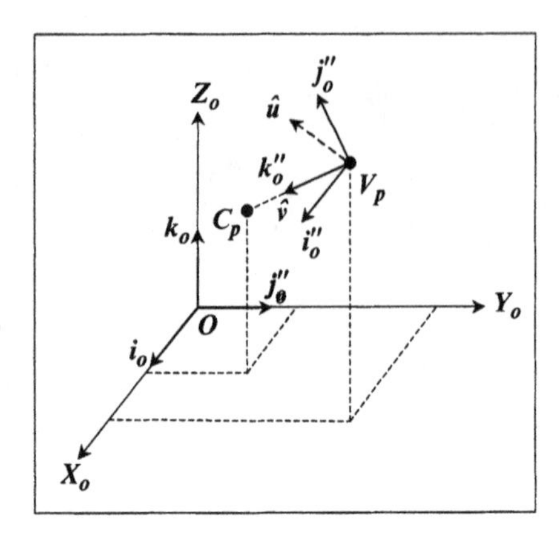

FIGURE 8.6. Aiming the $\boldsymbol{k}'_o$ vector along the $\hat{\boldsymbol{v}}$ vector.

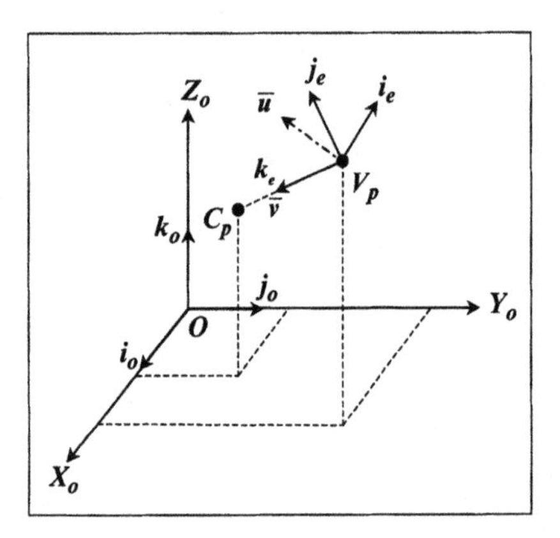

FIGURE 8.7. Reversing the direction of the $\boldsymbol{i}''_o$ vector.

So, to transform the frame $\{\boldsymbol{i}'_o, \boldsymbol{j}'_o, \boldsymbol{k}'_o, \boldsymbol{V_p}\}$ to the frame $\{\boldsymbol{i}_e, \boldsymbol{j}_e, \boldsymbol{k}_e, \boldsymbol{V_p}\}$ we require the following aiming transformation.

$$\boldsymbol{T_A} = \begin{bmatrix} x_{ie} & y_{ie} & z_{ie} & 0 \\ x_{je} & y_{je} & z_{je} & 0 \\ x_{ke} & y_{ke} & z_{ke} & 0 \\ 0 & 0 & 0 & 1 \end{bmatrix} \tag{8.5}$$

Above, for the purpose of producing clear illustrations, we have first translated the original frame to the viewing point and then reoriented it, while in reality we should have reoriented the frame first and then translated it. If we assume that the transformation $\boldsymbol{T_T}$ and $\boldsymbol{T_A}$ are local rather than global transformation, then we achieve the same result. By composing these two local transformations we obtain the composite transformation $(\boldsymbol{T_A} \cdot \boldsymbol{T_T})$ that converts the $\{\boldsymbol{i}_o, \boldsymbol{j}_o, \boldsymbol{k}_o, \boldsymbol{O}\}$ frame to the $\{\boldsymbol{i}_e, \boldsymbol{j}_e, \boldsymbol{k}_e, \boldsymbol{V_p}\}$ frame. We do not however wish to transform the frames of the coordinate systems, but to transform points from the object-space to the eye-space coordinate systems. Thus, we must use the inverse of the matrix product that would transform the frames. The inverse matrices for T_T and T_A are given by

$$T_T^{-1} = \begin{bmatrix} 1 & 0 & 0 & 0 \\ 0 & 1 & 0 & 0 \\ 0 & 0 & 1 & 0 \\ -V_{px} & -V_{py} & -V_{pz} & 1 \end{bmatrix} \tag{8.6}$$

and

$$T_A^{-1} = \begin{bmatrix} x_{ie} & x_{je} & x_{ke} & 0 \\ y_{ie} & y_{je} & y_{ke} & 0 \\ z_{ie} & z_{je} & z_{ke} & 0 \\ 0 & 0 & 0 & 1 \end{bmatrix} \tag{8.7}$$

as T_A is an orthogonal matrix and its inverse is the same as its transpose, i.e. $T_A^{-1} = T_A^{\mathrm{T}}$.

The viewing transformation can now be written as

$$T_V = (T_A \cdot T_T)^{-1} = T_T^{-1} \cdot T_A^{-1}$$

$$\therefore \quad T_V = \begin{bmatrix} 1 & 0 & 0 & 0 \\ 0 & 1 & 0 & 0 \\ 0 & 0 & 1 & 0 \\ -V_{px} & -V_{py} & -V_{pz} & 1 \end{bmatrix} \cdot \begin{bmatrix} x_{ie} & x_{je} & x_{ke} & 0 \\ y_{ie} & y_{je} & y_{ke} & 0 \\ z_{ie} & z_{je} & z_{ke} & 0 \\ 0 & 0 & 0 & 1 \end{bmatrix}$$

$$\therefore \quad T_V = \begin{bmatrix} x_{ie} & x_{je} & x_{ke} & 0 \\ y_{ie} & y_{je} & y_{ke} & 0 \\ z_{ie} & z_{je} & z_{ke} & 0 \\ -x_{ie} \cdot V_{px} - y_{ie} \cdot V_{py} - z_{ie} \cdot V_{pz} & -x_{je} \cdot V_{px} - y_{je} \cdot V_{py} - z_{je} \cdot V_{pz} & -x_{ke} \cdot V_{px} - y_{ke} \cdot V_{py} - z_{ke} \cdot V_{pz} & 1 \end{bmatrix}$$

$$T_V = \begin{bmatrix} \begin{bmatrix} \boldsymbol{i_e} \end{bmatrix} & \begin{bmatrix} \boldsymbol{j_e} \end{bmatrix} & \begin{bmatrix} \boldsymbol{k_e} \end{bmatrix} & \begin{matrix} 0 \\ 0 \\ 0 \end{matrix} \\ -\boldsymbol{i_e} \odot \boldsymbol{V_p} & -\boldsymbol{j_e} \odot \boldsymbol{V_p} & -\boldsymbol{k_e} \odot \boldsymbol{V_p} & 1 \end{bmatrix} \tag{8.8}$$

Hence

$$[x_e, y_e, z_e, 1] = [x_o, y_o, z_o, 1] \cdot T_V \tag{8.9a}$$

or alternatively

$$[x_e, y_e, z_e, 1]^{\mathrm{T}} = T_V^{\mathrm{T}} \cdot [x_o, y_o, z_o, 1]^{\mathrm{T}} \tag{8.9b}$$

where $\langle x_e, y_e, z_e \rangle$ is the point expressed in the *eye-space coordinate system.*

8.3 Projection Transformation

When we represent a 3D object on a 2D display we may attempt either to show the general appearance of the object, as in a photograph, or to depict the object so that its metric properties, such as distances and angles, can easily be derived from this image. Those methods of representation as well as the representations themselves are known as *projections*. Thus, a projection is defined to be both a mapping of a 3D coordinate system onto a 2D coordinate system, called the *projection coordinate system*, and the resulting image of applying such a mapping to an object.

The projected image of a 3D object is found by passing a line through each point of the object and finding the intersections of these lines with the *projection plane*, as shown in Figs. 8.8 and 8.9. These lines are called *projectors*, and emanate from a single point called the *centre of projection*. When the centre of projection is at infinity, so that the projectors are parallel to each other, the projection is known as a *parallel projection* (see Fig. 8.8). Alternatively, when the centre of projection is at a finite distance from the projection plane, then the

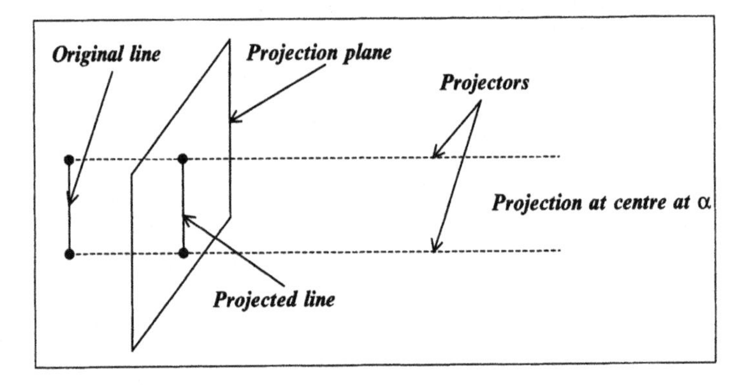

FIGURE 8.8. Parallel projections.

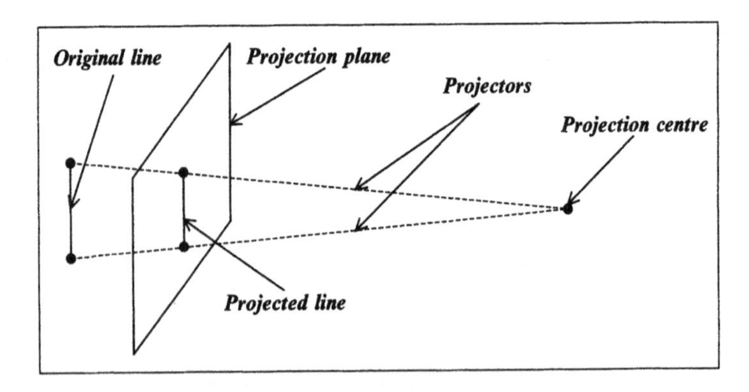

FIGURE 8.9. Perspective projections.

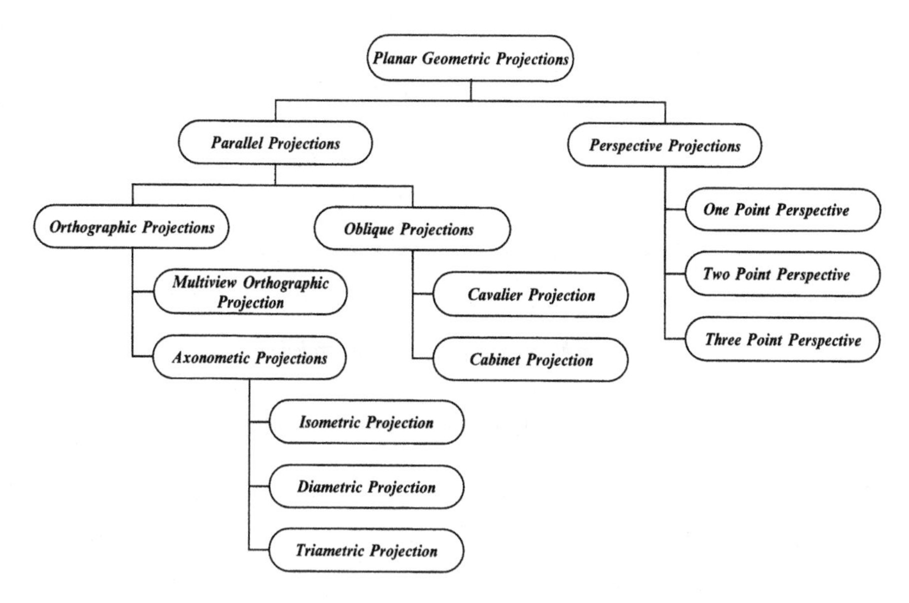

FIGURE 8.10. A classification of planar geometric projections.

projection is known as a *perspective projection* (see Fig. 8.9). Each of these types of projection has further sub-classifications, which are shown in Fig. 8.10.

The class of projections we have defined above is known as *planar geometric projections* because the projection is onto a planar rather than some curved surface and uses straight rather than curved projectors. Non-planar projections are possible but as they are not common in computer graphics we will not deal with such projections.

8.4 Projection Transformation Matrix

In the discussion that follows we will examine the individual projection transformations in terms of a 4×4 transformation matrix, which together with the viewing parameters uniquely defines the projection.

In our discussion of the *orthographic*, *oblique* and *perspective* projections we will assume that the viewing transformation has already been applied and that the coordinates of the 3D scene, to be projected, are expressed in the eye-space coordinate system. We will also assume that the projection plane is parallel to the $\boldsymbol{X_e Y_e}$ plane.

As we shall see later, certain projection transformations can only be achieved if we restrict the location and orientation of the viewing axis of our camera model. A distinct projection matrix will be derived for the orthographic, oblique and perspective projections.

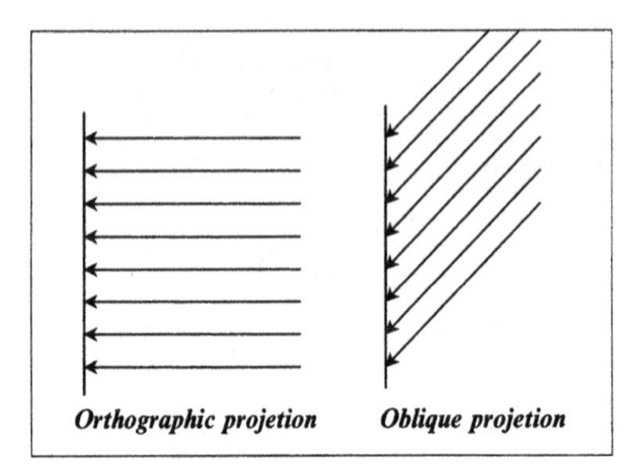

FIGURE 8.11. Orthographic and oblique projections.

8.5 Parallel Projections

Parallel projections are classified by the angle between the projectors and the projection plane. When the projectors are perpendicular to the projection plane, then the projection is known as an *orthographic projection*; otherwise it is known as an *oblique projection*. See Fig. 8.11.

8.5.1 Orthographic Projections

Orthographic projections are characterised by projectors that are perpendicular to the projection plane. These projections are therefore completely determined by the orientation of the projection plane. Orthographic projections are represented as *multi-view orthographic projections* or as *axonometric projections*.

8.5.1.1 Multi-View Orthographic Projections

A multi-view orthographic projection is not one but a collection of projections. These projections show, in one picture, two or more orthographic projections onto projection planes that are perpendicular to the coordinate axes, as shown in Fig. 8.12. Thus, only six distinct orthographic projections are possible.

A multi-view orthographic projection has the advantage that it illustrates the exact shape of two or more faces of an object, while at the same time it suffers from the disadvantage that the three-dimensional shape of the object may be hard to visualise from the separate views. The number of views required to adequately describe the dimensions of an object depends on the complexity of its shape. A simple symmetrical object with rectangular faces can often be described in only two or three views. These projections are often arranged relative to each other in a specific manner. Orthographic projections are often used because distances and angles can be directly measured from them.

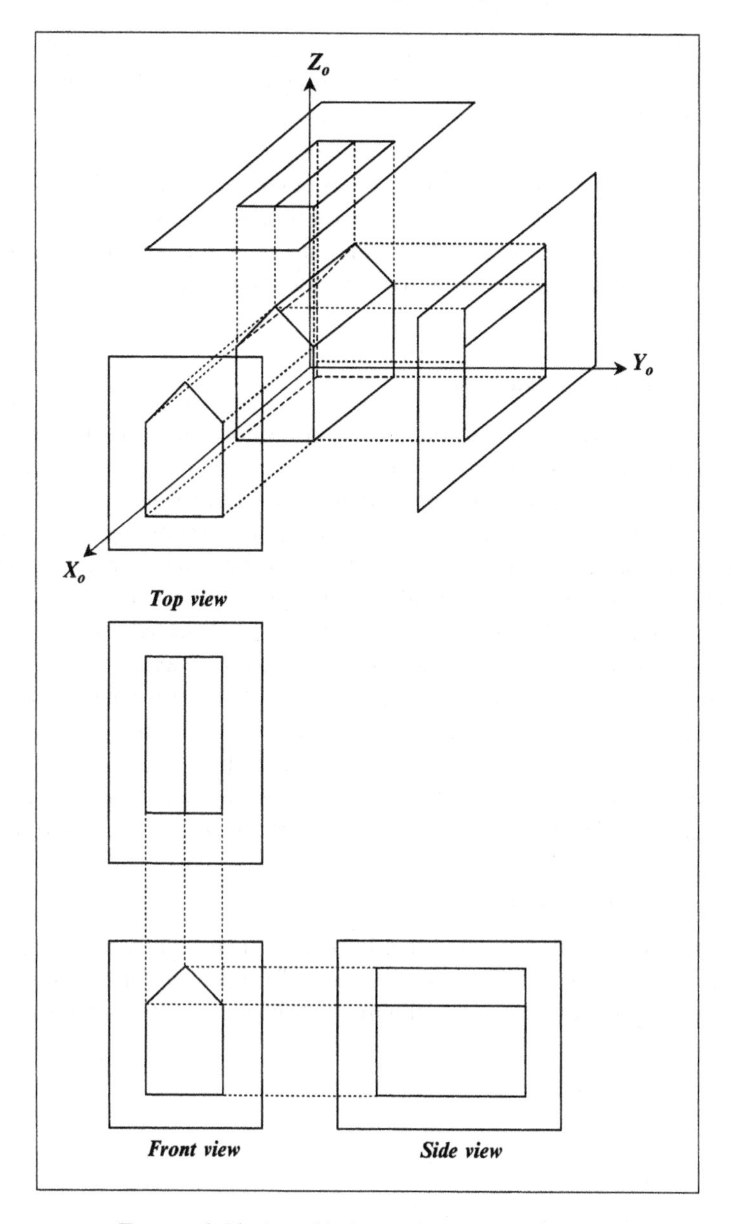

FIGURE 8.12. A multi-view orthographic projection.

In order to produce an orthographic projection we must restrict the location and the orientation of the camera as follows.

- For a *front view* (front elevation) the viewing axis of the camera must lie on the positive X_O-axis and point towards the origin (or be parallel to this axis).
- For a *back view* the viewing axis of the camera must lie on the negative X_O-axis and point towards the origin (or be parallel to this axis).

- For a *right side view* (side elevation) the viewing axis of the camera must lie on the positive Y_O-axis and point towards the origin (or be parallel to this axis).
- For a *left side view* (side elevation) the viewing axis of the camera must lie on the negative Y_O-axis and point towards the origin (or be parallel to this axis).
- For a top view (plan view) the viewing axis of the camera must lie on the positive Z_O-axis and point towards the origin (or be parallel to this axis).
- For a *bottom view* the viewing axis of the camera must lie on the negative Z_O-axis and point towards the origin (or be parallel to this axis).

8.5.1.2 Axonometric Projections

An axonometric projection is an orthographic projection onto the projection plane, where this plane is chosen in such a way that the general three-dimensional shape of the object is illustrated. Such projections usually represent an object so that three adjacent faces are visible, but the true shape and size of any of these faces are not shown unless a face is parallel to the projection plane. In an axonometric projection, parallel lines are equally foreshortened. In particular, axonometric projections produce uniform foreshortening along the projected coordinate axes; thus measurements can easily be made to scale along these axes.

Axonometric projections are classified according to the orientation of the projection plane and the coordinate axes (i.e. the angles between the projection plane and the coordinate axes), as seen in Fig. 8.13. If all three angles are equal, then the projection is called *isometric*. If only two angles are equal, then the projection is called *dimetric*. If all angles are different, then the projection is called *trimetric*.

8.5.1.2.1 Isometric Projections

In an isometric projection all three coordinate axes are equally foreshortened and the angles between the projected axes are equal. To obtain an isometric projection the projection plane must intersect all three coordinate axes at equal angles. This

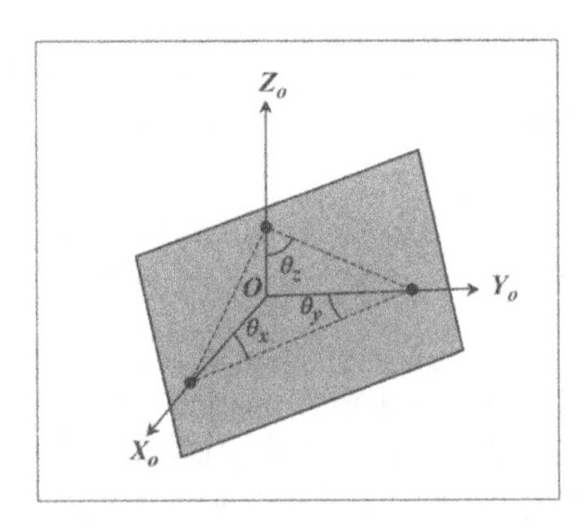

FIGURE 8.13. The angles between the projection plane and the coordinate axes.

is equivalent to requiring that the absolute values of all three direction cosines of the projection plane normal (i.e. the viewing axis) be equal.

Let

$$x = V_{px} - C_{px}$$
$$y = V_{py} - C_{py}$$
$$z = V_{pz} - C_{pz}$$

be the direction ratios of the projection plane normal, also let $l = \sqrt{x^2 + y^2 + z^2}$ be the length of this normal, then we require that

$$\frac{|x|}{l} = \frac{|y|}{l} = \frac{|z|}{l}$$

which implies that $\pm x = \pm y = \pm z$, i.e. the lines

$$+x = +y = +z \quad \text{(Case 1)}$$
$$-x = +y = +z \quad \text{(Case 2)}$$
$$+x = -y = +z \quad \text{(Case 3)}$$
$$-x = -y = +z \quad \text{(Case 4)}$$
$$+x = +y = -z \quad \text{(Case 5)}$$
$$-x = +y = -z \quad \text{(Case 6)}$$
$$+x = -y = -z \quad \text{(Case 7)}$$
$$-x = -y = -z \quad \text{(Case 8)}$$

The viewing axis of our camera must lie on or be parallel to one of the above eight lines and the camera must be looking towards the origin, as seen in Fig. 8.14. Thus, it is apparent that the isometric projection provides little freedom in the choice of the orientation of the projection plane and that an equal importance is given to all the coordinates axes. The right diagram of Fig. 8.14 depicts the isometric projection of a unit cube situated at the origin and aligned with the primary axes. In this projection, the viewing point of the camera is situated at the point labelled 4 on the left diagram of the figure.

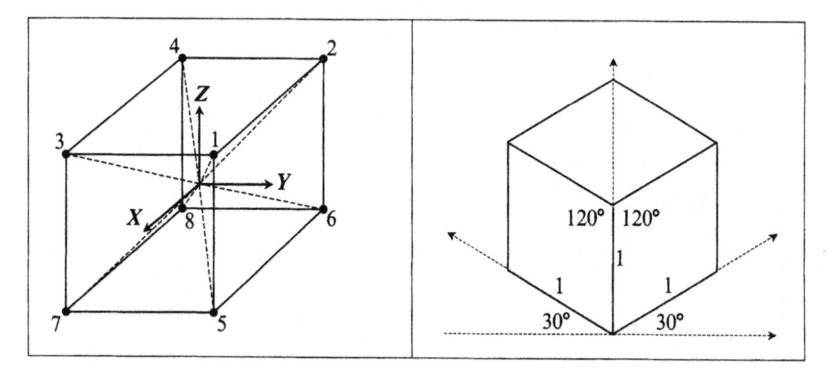

FIGURE 8.14. An isometric projection.

8.5.1.2.2 *Dimetric Projections*

In a dimetric projection only two coordinate axes are equally foreshortened and only two of the angles between the projected axes are equal. In order to obtain a dimetric projection, the projection plane must intersect two of the coordinate axes at equal angles. This is equivalent to requiring the absolute value of two of the direction cosines of the projection plane normal to be equal. Thus we require

$$\frac{|x|}{l} = \frac{|y|}{l} \quad \text{or} \quad \frac{|x|}{l} = \frac{|z|}{l} \quad \text{or} \quad \frac{|y|}{l} = \frac{|z|}{l}$$

which implies that the projection plane normal lies on a plane parallel to one of the following six planes:

$$\begin{aligned} x &= \pm y \\ x &= \pm z \\ y &= \pm z \end{aligned}$$

which means that the viewing axis of our camera must lie on one of these planes or on a plane parallel to one of these planes, as seen in Fig. 8.15.

A dimetric projection affords us more freedom in the choice of the projection plane orientation, thus allowing us to emphasise the face of the object we are most interested in.

To best visualise the above six planes, think of them as dissecting opposite faces of the solid cube of the left diagram of Fig. 8.15. For example, the points 1, 4, 8 and 5 define the $x = +y$ plane and points 2, 3, 7 and 6 define the $x = -y$ plane.

8.5.1.2.3 *Trimetric Projections*

A trimetric projection is the most general form of an axonometric projection. It produces different foreshortening for each of the coordinate axes and no two angles between the projected axes are equal. A trimetric projection affords us almost complete freedom in the choice of projection plane orientation and if the

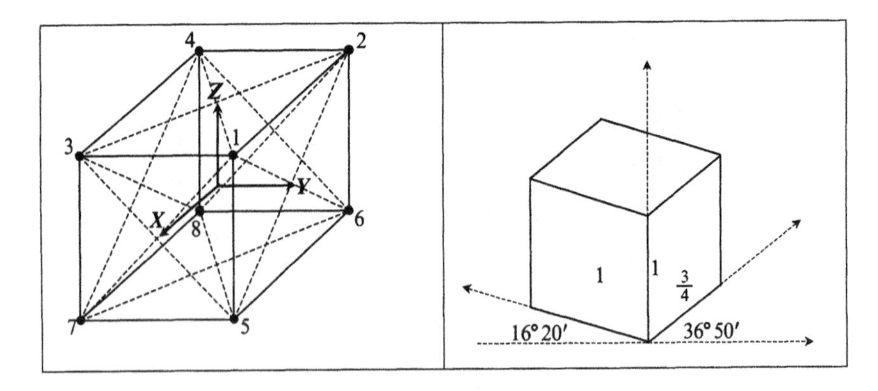

FIGURE 8.15. A dimetric projection.

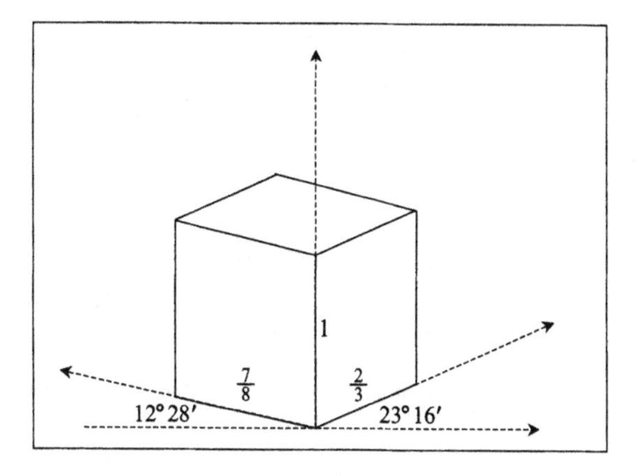

FIGURE 8.16. A trimetric projection.

projection plane is appropriately chosen, it gives the most realistic appearance. See Fig. 8.16.

8.5.1.3 Orthographic Projection Matrix

An orthographic projection onto the $\boldsymbol{X_p Y_p}$ plane, which is parallel to the $\boldsymbol{X_e Y_e}$ plane, is achieved by simply ignoring the z_e coordinate of the point $\langle x_e, y_e, z_e \rangle$. Thus the projection matrix is

$$\boldsymbol{P} = \begin{bmatrix} 1 & 0 & 0 & 0 \\ 0 & 1 & 0 & 0 \\ 0 & 0 & 0 & 0 \\ 0 & 0 & 0 & 1 \end{bmatrix} \tag{8.10}$$

Hence

$$\left[x_p, y_p, 0, 1\right] = [x_e, y_e, z_e, 1] \cdot \boldsymbol{P} \tag{8.11a}$$

or alternatively

$$\left[x_p, y_p, 0, 1\right]^{\mathrm{T}} = \boldsymbol{P}^{\mathrm{T}} \cdot [x_e, y_e, z_e, 1]^{\mathrm{T}} \tag{8.11b}$$

where $\langle x_p, y_p, 0 \rangle$ is the projected point expressed in the *projection-space coordinate system*.

8.5.2 *Oblique Projections*

Oblique projections combine the properties of the multi-view orthographic projections and those of the axonometric projections. An oblique projection normally represents the exact shape of one face of an object, and at the same time illustrates the general three-dimensional appearance of the object.

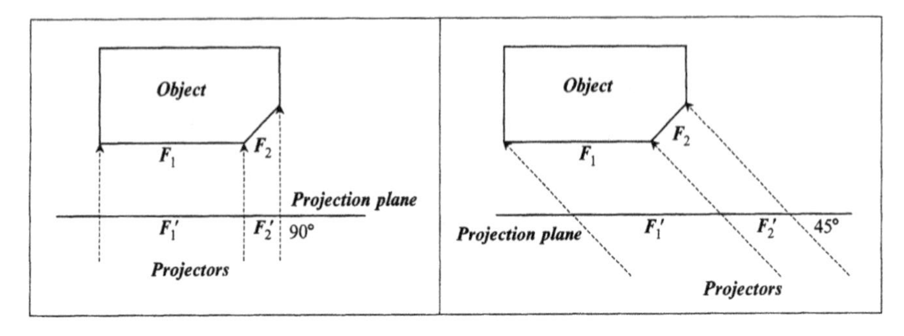

FIGURE 8.17. Comparison of an orthographic and an oblique projection.

Figure 8.17 allows us to compare two projections of an object with faces $\boldsymbol{F}_1$ and $\boldsymbol{F}_2$, the details of which we wish to emphasise; an orthographic projection (on the left) and an oblique projection (on the right). Observe that in both projections the relative length of the projection $\boldsymbol{F}_1'$ of face $\boldsymbol{F}_1$ remains unchanged. This is so because this face is parallel to the projection plane. In contrast, the relative length of the projection $\boldsymbol{F}_2'$ of the face $\boldsymbol{F}_2$ is larger in the oblique projection, thus allowing us to show more detail in the projection of this face. This has been achieved by altering the projection angle alone in this case.

The oblique projection is characterised by projectors that meet the projection plane at an oblique angle and is determined by

- the orientation of the projection plane (which is determined by the orientation of the viewing axis);
- the projection angle α_1, i.e. the angle between the projectors and the projection plane;
- the orientation angle α_2, i.e. the orientation of the projectors with respect to the projection plane normal. This angle is usually measured from the horizontal axis of the projection-space coordinate system (i.e. the $\boldsymbol{X}$-axis).

See Figs. 8.18 and 8.19.

The projection plane of an oblique projection is usually positioned either

- parallel to the largest face of the object or
- parallel to the face of the object with the most detail,

so that this face is projected without distortion.

The orientation of the projectors is chosen to best illustrate the third dimension of the object.

An oblique projection is classified by its projection angle either as a *cavalier projection* when $\alpha_1 = 45°$ or as a *cabinet projection* when $\alpha_1 \approx 64°$.

The projection angle α_1 determines the thickness of the projected object, while the orientation angle α_2 determines the relative emphasis of the receding planes. The orientation angle is often chosen such that the projection plane normal is projected at 30° or at 45° with respect to the horizontal projection-space coordinate axis, $\boldsymbol{X_p}$.

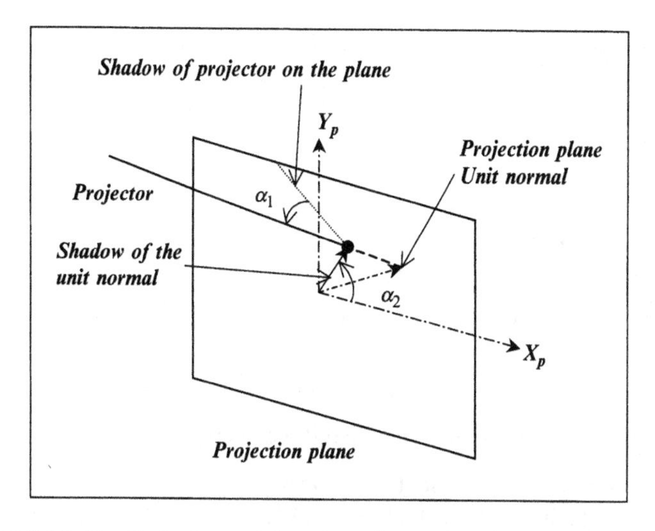

FIGURE 8.18. The Projection and the orientation angles of the oblique projection.

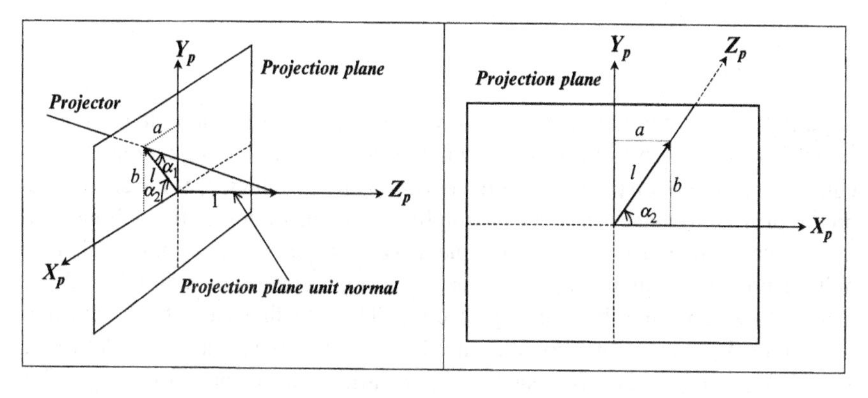

FIGURE 8.19. The oblique projection parameters.

The parameters of an oblique projection are best defined with respect to a coordinate system with two axes on the projection plane and the third along the projection plane normal. See Fig. 8.19.

These parameters are

l the length of the projected unit normal on the projection plane (i.e. the foreshortening ratio);

a and b the coordinates of the tip of the projected unit normal on the projection plane;

α_1 the projection angle;

α_2 the orientation angle.

From Fig. 8.19 we have

$$\tan(\alpha_1) = \frac{1}{l}$$

$$\therefore \quad l = \frac{1}{\tan(\alpha_1)}$$

Also from Fig. 8.19 we have

$$\cos(\alpha_2) = \frac{a}{l} \quad \text{and} \quad \sin(\alpha_2) = \frac{b}{l}$$

$$\therefore \quad a = l \cdot \cos(\alpha_2) \quad \text{and} \quad b = l \cdot \sin(\alpha_2)$$

Thus an oblique projection is specified if we know

α_1 and α_2 in which case we calculate l, a and b;
or l and α_2 in which case we calculate a and b;
or a and b directly.

In oblique projections there are no restrictions placed on the location and orientation of the camera, provided that $0° < |\alpha_1| < 90°$.

8.5.2.1 Oblique Projection Matrix

Usually when we think of a projection we envisage the projection plane as being in front of us (i.e., in front of the viewing point). In parallel projections the distance of the projection plane from the viewing point is irrelevant, as it has no effect on the projected image. In order to simplify the computation of the oblique projection we assume that the projection plane $\boldsymbol{X_p Y_p}$ lies on the $\boldsymbol{X_e Y_e}$ plane. Then an oblique projection can be thought of as a shearing transformation along the $\boldsymbol{X_e Y_e}$ plane followed by an orthographic projection. Thus, we first shear the $\boldsymbol{Z_e}$-axis parallel to the $\boldsymbol{X_e}$-axis by an amount a and then we shear the $\boldsymbol{Z_e}$-axis parallel to the $\boldsymbol{Y_e}$-axis by an amount b. To visualise this shearing transformation see Fig. 8.20, which depicts a unit cube situated at the origin of the eye-space coordinate system, starting from its neutral position and undergoing the two shearing transformations in sequence.

The combined shearing transformation can be expressed in matrix form as

$$\boldsymbol{S} = \begin{bmatrix} 1 & 0 & 0 & 0 \\ 0 & 1 & 0 & 0 \\ a & b & 1 & 0 \\ 0 & 0 & 0 & 1 \end{bmatrix}$$

Now the projection transformation for the oblique projection becomes

$$\boldsymbol{P} = \begin{bmatrix} 1 & 0 & 0 & 0 \\ 0 & 1 & 0 & 0 \\ a & b & 1 & 0 \\ 0 & 0 & 0 & 1 \end{bmatrix} \cdot \begin{bmatrix} 1 & 0 & 0 & 0 \\ 0 & 1 & 0 & 0 \\ 0 & 0 & 0 & 0 \\ 0 & 0 & 0 & 1 \end{bmatrix} = \begin{bmatrix} 1 & 0 & 0 & 0 \\ 0 & 1 & 0 & 0 \\ a & b & 0 & 0 \\ 0 & 0 & 0 & 1 \end{bmatrix} \tag{8.12}$$

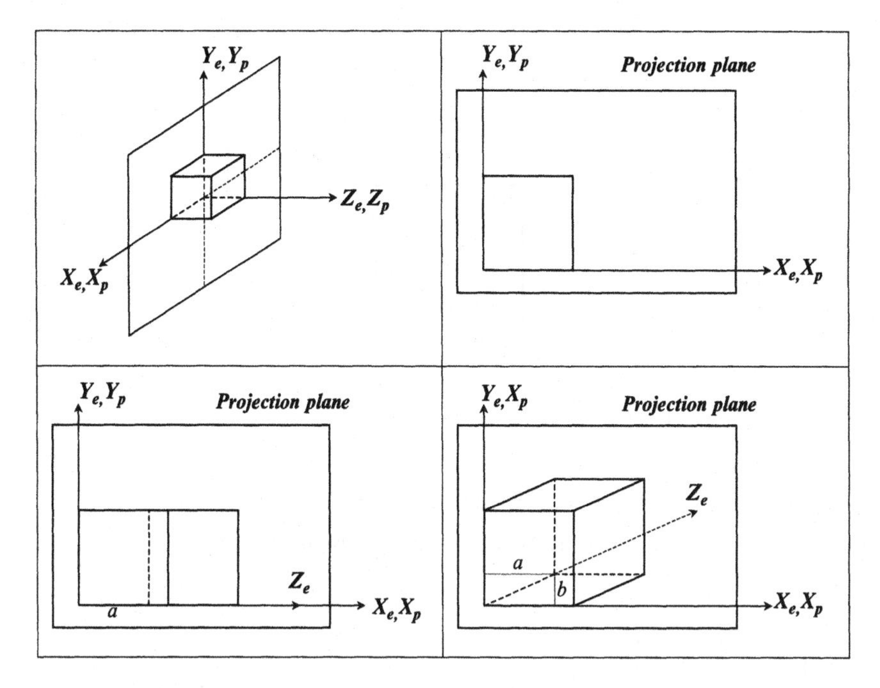

FIGURE 8.20. The oblique projection of a unit cube.

Hence

$$\left[x_p, y_p, 0, 1\right] = [x_e, y_e, z_e, 1] \cdot \boldsymbol{P} \tag{8.13a}$$

or alternatively

$$\left[x_p, y_p, 0, 1\right]^{\mathrm{T}} = \boldsymbol{P}^{\mathrm{T}} \cdot [x_e, y_e, z_e, 1]^{\mathrm{T}} \tag{8.13b}$$

where $\langle x_p, y_p, 0\rangle$ is the projected point expressed in the *projection-space coordinate system.*

8.6 Perspective Projections

A perspective projection gives a natural appearance to the image of the object. Perspective projections do not preserve the shape of the object and measurements can only be made to scale on faces of the object that lie on the projection plane. A perspective projection is distinguished from a parallel projection by

- the convergence of parallel lines;
- the diminution of size (i.e. foreshortening); and
- non-uniform foreshortening.

Lines that are parallel to the projection plane remain parallel after the projection, all other parallel lines converge to one, two or three *vanishing points.* See Figs. 8.21–8.23. The convergence of parallel lines results in the diminution of size

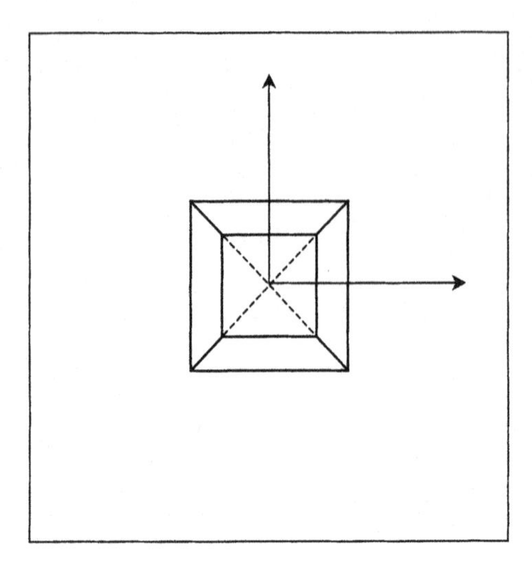

FIGURE 8.21. A one-point perspective projection.

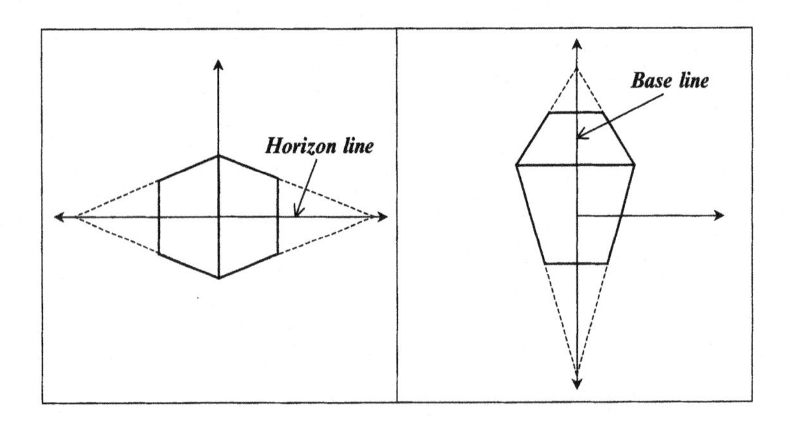

FIGURE 8.22. Two two-point perspective projections.

and non-uniform foreshortening of objects. Objects of equal size appear smaller as their distance from the eye increases. Only faces on the projection plane retain their true size.

Perspective projections are classified according to the number of vanishing points they produce. A perspective projection has the same number of vanishing points as the number of coordinate axes the projection plane intersects. Axes tangential to the projection plane are not counted. There are three classes of perspective projection:

- *one-point perspective* or *parallel perspective*, with one vanishing point;
- two-point perspective or *angular perspective*, with two vanishing points; and
- *three-point perspective*, with three vanishing points.

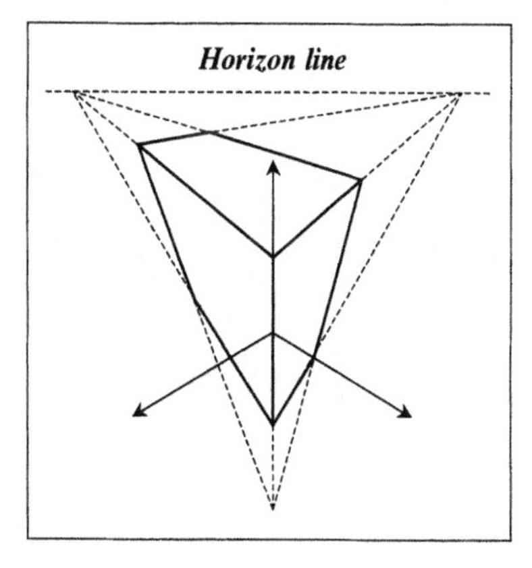

FIGURE 8.23. A three-point perspective projection.

To obtain a one-point perspective projection we must restrict the viewing axis of our camera to coincide with one of the coordinate axes $\boldsymbol{X_o}$, $\boldsymbol{Y_o}$ or $\boldsymbol{Z_o}$ (i.e. one of the *object-space* axes) or to be parallel to it. To obtain a two-point perspective projection we must restrict the viewing axis of our camera to lie on one of the object-space major planes $\boldsymbol{X_o Y_o}$, $\boldsymbol{X_o Z_o}$ or $\boldsymbol{Y_o Z_o}$, or to lie on a plane parallel to one of these planes. No restrictions are imposed on the viewing axis of our camera if a three-point perspective projection is required. See Figs. 8.21–8.23.

8.6.1 Perspective Projection Matrix

A perspective projection on the $\boldsymbol{X_p Y_p}$ plane, which is parallel to the $\boldsymbol{X_e Y_e}$ plane at a distance d_V (*viewing distance*) from it, is defined as outlined below. See Fig. 8.24.

A point $p = \langle x_e, y_e, z_e \rangle$ projects onto a point $p' = \langle x_p, y_p, d_V \rangle$ on the projection plane. From Fig. 8.24 we have

$$\frac{x_p}{x_e} = \frac{d_V}{z_e} \quad \text{and} \quad \frac{y_p}{y_e} = \frac{d_V}{z_e}$$

$$\therefore \quad x_p = x_e \frac{d_V}{z_e} = \frac{x_e}{(z_e/d_V)} \quad \text{and} \quad y_p = y_e \frac{d_V}{z_e} = \frac{y_e}{(z_e/d_V)}$$

Thus the analytical form of the perspective projection transformation is given by

$$\begin{aligned} x_p &= x_e \cdot \frac{1}{(z_e/d_V)} \\ y_p &= y_e \cdot \frac{1}{(z_e/d_V)} \end{aligned} \tag{8.14}$$

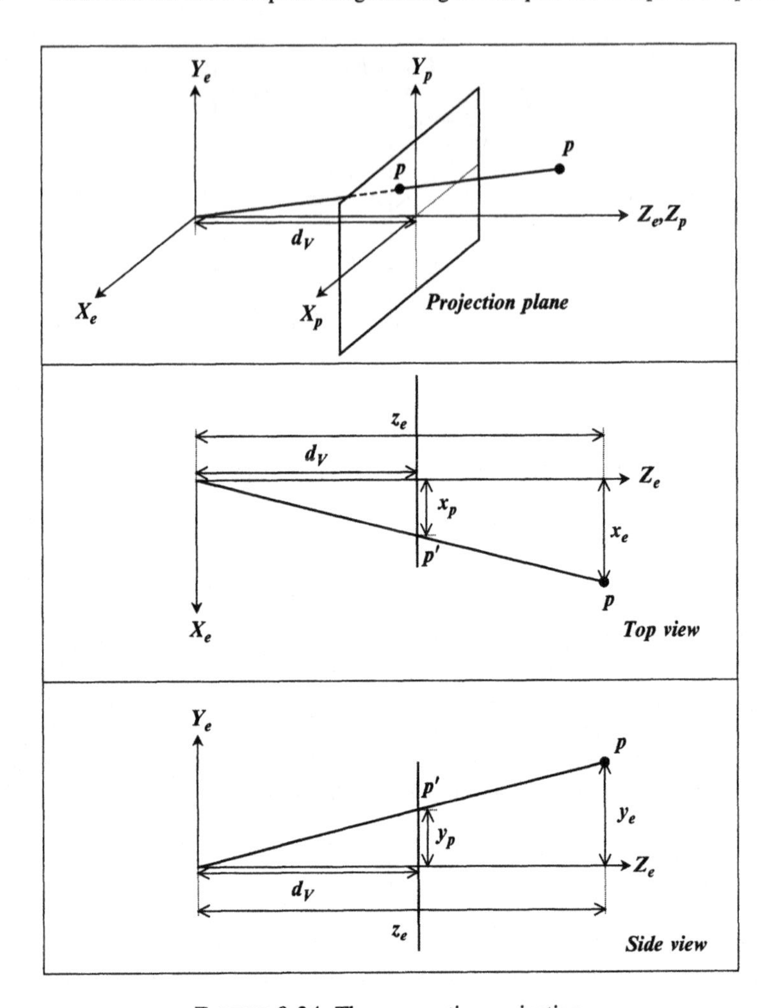

FIGURE 8.24. The perspective projection.

and the matrix form of this transformation is given by

$$P = \begin{bmatrix} 1 & 0 & 0 & 0 \\ 0 & 1 & 0 & 0 \\ 0 & 0 & 1 & \dfrac{1}{d_V} \\ 0 & 0 & 0 & 0 \end{bmatrix} \tag{8.15}$$

Let us verify that this is so. Post-multiplying the homogeneous point $p^{\mathrm{h}} = \langle x_e, y_e, z_e, 1\rangle$ by this matrix yields the general homogeneous point $\langle x, y, z, w\rangle$, i.e.

$$\begin{bmatrix} x & y & z & w \end{bmatrix} = \begin{bmatrix} x_e & y_e & z_e & 1 \end{bmatrix} \cdot P \tag{8.16a}$$

$$\text{or} \quad \begin{bmatrix} x & y & z & w \end{bmatrix} = \begin{bmatrix} x_e & y_e & z_e & \dfrac{z_e}{d_V} \end{bmatrix}$$

$$\therefore \quad x = x_e, y = y_e, z = z_e, w = \frac{z_e}{d_V}$$

Dividing both sides of the above vector equation by w (i.e. homogenising the point) we obtain

$$\begin{bmatrix} \dfrac{x}{w} & \dfrac{y}{w} & \dfrac{z}{w} & 1 \end{bmatrix} = \begin{bmatrix} \dfrac{x_e}{(z_e/d_V)} & \dfrac{y_e}{(z_e/d_V)} & d_V & 1 \end{bmatrix}$$

which is equivalent to

$$\begin{bmatrix} x_p & y_p & d_V & 1 \end{bmatrix} = \begin{bmatrix} \dfrac{x_e}{(z_e/d_V)} & \dfrac{y_e}{(z_e/d_V)} & d_V & 1 \end{bmatrix}$$

Thus we arrive at the same result as in Eq. (8.14).

$$x_p = \frac{x_e}{(z_e/d_V)}$$
$$y_p = \frac{y_e}{(z_e/d_V)}$$

We can arrive at the same result using the alternative form

$$\begin{bmatrix} x & y & z & w \end{bmatrix}^{\mathrm{T}} = \boldsymbol{P}^{\mathrm{T}} \cdot \begin{bmatrix} x_e & y_e & z_e & 1 \end{bmatrix}^{\mathrm{T}} \tag{8.16b}$$

Note that z_e can assume any value apart from zero as this would cause the values of x_p and y_p to become infinite and will therefore cause a floating point error in a computer program that implements this projection. Later, we will see how we resolve this problem through clipping.

8.7 Screen or Device Coordinate System

Having performed the perspective projection, the image of our scene lies on the projection plane and is centred about the origin of the projection coordinate system. Let us assume that there is a window defined on the projection plane of size $2s_x \times 2s_y$ that is centred at the origin, as shown in Fig. 8.25.

In order to display the contents of this window on the screen (or any other graphics device), inside a viewport of size $v_{sx} \times v_{sy}$ centred at $\langle v_{cx}, v_{cy} \rangle$, we must scale and offset each projected point, thus transforming it into the *screen coordinate system*. This transformation is known as *the viewporting transformation* and is computed as

$$x_s = x_p \cdot \left(\frac{v_{sx}/2}{s_x} \right) + v_{cx} \quad \text{and} \quad y_s = y_p \cdot \left(\frac{v_{sy}/2}{s_y} \right) + v_{cy} \tag{8.17}$$

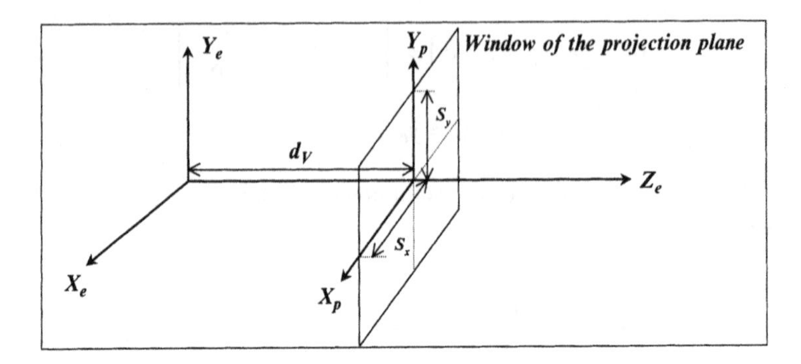

FIGURE 8.25. The projection plane window.

$$\text{i.e.}\quad x_s = \frac{x_e}{(z_e/d_V)} \cdot \left(\frac{v_{sx}/2}{s_x}\right) + v_{cx} \quad \text{and} \quad y_s = \frac{y_e}{(z_e/d_V)} \cdot \left(\frac{v_{sy}/2}{s_y}\right) + v_{cy}$$

which can be rewritten as

$$x_s = \left(\frac{x_e}{z_e}\right)\cdot\left(\frac{d_V}{s_x}\right)\cdot\left(\frac{v_{sx}}{2}\right)+v_{cx} \quad \text{and} \quad y_s = \left(\frac{y_e}{z_e}\right)\cdot\left(\frac{d_V}{s_y}\right)\cdot\left(\frac{v_{sy}}{2}\right)+v_{cy} \qquad (8.18)$$

See Fig. 8.25.

If $s = \max\left(s_x, s_y\right)$, then the ratio (d_V/s) is the cotangent (i.e. the reciprocal of the tangent) of half the viewing angle and it thus allows us to simulate different types of camera lenses. If this ratio is small, then the viewing angle will be large thus producing an image similar to that of a wide-angle lens (or a fish-eye lens). Conversely, if this ratio is large, then the viewing angle will be small thus producing an image similar to that of a telephoto lens. See Fig. 8.26.

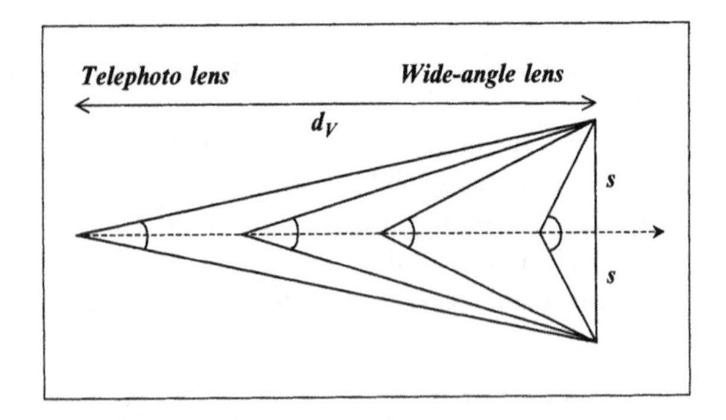

FIGURE 8.26. Simulating different types of lenses by varying the viewing distance.

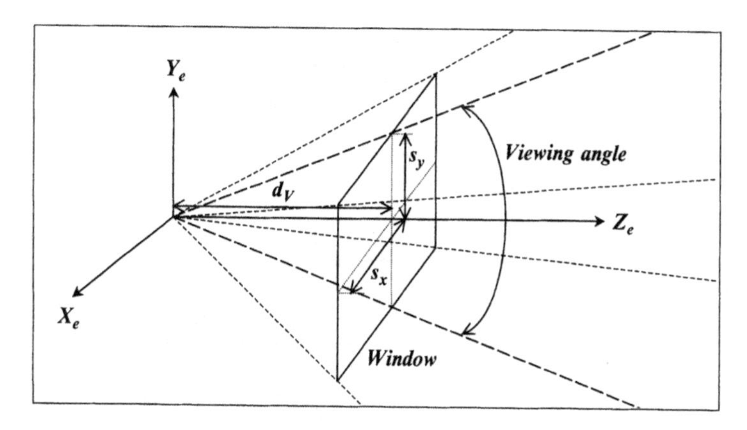

FIGURE 8.27. The viewing pyramid.

8.8 3D Line Clipping

If we join the origin of the eye-space coordinate system with the corners of the projection plane window (that will be mapped onto the viewport of the screen) we define the *viewing pyramid* which determines the visible portion of eye-space. Each edge of the window together with the origin of the eye-space defines a *clipping plane*. Thus, the top, bottom, right and left edges of the window define the top, bottom, right and left clipping planes, respectively. See Figs. 8.27 and 8.28.

All the edges of an object (defined in object-space) must first be transformed to the eye-space, then clipped against the viewing pyramid and finally projected onto the viewport of the screen.

From Fig. 8.28 it can be seen that the clipping planes are defined as

Top plane: $\left(\frac{d_V}{s_y}\right) \cdot y_e = +z_e$

Bottom plane: $\left(\frac{d_V}{s_y}\right) \cdot y_e = -z_e$

Right plane: $\left(\frac{d_V}{s_x}\right) \cdot x_e = +z_e$

Left plane: $\left(\frac{d_V}{s_x}\right) \cdot x_e = -z_e$

Thus, for a point to lie within the viewing pyramid the following conditions must hold true

$$-z_e \leq \left(\frac{d_V}{s_x}\right) \cdot x_e \leq +z_e \quad \text{and} \quad -z_e \leq \left(\frac{d_V}{s_y}\right) \cdot y_e \leq +z_e \tag{8.19}$$

In order to simplify the clipping algorithm we now introduce a new coordinate system, called the *clip-space coordinate system*. To effect this transformation we multiply both the x_e and y_e coordinates of the eye-space point by the scaling

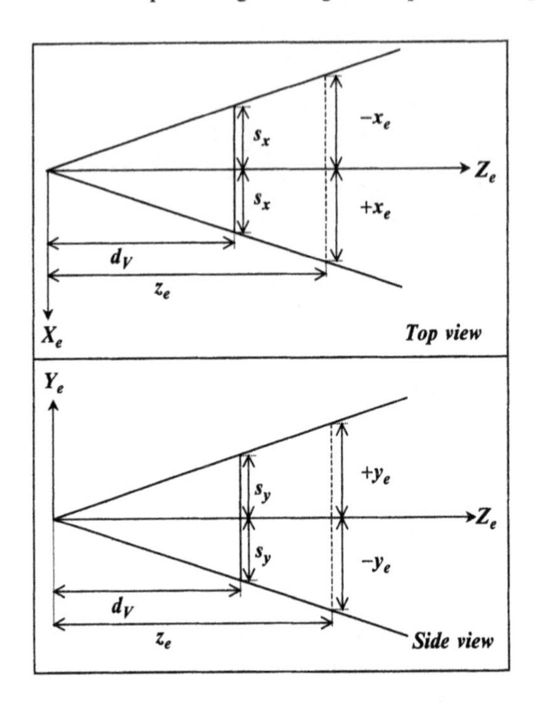

FIGURE 8.28. Two views of the eye-space viewing pyramid.

factors (d_V/s_x) and (d_V/s_y). In matrix form this transformation is expressed as

$$C = \begin{bmatrix} \left(\frac{d_V}{s_x}\right) & 0 & 0 & 0 \\ 0 & \left(\frac{d_V}{s_y}\right) & 0 & 0 \\ 0 & 0 & 1 & 0 \\ 0 & 0 & 0 & 1 \end{bmatrix} \tag{8.20}$$

Thus

$$[x_c, y_c, z_c, 1] = [x_e, y_e, z_e, 1] \cdot \boldsymbol{C} \tag{8.21a}$$

or alternatively as

$$[x_c, y_c, z_c, 1]^{\mathrm{T}} = \boldsymbol{C}^{\mathrm{T}} \cdot [x_e, y_e, z_e, 1]^{\mathrm{T}} \tag{8.21b}$$

This has the effect of transforming the clipping planes to

Top plane: $y_c = +z_c$
Bottom plane: $y_c = -z_c$
Right plane: $x_c = +z_c$
Left plane: $x_c = -z_c$

The four conditions for a point being visible now become

$$-z_c \le x_c \le +z_c \quad \text{and} \quad -z_c \le y_c \le +z_c \tag{8.22}$$

The geometric significance of this is that in the clip-space coordinate system the viewing angle is transformed to a 90° angle as shown in Fig. 8.29.

We can now derive a 3D line-clipping algorithm by extending the 2D Choen and Sutherland line-clipping algorithm. This algorithm determines whether the endpoints of a line segment lie inside or outside the limits of the viewing pyramid by testing the above four conditions and by computing a 4-bit code for each endpoint of the line segment. A given bit being set has the following meaning.

1st bit:	$\boldsymbol{x_c}$	is to the *left* of the *left clipping plane*, i.e.	$x_c < -z_c$
2nd bit:	$\boldsymbol{x_c}$	is to the *right* of the *right clipping plane*, i.e.	$x_c > +z_c$
3rd bit:	$\boldsymbol{y_c}$	is to the *bottom* of the *bottom clipping plane*, i.e.	$y_c < -z_c$
4th bit:	$\boldsymbol{y_c}$	is to the *top* of the *top clipping plane*, i.e.	$y_c > +z_c$

4-Bit codes are computed for both endpoints of the line segment to be clipped in order to speed up the process of *trivially accepting* the segment (as being totally visible) or *trivially rejecting* the segment (as being totally invisible). If the codes of both endpoints are zero, then both endpoints must lie within the viewing pyramid and the segment can be **trivially accepted**. If the *logical intersection* (logical ANDing) of the codes is non-zero, then both endpoints must lie on the invisible side of at least one of the clipping planes and the segment can be

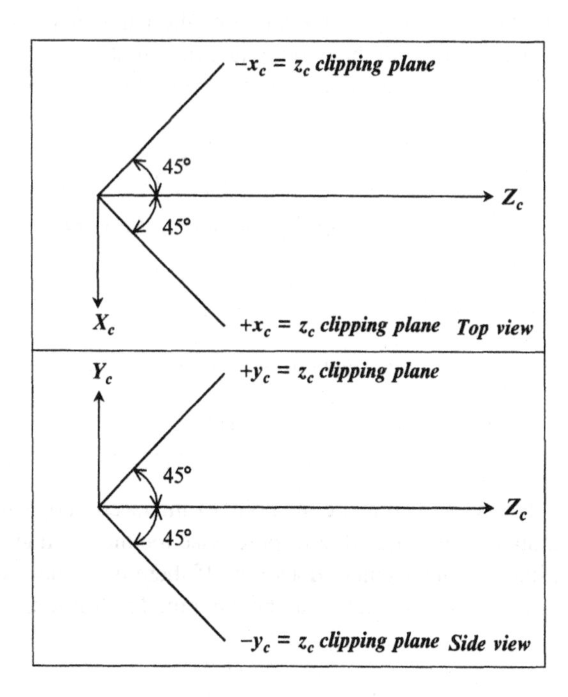

FIGURE 8.29. Two views of the clip-space viewing pyramid.

trivially rejected. If both the above conditions fail, then we must consider the segment further as it might be partially visible. In order to determine the visible portion of the segment (if any), we must compute the point of intersection of the segment with each of the clipping planes.

The intersection computation makes use of the parametric form of the line. Say a line segment is defined by the points $\boldsymbol{p}_1 = \langle x_1, y_1, z_1 \rangle$ and $\boldsymbol{p}_2 = \langle x_2, y_2, z_2 \rangle$, then the general point on the line is given as

$$\begin{aligned} x &= t \cdot (x_2 - x_1) + x_1 \\ y &= t \cdot (y_2 - y_1) + y_1 \\ z &= t \cdot (z_2 - z_1) + z_1 \end{aligned} \tag{8.23}$$

where $0 \leq t \leq 1$ if the point lies on the line segment $\overrightarrow{\boldsymbol{p}_1\boldsymbol{p}_2}$.

To compute the point of intersection of the line with one of the clipping planes we first compute the value of the parameter t at the point of intersection. This is done as follows.

Let us assume that we wish to compute the point of intersection of the line with the *right clipping plane* (i.e. the $x_c = +z_c$ plane). Using Eq. (8.23) we have

$$t \cdot (x_2 - x_1) + x_1 = t \cdot (z_2 - z_1) + z_1$$

$$\therefore \quad t \cdot (x_2 - x_1) - t \cdot (z_2 - z_1) = z_1 - x_1$$

$$\therefore \quad t = \frac{z_1 - x_1}{(x_2 - x_1) - (z_2 - z_1)} \tag{8.24a}$$

In a similar fashion we compute the value of the parameter t at the point of intersection of the segment with the remaining clipping planes.

At the intersection with the *left clipping plane* (i.e. the $x_c = -z_c$ plane)

$$t = \frac{-(z_1 + x_1)}{(x_2 - x_1) + (z_2 - z_1)} \tag{8.24b}$$

At the intersection with the *top clipping plane* (i.e. the $y_c = +z_c$ plane)

$$t = \frac{z_1 - y_1}{(y_2 - y_1) - (z_2 - z_1)} \tag{8.24c}$$

At the intersection with the *bottom clipping plane* (i.e. the $y_c = -z_c$ plane)

$$t = \frac{-(z_1 + y_1)}{(y_2 - x_1) + (y_2 - z_1)} \tag{8.24d}$$

If $0 \leq t \leq 1$, then we can compute the point of intersection by substituting the value of the parameter t into Eq. (8.23); otherwise the infinite line intersects the clipping plane, but the line segment does not. If there is a point of intersection with the line segment, its coordinates can be computed as follows.

$$\begin{aligned} x_i &= t \cdot (x_2 - x_1) + x_1 \\ y_i &= t \cdot (y_2 - y_1) + y_1 \\ z_i &= t \cdot (z_2 - z_1) + z_1 \end{aligned} \tag{8.25}$$

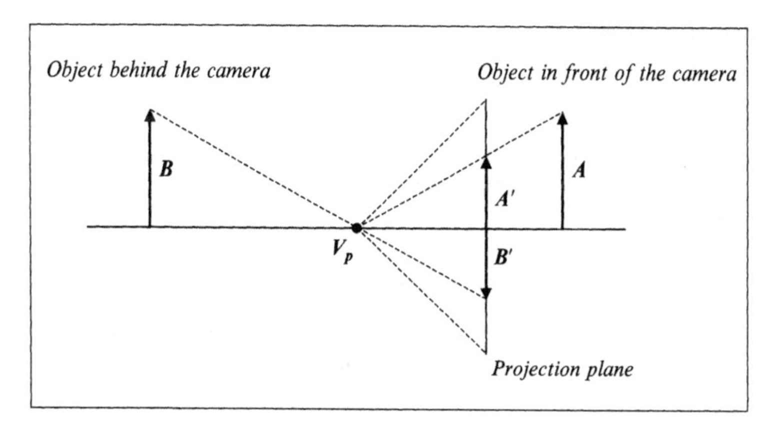

FIGURE 8.30. An unclipped perspective projection.

If the clipping process yields a visible line segment, then the *viewporting transformation* must be applied to both its endpoints before the segment can be displayed on the screen.

$$
\begin{aligned}
x_s &= \left(\frac{x_c}{z_c}\right) \cdot \left(\frac{v_{sx}}{2}\right) + v_{cx} \\
y_s &= \left(\frac{y_c}{z_c}\right) \cdot \left(\frac{v_{sy}}{2}\right) + v_{cy}
\end{aligned}
\tag{8.26}
$$

It is important to observe that clipping *must* take place *before* the perspective division. Failure to do so will produce incorrect results with parts of the scene that lie behind the viewing point (and should therefore be invisible) being drawn upside-down on the screen, as seen in Fig. 8.30

An outline of the 3D Cohen and Sutherland line-clipping algorithm is given in Algorithm 8.1.

```
1.  c1 ← compute the 4-bit code for p1 ;
    c2 ← compute the 4-bit code for p2 ;

2.  if (c1 = 0) and (c2 = 0) then
    {
      trivially accept the line segment;
      add a small displacement dz to z1 and z2;
      done;
    }

3.  if (c1 bit-wise AND c2) then
    {
      trivially reject the line segment;
      done;
    }
```

```
4.  if (c1 = 0) then c = c2;
               else c = c1 ;

5.  if the first bit of c is set then
    {
      compute the point of intersection (xi, yi, zi) of the line
      segment with the left clipping plane;
      goto step 9;
    }

6.  if the second bit of c is set then
    {
      compute the point of intersection (xi, yi, zi) of the line
      segment with the right clipping plane;
      goto step 9;
    }

7.  if the third  bit of c is set then
    {
      compute the point of intersection (xi, yi, zi) of the line
      segment with the bottom clipping plane;
      goto step 9;
    }
8.  if the fourth  bit of c is set then
    {
      compute the point of intersection (xi, yi, zi) of the line
      segment with the top clipping plane;
      goto step 9;
    }

9.  if (c = c1) then
    {
      x1 ← xi ;
      y1 ← yi ;
      z1 ← zi ;
      c1 ← recompute the 4-bit code for this point;
      goto step 2;
    }
      else
    {
      x2 ← xi ;
      y2 ← yi ;
      z2 ← zi ;
      c2 ← recompute the 4-bit code for this point;
      goto step 2;
    }
```

Algorithm 8.1 The 3D Choen and Sutherland line-clipping algorithm.

The following **C** functions implement a 3D point-clipping routine and the Cohen and Sutherland 3D line-clipping algorithm.

```
/*
 * Common Constants and typedefs.
 */

typedef unsigned char boolean_t;

#define False (boolean_t) 0
#define True  (boolean_t) 1

/*
 * Constants and typedef for Choen and Sutherland 3D line-clipping
routine.
 */

#define small_t        0.000000005
#define z_displacement 0.0005

typedef unsigned char plane_code_t;

#define left_plane   (plane_code_t) 1)
#define right_plane  (plane_code_t) 2)
#define bottom_plane (plane_code_t) 4)
#define top_plane    (plane_code_t) 8)

/*-----------------------------------------------------------------*/

boolean_t clip_point_3d(double *x, /* Test Point (In/Out) */
                        double *y,
                        double *z
                       )
{
 boolean_t inside;

 inside = (
           (-(*z) <= (*x)) &&
           ( (*x) <= (*z)) &&
           (-(*z) <= (*y)) &&
           ( (*y) <= (*z))
          );

  /*
   * If the point is visible add a small displacement to its
   * z-coordinate in order to prevent a potential zero divide
   * in the perspective division.
   */

 if (inside) (*z) += z_displacement;

 return(inside);

} /* clip_point_3d */

/*-----------------------------------------------------------------*/
```

```
void cs_get_plane_code(double        x,
                       double        y,
                       double        z,
                       plane_code_t *c
                      )
{
 if (x < -z) *c = left_plane; else
 if (x >  z) *c = right_plane;

 if (y < -z) *c |= bottom_plane; else
 if (y >  z) *c |= top_plane;

} /* cs_get_plane_code */

/*------------------------------------------------------------*/

boolean_t cs_clip_line_3d(double *x1, /* Test Line (In/Out) */
                          double *y1,
                          double *z1,
                          double *x2,
                          double *y2,
                          double *z2
                         )
{

 /*
  * Clip a 3D line segment in the clip-space coordinate system.
  */

 double       x, y, z, t;
 plane_code_t c, c1, c2, done;
 boolean_t result;

 result = False;
 done   = 0;

 cs_get_plane_code(*x1, *y1, *z1, &c1);
 cs_get_plane_code(*x2, *y2, *z2, &c2);

 while ((c1 != 0) || (c2 != 0))
  {
   if (c1 & c2) return(False); /* Trivial rejection */

   /*
    * The line segment is at least partially outside the clipping
    * pyramid.
    */

   if (c1 == 0) c = c2;
   else         c = c1;

   if (c & left_plane)
    {
     /*
```

```
  * Compute the intersection with x = -z clipping plane.
  */

 if (done & left_plane) return(result);
 else                   done |= left_plane;

 t = (*x2 - *x1) + (*z2 - *z1);

 if (fabs(t) < small_t) t = small_t;

  t = -(*z1 + *x1) / t;
  x = -z;
  y = t * (*y2 - *y1) + *y1;
  z = t * (*z2 - *z1) + *z1;
 }
else

if (c & right_plane)
 {

  /*
   * Compute the intersection with x = +z clipping plane.
   */

  if (done & right_plane) return(result);
  else                    done |= right_plane;

  t = (*x2 - *x1) - (*z2 - *z1);

  if (fabs(t) < small_t) t = small_t;

  t = (*z1 - *x1) / t;
  x = z;
  y = t * (*y2 - *y1) + *y1;
  z = t * (*z2 - *z1) + *z1;
 }
else

if (c & bottom_plane)
 {
  /*
   * Compute the intersection with y = -z clipping plane.
   */

  if (done & bottom_plane) return(result);
  else                     done |= bottom_plane;

  t = (*y2 - *y1) + (*z2 - *z1);

  if (fabs(t) < small_t) t = small_t;

  t = -(*z1 + *y1) / t;
  x = t * (*x2 - *x1) + *x1;
  y = -z;
```

```
      z = t * (*z2 - *z1) + *z1;
     }
    else

    if (c & top_plane)
     {
      /*
       * Compute the intersection with y = +z clipping plane.
       */

      if (done & top_plane) return(result);
      else                  done |= top_plane;

      t = (*y2 - *y1) - (*z2 - *z1);

      if (fabs(t) < small_t) t = small_t;

      t = (*z1 - *y1) / t;
      x = t * (*x2 - *x1) + *x1;
      y = z;
      z = t * (*z2 - *z1) + *z1;
     }

    /*
     * Recompute the code.
     */

    if (c == c1)
     {
      *x1 = x;
      *y1 = y;
      *z1 = z;
      cs_get_plane_code(x, y, z, &c1);
     }
    else
     {
      *x2 = x;
      *y2 = y;
      *z2 = z;
      cs_get_plane_code(x, y, z, &c2);
     }
   } /* while loop */

 /*
  * Add a small displacement to the z-coordinates of the
  * line segment in order to prevent a potential zero
  * divide in the perspective division.
  */

 *z1 += z_displacement;
 *z2 += z_displacement;

 return(True);

} /*cs_clip_line_3d */
```

8.9 Perspective Depth

The process of removing hidden lines or hidden surfaces (when displaying a scene) requires a perspective transformation with special properties. Here, we wish to produce a perspective view and at the same time we require the depth, z_s, of each point in the perspective image, so that we can make decisions about which surfaces hide lines and other surfaces. We must therefore augment the screen coordinate system to be a 3D system with coordinates $\langle x_s, y_s, z_s \rangle$, where x_s and y_s are the coordinates of the point on the screen, as before, and z_s to be calculated so as to retain the depth information, without altering x_s and y_s. Thus, to display a point $\langle x_s, y_s, z_s \rangle$ on the screen we use x_s and y_s, and ignore z_s, as shown in Fig. 8.31.

In this figure, the perspective image $\boldsymbol{p}'$ of points $\boldsymbol{p}_1$ and $\boldsymbol{p}_2$ is identical because they have the same x_s and y_s coordinates, but point $\boldsymbol{p}_1$ is closer to the eye than point $\boldsymbol{p}_2$.

The screen coordinates can now be computed as

$$
\begin{aligned}
x_s &= \left(\frac{x_c}{z_c}\right) \cdot \left(\frac{v_{sx}}{2}\right) + v_{cx} \\
y_s &= \left(\frac{y_c}{z_c}\right) \cdot \left(\frac{v_{sy}}{2}\right) + v_{cy} \\
z_s &= -\frac{1}{z_c}
\end{aligned}
\tag{8.27}
$$

This expression for z_s has the advantage that it preserves the intuitive notion of depth. If a point has a larger z_e coordinate than another point, then it will also have a larger z_s coordinate, i.e.

$$
z_{e2} > z_{e1} \Leftrightarrow z_{s2} > z_{s1}
$$

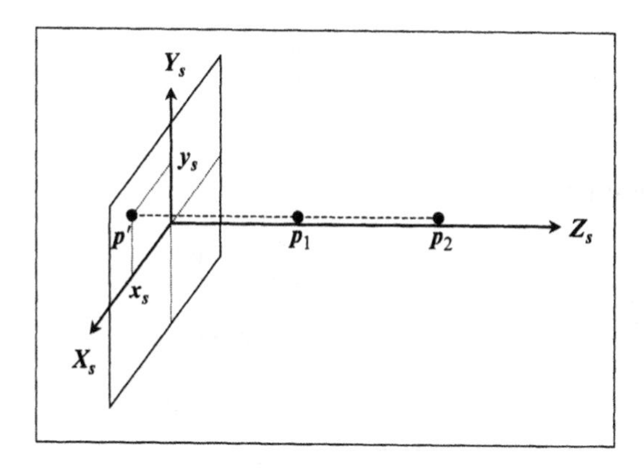

FIGURE 8.31. The screen coordinate system.

In matrix form the above transformation is expressed as

$$S = \begin{bmatrix} \left(\frac{v_{sx}}{2}\right) & 0 & 0 & 0 \\ 0 & \left(\frac{v_{sy}}{2}\right) & 0 & 0 \\ v_{cx} & v_{cy} & 0 & 1 \\ 0 & 0 & -1 & 0 \end{bmatrix} \tag{8.28}$$

Now multiplying the homogeneous point $\boldsymbol{p}^{\mathrm{h}} = \langle x_c, y_c, z_c, 1\rangle$ by this matrix yields the general homogeneous point $\langle x, y, z, w\rangle$, i.e.

$$[x, y, z, w] = [x_c, y_c, z_c, 1] \cdot \boldsymbol{S} \tag{8.29a}$$

or alternatively as

$$[x, y, z, w]^{\mathrm{T}} = \boldsymbol{S}^{\mathrm{T}} \cdot [x_c, y_c, z_c, 1]^{\mathrm{T}} \tag{8.29b}$$

Thus

$$\begin{aligned} x &= x_c \cdot \left(\frac{v_{sx}}{2}\right) + z_c \cdot v_{cx} \\ y &= y_c \cdot \left(\frac{v_{sy}}{2}\right) + z_c \cdot v_{cy} \\ z &= -1 \\ w &= z_c \end{aligned}$$

To recover the 3D point we homogenise the 4D point to obtain

$$\begin{aligned} x_s &= \frac{x}{w} \\ y_s &= \frac{y}{w} \\ z_s &= \frac{z}{w} \end{aligned}$$

which after substitution yields

$$\begin{aligned} x_s &= \frac{x_c}{z_c} \cdot \left(\frac{v_{sx}}{2}\right) + \frac{z_c \cdot v_{cx}}{z_c} \\ y_s &= \frac{y_c}{z_c} \cdot \left(\frac{v_{sy}}{2}\right) + \frac{z_c \cdot v_{cy}}{z_c} \\ z_s &= -\frac{1}{z_c} \end{aligned}$$

and after simplification we obtain the same result as in Eq. (8.27):

$$x_s = \left(\frac{x_c}{z_c}\right) \cdot \left(\frac{v_{sx}}{2}\right) + v_{cx}$$

$$y_s = \left(\frac{y_c}{z_c}\right) \cdot \left(\frac{v_{sy}}{2}\right) + v_{cy}$$

$$z_s = -\frac{1}{z_c}$$

8.10 Simple C Library for 3D Transformations

See *Appendix 4*.

9

3D Rendering

9.1 Introduction

Computer image synthesis is the process of generating a two-dimensional pictorial representation of a mathematically defined three-dimensional object or scene. Sometimes this process is referred to as *rendering* the scene and the computer program that caries out this process is called a *renderer*.

Given the mathematical description of a collection of objects or curved surfaces in a scene, there are three main ways in which we may render this scene. We may render it as

- a *wire-frame drawing*, depicting all the edges or curves, as shown in Fig. 9.1a;
- a *hidden-line drawing*, depicting only the visible edges or curves (i.e. all the edges or curves not obscured by any faces or surfaces closer to the observer), as shown in Fig. 9.1b;
- a *shaded image*, which attempts to capture the photographic appearance of the scene by simulating how light is reflected off the surface of its objects or curved surfaces, as shown in Fig. 9.1c.

There are clear distinctions between these three techniques of rendering, which we shall examine shortly.

To simplify our discussion, we will only deal with scenes that are described as a collection of polygons. In this scheme, curved surfaces are assumed to have been decomposed into polygonal meshes which approximate their shape. This is common practice in most rendering schemes.

In our scheme, a scene is defined as a set (collection) of objects, i.e. $\boldsymbol{S} = \{\boldsymbol{O}_1, \boldsymbol{O}_2, \ldots, \boldsymbol{O}_k\}$. Each object is defined as a set of polygons, faces or facets $\boldsymbol{F}$, a set of edges $\boldsymbol{E}$ and a set of vertices $\boldsymbol{V}$. Thus an object $\boldsymbol{O}$ is defined as $\boldsymbol{O} = \{\boldsymbol{F}, \boldsymbol{E}, \boldsymbol{V}\}$, where $\boldsymbol{F} = \left\{\boldsymbol{f}_1, \boldsymbol{f}_2, \ldots, \boldsymbol{f}_l\right\}$, $\boldsymbol{E} = \{\boldsymbol{e}_1, \boldsymbol{e}_2, \ldots, \boldsymbol{e}_m\}$ and $\boldsymbol{V} = \{\boldsymbol{v}_1, \boldsymbol{v}_2, \ldots, \boldsymbol{v}_n\}$. Each vertex $\boldsymbol{v}_i$ is defined by its three coordinates $\boldsymbol{v}_i = \{x_i, y_i, z_i\}$, each edge $\boldsymbol{e}_j$ is defined by its two endpoints $\boldsymbol{e}_j = \{\boldsymbol{v}_s, \boldsymbol{v}_e\}$ and each face $\boldsymbol{f}_p$ may be defined by the set of edges $\boldsymbol{f}_p = \{\boldsymbol{e}\}$ or by the set of vertices $\boldsymbol{f}_p = \{\boldsymbol{v}\}$ that describe it.

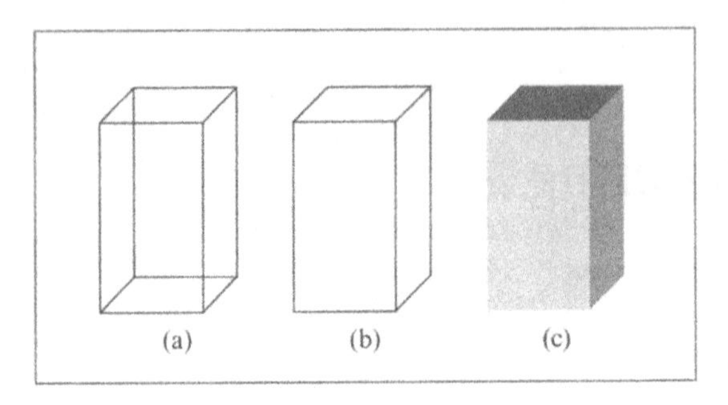

FIGURE 9.1. Three rendering techniques.

Figure 9.2 depicts a cuboid with its faces, edges and vertices labelled, and Table 9.1 shows the corresponding description of this object. In Table 9.1, the face descriptions $f = \{v\}$ are given in counter-clockwise order when seen from the outside of the object. The reason for this is simple. If we use three consecutive vertices in a face description, we can construct two vectors lying on the face. We can then compute the cross product of these two vectors to determine the normal of the face, which points outwards from the object. This only works with convex polygonal facets. For convex or concave polygonal facets with n vertices we have to use a more sophisticated method developed by Martin Newell. His method also works with non-flat facets, where it calculates the average plane equation of the facet. Recall that for a point to lie on a plane it must satisfy its equation $a \cdot x + b \cdot y + c \cdot z + d = 0$, where a, b, c, d are the coefficients of the plane equation and x, y, z are the coordinates of the general point. Newell computes the coefficients of this equation as follows.

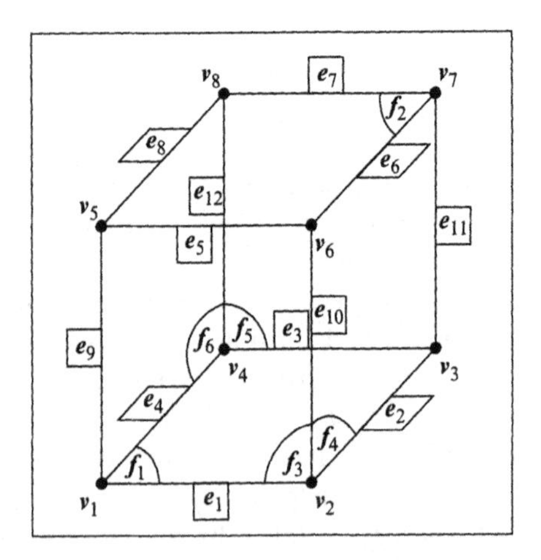

FIGURE 9.2. A cuboid with its faces, edges and vertices labelled.

TABLE 9.1. The vertex, edge and face lists of the cuboid depicted in Fig. 9.2.

Vertex list	Edge list	Face list	
$v = \{x, y, z\}$	$e = \{v\}$	$f = \{v\}$	$f = \{e\}$
$v_1 = \{x_1, y_1, z_1\}$	$e_1 = \{v_1, v_2\}$	$f_1 = \{v_4, v_3, v_2, v_1\}$	$f_1 = \{e_4, e_3, e_2, e_1\}$
$v_2 = \{x_2, y_2, z_2\}$	$e_2 = \{v_2, v_3\}$	$f_2 = \{v_5, v_6, v_7, v_8\}$	$f_2 = \{e_5, e_6, e_7, e_8\}$
$v_3 = \{x_3, y_3, z_3\}$	$e_3 = \{v_3, v_4\}$	$f_3 = \{v_1, v_2, v_6, v_5\}$	$f_3 = \{e_1, e_{10}, e_5, e_9\}$
$v_4 = \{x_4, y_4, z_4\}$	$e_4 = \{v_4, v_1\}$	$f_4 = \{v_2, v_3, v_7, v_6\}$	$f_4 = \{e_2, e_{11}, e_6, e_{10}\}$
$v_5 = \{x_5, y_5, z_5\}$	$e_5 = \{v_5, v_6\}$	$f_5 = \{v_3, v_4, v_8, v_7\}$	$f_5 = \{e_3, e_{12}, e_7, e_{11}\}$
$v_6 = \{x_6, y_6, z_6\}$	$e_6 = \{v_6, v_7\}$	$f_6 = \{v_4, v_1, v_5, v_8\}$	$f_6 = \{e_4, e_9, e_8, e_{12}\}$
$v_7 = \{x_7, y_7, z_7\}$	$e_7 = \{v_7, v_8\}$		
$v_8 = \{x_8, y_8, z_8\}$	$e_8 = \{v_8, v_5\}$		
	$e_9 = \{v_1, v_5\}$		
	$e_{10} = \{v_2, v_6\}$		
	$e_{11} = \{v_3, v_7\}$		
	$e_{12} = \{v_4, v_8\}$		

$$a = \sum_{i=1}^{n} (y_i - y_j) \cdot (z_i + z_j)$$

$$b = \sum_{i=1}^{n} (z_i - z_j) \cdot (x_i + x_j)$$

$$c = \sum_{i=1}^{n} (x_i - x_j) \cdot (y_i + y_j)$$

$$\text{and} \quad d = -(a \cdot x_1 + b \cdot y_1 + c \cdot z_1) \tag{9.1}$$

where if $(i < n)$ then $j = i + 1$ else $j = 1$

As we have seen in Chapter 2 (Section 2.24.8), the normal of this plane is given by $n = [a, b, c]$ and d is the signed distance of the plane from the origin.

Incidentally, the signed distance of a point $\boldsymbol{P} = \langle x, y, z \rangle$ from the plane is given by

$$d_p = \frac{(a \cdot x + b \cdot y + c \cdot z)}{|\boldsymbol{n}|} = \frac{(a \cdot x + b \cdot y + c \cdot z)}{\sqrt{a^2 + b^2 + c^2}} \tag{9.2}$$

Here, if $d_p > 0$, then this point lies in front of the plane (i.e. in the half-space towards which the plane normal points). If $d_p < 0$, then this point lies behind the plane (i.e. in the half-space opposite to the one towards which the plane normal points). Finally, if $d_p = 0$, then this point lies on the plane.

Another interesting fact is that, the coefficients a, b, c of the plane equation (i.e. the components of its normal) are proportional to the area of the projections of the polygon on the ***YZ***, the ***XZ*** and the ***XY*** planes, respectively. The area of the projected polygon on the ***YZ*** plane is given by $|a|/2$, on the ***XZ*** plane by $|b|/2$ and on the ***XY*** by $|c|/2$.

In the above discussion, it is assumed that the polygon is described in counter-clockwise fashion when seen from the front (i.e. from the half-space towards which its normal points).

9.2 Rendering Algorithms

Producing a wire-frame drawing of the objects in a scene is fairly simple. First we must transform all the vertices of the scene into the eye-space coordinate system. Then, we clip each edge of the scene against the viewing pyramid and if the clipped edge is at least partially visible, we transform its endpoints into the image-space coordinate system and we draw a 2D line segment using the available graphics package.

Producing a hidden-line drawing or a shaded image of the objects in a scene is slightly more complex. In both cases, we first transform all the vertices of the scene into the eye-space coordinate system and we clip all the polygons of the scene against the viewing pyramid using a 3D polygon-clipping algorithm.

To produce a hidden-line drawing of the scene we need to use a *hidden-line elimination algorithm*. This algorithm examines each edge in the scene and compares it against each face in the scene in order to determine if it is obscured by any face that is closer to the observer. If the algorithm determines that an edge is totally obscured by a face, then it eliminates this edge and proceeds with the next one. If an edge is partially obscured by a face, then the algorithm removes the invisible part and continues checking the visible part of the edge against the remaining faces. If an entire edge or a part of an edge is found to be obscured by none of the faces, then its endpoints are transformed into the image-space coordinate system and it is displayed as before.

To produce a shaded image of the scene we need to use a *hidden-surface elimination algorithm* and a *shading algorithm*. In this case our image can only be displayed on a *raster scan display* device. Such devices are capable of displaying a number of *scan-lines* on the screen. A scan-line is a horizontal line on the screen which is composed of a number of *pixels* (picture elements) that appear as dots on the screen. A typical display may consist of 1024 scan-lines each of which has 1280 pixels, which gives us a total of 1,310,720 pixels. These pixels are commonly stored in a rectangular array in memory called the *frame-buffer*.

At each pixel of the frame-buffer, the hidden-surface elimination algorithm has to examine each polygon in the scene in order to determine the polygon that is closest to the observer thus obscuring all other polygons at that pixel. While performing this determination, the algorithm needs to store the depth of the polygon found to be the closest to the observer so far. These depths are stored in a rectangular array, called the *depth-buffer* or *z-buffer*, which has the same dimensions as the frame-buffer.

After the hidden-surface elimination algorithm has determined the closest polygon at a given pixel, the shading algorithm must determine the intensity and colour of the light reflected by the polygon covering the pixel. This intensity and colour will depend on the surface finish and colour texture of the polygon, on the lighting characteristics of the light source, on whether the polygon is lit directly or is in shadow and on whether the polygon is opaque or transparent. The more realistic the rendering is required to be the more complex and computationally expensive the rendering algorithm becomes.

The removal of the hidden parts from an image of solid objects is one of the most challenging problems in computer graphics and has occupied researchers since the early 1960s. Although there are many hidden-line and hidden-surface elimination algorithms there is no single answer to the hidden-line and hidden-surface elimination problem. There is no best algorithm. In an excellent paper, Sutherland, Sproull and Schumacker categorise the hidden-line and hidden-surface elimination algorithms into two basic classes: *object-space* algorithms and *image-space* algorithms [Sutherland 74]. Although some algorithms fit partially in each class, namely the *list-priority* algorithms. This classification also happens to separate the hidden-line and hidden-surface elimination algorithms. Three-dimensional objects are considered to be a collection of n_f polygonal facets by both categories of algorithm.

Object-space algorithms seek to compute exactly what the image should be by discovering what parts of the scene are hidden by other parts. Each of the n_f facets must be compared to the remaining $(n_f - 1)$ facets in order to determine the facets or the portions of facets that are not visible. Thus the computations required for this determination are of the order $n_f \times (n_f - 1)$.

Image-space algorithms, on the other hand, seek to determine what the image will be at each of the pixels of the display screen. For each of the n_p pixels of the display the algorithm must examine all n_f facets in order to determine the facet that is closest to the observer. Thus the computations required for this determination are of the order $n_f \times n_p$. As we have seen above, the number of pixels n_p is very large. Thus, when $(n_f - 1) < n_p$ object-space algorithms tend to be more efficient than image-space algorithms. This is only a rule of thumb since the computational steps required for image-space algorithms tend to be simpler than those required for object-space algorithms.

Next we examine a collection of simple hidden surface elimination algorithms.

9.2.1 A Simple Rendering Algorithm

A simple, but inefficient, algorithm is one that looks at all surfaces in a scene at each pixel and declares the one closest to the viewer as the visible surface. If the image has a resolution of $(x_{max} + 1) \times (y_{max} + 1)$ pixels, the scene contains $(p_{max} + 1)$ polygons and the depth (distance from the viewer) is computed by the function **compute_depth(p,x,y)**, then this simple algorithm can be expressed by the following pseudo-code.

```
#define background_depth      1e30
#define background_intensity 0

for (y = Y_max; y >=0; y--)
 {
  /* For every scan-line */

  for (x = 0; x <= X_max; x++)
   {
    /* For every pixel in this scan-line */
```

```
    depth_buffer[x][y]  = background_depth;
    frame_buffer[x][y]  = background_intensity;

    for (p = 0; p <= P_max; p++)
     {
      /* For every polygon in the scene */

      if (point_inside_polygon(p, x, y))
       {
        /* Pixel is inside the perimeter of the polygon */

        z = compute_depth(p, x, y);

        if (z < depth_buffer[x][y])
         {
          /* Store the depth and intensity of the polygon */

          depth_buffer[x][y]  = z;
          frame_buffer[x][y]  = compute_intensity(p, x, y, z);

         }
       }
     }
  }
}
```

One way the function **compute_depth(p,x,y)** may compute the depth of the polygon is to trace a ray from the point $\langle x, y, 0\rangle$ parallel to the $-\mathbf{Z}$-axis and intersect it with each polygon in the scene, hence the term *ray-tracing*. The visibility calculation must be performed $(x_{\max}+1)\times(y_{\max}+1)\times(p_{\max}+1)$ times per image, which makes this algorithm very expensive. The function **compute_intensity(p,x,y,z)** implements the shading algorithm (shader routine).

The reason why this simple, brute force, ray-tracing algorithm is wasteful is primarily because it operates on a pixel-by-pixel basis. A more fruitful approach is to attempt to determine visibility over large regions of the image exploiting some of the coherence properties of the image.

9.2.2 Warnock (Screen Subdivision) Algorithm

Warnock developed one of the first *area-subdivision* algorithms that takes advantage of the property of *area-coherence* [Warnock 69]. Area-coherence is based on the observation that the image of a typical polygon has similar extents in both the x and y directions. Thus, the pixels within such an area are coherent in that they depict a single polygon.

The Warnock algorithm first tries to resolve the visibility problem (hidden-surface problem) for a window that covers the entire screen. If the visibility problem is simple to resolve, then the algorithm displays the contents of the window, otherwise it recursively subdivides the current window into four sub-windows and repeats the process until the visibility problem can be resolved for each sub-window. If the visibility problem continues to prove intractable with successive subdivisions, then this recursive subdivision process is forced to terminate when the size of the sub-window reaches the size of a pixel. See Fig. 9.3.

At each stage of the recursive subdivision, the projection of each polygon and the window have one of the following relationships (shown in Fig. 9.4).

1. The polygon surrounds the window. Such a polygon is called a *surrounder* polygon.
2. The polygon intersects the window. Such a polygon is called an *intersector* polygon.
3. The polygon is contained in the window.
4. The polygon is disjoint from the window.

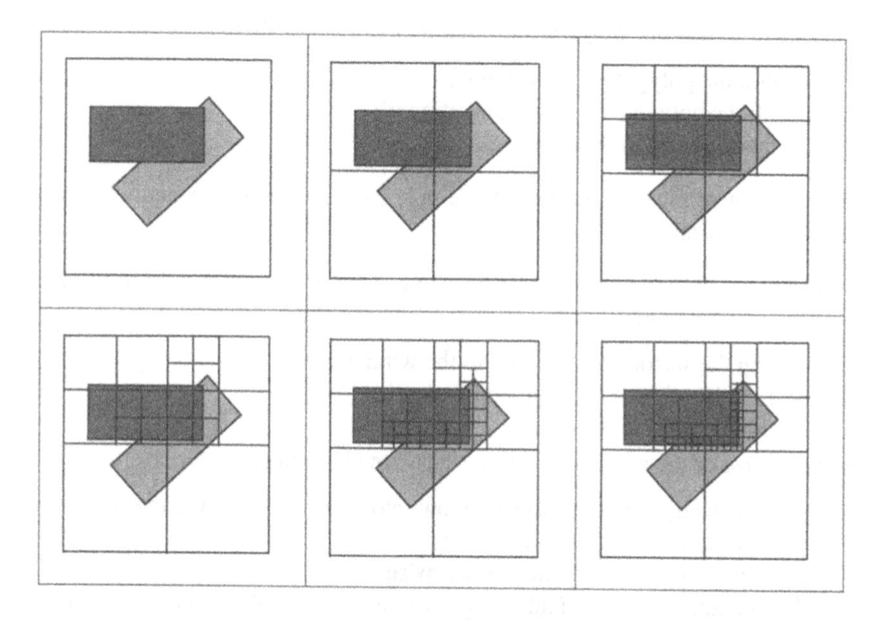

FIGURE 9.3. The recursive subdivision of the Warnock algorithm.

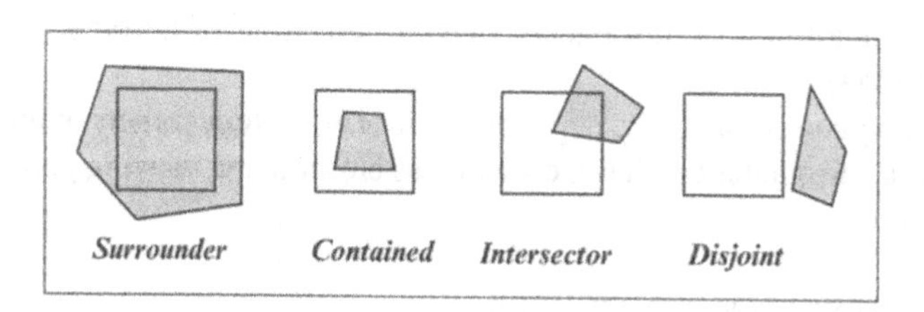

FIGURE 9.4. The classification of polygons.

Decisions about the visibility problem at a given window are made as follows.

1 **If** all the polygons are disjoint from the window, **then**

  1.1 fill the window with the background colour;
  1.2 **done** (return to the caller).

2 **If** there is only one polygon contained in the window, **then**

  2.1 fill the window with the background colour;
  2.2 fill the polygon (by scan-converting it);
  2.3 **done** (return to the caller).

3 **If** there is only one polygon intersecting the window **then**

  3.1 fill the window with the background colour;
  3.2 clip the polygon to the window;
  3.3 fill the polygon (by scan-converting it);
  3.4 **done** (return to the caller).

4 **If** there is only one polygon surrounding the window **then**

  4.1 clip the polygon to the window;
  4.2 fill the polygon (by scan-converting it);
  4.3 **done** (return to the caller).

otherwise there must be more than one polygons intersecting, contained or surrounding the window.

5 **If** there is only one polygon surrounding the window and it is in front of all other polygons **then**

  5.1 clip the surrounder polygon to the window;
  5.2 fill this polygon (by scan-converting it);
  5.3 **done** (return to the caller).

6 **If** the size of the window is greater than one pixel **then**

  6.1 recursively subdivide this window into four sub-windows and repeat the algorithm for each sub-window;
  6.2 **done** (return to the caller). **otherwise**
  6.3 compute the depth of all polygons at the centre of the window and use the closest polygon to colour the pixel;
  6.4 **done** (return to the caller).

Determining that the surrounder polygon is in front of all other polygons is done as follows.
for each polygon compute the depth of its plane at the four corners of the window.
If there is a surrounder that has z coordinates closer to the viewing point than any other polygon

then the test succeeds (see Fig. 9.5a);
else the test fails (see Fig. 9.5b).

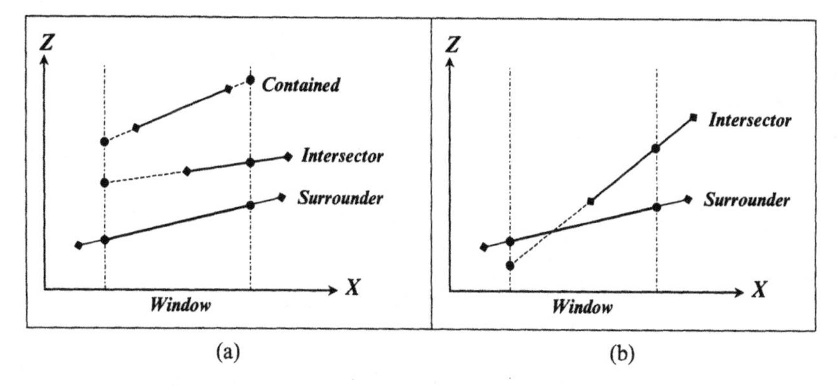

FIGURE 9.5 (a) Successful surrounder in front test. (b) Unsuccessful surrounder in front test.

So the Warnock algorithm exhibits the following characteristics.

- It exploits the property of area coherence.
- It is useful for a scene with many large polygons.
- It is relatively easy to implement (though not the simplest) using recursive calls.
- Its memory costs can be large. For a 1024 × 1024 display, the algorithm may recur to *at most* 10 levels of subdivision (as $2^{10} = 1024$).
- It is not the fastest available algorithm.

9.2.3 Newell, Newell and Sancha Algorithm

The Newell, Newell and Sancha algorithm is a *list-priority* algorithm [Newell 72]. All list-priority algorithms have to determine, in eye space, a *display-priority* (or *visibility-ordering*) for all polygonal facets in the scene, before they are able to generate a picture, in the screen space. The polygons are sorted in such a way that if two polygons are compared, the one with the lower display priority is the most visible one (i.e. not obscured by a higher display priority polygon).

The allocation of priorities is achieved through what is known as a *topological sort.* The topological sort begins by sorting all the polygons in the scene in descending order of their z_{max}, i.e. the depth of the point in each polygon that is furthest away from the viewing point. This sort orders the polygons from the deepest polygon (which is assigned the highest display priority) to the shallowest polygon (which is assigned the lowest priority).

If after this depth sort, no two polygons (that are adjacent in the depth-sorted list) overlap in z, then the list of polygons is sorted in the correct priority order (see Fig. 9.6a). Such a simple priority sort is only guarantied to work with scenes consisting of a series of polygons nearly perpendicular to the viewing axis.

If two polygons, in the depth-sorted list, overlap in depth but do not have an $x - y$ bounding box overlap, then their priority order is correct (see Fig. 9.6b).

If two polygons, in the depth-sorted list, overlap in depth and have an $x - y$ bounding box overlap, then their priority order may be incorrect and we must determine if the deeper (higher priority) polygon $\boldsymbol{P}$ obscures the shallower (lower

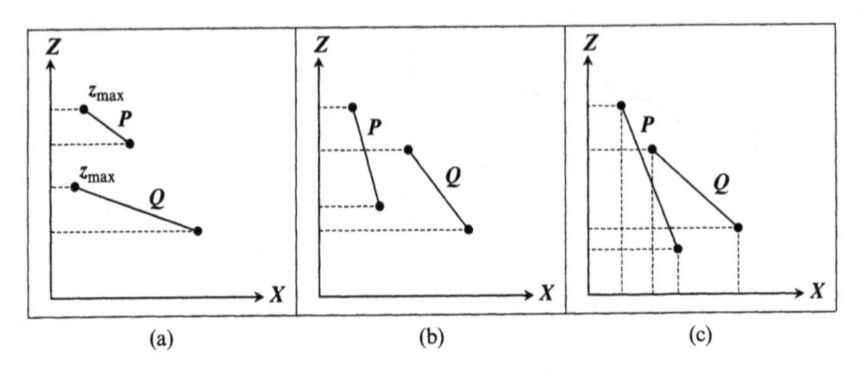

FIGURE 9.6 (a) No depth-overlap. (b) Depth-overlap with no x–y bounding box overlap. (c) Depth-overlap and x–y bounding box overlap.

priority) polygon ***Q*** (see Fig. 9.6c). In this example the priority of the two polygons is incorrect and must be reversed, i.e. ***Q*** must be displayed before ***P***.

Polygon ***P*** may have a depth-overlap with a set of polygons $\{Q_i\}$ that immediately follow it in the list. If ***P*** does not obscure any of the polygons Q_i, then polygon ***P*** has the correct display priority (see Fig. 9.7a), otherwise Q_i must be given a higher display priority than ***P*** (see Fig. 9.7b). In this example after the

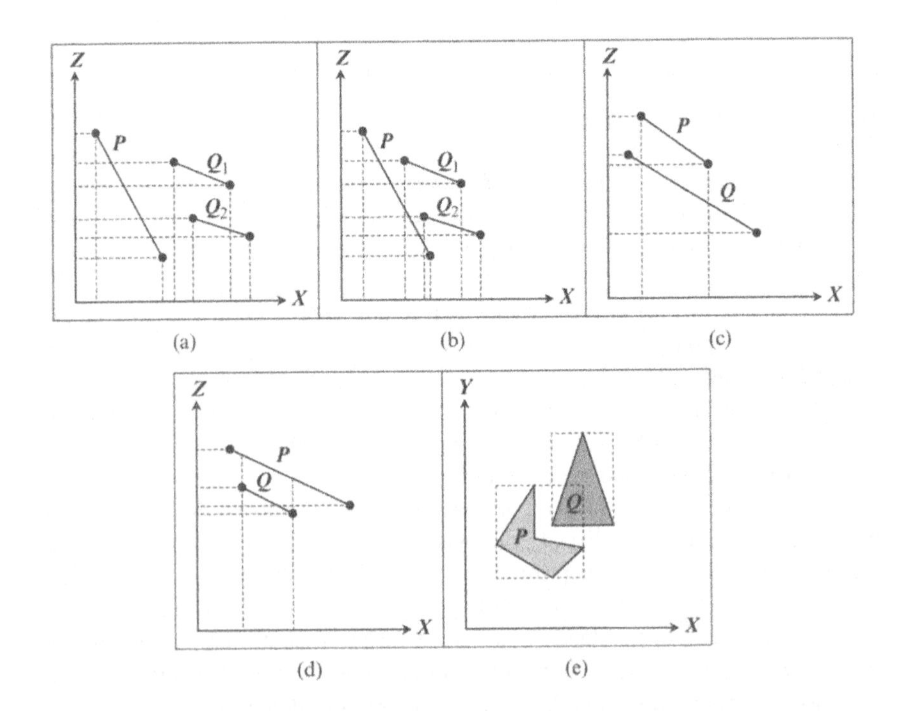

FIGURE 9.7 (a) ***P*** does not obscure polygons Q_i. (b) ***P*** obscures polygons Q_i. (c) ***P*** behind ***Q***. (d) ***Q*** in front of ***P***. (e) No proper x–y overlap.

depth-sort the initial priority order is P, Q_1, Q_2 but after the ***P-obscures-Q*** test the priority order becomes Q_1, Q_2, P.

The ***P-obscures-Q*** test is *false* if any of the following tests is *true*:

1. ***P*** and ***Q*** do not overlap in depth and ***Q*** is closer to the viewing point than ***P*** (see Fig. 9.6a). This test is done by the initial depth-sort.
2. ***P*** and ***Q*** do not have an $x - y$ bounding box overlap (see Fig. 9.7a). This test is done above.
3. All vertices of ***P*** are further away from the viewing point than the plane of ***Q*** (i.e. all the vertices of ***P*** are in the half-space that lies behind the plane of ***Q***). See Fig. 9.7c. This test is done by substituting the coordinates of the vertices of ***P*** into the plane equation of ***Q***.
4. All vertices of ***Q*** are closer to the viewing point than the plane of ***P*** (i.e. all the vertices of ***Q*** are in the half-space that lies in front of the plane of ***P***). See Fig. 9.7d. This test is done by substituting the coordinates of the vertices of ***Q*** into the plane equation of ***P***.
5. There is no proper overlap of ***P*** and ***Q*** on the ***XY*** plane (i.e. no vertices of ***P*** are in ***Q*** and no vertices of ***Q*** are in ***P***, and the edges of ***P*** and ***Q*** do not intersect). See Fig. 9.7e.

Tests 1–5 are performed in the specified order as they become progressively more expensive to compute. If ***P-obscures-Q***, then we reverse the order of ***P*** and ***Q*** in the display priority list. This is not the end of the story however since there are some problem cases:

1. Penetrating polygons cannot be ordered by the above procedure and one of them must be split about the plane of the other before we can sort them topologically. See Fig. 9.8a.
2. Cyclically overlapping polygons cannot be ordered by the above procedure and one or more of them must be split before we can sort them topologically. See Fig. 9.8b and c.

The Newell, Newell and Sancha algorithm calculates the visibility of the scene using geometric criteria rather than using pixel-by-pixel depth comparisons. It

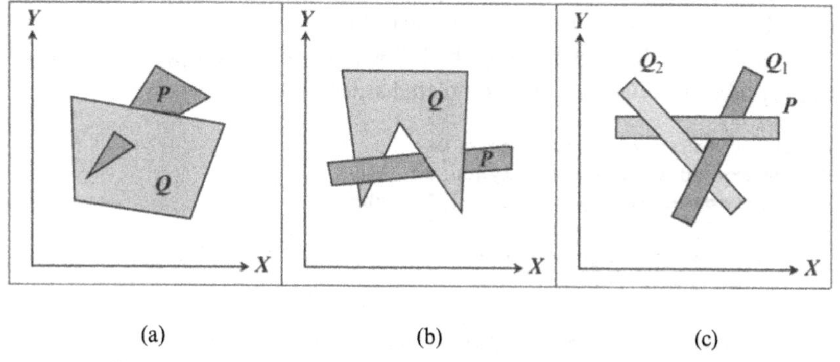

(a) (b) (c)

FIGURE 9.8. (a) Penetrating polygons. (b and c) Cyclic overlap.

thus takes advantage of the *depth coherence* of polygons to make visibility decisions about entire polygons. This algorithm performs the priority sort in eye space and the resulting priority list is only valid for a single camera position.

9.2.4 Single Scan-Line Depth-Buffer Algorithm

This algorithm consists of two parts. In the first part, the algorithm sets up the *y-bucket list*, which sorts each edge of each polygon in the scene according to its maximum y coordinate. This operation is known as the *y-bucket sort*. The y-bucket list is an array whose elements correspond to the scan-lines of the display. At the end of this pre-processing step, each y-bucket entry contains either a null pointer (indicating that there are no polygon edges starting at this scan-line) or points to a linked list of *edge records* (each representing a polygon edge starting at this scan-line). Horizontal or near horizontal edges are omitted from the y-bucket list. Edge records contain the following fields.

count containing minus the number of scan-lines crossed by this edge.

current_x containing the x coordinate of the point of intersection of the edge with the current scan-line.

current_z containing the z coordinate of the point of intersection of the edge with the current scan-line.

delta_x containing the amount by which the current x coordinate of the point of intersection of the edge is to be incremented or decremented by on each subsequent scan-line.

delta_z containing the amount by which the current z coordinate of the point of intersection of the edge is to be incremented or decremented by on each subsequent scan-line.

poly_idx containing the index of the polygon to which the edge belongs.

next_ptr containing a pointer to the next edge record in the list of this y-bucket entry.

To build the y-bucket list the algorithm examines each edge of each facet of the geometric model. As each non-horizontal edge is processed, its maximum y coordinate is determined, its record is completed and it is inserted into the appropriate linked list of edges starting at the same scan-line.

Given that an edge is defined by the following information **{x1, y1, z1, x2, y2, z2, poly_idx}**, its edge record is completed as follows.

```
for every polygon in the scene do
 for every edge in this polygon do
  if (truncate(y1) != truncate(y2)) then
   {
    create a new edge record;
    if (y1 > y2) then
     {
       x_max = x1; y_max = y1; z_max = z1;
       x_min = x2; y_min = y2; z_min = z2;
```

```
    }
   else
    {
     x_max = x2; y_max = y2; z_max = z2;
     x_min = x1; y_min = y1; z_min = z1;
    }

   count = (y_min - y_max);
   current_x = x_max;
   current_z = z_max;
   delta_x = (x_min - x_max) / (y_max - y_min);
   delta_z = (z_min - z_max) / (y_max - y_min);
   link this record in the appropriate y-bucket entry list;
  }
```

The second part of this algorithm maintains a list of *active edges* (i.e. edges that intersect the current scan-line). This list, which is known as the *active edge list*, is a sorted list of all the edges that intersect the scan-line under consideration. This algorithm also maintains an *intensity array* (or *frame-buffer*) and a *depth buffer array* (or *z-buffer*) for the current scan-line. These arrays have as many pixels as the scan-line.

The image is generated starting from the top scan-line of the display and proceeding towards the bottom. Initially, the active edge list is empty and it is updated before processing each scan-line. Each scan-line is processed by first updating the active edge list. At this stage the algorithm removes from the active edge list any edges that have ceased to be active (as they no longer intersect the current scan-line) and it inserts the newly activated edges from the corresponding y-bucket entry using an insertion sort procedure. This procedure inserts a newly activated edge in ascending polygon-index order. If two edges have the same polygon-index, then it inserts the new edge in ascending current-x order. If two edges have the same current-x coordinate, then it inserts the new edge in ascending delta-x order.

At the end of the updating stage, the algorithm initialises each pixel of the frame-buffer to the background colour and each pixel of the z-buffer to some large positive value (representing the depth of the background). Next, pairs of neighbouring edges, known as *polygon spans*, are *scan-converted* in sequence. Scan-converting a span involves determining the pixels of the current scan-line that are covered by this span. At each pixel the algorithm determines if the depth of the polygon at that pixel is closer to the observer than the current depth value stored in this pixel. If it is, then that means that this polygon is visible at this pixel and its shade must be computed and stored in this frame-buffer pixel.

When all the polygon spans have been processed in this way, the z-buffer contains the depth and the frame-buffer contains the shade of the nearest polygon at each pixel of the scan-line.

Because visibility must be resolved at each pixel covered by one or more polygons, and because all polygon spans must be processed whether they are visible or not, the single scan-line depth buffer algorithm is inefficient compared to some other algorithms. However, the depth comparisons are very simple operations that can be executed quickly, giving this algorithm the potential for good performance. This algorithm becomes relatively more efficient than other algorithms when the number of polygons in the scene is greater than the total number of pixels on the screen.

Part two of the algorithm has the following steps.

```
1. Set the active edge list to empty;

2. current_y = maximum scan-line index;

while (current_y >= 0) do
{
  3. if the active edge list is not empty then
      {
       Update all the active edges as follows:

       if the edge has expired (i.e. count == 0) then
        {
         Remove it from the active edge list;
        }
       else
        {
         Update the edge record:
         current_x += delta_x;
         current_z += delta_z;
         count ++;
        }
      }

  4. if the y-bucket entry for the current scan-line is not empty then
      {
       Insert all the newly activated edges into the active
       edge list. This involves an insertion sort by
       ascending poly_idx order, if two entries have the
       same poly_idx, then by ascending current_x order and
       if two entries have the same current_x then by
       ascending delta_x order.
      }

  5. if the active edge list is not empty then
      {

    5.1 Initialise the z-buffer array to a large positive
        number (representing the background depth) and the
        intensity array to the background colour.

    5.2 while (there are more entries in the active edge list) do
        {
         5.2.1 Pick a pair of entries from the active
               edge list. (each pair of edges defines an
               active polygon span).
```

```
        5.2.2 xl = the current_x of the left edge of
                   the polygon span;
              zl = the current_z of the left edge of the
                   polygon span;
              xr = the current_x of the right edge of the
                   polygon span;
              zr = the current_z of the right edge of the
                   polygon span;
              dz = (zr - zl) / (xr - xl);
                   (where dz is the amount by which the depth of
                    the span changes from pixel to pixel as we
                    move along the span).

        5.2.3 for xc = xl to xr do
              {
               if (zl < z_buffer[xc]) then
                {
                 z_buffer[xc] = zl;
                 intensity[xc] =
                 compute_intencity(poly_idx,xc,current_y,zl);
                }

               zl += dz;
              }
           }

       5.3 Display the contents of the intensity buffer.
       }

  6. current_y --;
 }

7. Done.
```

9.3 Reflection Models and Shading Techniques

Having determined which polygon (or surface) is visible at any given pixel of the display device we must determine what colour to paint it with. The determination of this colour depends on the shading technique and on the reflection model used. Additionally, texturing and transparency effects can influence the colour of a pixel. In this section we will examine a number of simple *reflection models* and *shading techniques*.

Briefly put, a shading technique determines how and at what instance we calculate the normal vector of a polygon while scan converting it. Shading tends to work either entirely in eye-space or partly in eye-space and partly in screen-space. We will examine three different shading techniques: *flat polygon shading*, *Gouraud smooth shading* and *Prong smooth shading*.

In contrast, a reflection model determines how the light that reaches a surface is reflected off it. Our discussion will concentrate on three simple light reflection models: *ambient light reflection*, *diffuse light reflection* and *specular light reflection*.

In our discussion we will assume that the scene is composed of a number of polygons that posses a number of attributes such as colour and surface finish. We will also assume that our scene is lit by one or more point light sources. Let us start by examining these three light reflection models.

9.3.1 Ambient Light Reflection

In most environments there is ambient light from a variety of sources that is repeatedly reflected off various surfaces and eventually reaches the surface of interest. Examples of ambient light are: the light that is reflected off the walls, the ceiling and the floor of a room or sunlight on an overcast day. Rather than model such light sources individually, we make the simplifying assumption that ambient light comes equally from all directions. Then we can model the reflection of this ambient light as a constant term that has no dependence on the viewer, the light source, or the surface normal directions. The model for ambient light reflection can be expressed as follows.

$$I_a = L_a \cdot k_a \tag{9.3}$$

where I_a is the intensity of the reflected ambient light,
L_a is the intensity of the incident ambient light, i.e. the ambient light reaching the surface, and
k_a is the coefficient of ambient reflection (i.e. the fraction of ambient light that is reflected off the surface).

If an object is lit only with ambient light, it will appear dull and featureless and with no discernible surface texture.

9.3.2 Diffuse Light Reflection

Matt surfaces, such as chalk, exhibit a rough surface finish that is composed of randomly distributed micro-facets. When parallel light rays hit such a surface they are reflected in a random fashion thus scattering in all directions, as shown in Fig. 9.9. Such surfaces appear to be equally bright when viewed from any direction, thus the direction of the viewer is unimportant. Such reflection is known as *diffuse* or *Lambertian* reflection. According to Lambert's law what determines the brightness of such surfaces is the *incidence angle* i, i.e. the angle between the incoming light ray and the surface normal vector, as shown in Fig. 9.10. When the incidence angle is small then the surface appears brighter and when this angle is large the surface appears darker. To be more precise, the surface brightness is proportional to the cosine of the incidence angle, $\cos i$.

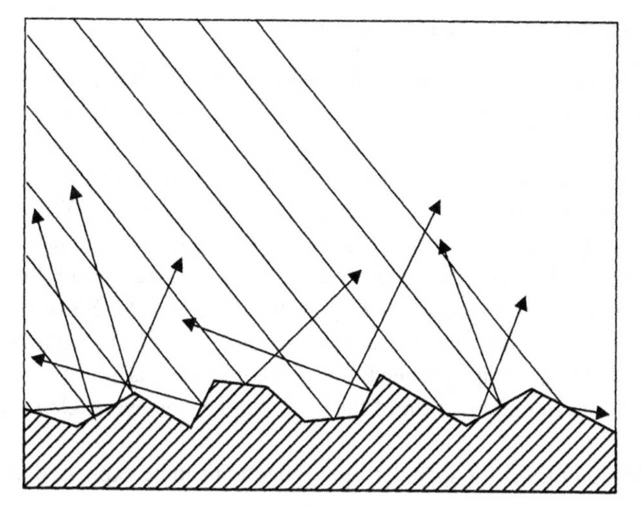

FIGURE 9.9. The surface micro-facets.

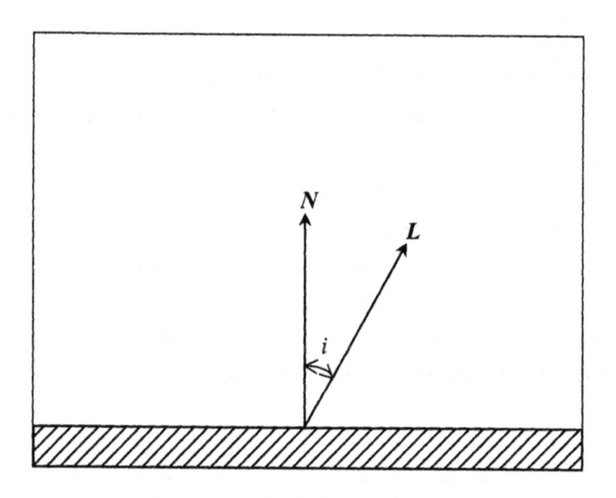

FIGURE 9.10. Diffuse reflection.

The model for diffuse light reflection can be expressed as follows.

$$I_d = L_p \cdot k_d \cdot \cos i = L_p \cdot k_d \cdot (\boldsymbol{L} \odot \boldsymbol{N}) \tag{9.4}$$

where I_d is the intensity of the reflected diffuse light,

L_p is the intensity of the light of a point source incident on the surface,

$\boldsymbol{L}$ is the unit vector in the direction of the light source,

$\boldsymbol{N}$ is surface unit normal vector, and

k_d is the coefficient of diffuse reflection (i.e. the fraction of light that is reflected diffusely off the surface).

9.3.3 Specular Light Reflection

Glossy surfaces, such as mirrors and polished metal surfaces, exhibit a smooth surface finish. When parallel light rays hit such a surface and are reflected off it, they remain parallel. The angle of reflection r of a ray is equal to its angle of incidence i, as shown in Fig. 9.11. The best visual indicators of glossiness are highlights, i.e. the specular reflection of the light source. Were the surface a perfect mirror, light would only reach the viewer if the surface normal pointed halfway between the light source direction vector $\boldsymbol{L}$ and the viewer direction vector $\boldsymbol{E}$. We will call the vector pointing in this direction $\boldsymbol{H}$. Since most glossy surfaces are not perfectly smooth mirrors, the intensity of the highlight falls off smoothly rather than abruptly when the angle of incidence is not equal to the angle of a ray reflected in the direction of the observer, i.e. the eye angle e as depicted in Fig. 9.12.

The reflection model for specular light reflection can be expressed as follows.

$$I_s = L_p \cdot k_s \cdot F_s(\boldsymbol{N}, \boldsymbol{L}, \boldsymbol{E}, n) \tag{9.5}$$

where I_s is the intensity of the reflected specular light,
L_p is the intensity of the light of a point source incident on the surface,
$\boldsymbol{L}$ is the unit vector in the direction of the light source,
$\boldsymbol{E}$ is the unit vector in the direction of the viewer,
$\boldsymbol{N}$ is the surface unit normal vector,
n is the *specular sharpness*, i.e. the sharpness of the specular peak,
F_s is the specular reflection function, and
k_s is the coefficient of specular reflection (i.e. the fraction of light that is reflected off the surface in a specular fashion).

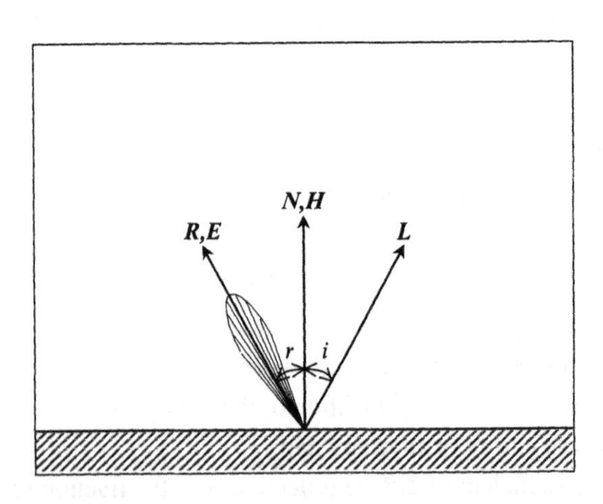

FIGURE 9.11. The direction of maximum highlights.

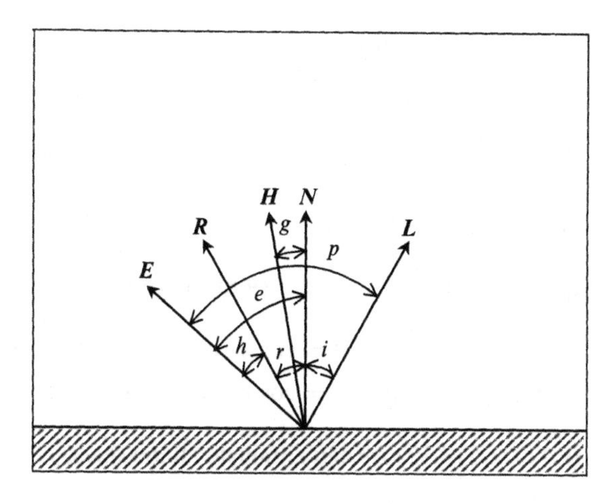

FIGURE 9.12. Specular reflection.

The meaning of the remaining vectors and angles in Fig. 9.12 is as follows.

R is the unit vector in the direction of the reflected ray, i.e. the direction of maximum highlights.

H is the vector which points halfway between the light source direction and the eye direction. This vector points in the direction in which the surface normal should point if the viewer were to receive the maximum amount of highlights.

i is the *incidence angle.*

r is the *reflection angle*. Observe that $r = i$.

e is the *eye angle*, i.e. the angle between vectors **N** and **E**.

p is the *phase angle*, i.e. the angle between vectors **L** and **E**.

h is the *highlight angle*, i.e. angle between vectors **E** and **R**.

g is the angle between vectors **N** and **H**. Observe that $g = h/2$.

The specular reflection function F_s ( ) can be calculated using either Horn's or Blinn's method.

9.3.3.1 Horn's Method for Computing the Specular Reflection Function

In Horn's method the specular reflection function is given by

$$F_s(\mathbf{N}, \mathbf{L}, \mathbf{E}, n) = \cos^n h \tag{9.6}$$

From Fig. 9.12 we have the following relationships between the angles:

$$h = e - r$$

and since $r = i$ we have

$$h = e - i \tag{9.7}$$

$$\text{also } p = e + i \tag{9.8}$$

Given the trigonometric identities:

$$\cos(\alpha - \beta) = \cos\alpha \cdot \cos\beta + \sin\alpha \cdot \sin\beta \tag{9.9}$$

$$\text{and} \quad \cos(\alpha + \beta) = \cos\alpha \cdot \cos\beta - \sin\alpha \cdot \sin\beta \tag{9.10}$$

Combining Eqs. (9.7) and (9.9), and Eqs. (9.8) and (9.10) we get

$$\cos h = \cos(e - i) = \cos e \cdot \cos i + \sin e \cdot \sin i \tag{9.11}$$

$$\text{and} \quad \cos p = \cos(e + i) = \cos e \cdot \cos i - \sin e \cdot \sin i \tag{9.12}$$

Adding Eqs. (9.11) and (9.12) we can eliminate all the sines:

$$\cos h + \cos p = 2 \cdot \cos e \cdot \cos i$$

$$\therefore \quad \cos h = 2 \cdot \cos e \cdot \cos i - \cos p$$

$$\therefore \quad \cos h = 2 \cdot (\boldsymbol{N} \odot \boldsymbol{E}) \cdot (\boldsymbol{N} \odot \boldsymbol{L}) - (\boldsymbol{E} \odot \boldsymbol{L}) \tag{9.13}$$

Therefore the specular reflection function can be computed as

$$F_s(\boldsymbol{N}, \boldsymbol{L}, \boldsymbol{E}, n) = [2 \cdot (\boldsymbol{N} \odot \boldsymbol{E}) \cdot (\boldsymbol{N} \odot \boldsymbol{L}) - (\boldsymbol{E} \odot \boldsymbol{L})]^n \tag{9.14}$$

where $n > 4$ for glossy surfaces and $n < 2$ for less shinny surfaces.

9.3.3.2 Blinn's Method for Computing the Specular Reflection Function

In Blinn's method the specular reflection function is given by

$$F_s(\boldsymbol{N}, \boldsymbol{L}, \boldsymbol{E}, n) = \cos^n g \tag{9.15}$$

This is a simplification of Horn's function based on the observation that $g = h/2$.

Now, from Fig. 9.12 we have

$$\cos g = \boldsymbol{N} \odot \boldsymbol{H}$$

where $\boldsymbol{N}$ is known and $\boldsymbol{H}$ is easy to compute as it is the vector halfway between the vectors $\boldsymbol{E}$ and $\boldsymbol{L}$. Thus

$$\boldsymbol{H} = \frac{\boldsymbol{E} + \boldsymbol{L}}{|\boldsymbol{E} + \boldsymbol{L}|}$$

Thus the specular reflection function can be computed as

$$F_s(\boldsymbol{N}, \boldsymbol{L}, \boldsymbol{E}, n) = (\boldsymbol{N} \odot \boldsymbol{H})^n \tag{9.16}$$

where $n > 20$ for glossy surfaces and $n < 8$ for less shinny surfaces.

Both Horn's and Blinn's methods are essentially the same as $g = h/2$, but Horn's highlights decay faster and are therefore sharper and look more realistic.

9.3.4 Phong's Lighting Model

Phong's lighting model is based on the observation that "real life" surfaces are sometimes glossy or chalky or something in-between, and incorporates all three reflection models that we have examined so far. Thus

$$I = I_a + I_d + I_s$$

$$\text{i.e.} \quad I = L_a \cdot k_a + L_p \cdot [k_d \cdot (\boldsymbol{L} \odot \boldsymbol{N}) + k_s \cdot F_s(\boldsymbol{N}, \boldsymbol{L}, \boldsymbol{E}, n)] \tag{9.17}$$

9.3.4.1 Simulating Multiple Light Sources

Phong's lighting model can farther be improved by allowing for multiple light sources.

$$I = L_a \cdot k_a + \sum_{j=1}^{m} I_j \tag{9.18}$$

where I is the total intensity of the reflected light,
L_a is the incident ambient light intensity,
k_a is the ambient reflection coefficient of the surface, $0 \leq k_a \leq 1$,
I_j is the intensity of reflected light contributed by the jth light source, and
m is the number of light sources.

The intensity of reflected light contributed by the jth light source is given by

$$I_j = L_{pj} \cdot \left[k_d \cdot \left(\boldsymbol{L}_j \odot \boldsymbol{N}\right) + k_s \cdot F_s\left(\boldsymbol{N}, \boldsymbol{L}_j, \boldsymbol{E}, n\right)\right] \tag{9.19}$$

where L_{pj} is the incident intensity of the jth light source,
k_d is the coefficient of diffuse reflection of the surface, $0 \leq k_d \leq 1$,
k_s is the coefficient of specular reflection of the surface, $0 \leq k_s \leq 1$,
$\boldsymbol{N}$ is the surface unit normal vector,
$\boldsymbol{L}_j$ is the unit vector in the direction of the jth light source,
$\boldsymbol{E}$ is the unit vector in the direction of the viewer,
n is the *specular sharpness*, and
F_s is the specular reflection function.

9.3.4.2 Simulating Distant Light Sources

This model can further be improved by attenuating the intensity of reflected light contributed by the jth light source, I_j, by the distance of this light source from the surface. Thus

$$I_j = A\left(d_j\right) \cdot L_{pj} \cdot \left[k_d \cdot \left(\boldsymbol{L}_j \odot \boldsymbol{N}\right) + k_s \cdot \boldsymbol{F}_s\left(\boldsymbol{N}, \boldsymbol{L}_j, \boldsymbol{E}, n\right)\right] \tag{9.20}$$

where d_j is the distance of the jth light source from the surface, and
A is the intensity attenuation function.

A number of intensity attenuation functions may be constructed. One such function is

$$A(d) = \frac{1}{d^2 + 1} \tag{9.21}$$

Although this function is based on Lambert's law and it is physically correct, it causes the intensity to decay too dramatically and does not produce good CG results. Observe that we have augmented Lambert's law by adding 1 to the denominator of the fraction to avoid a division by zero when the light source is very near the surface.

An alternative attenuation function is

$$A(d) = \frac{1}{d + k} \tag{9.22}$$

where k is an arbitrary constant. This function is based on empirical observation.

Another alternative attenuation function is

$$A(d) = k^d \tag{9.23}$$

where k is an arbitrary constant, $0 \leq k \leq 1$. This is an empirical function, which is easy to control, and produces good visual results. If, say, we require the intensity contributed by a given light source to fall to half the original intensity at a distance of 20 units from the light source, then we set the value of the constant to $k = 0.5^{1/20}$. In general, if we require the intensity contributed by the light source to fall to a fraction f of the original intensity at a distance of d_f units from the light source, then we set the value of the constant to $k = f^{(1/d_f)}$.

9.3.4.3 Coloured Light Sources

Phong's lighting model, as presented above, only deals with white ambient light and white light sources. We can easily modify this lighting model to deal with coloured lights. To do this we have to rewrite Eq. (9.18) as a set of three equations, one for each of the primary colour components of light.

$$\begin{aligned} I_R &= L_{aR} \cdot k_a + \sum_{j=1}^{m} I_{jR} \\ I_G &= L_{aG} \cdot k_a + \sum_{j=1}^{m} I_{jG} \\ I_B &= L_{aB} \cdot k_a + \sum_{j=1}^{m} I_{jB} \end{aligned} \tag{9.24}$$

where I_R, I_G, I_B are the RGB components of the total intensity of the reflected light,

L_{aR}, L_{aG}, L_{aB} are the RGB components of the incident ambient light intensity,

I_{jR}, I_{jG}, I_{jB} are the RGB components of the intensity of reflected light contributed by the jth light source,

and all other terms retain their original meaning.

Equation (9.20) will also have to be rewritten as a set of three equations:

$$I_{jR} = A\left(d_j\right) \cdot L_{pjR} \cdot \left[k_d \cdot \left(\boldsymbol{L}_j \odot \boldsymbol{N}\right) + k_s \cdot F_s\left(\boldsymbol{N}, \boldsymbol{L}_j, \boldsymbol{E}, n\right)\right]$$
$$I_{jG} = A\left(d_j\right) \cdot L_{pjG} \cdot \left[k_d \cdot \left(\boldsymbol{L}_j \odot \boldsymbol{N}\right) + k_s \cdot F_s\left(\boldsymbol{N}, \boldsymbol{L}_j, \boldsymbol{E}, n\right)\right] \quad (9.25)$$
$$I_{jB} = A\left(d_j\right) \cdot L_{pjB} \cdot \left[k_d \cdot \left(\boldsymbol{L}_j \odot \boldsymbol{N}\right) + k_s \cdot F_s\left(\boldsymbol{N}, \boldsymbol{L}_j, \boldsymbol{E}, n\right)\right]$$

where $L_{pjR}, L_{pjG}, L_{pjB}$ are the RGB components of the incident intensity of the jth light source, and all other terms retain their original meaning.

9.4 Shading Techniques

Having developed a number of reflection models, let us now examine how shading of a given polygon is achieved. We will examine three simple shading techniques: *flat polygon shading*, *Gouraud smooth shading* and *Prong smooth shading*.

Let us start with the simplest of these techniques.

9.4.1 Flat Polygon Shading Technique

In the simplest case, each polygon is assigned a fixed *shade value* (i.e. reflected colour value). Then whenever the polygon is visible, this fixed shade value is written into the appropriate pixel of the frame buffer. To calculate the polygon's shade value we use the polygon's normal to calculate the RGB colour components of the intensity of the reflected light and then we use the components of the polygon colour to modulate the colour components of this intensity. Thus

$$\begin{aligned} S_R &= P_R \cdot I_R \\ S_G &= P_G \cdot I_G \\ S_B &= P_B \cdot I_B \end{aligned} \quad (9.26)$$

where S_R, S_G, S_B are the RGB components of the shade value,
P_R, P_G, P_B are the RGB components of the polygon colour, and
I_R, I_G, I_B are the RGB components of the total intensity of light reflected off the polygon.

If a polyhedral object is flat polygon shaded, its individual facets will be discernible and it will appear faceted.

9.4.2 Gouraud Smooth Shading Technique

If the polygons to be displayed are meant to approximate a curved surface, the flat polygon shading technique will produce an object with a faceted appearance, thus betraying the construction method of the geometric model. A possible solution to this problem is to approximate the curved surface with a large number of polygons. Then the differences in shade between adjacent polygons will be too small for the viewer to see. A less expensive approach is to use a *smooth shading technique*.

For most surfaces of interest it is possible to calculate the exact surface normal at each sample point on the surface. If these points correspond to the vertices of the polygonal approximation to the surface, then a sample surface normal can be associated with each vertex. If we cannot calculate the exact surface normal at these vertices, then we compute the average normal at each vertex by averaging the surface normals of all the faces meeting at this vertex. In the example depicted in Fig. 9.13, the averaged normal is given by

$$\boldsymbol{N}_a = \frac{\sum_{i=1}^{5} \boldsymbol{N}_i}{\left|\sum_{i=1}^{5} \boldsymbol{N}_i\right|}$$

$$\text{i.e.}\quad \boldsymbol{N}_a = \frac{\boldsymbol{N}_1 + \boldsymbol{N}_2 + \boldsymbol{N}_3 + \boldsymbol{N}_4 + \boldsymbol{N}_5}{|\boldsymbol{N}_1 + \boldsymbol{N}_2 + \boldsymbol{N}_3 + \boldsymbol{N}_4 + \boldsymbol{N}_5|}$$

The trick to produce a smooth shaded image from a polyhedral model with only a few vertices is to calculate a shade value for each vertex (instead of each polygon) and then to interpolate these shade values over the entire polygon. Henri Gouraud devised this technique in 1971 while studying for his Ph.D. at the Utah State University. This technique is known as *Gouraud smooth shading* [Gouraud 71].

The interpolation is accomplished by first linearly interpolating along each edge, and then by linearly interpolating along the scan-line, as shown in Fig. 9.14 and Eq. (9.27). This, of course, is exactly how z is interpolated in ordered-edge

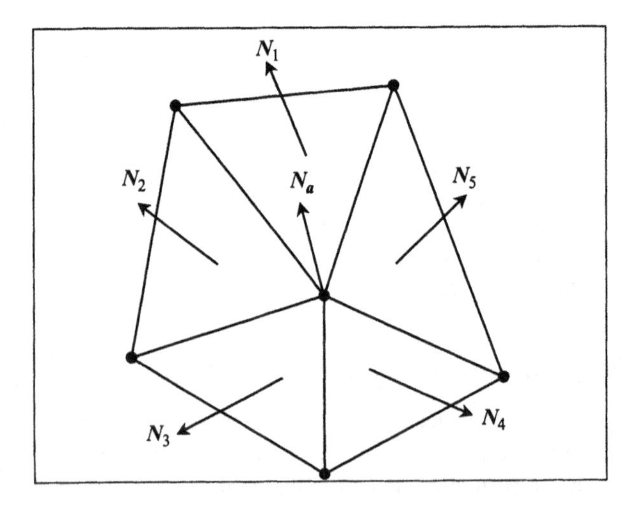

FIGURE 9.13. The average vertex normal vector.

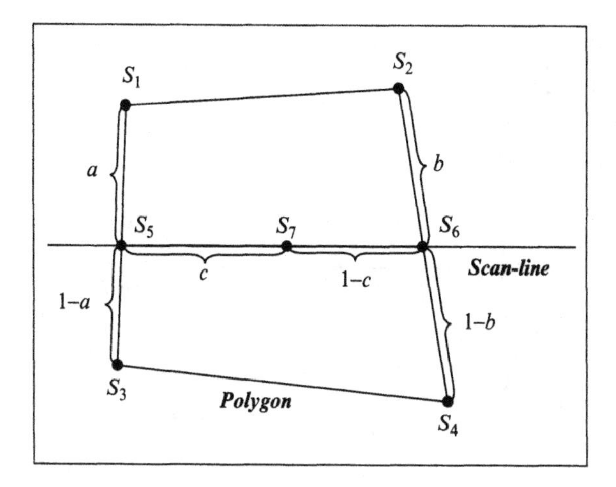

FIGURE 9.14. Gouraud smooth shading.

scan-line algorithms. Consequently, scan-line algorithms can be modified almost trivially to handle Gouraud smooth shading.

In Gouraud smooth shading, the shade value interpolation is done as follows.

$$\begin{aligned} S_5 &= (1-a)\cdot S_1 + a\cdot S_3 \\ S_6 &= (1-b)\cdot S_2 + b\cdot S_4 \quad \text{where} \quad 0 \le a,b,c \le 1 \\ S_7 &= (1-c)\cdot S_5 + c\cdot S_6 \end{aligned} \tag{9.27}$$

where a and b represent the fractions of the left and right edges traversed so far and c represents the fraction of the polygon span traversed so far.

9.4.3 Phong Smooth Shading Technique

If one attempts to apply the Phong illumination model to each vertex of a polyhedral model and then uses Gouraud's smooth shading technique to interpolate the resulting shade values, the result will not be satisfactory, as it will produce peculiar looking highlights. For instance, unless a highlight is centred on a vertex, maximum highlights will not result from the shade value interpolation. In an animation sequence the highlights of a rotating object will appear to be jumping from vertex to vertex in an erratic manner. To overcome this problem Phong Bui-Tuong devised a new shading technique in 1973 while studying for his Ph.D. at the Utah State University. This technique is known as *Phong smooth shading* [Phong 73].

Phong's solution was to interpolate the surface normal vectors of the vertices of the polygon rather than the shade values of these vertices. See Fig. 9.15. The illumination model is then applied at each pixel instead of just at the vertices of the polygon. Here, the surface normal is interpolated just like the z values or the shade values in Gouraud smooth shading with one important difference. The illumination model requires that the surface normal be unit length. Consequently the

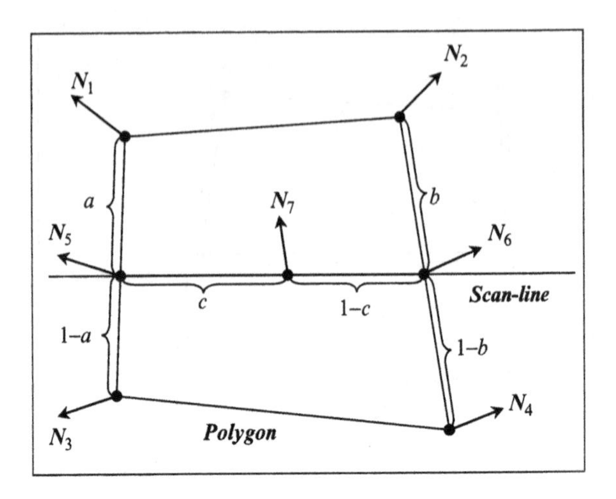

FIGURE 9.15. Phong smooth shading.

interpolated surface normal must be normalised at every point in the interpolation process which is computationally expensive. Slow as it may be, Phong shading can produce realistic looking results.

In Phong smooth shading, the normal vector interpolation is done as follows.

$$\begin{aligned} N_5 &= (1-a)\cdot N_1 + a\cdot N_3 \\ N_6 &= (1-b)\cdot N_2 + b\cdot N_4 \quad \text{where} \quad 0 \le a, b, c \le 1 \\ N_7 &= (1-c)\cdot N_5 + c\cdot N_6 \end{aligned} \tag{9.28}$$

where a and b represent the fractions of the left and right edges traversed so far and c represents the fraction of the polygon span traversed so far.

References

[Gouraud 71] Gouraud, H. Computer display of curved surfaces. University of Utah Computer Science Department, UTEC-Cs-71-113, June 1971.

[Newell 72] Newell, M. E., Newell, R. G., and Sancha T. L. A new approach to the shaded picture problem. *Proceedings of the ACM National Conference*, p. 443, 1972.

[Phong 73] Bui-Tuong, Phong. Illumination for computer-generated images. University of Utah Computer Science Department, UTEC-Cs-73-129, July 1973.

[Sutherland 74] Sutherland, I. E., Sproull, R. F., and Schumacker, R. A. A characterization of ten hidden surface algorithms. *ACM Computing Surveys* 6 (1), 1–55, 1974.

[Warnock 69] Warnock, J. E. A hidden-surface algorithm for computer generated half-tone pictures. University of Utah Computer Science Department, Technical Report: TR 4-15, 1969.

10

Physically Based Lighting and Shading Models and Rendering Algorithms

In the last chapter we have examined a simple model of how light, emanating from light sources, interacts with the surfaces of objects in a scene. This model is purely empirical and bears very little relation to what actually takes place in reality. In this chapter we will introduce a much more sophisticated physically based model of light and examine how it interacts with the surfaces, volume and substance of objects in a computer-generated scene. To do that we must first examine the nature of light and of the materials lit by it. We start by examining how our understanding of light and its perception has evolved over the ages.

10.1 Evolution of the Theory of Light

Much of our interaction with the physical universe is mediated by light or through images generated by the mediation of light. So understanding what light is and how we perceive it has preoccupied human beings since the dawn of history.

The early Greeks believed that Aphrodite, the goddess of love, made the human eye using the four elements (of water, earth, wind and fire) and lit the fire that shines out of the eye as light rays, thus making sight possible. This simple belief does not explain why we cannot see at night and why we see shadows during the day. In computer graphics terms this is equivalent to ray tracing a scene which is lit by a single point light source situated at the eye.

The Greek philosopher Empedocles of Acrages in Sicily (492–432 BC) questioned this belief and postulated an interaction between the rays from the eyes and the rays from an external light source such as the sun. In computer graphics terms this is equivalent to ray tracing a scene which is lit by external light sources.

The Greek mathematician Euclid of Alexandria (325–265 BC) wrote his book "Optica" around 300 BC, in which he presented a mathematical study of light and optics. Euclid postulated that light travels in straight lines and developed the mathematics for the law of reflection.

In 50 BC, the Roman poet Titus Lucratius Charus (99–55 BC) wrote a poem entitled "On the Nature of the Universe" in which he wrote "the light and the heat of the sun are composed of minute atoms which, when they are shoved off, lose no time in shooting right across the inter-space of air in the direction imparted by the shove". He was thus the first person to express the idea that light is composed of particles.

Following the Greeks, the next big breakthrough in our understanding of light came from the Arab mathematician Abu Ali al-Hasan ibn al-Haytham who was born in Basra (Iraq) in AD 965 and died in Cairo (Egypt) in AD 1040. He wrote that light emanates from light sources, is reflected off surfaces and enters the eye, thus making sight possible. He used a pinhole camera to prove his point. In this type of camera, light enters the camera chamber through a pinhole situated on the front wall of the chamber and the image is formed upside-down on the back wall of the chamber. His experiment showed that rays emanating from the eye were not required in order to generate an image. Al-Haytham believed that light is composed of tiny particles that travel in straight lines. He further believed that these particles must travel at a very high but finite speed and that their speed varied depending on the medium through which they travelled, and he thought that this was the reason why refraction took place. Unfortunately, his writings were not available to European readers until the end of the sixteenth century, his work went unnoticed and European thinkers continued to subscribe to the Greek explanation of sight.

At the beginning of the seventeenth century the German astronomer Johannes Kepler (1571–1630) worked on optics and was the first to develop a correct mathematical explanation of the camera obscura and to correctly explain how the human eye works.

In the last third of the seventeenth century two opposing major theories of light were put forward. The first theory was put forward by the English mathematician Isaac Newton (1643–1727). The second theory was developed independently by the Dutch astronomer Christian Huygens (1629–1695) and by the English scholar Robert Hook (1635–1703). Newton's theory was corpuscular in nature (i.e. it was a particle theory), while the Huygens–Hook theories were wave theories. For Newton's theory to be correct it was necessary for light to travel faster in denser materials (which we now know is false) and for the Huygens–Hook theories to be correct it was necessary for light to travel slower in denser materials (which we now know is true).

During the eighteenth century most mathematicians and scientists sided with Newton, but in 1799 the English philosopher Thomas Young (1773–1829) produced some experimental results in favour of the wave theory of light.

In 1817, the work of the French mathematician Augustin Jean Fresnel (1788–1827) on light diffraction provided further proof of the validity of the wave theory of light.

In 1849, the French physicist Armand Hyppolyte Louis Fizeau (1819–1896) determined experimentally the speed of light to be 300,000 km/s.

In 1862, the English natural philosopher (physicist) James Clark Maxwell (1831–1879) realised that electromagnetic phenomena are related to light when he discovered that electromagnetic waves travelled at the same speed as light. Maxwell is considered to be both the person that completed the classical description of light and that instigated some of the modern developments in the theory of light.

In 1887, the German physicist Heinrich Hertz (1857–1894) discovered the *photoelectric effect*. By experimenting he noticed that when he shone ultraviolet light on two metallic electrodes this lowered the voltage at which a spark would be produced between the two electrodes. This effect could only be explained using the particle theory of light and we now know that it is caused by electrons being ejected from the surface of the electrodes when they are struck by light rays.

In 1905, the German physicist Albert Einstein (1879–1955) published his *special relativity* theory in which he suggests that the speed of light remains constant for all observers independent of their relative velocities. Despite the fact that this notion is difficult to accept, it is less paradoxical than the alternative proposed by the classical theory – according to which an observer travelling faster than the speed of light could arrive at his destination in time to be able to observe himself setting off for the journey. In his paper Einstein showed how the photoelectric effect could be explained if we accept that light is composed of discrete particles which are energy quanta. These particles are called *photons* and derive their name from the Greek word $\varphi\omega\tau o$s (meaning of the light).

In 1915, Einstein published his *general relativity* theory in which he predicted that light rays can be bent when passing through a gravitational field.

In 1924, the Indian physicist Satyendranath Bose (1894–1974) put forward the hypothesis that light consists of particles that obeyed certain statistical laws.

In the same year, the French physicist Pierre Raymond duc de Broglie (1892–1887) put forward his *wave-particle duality* theory that states that matter has both the properties of waves and particles. Thus, not only photons act as waves but electrons could as well. Such particles are often referred to as *wavicles* (short for wave-particles).

In 1926, the American physicist Albert Abraham Michelson (1852–1931) accurately established the speed of light to be 299,796 km/s.

By 1930 the stage was set for the Danish physicist Neils Henrik David Bohr (1885–1962) and his group of collaborators to complete the *Copenhagen interpretation* of *quantum* theory. This interpretation of quantum theory attempts to explain the dual wave-particle nature of light by stating that light "travels as a wave but departs and arrives as a particle". The main idea behind the Copenhagen interpretation is that observing light waves causes them to change into particles. This interpretation, which seems to indicate that the observer affects the way light behaves, although difficult to accept seems to be supported by experimental evidence. In 1928, Bohr put forward the *complementarity principle* which states that photons could behave either as waves or as particles but that it is impossible to observe both these aspects of their behaviour at the same time. Some recent experiments, however, have created situations where light behaves both as waves and particles at the same time, which contradicts the complementarity principle.

10.2 Nature of Light

Let us now put together what we know about the nature of light from physics. Light is *electromagnetic radiation* in a range of wavelengths that can be detected by the human visual system.

According to the wave-particle duality principle of quantum theory, light exhibits properties of both waves and particles. Alternatively, light consists of *quanta* (small packets) of excitation of a quantised electromagnetic field, called photons. Thus a photon is both a fundamental particle and a small packet of electromagnetic radiation (i.e. an electromagnetic wave). In this sense a photon is the smallest building block of electromagnetic radiation and all electromagnetic radiation (from radio waves to gamma rays) is quantised as photons.

Photons have an infinite lifetime, although they can be created and destroyed. Photons can be created in a variety of ways. For instance, a photon can be created when an electron changes its energy state, during a nuclear transition in a particle–antiparticle collision or due to the fluctuation of an electromagnetic field.

Table 10.1 represents the entire spectrum of electromagnetic radiation. In this table the various types of electromagnetic waves are tabulated in ascending order of wavelength. The frequency of each type of wave and the associated energy of the corresponding photon is also shown. A more detailed explanation of these terms will follow shortly. The wavelength of an electromagnetic wave is measured in meters (m), its frequency is measured in Hertz (Hz) and the energy of the corresponding photon is measured in electron-volts (eV). A Hertz is the number of events per second; thus a wave that completes 10 complete undulations in a second is said to have a frequency of 10 Hz. An electron-volt is the very small amount of energy that a free electron gains when it falls through an electrostatic potential difference of 1 V. Note that $1\ \text{eV} = 1.602176462 \times 10^{-19}$ J (joules).

At the top of this table we find γ (gamma) rays. These correspond to the most energetic photons with energies in the range of mega-electron-volts, wavelengths in the range of pico-meters and frequencies in the range of a few hexa-Hertz. While, at the bottom of this table we find extremely low frequency waves that can be detected as very base sounds reproduced by subwoofer speakers or in the transmission of electrical currents. The photons corresponding to such waves have energies in the range of pico-electron-volts, wavelengths in the range of mega-meters and frequencies in the range of a few Hertz. Visible light represents a very narrow electromagnetic spectrum band that lies in the wavelength range between 380 and 740 nm and the frequency range of 789 and 480 tera-Hz. Most of the Sun's radiation is emitted at this wavelength range and thus the human visual system has evolved to detect radiation at this range. Similarly skin sensors have evolved to detect infrared radiation as heat.

TABLE 10.1. The electromagnetic spectrum.

Wavelength (m)	Frequency (Hz)	Energy (eV)	Type of radiation
1.00×10^{-12}	3.00×10^{20}	1.24×10^{6}	
			Gamma rays
1.00×10^{-11}	3.00×10^{19}	1.24×10^{5}	
			X-rays
1.00×10^{-9}	3.00×10^{17}	1.24×10^{3}	
			Ultraviolet rays
3.80×10^{-7}	7.89×10^{14}	3.26×10^{0}	
			Violet (visible light)
4.40×10^{-7}	6.81×10^{14}	2.82×10^{0}	
			Blue (visible light)
5.10×10^{-7}	5.88×10^{14}	2.43×10^{0}	
			Cyan (visible light)
5.20×10^{-7}	5.77×10^{14}	2.38×10^{0}	
			Green (visible light)
5.65×10^{-7}	5.31×10^{14}	2.19×10^{0}	
			Yellow (visible light)
5.90×10^{-7}	5.08×10^{14}	2.10×10^{0}	
			Orange (visible light)
6.25×10^{-7}	4.80×10^{14}	1.98×10^{0}	
			Red (visible light)
7.40×10^{-7}	4.05×10^{14}	1.68×10^{0}	
			Infrared rays
1.00×10^{-3}	3.00×10^{11}	1.24×10^{-3}	
			Microwaves
1.00×10^{-1}	3.00×10^{9}	1.24×10^{-5}	
			Ultra high frequency waves (TV)
1.00×10^{0}	3.00×10^{8}	1.24×10^{-6}	
			Very high frequency waves (FM radio)
1.00×10^{1}	3.00×10^{7}	1.24×10^{-7}	
			Radio waves
1.00×10^{5}	3.00×10^{3}	1.24×10^{-11}	
			Voice frequency waves (Telephony)
1.00×10^{6}	3.00×10^{2}	1.24×10^{-12}	
			Extremely low frequency waves (Electrical power)
1.00×10^{7}	3.00×10^{1}	1.24×10^{-13}	

Electromagnetic radiation is a combination of an electrical field and a magnetic field moving in unison through a medium and transferring energy from one place to another. These two fields are mutually perpendicular to each other and to the direction of the propagation of the wave. As seen in Fig. 10.1, the $\boldsymbol{E}$-axis represents the direction of the electrical field and the $\boldsymbol{B}$-axis represents the direction of the magnetic field. Such a wave is called a *transverse wave*, as it oscillates in a direction perpendicular to which it advances.

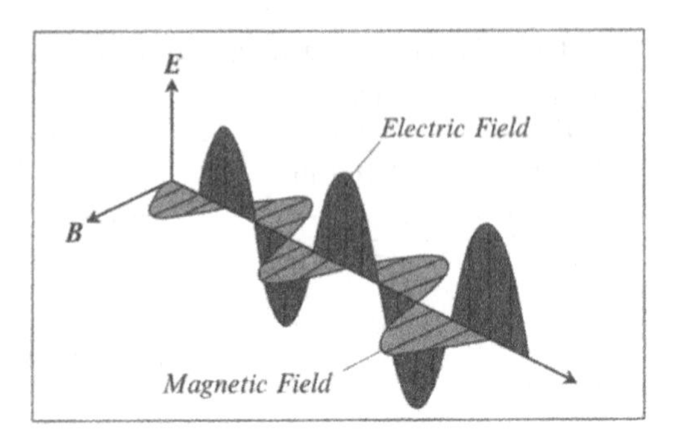

FIGURE 10.1. An electromagnetic wave.

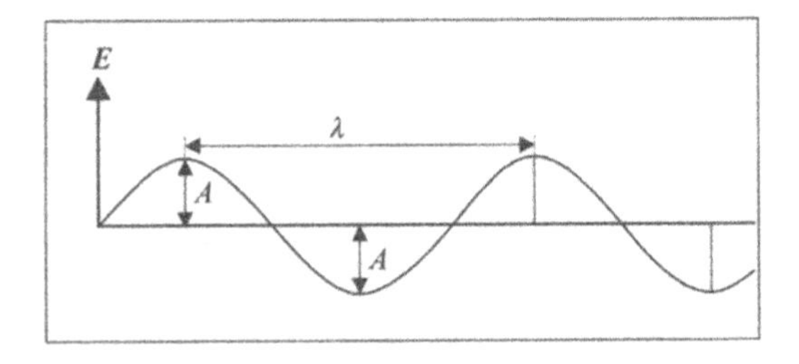

FIGURE 10.2. The amplitude and wavelength of a wave.

Any wave is determined by its *wavelength*, *frequency*, *amplitude* and *period*.

The amplitude of a wave, usually denoted by the letter A, is the magnitude of the maximum displacement in the medium in one wave cycle (see Fig. 10.2). The magnitude of an electromagnetic wave is measured in volts per meter (V/m).

The wavelength of a wave, usually denoted by λ (the Greek letter *lambda*), is the length of one entire oscillation (cycle) of the wave (see Fig. 10.2). The wavelength of a wave is usually measured in meters (m) and in the case of visible light in nanometers (nm).

The period of a wave, usually denoted by the letter T, is the time taken to complete one entire oscillation of the wave. The period of a wave is usually measured in seconds (s).

The frequency of a wave, usually denoted by the letter f, is the number of periods of the wave per unit time. The frequency of a wave is usually measured in Hertz (Hz), and is given by

$$f = \frac{1}{T} \tag{10.1}$$

or alternatively by

$$f = \frac{c}{\lambda} \tag{10.2}$$

where c is the speed of light in vacuum ($c = 299,792,458\,\mathrm{m/s}$).

Sometimes the frequency is expressed in terms of the *angular frequency* or *angular velocity* of the wave as follows.

$$f = \frac{\omega}{2\pi} \tag{10.3}$$

where ω (the Greek letter *omega*) is the angular frequency of the wave, which is a measure of its rotation rate.

The angular frequency is defined as

$$\omega = \frac{2\pi}{T} = 2\pi f \tag{10.4}$$

and is measured in radians per second (rad/s).

The *wave number* of a wave, usually denoted by the letter k, is defined to be the number of wavelengths (or wave crests) in a set distance and is given as

$$k = \frac{2\pi}{\lambda} = \frac{2\pi f}{c} = \frac{\omega}{c} \tag{10.5}$$

In general, the *speed* of a wave is given by

$$v = \frac{\omega}{k} = \lambda f \tag{10.6}$$

Photons have a definite finite *energy*, which is given by

$$e = hf = \frac{hc}{\lambda} \tag{10.7}$$

where h is Planck's constant, which is measured in joules by seconds (J·s) ($h \approx 6.63 \times 10^{-34}$ J s). Planck's constant is the ratio between the angular velocity and the group velocity of photons or, equivalently, the ratio between their *momentum* and their energy. The momentum (mass $\times$ speed) of a photon is usually denoted by ρ (the Greek letter *rho*) and is given by

$$\rho = \frac{h}{\lambda} = \frac{hf}{c} \tag{10.8}$$

A question that often arises is the following. What is the mass of a photon? Do photons have mass or are they "massless"? This is a thorny question and the answer depends on how we define mass. Using Einstein's equation $E = mc^2$, we can derive the *relativistic mass* of a particle as

$$m = \frac{E}{c^2} \tag{10.9}$$

This definition gives every particle a speed related mass. Modern physics, however, assigns a *speed invariant mass* to each particle that is given as

$$m = \frac{E_0}{c^2} \tag{10.10}$$

where E_0 is the energy of a particle at rest (called the *rest energy*). This mass is called the *rest mass* or *invariant mass* of the particle. Photons have zero invariant mass.

In vacuum, photons of all wavelengths travel with the speed of light and in the absence of a gravitational field they travel in a straight line.

In a material, photons behave in a more complex manner. When a photon, which as we have seen is an electromagnetic wave, enters a material, its electrical field causes a *disturbance* to the charges of the electrons of the atoms of the material (causing them to oscillate). This disturbance is proportional to the *permittivity* of the material. The permittivity, ε, of a material (medium) is a measure of how much the medium changes to absorb energy when subjected to an electrical field. This oscillation of the charges of the electrons in turn causes the radiation of an electromagnetic wave that is slightly out of phase with the electromagnetic wave that the photon represents. The sum of these two electromagnetic waves is now a wave with the same frequency but a shorter wavelength than the original electromagnetic wave of the photon. This explains why photons travel slower in materials than they do in the vacuum.

In quantum physics this electromagnetic disturbance caused by the photon entering the material is called an *excitation* and is represented by quasi-particles called *excitons*. Thus when a photon enters a medium it *couples* with it. This means that the photon gets absorbed and the medium gets excited, which in turn means that the photon gets transformed into an exciton. When in this state, it either gets absorbed by the medium and its energy is stored as heat in the medium (which is likely to occur in opaque materials) or it gets transformed back into a photon that re-emerges from the surface of the medium into space (which is likely to occur in transparent materials). This transformation from an exciton back into a photon is due to the medium relaxing and re-emitting the stored energy in the form of a photon.

This is a simplified explanation of what is believed to occur, but will suffice for the purposes of our discussion.

10.3 Interaction of Light with Various Materials

There are two distinct illumination phenomena that we will examine in more detail below. The interaction of light with the boundaries between different types of materials and the scattering and absorption of light as it is reflected from the surface of and transmitted through the volume of different types of materials.

We can categorise the materials found in nature as being *homogenous* and *non-homogenous*. Homogenous materials have a constant composition and the same optical properties/qualities throughout their volume, while non-homogenous materials are composed of two or more different types of homogenous materials (where one type is embedded within another) and their optical properties may vary widely throughout their volume (depending on the concentration of the various homogenous materials that make up their volume).

Homogenous materials can be further subdivided into *opaque* and *transparent* materials.

Opaque materials are frequently called *conductors* or *conducting materials*, as they tend to be good conductors of electrical currents. Examples of conducting materials are iron and copper. Some opaque materials, however, are bad conductors, such as plastics and wood. Opaque materials prevent light from passing through their volume.

Transparent materials are frequently called *dielectric materials*, as they tend to be *insulators* (i.e. they do not conduct electrical currents). Examples of dielectric materials are gases and glass. Some transparent materials, however, are good conductors, such as water and other liquids. Transparent materials allow light to pass through their volume.

Some materials are called *translucent*, because they are semitransparent. They allow light to pass through them only diffusely. Thus, we can see light passing though such materials but we cannot see a clear image. Examples of translucent materials are smoke, steam, paper, frosted glass and some plastics.

In computer graphics we are usually interested in the behaviour of light at the surface between two different types of material. We shall call such a surface the *interface* and the two materials either side of it the *participating materials* or *participating media*. Of most interest are the interface between two transparent materials and between a transparent and an opaque material.

When a beam of light, consisting of many photons, collides with the interface between two participating materials, we observe the phenomena of *reflection*, *refraction* and *transmission*, *scattering*, *absorption* and *subsurface scattering*.

10.3.1 Light Reflection

Light reflection is a phenomenon observed when a light beam travels through the first participating medium, collides with the interface surface, its direction of travel changes abruptly and continues to travel in the first participating medium. We can distinguish two types of reflections, namely *specular reflections* and *diffuse reflections*. Specular or mirror-like reflections are most pronounced when the interface surface between the two participating media is smooth (shiny). In this case an incoming ray of light is reflected in one direction as shown in Fig. 10.3. In this figure, $\boldsymbol{N}$ represents the unit normal vector of the surface, the angle θ_i (known as the angle of *incidence*) is the angle between the incoming ray and the surface normal, and the angle θ_r (known as the angle of *reflection*) is the angle between the reflected (outgoing) ray and the surface normal. The law of reflection states that $\theta_i = \theta_r$. If the light beam collides with the interface surface at right angles, then the light beam bounces in the direction from which it came. This kind if reflection is called *retro-reflection*.

Diffuse reflections are most pronounced when the interface surface between the two participating media is rough (matt) and can be seen to be composed of many micro-facets, as shown in Fig. 10.4a. In this case an incoming ray of light is reflected in many different directions, as shown in Fig. 10.4. In Fig. 10.4b, $\boldsymbol{N}$

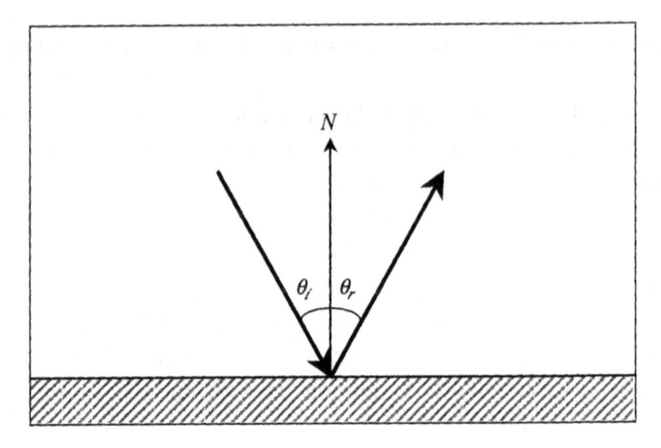

FIGURE 10.3. Specular reflection.

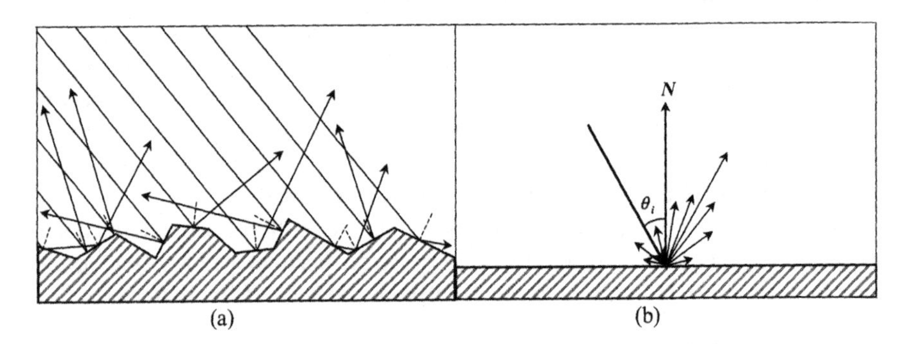

FIGURE 10.4. (a) The surface micro-facets. (b) Diffuse reflection.

represents the average unit normal vector of the interface surface. The dotted lines in Fig. 10.4a represent the normals of the micro-facets of the interface surface.

10.3.2 Light Refraction and Transmission

Light refraction and transmission are phenomena observed when a light beam travels through the first participating medium (which is transparent), collides with the interface surface, penetrates this surface changing direction abruptly and continues to travel through the second participating medium (which is transparent). Transmission is the phenomenon of light travelling through a transparent material and refraction is phenomenon of the abrupt change of direction when the beam enters the second participating medium. This phenomenon is observed at the interface between two transparent materials with different *refractive indices*. It occurs because at the interface surface the electromagnetic wave changes velocity and its wavelength increases or decreases but its frequency remains the same. Figure 10.5 shows an incoming light ray that collides with the interface surface at an incidence angle θ_i and it is refracted and transmitted in the second medium. Its *transmission*

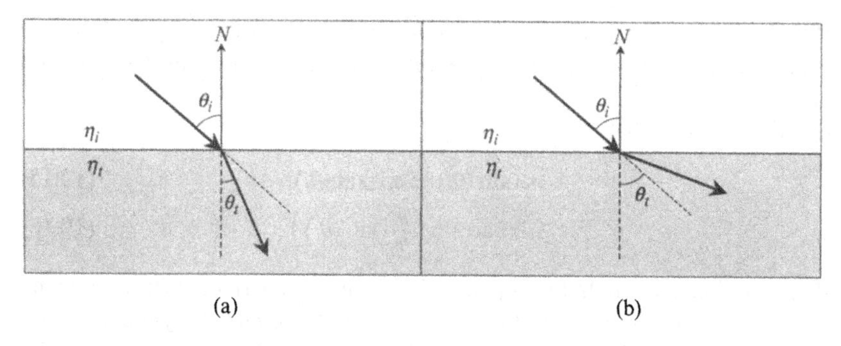

FIGURE 10.5. (a) Refraction with $\eta_i < \eta_t$. (b) Refraction with $\eta_i > \eta_t$.

angle is θ_t. In general $\theta_i \neq \theta_t$, unless the two participating media have the same refractive index.

The refractive index of a material at a given frequency is the rate by which an electromagnetic wave, of that frequency, is slowed down with respect to its speed in vacuum. The refractive index of a material at a given frequency, which is denoted by η (the Greek letter *eta*), is given as

$$\eta = \frac{c}{v} \tag{10.11}$$

where c is the speed of light in vacuum and v is the speed of light of that frequency in the material. The refractive index is typically larger than 1. The denser the material is the larger its refractive index.

The refractive index of a material is also defined as

$$\eta = \sqrt{\varepsilon_r \mu} \tag{10.12}$$

where ε_r is the *dielectric constant* or *relative permittivity* of the material and μ (the Greek letter *mu*) is the *permeability* of the material.

The relative permittivity is defined as

$$\varepsilon_r = \frac{\varepsilon}{\varepsilon_0} \tag{10.13}$$

where ε is the *permittivity* of the material and ε_0 is the permittivity of the material in vacuum. As we have seen earlier, the permittivity of a material is a measure of how much the material changes to absorb energy when subjected to an electrical field.

The permeability, μ, of a material is the degree of magnetisation of a material in response to a magnetic field. The product of the permittivity, ε, and the permeability, μ, of a material is inversely proportional to the square of the speed with which electromagnetic radiation travels through the material. Thus

$$\varepsilon\mu = \frac{1}{v^2} \quad \text{or} \quad v = \frac{1}{\sqrt{\varepsilon\mu}} \tag{10.14}$$

If we know the refractive indices η_i and η_t for the two participating media either side of the interface surface, we can use Snell's law to determine the transmission angle θ_t.

Snell's law states that

$$\eta_i \cdot \sin(\theta_i) = \eta_t \cdot \sin(\theta_t) \tag{10.15}$$

$$\therefore \quad \theta_t = \arcsin\left(\frac{\eta_i}{\eta_t} \cdot \sin(\theta_i)\right) \tag{10.16}$$

If $\theta_i = 0$ (i.e. if the light ray is perpendicular to the interface surface), then no refraction takes place and $\theta_t = 0$. If $\eta_i < \eta_t$, then the refracted ray moves closer to the direction normal to the interface surface and $\theta_i > \theta_t$, as seen in Fig. 10.5a. Alternatively if $\eta_i > \eta_t$, then the refracted ray moves further away from the direction normal to the interface surface and $\theta_i < \theta_t$, as seen in Fig. 10.5b.

10.3.3 Total Internal Reflection

When $\eta_i > \eta_t$ (i.e. when moving from a denser medium to a less dense medium), Eq. (10.16) has no solution if $(\eta_i/\eta_t)\sin(\theta_i) > 1$. Thus when the incidence angle θ_i becomes greater than some *critical angle* θ_c, then no refraction or transmission takes place and the incident ray is reflected off the interface surface and continues to travel in the first participating medium. This phenomenon is known as *total internal reflection*. See Figs. 10.6 and 10.36. The critical angle is computed as follows.

$$\theta_c = \arcsin\left(\frac{\eta_i}{\eta_t}\right) \tag{10.17}$$

Figure 10.6 shows a light ray $\boldsymbol{V}$ that strikes an interface surface, where $\eta_i > \eta_t$. Figure 10.6a shows that when $\theta_i < \theta_c$, some of the incident light is reflected in the direction of $\boldsymbol{V}_r$ and some is refracted and transmitted in the direction of $\boldsymbol{V}_t$. Figure 10.6b shows that when $\theta_i \geq \theta_c$, all the incident light undergoes total internal reflection and no light is transmitted through the second participating medium.

As we have seen above, the refractive index of any material varies with frequency. This means that not all colours (frequencies) of light travel with the same

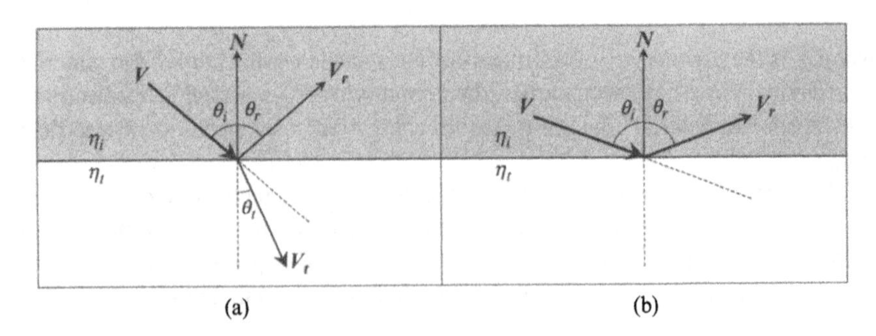

FIGURE 10.6. (a) When $\theta_i < \theta_c$, reflection and transmission take place. (b) When $\theta_i \geq \theta_c$, only total internal reflection takes place.

speed through a material and that the transmission angles of different colours of light are different. This phenomenon, which is known as *dispersion*, can be observed when a beam of light passes through a prism and divides into its constituent spectral colours. A rainbow is the most common manifestation of this phenomenon in nature. The creation of a rainbow can be explained as follows. As a ray of white light from the sun enters a near spherical droplet of rainwater it is refracted and dispersed into a number of coloured rays. These rays continue to travel through the droplet until they reach the back surface of the droplet where they undergo total internal reflection. They continue to travel through the droplet until they reach the front surface of the droplet where they exit the droplet undergoing further refraction. The angle between the entering ray of white light and the exiting coloured rays is between 40 ° and 42 °.

10.3.4 Light Scattering and Absorption

Light scattering is a phenomenon observed when a light beam travels through a non-homogenous medium where small particles of one material are suspended in a second material with a different refractive index. When the photons of the light beam collide with these particles they are scattered in all directions, resulting in the diffusion of the light beam. Light scattering varies as a function of the wavelength of the individual photons and the radius of the suspended particles. See Fig. 10.7.

Light absorption is the phenomenon observed when a photon travelling through a material collides with an atom of this material. This collision results in the distraction of the photon, whose energy is stored as heat in the material.

The light scattering and absorption properties of a material are frequency dependent. For instance, when a beam of white light passes trough a yellow filter, the photons corresponding to the blue frequencies of light are absorbed by the material while the photons corresponding to the red and green frequencies are allowed to pass through the filter. Thus the light leaving the filter is yellow. Similarly,

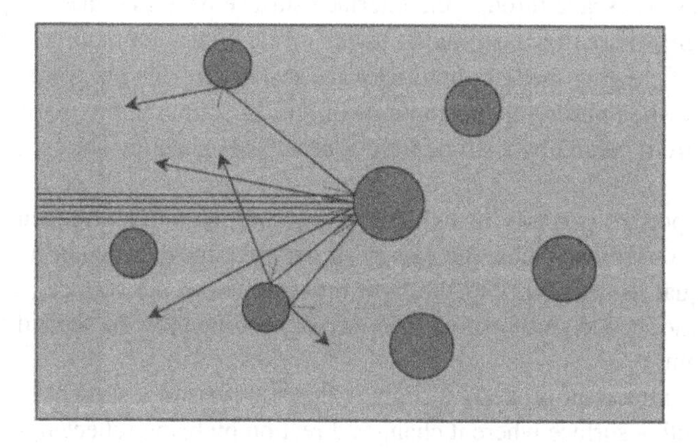

FIGURE 10.7. Light scattering.

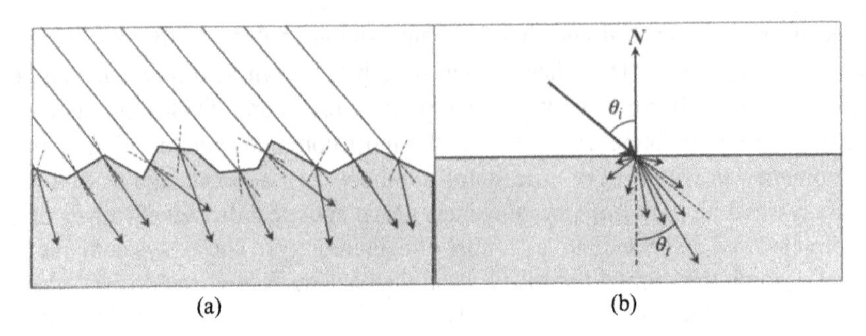

FIGURE 10.8. (a) The interface surface micro-facets. (b) Refraction with scattering.

when sunlight passes through the earth's atmosphere the photons corresponding to the blue frequencies of light are scattered while the photons corresponding to the red and green frequencies are allowed to pass through the atmosphere. This causes the sky to appear blue and since the blue component of the sunlight is scattered the sun appears yellow-orange depending on the time of day.

A form of scattering also occurs when a light beam collides with the interface between the two participating media, which is rough and composed of many micro-facets. If the two participating media are transparent and have different refractive indices, refraction also takes place. This phenomenon is illustrated in Fig. 10.8.

10.3.5 Subsurface Scattering

Subsurface scattering is the phenomenon observed when a light beam travels through the first participating medium (which is transparent), collides with the interface surface between this medium and an opaque or semitransparent non-homogenous medium. Upon entering the second participating medium, the individual photons of the light beam are refracted and scattered. The photons continue to travel in this medium for a short distance under the interface surface until they collide with some suspended particles that cause them to scatter once again. Some of the photons escape through the interface surface back into the first participating medium where they continue to travel, while others continue to travel in the second participating medium until they are absorbed. This phenomenon is most noticeable when observing the contours of human skin or a thin marble structure that is lit from behind by a strong light source. This phenomenon is illustrated in Fig. 10.9.

One important property of light travelling through an environment is that of *energy preservation*. Thus, the energy of the light incident on an interface surface is equal to the energy of the light reflected (from the surface), transmitted (through the second participating medium) or absorbed (by the second participating medium).

Another important property of light is that it travels in a straight line unless it collides with a surface where it changes direction by being reflected, refracted or scattered.

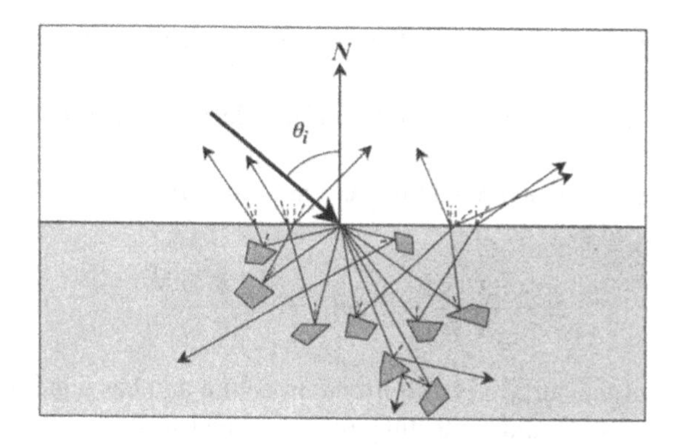

FIGURE 10.9. Subsurface scattering.

10.4 Some Useful Concepts, Definitions and Conventions

Next we will introduce some notation in order to best explain the above-mentioned phenomena. Let us concentrate on the behaviour of light at a point s which lies on the interface surface between two participating media. At that point the surface unit normal vector is $\boldsymbol{N}$ which points from the surface and into the first participating medium. By convention all vectors that we use in our explanations are unit vectors and point away from point s. Thus, as seen in Fig. 10.10, the unit vector $\boldsymbol{L}$ points from the surface point s towards the light source (i.e. it points in the opposite direction from which the light is coming).

Vector $\boldsymbol{E}$ points in the direction of the viewer (the eye). Vector $\boldsymbol{H}$ points halfway between vectors $\boldsymbol{L}$ and $\boldsymbol{E}$. Vectors $\boldsymbol{R_L}$ and $\boldsymbol{R_E}$ are the directions of the reflection of vectors $\boldsymbol{L}$ and $\boldsymbol{E}$, respectively, relative to the surface normal $\boldsymbol{N}$. While, vectors $\boldsymbol{T_L}$ and $\boldsymbol{T_E}$ are the directions of the transmission of vectors $\boldsymbol{L}$ and $\boldsymbol{E}$, respectively, relative to the surface normal $\boldsymbol{N}$.

It is convenient to define a *local surface Cartesian frame* $\{\boldsymbol{T}, \boldsymbol{B}, \boldsymbol{N}, s\}$ with its origin at the surface point s, where $\boldsymbol{N}$ is the unit normal vector of the surface, $\boldsymbol{T}$

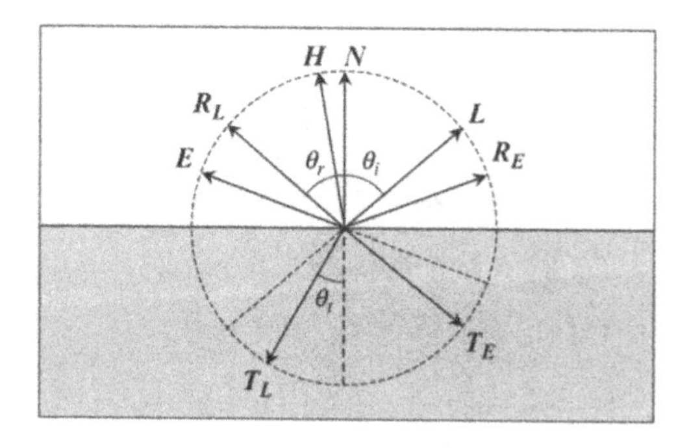

FIGURE 10.10. The reflection and refraction vectors.

is the unit *tangent* vector of the surface and $\boldsymbol{B}$ is the unit *binormal* or *bitangent* vector of the surface.

If the surface is *isotropic* in nature (i.e. if it reflects, refracts and scatters light in a uniform manner), then the tangent vector $\boldsymbol{T}$ can be assumed to lie on the plane defined by the vectors $\boldsymbol{L}$ and $\boldsymbol{N}$. In this case, as seen in Fig. 10.11, the unit vectors $\boldsymbol{B}$ and $\boldsymbol{T}$ are defined as

$$\boldsymbol{B} = \frac{\boldsymbol{N} \otimes \boldsymbol{L}}{|\boldsymbol{N} \otimes \boldsymbol{L}|} \quad \text{and} \quad \boldsymbol{T} = \frac{\boldsymbol{B} \otimes \boldsymbol{N}}{|\boldsymbol{B} \otimes \boldsymbol{N}|} \tag{10.18}$$

Alternatively if the surface is *anisotropic* in nature and has a grain pointing in a particular direction $\boldsymbol{G}$ (thus reflecting, refracting and scattering light in a biased manner), then the tangent unit vector can be taken in the direction of the grain. Thus, as seen in Fig. 10.12, the unit vectors $\boldsymbol{T}$ and $\boldsymbol{B}$ are defined as

$$\boldsymbol{T} = \frac{\boldsymbol{G}}{|\boldsymbol{G}|} \quad \text{and} \quad \boldsymbol{B} = \frac{\boldsymbol{N} \otimes \boldsymbol{T}}{|\boldsymbol{N} \otimes \boldsymbol{T}|} \tag{10.19}$$

Next, we define a spherical coordinate system with its origin at the surface point s, as seen in Fig. 10.13. The spherical coordinates θ and ϕ, of a given vector $\boldsymbol{V}$, are measured from the normal vector $\boldsymbol{N}$ and from the plane defined by the primary axis $\boldsymbol{N}$ and the secondary axis $\boldsymbol{T}$. Thus, for isotropic surfaces the ϕ

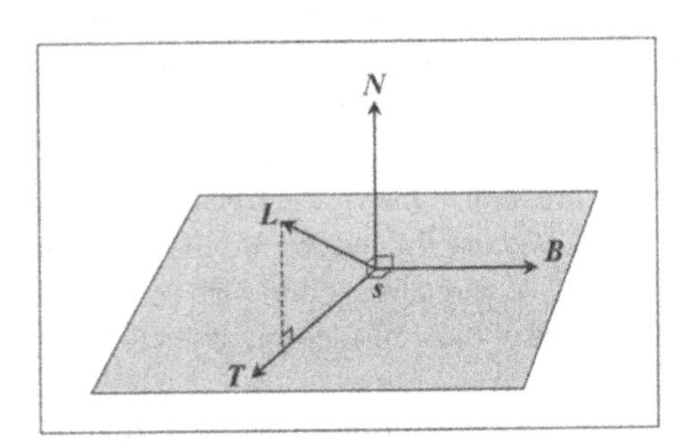

FIGURE 10.11. The cartesian frame of an isotropic surface.

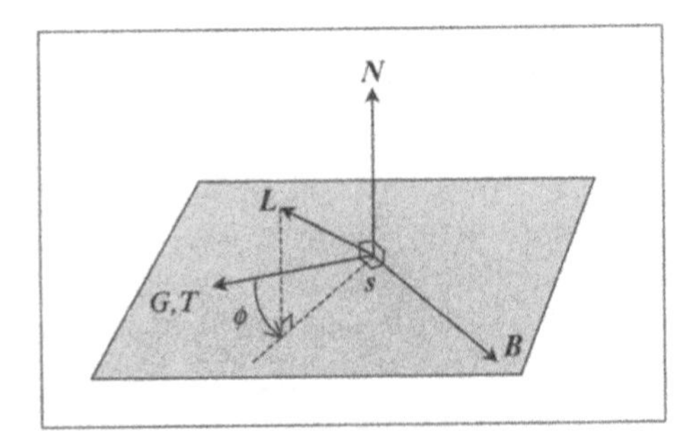

FIGURE 10.12. The cartesian frame of an anisotropic surface.

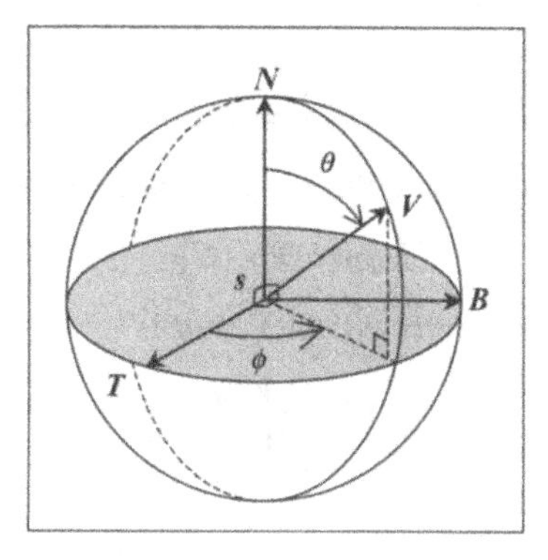

FIGURE 10.13. A spherical coordinate system with its origin at surface point $\boldsymbol{s}$.

spherical coordinate of vector $\boldsymbol{L}$ is zero and for anisotropic surfaces ϕ of $\boldsymbol{L}$ is generally non-zero.

The components of the direction vectors $\boldsymbol{T}$, $\boldsymbol{B}$, $\boldsymbol{N}$ and the position vector $\boldsymbol{s}$ are defined relative to the *global Cartesian frame* $\{\boldsymbol{X}, \boldsymbol{Y}, \boldsymbol{Z}, \boldsymbol{O}\}$, where $\boldsymbol{O}$ is the origin of this frame.

10.4.1 Spherical Coordinates of a Vector

Given an arbitrary unit vector $\boldsymbol{V} = \left[v_x, v_y, v_z\right]$ defined relative to the global Cartesian frame $\{\boldsymbol{X}, \boldsymbol{Y}, \boldsymbol{Z}, \boldsymbol{O}\}$ we can calculate its spherical coordinates (θ_V, ϕ_V) relative to the local surface Cartesian frame $\{\boldsymbol{T}, \boldsymbol{B}, \boldsymbol{N}, \boldsymbol{s}\}$ by referring to Fig. 10.14 and reasoning as follows.

The projection of vector $\boldsymbol{V}$ onto vector $\boldsymbol{N}$ is

$$\boldsymbol{C_V} = (\boldsymbol{V} \odot \boldsymbol{N}) \cdot \boldsymbol{N} = \cos(\theta_V) \cdot \boldsymbol{N}$$

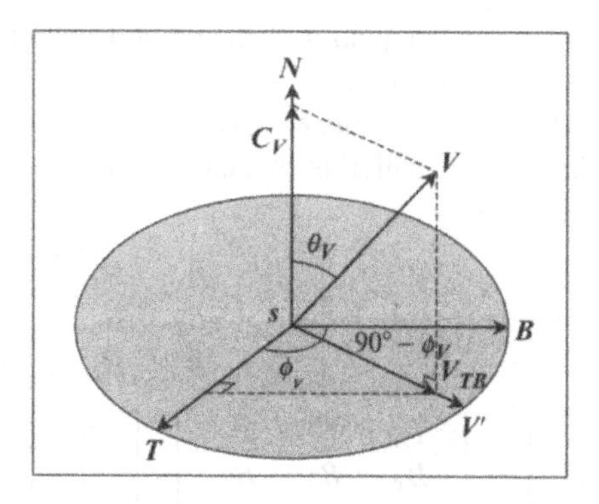

FIGURE 10.14. Computing the spherical coordinates of an arbitrary unit vector.

and the length of this projection is

$$\cos(\theta_V) = \boldsymbol{N} \odot \boldsymbol{V} \tag{10.20}$$

The projection of vector $\boldsymbol{V}$ onto the $\boldsymbol{TB}$ plane is given by

$$\boldsymbol{V_{TB}} = \boldsymbol{V} - \boldsymbol{C}_V$$

and after normalisation the projected unit vector is

$$\boldsymbol{V}' = \frac{\boldsymbol{V_{TB}}}{|\boldsymbol{V_{TB}}|}$$

Now,

$$\cos(\phi_V) = \boldsymbol{T} \odot \boldsymbol{V}' \tag{10.21}$$

$$\sin(\phi_V) = \cos\left(90^0 - \phi_V\right) = \boldsymbol{B} \odot \boldsymbol{V}' \tag{10.22}$$

Since $0 \le \theta_V \le \pi$ and $0 \le \phi_V \le 2\pi$ we can use $\arccos(\cos(\theta_V))$ to determine θ_V and, $\arccos(\cos(\phi_V))$ and $\arcsin(\sin(\phi_V))$ to determine ϕ_V.

To convert the spherical coordinates (θ_V, ϕ_V) of the vector to its components $[v_t, v_b, v_n]$, relative to the local surface Cartesian basis, is simple, as

$$\begin{aligned} v_t &= \sin(\theta_V) \cdot \cos(\phi_V) \\ v_b &= \sin(\theta_V) \cdot \sin(\phi_V) \\ v_n &= \cos(\theta_V) \end{aligned} \tag{10.23}$$

$$\text{and} \quad \boldsymbol{V} = [v_t, v_b, v_n] = v_t \cdot \boldsymbol{T} + v_b \cdot \boldsymbol{B} + v_n \cdot \boldsymbol{N} \tag{10.24}$$

To convert the vector components from the local frame to the global frame we use the change of basis matrix.

$$\left[v_x, v_y, v_z\right] = [v_t, v_b, v_n] \cdot \begin{bmatrix} T_x & T_y & T_z \\ B_x & B_y & B_z \\ N_x & N_y & N_z \end{bmatrix} \tag{10.25a}$$

$$\text{or} \quad \begin{bmatrix} v_x \\ v_y \\ v_z \end{bmatrix} = \begin{bmatrix} T_x & B_x & N_x \\ T_y & B_y & N_y \\ T_z & B_z & N_z \end{bmatrix} \cdot \begin{bmatrix} v_t \\ v_b \\ v_n \end{bmatrix} \tag{10.25b}$$

The choice of the appropriate equation depends on whether we use the row or the column representation for vectors.

Similarly, to convert the components of a vector from the global frame to the local frame we use the inverse (which in this case is the transpose) matrix of the change of basis matrix.

$$[v_t, v_b, v_n] = \left[v_x, v_y, v_z\right] \cdot \begin{bmatrix} T_x & B_x & N_x \\ T_y & B_y & N_y \\ T_z & B_z & N_z \end{bmatrix} \tag{10.26a}$$

$$\text{or} \quad \begin{bmatrix} v_t \\ v_b \\ v_n \end{bmatrix} = \begin{bmatrix} T_x & T_y & T_z \\ B_x & B_y & B_z \\ N_x & N_y & N_z \end{bmatrix} \cdot \begin{bmatrix} v_x \\ v_y \\ v_z \end{bmatrix} \tag{10.26b}$$

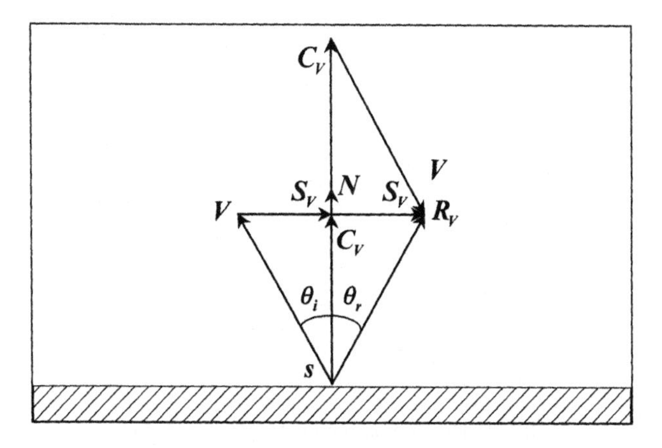

FIGURE 10.15. Computing the reflection vector.

10.4.2 Determining the Reflection Vector

Now, given a unit vector $\boldsymbol{V}$ incident at a point $\boldsymbol{s}$ of an interface surface with unit normal vector $\boldsymbol{N}$, we wish to determine the reflection vector $\boldsymbol{R}_V$. This vector lies on the plane defined by the vectors $\boldsymbol{V}$ and $\boldsymbol{N}$. To compute this vector we refer to Fig. 10.15 and we reason as follows.

Let vector $\boldsymbol{C}_V$ be the projection of vector $\boldsymbol{V}$ onto vector $\boldsymbol{N}$. In the right-angled triangle containing the angle θ_i $|\boldsymbol{C}_V| = |\cos(\theta_i)|$ and $|\boldsymbol{S}_V| = |\sin(\theta_i)|$. Thus,

$$\boldsymbol{C}_V = (\boldsymbol{V} \odot \boldsymbol{N}) \cdot \boldsymbol{N} = \cos(\theta_i) \cdot \boldsymbol{N}$$

and

$$\cos(\theta_i) = \boldsymbol{V} \odot \boldsymbol{N} \tag{10.27}$$

Now

$$\boldsymbol{V} = \boldsymbol{C}_V + (-\boldsymbol{S}_V) = \boldsymbol{C}_V - \boldsymbol{S}_V$$
$$\therefore \quad \boldsymbol{S}_V = \boldsymbol{C}_V - \boldsymbol{V}$$

Since $\theta_i = \theta_r$, from the above diagram we have

$$\begin{aligned} \boldsymbol{R}_V &= \boldsymbol{V} + 2 \cdot \boldsymbol{S}_V \\ &= \boldsymbol{V} + 2 \cdot \boldsymbol{C}_V - 2 \cdot \boldsymbol{V} \\ &= 2 \cdot \boldsymbol{C}_V - \boldsymbol{V} \end{aligned}$$

$$\therefore \quad \boldsymbol{R}_V = 2 \cdot (\boldsymbol{V} \odot \boldsymbol{N}) \cdot \boldsymbol{N} - \boldsymbol{V} \tag{10.28}$$

$$\text{or} \quad \boldsymbol{R}_V = 2 \cdot \cos(\theta_i) \cdot \boldsymbol{N} - \boldsymbol{V} \tag{10.29}$$

10.4.3 Determining the Transmission Vector

Next we wish to determine the transmission vector $\boldsymbol{T}_V$ after refraction takes place. We will assume that the refractive indices of the two participating media are η_i

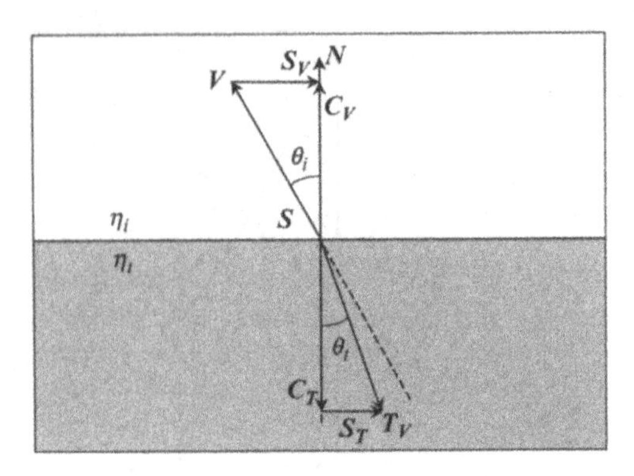

FIGURE 10.16. Computing the transmission vector.

and η_t, respectively. The transmission vector also lies on the plane defined by the vectors $\boldsymbol{V}$ and $\boldsymbol{N}$. To compute this vector we refer to Fig. 10.16 and we reason as follows.

Let vector $\boldsymbol{C_T}$ be the projection of vector $\boldsymbol{T_V}$ onto vector $\boldsymbol{N}$. In the right-angled triangle containing the angle θ_t $|\boldsymbol{C_T}| = |\cos(\theta_t)|$ and $|\boldsymbol{S_T}| = |\sin(\theta_t)|$. Using Snell's law we have

$$\eta_i \cdot |\boldsymbol{S_V}| = \eta_t \cdot |\boldsymbol{S_T}|$$

$$\therefore \quad |\boldsymbol{S_T}| = \frac{\eta_i}{\eta_t} \cdot |\boldsymbol{S_V}|$$

$$\text{and} \quad \boldsymbol{S_T} = \frac{\eta_i}{\eta_t} \cdot \boldsymbol{S_V} \tag{10.30}$$

From the right-angled triangle containing the angle θ_t we have

$$|\boldsymbol{C_T}|^2 + |\boldsymbol{S_T}|^2 = 1 \tag{10.31}$$

$$\therefore \quad |\boldsymbol{C_T}| = \sqrt{1 - |\boldsymbol{S_T}|^2}$$

Note that the length of the $\boldsymbol{S_T}$ vector may not exceed the value of one, as when this occurs total internal reflection takes place instead of refraction and transmission.

The vector $\boldsymbol{C_T}$ can be defined as

$$\boldsymbol{C_T} = -\boldsymbol{N} \cdot |\boldsymbol{C_T}|$$

$$\therefore \quad \boldsymbol{C_T} = \left(-\sqrt{1 - |\boldsymbol{S_T}|^2}\right) \cdot \boldsymbol{N}$$

$$\text{or} \quad \boldsymbol{C_T} = -\cos(\theta_t) \cdot \boldsymbol{N} \tag{10.32}$$

Thus we can compute the transmission vector $\boldsymbol{T_V}$ as

$$\boldsymbol{T_V} = \boldsymbol{C_T} + \boldsymbol{S_T}$$

Using Eqs. (10.32) and (10.30) we can expand this to

$$\begin{aligned}
\boldsymbol{T}_V &= \frac{\eta_i}{\eta_t}\cdot \boldsymbol{S}_V - \cos(\theta_t)\cdot \boldsymbol{N} \\
\therefore\quad \boldsymbol{T}_V &= \frac{\eta_i}{\eta_t}\cdot (\boldsymbol{C}_V - \boldsymbol{V}) - \cos(\theta_t)\cdot \boldsymbol{N} \\
\therefore\quad \boldsymbol{T}_V &= \frac{\eta_i}{\eta_t}\cdot [\cos(\theta_i)\cdot \boldsymbol{N} - \boldsymbol{V}] - \cos(\theta_t)\cdot \boldsymbol{N} \\
\therefore\quad \boldsymbol{T}_V &= \left[\frac{\eta_i}{\eta_t}\cdot \cos(\theta_i) - \cos(\theta_t)\right]\cdot \boldsymbol{N} - \frac{\eta_i}{\eta_t}\cdot \boldsymbol{V}
\end{aligned} \tag{10.33}$$

To compute the cosine of the transmission angle we start from Eq. (10.31) and proceed as follows.

$$\begin{aligned}
&|\boldsymbol{C}_T|^2 = 1 - |\boldsymbol{S}_T|^2 \\
\therefore\quad &\cos^2(\theta_t) = 1 - \sin^2(\theta_t) \text{ and} \\
&\cos(\theta_t) = \sqrt{1 - \sin^2(\theta_t)} \\
&\qquad = \sqrt{1 - (\sin(\theta_t))^2} \\
&\qquad = \sqrt{1 - \left[\frac{\eta_i}{\eta_t}\sin(\theta_i)\right]^2} \\
&\qquad = \sqrt{1 - \left(\frac{\eta_i}{\eta_t}\right)^2\cdot \sin^2(\theta_i)} \\
\therefore\quad &\cos(\theta_t) = \sqrt{1 - \left(\frac{\eta_i}{\eta_t}\right)^2\cdot \left[1 - \cos^2(\theta_i)\right]}
\end{aligned} \tag{10.34}$$

Thus collecting all the above results we have

$$\begin{aligned}
&\cos(\theta_i) = \boldsymbol{V}\odot \boldsymbol{N} \\
&\cos(\theta_t) = \sqrt{1 - \left(\frac{\eta_i}{\eta_t}\right)^2\cdot \left[1 - \cos^2(\theta_i)\right]} \\
&\boldsymbol{R}_V = 2\cdot \cos(\theta_i)\cdot \boldsymbol{N} - \boldsymbol{V} \\
&\boldsymbol{T}_V = \left[\frac{\eta_i}{\eta_t}\cdot \cos(\theta_i) - \cos(\theta_t)\right]\cdot \boldsymbol{N} - \frac{\eta_i}{\eta_t}\cdot \boldsymbol{V}
\end{aligned}$$

10.4.4 Illuminating Hemisphere and Solid Angles

Given a synthetic scene that we wish to render, we will assume that its geometry consists of a finite set of surfaces that exist in $\mathcal{E}^3$. Each surface is assumed to be a two-dimensional differentiable manifold $\mathcal{M}_i$ with boundary $\partial\mathcal{M}_i$. A 2D manifold is an entity that is topologically equivalent to a 2D Euclidean space. Thus a polygon and its outline or a curved surface patch and its boundary curves, although they may exist in a 3D Euclidean space, they are topologically equivalent to a subset of the 2D Euclidean space (i.e. they can be mapped/flattened onto a 2D plane through a parameter mapping). A differentiable manifold is a manifold on which we apply the rules of Calculus, such as differentiation. Associated with every point of a differentiable manifold is a 2D *tangent space*, which is a vector space in which we define the directional derivatives of the manifold (i.e. its tangent vectors), and a 2D *cotangent space*, which is also a vector space in which we define the differentials of the manifold (i.e. its cotangent vectors). The entire scene is assumed to be the set $\mathcal{M}$, which is the union of n such manifolds $\mathcal{M}_1, \mathcal{M}_1, \ldots, \mathcal{M}_n$. A region $\mathcal{D}$ is defined to be a subset of the set of all the manifolds in the scene, thus $\mathcal{D} \subseteq \mathcal{M}$. We define an area measure A on $\mathcal{M}$, so that $A(\mathcal{D})$ is the area of the region $\mathcal{D}$.

Given that directions in $\mathcal{E}^3$ are represented by unit vectors $\boldsymbol{\omega} \in \mathcal{E}^3$, the set of all directions in a unit sphere centred at a surface point $\boldsymbol{x}$ is denoted by $\mathcal{S}^2$.

A *solid angle* is the three-dimensional equivalent of a two dimensional (planar) angle. The solid angle $\sigma(\boldsymbol{\omega}_x)$ subtended by a surface patch $\mathcal{P}$ (in the direction $\boldsymbol{\omega}_x$) at a point $\boldsymbol{x}$, is defined as the proportion of the area of the unit sphere (centred at point $\boldsymbol{x}$) that is covered by the projection $\mathcal{P}'$ of the patch $\mathcal{P}$ onto the unit sphere. See Fig. 10.17, which shows how the solid angle $\sigma(\boldsymbol{\omega}_x)$ is computed. In a sphere of radius r, the solid angle subtended by a spherical area a is a/r^2. Solid angles are measured in *steradians* (sr), i.e. radians squared. The steradian derives its name from the Greek word "stereos" (meaning solid). A sphere has a total of $4\pi r$ steradians.

The solid angle $\sigma(\boldsymbol{\omega}_x)$ can be computed as

$$\sigma(\boldsymbol{\omega}_x) = \frac{A(\mathcal{P})}{r^2} = \frac{A(\mathcal{P}')}{1^2} \tag{10.35}$$

Now starting with the direction vector $\boldsymbol{\omega}_x$ with spherical coordinates $(\theta_\omega, \phi_\omega)$ we define a differential patch on a unit sphere, which subtends angles $\partial\theta_\omega$ in the longitudinal direction and $\partial\phi_\omega$in the latitudinal direction (as shown in Fig. 10.18).

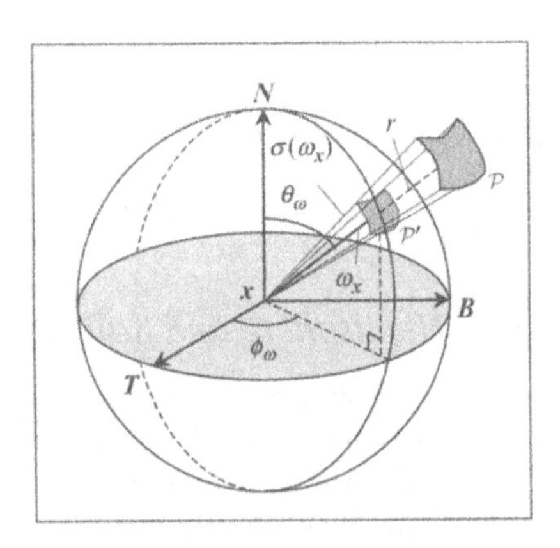

FIGURE 10.17. The definition of a solid angle $\sigma(\boldsymbol{\omega}_x)$.

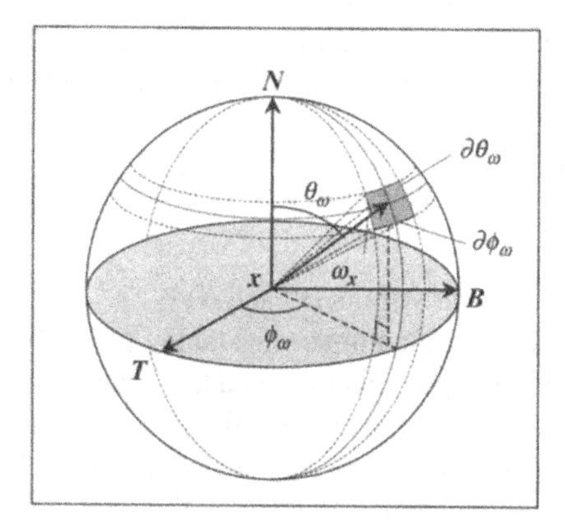

FIGURE 10.18. The definition of a differential solid angle $\partial\sigma(\boldsymbol{\omega}_x)$.

The length of the longitudinal arc of this patch is $(1\cdot\partial\theta_\omega)$, as the longitudinal angle changes from $\theta_\omega - \partial\theta_\omega/2$ to $\theta_\omega + \partial\theta_\omega/2$, and the length of the latitudinal arc of this patch is $\sin(\theta_\omega)\cdot\partial\phi_\omega$, as the latitudinal angle changes from $\phi_\omega - \partial\phi_\omega/2$ to $\phi_\omega + \partial\phi_\omega/2$.

Thus, the area of this differential spherical patch is

$$\partial A_x = (1\cdot\partial\theta_\omega)\cdot[\sin(\theta_\omega)\cdot\partial\phi_\omega] = \sin(\theta_\omega)\cdot\partial\theta_\omega\cdot\partial\phi_\omega \tag{10.36}$$

and the differential solid angle $\partial\sigma(\boldsymbol{\omega}_x)$ of this differential spherical patch is

$$\partial\sigma(\boldsymbol{\omega}_x) = \frac{\partial A_x}{1^2} = \sin(\theta_\omega)\cdot\partial\theta_\omega\cdot\partial\phi_\omega \tag{10.37}$$

since the distance of the spherical patch from the origin of the unit sphere is 1.

It is very convenient to think of the differential solid angle $\partial\sigma(\boldsymbol{\omega}_x)$ as a vector, $\partial\boldsymbol{\omega}_x$. Where the vector $\partial\boldsymbol{\omega}_x$ points to a point on the unit sphere with spherical coordinates $(\theta_\omega, \phi_\omega)$ and its length is equal to the size of the differential solid angle $\partial\sigma(\boldsymbol{\omega}_x)$ in the direction of the vector $\boldsymbol{\omega}_x$.

As we have seen in Section 10.4.1, given the spherical coordinates of a vector $\boldsymbol{\omega}_x$ relative to the local surface Cartesian frame $\{\boldsymbol{T}, \boldsymbol{B}, \boldsymbol{N}, \boldsymbol{x}\}$ we can compute its direction by combining Eqs. (10.23) and (10.24). Thus,

$$\boldsymbol{\omega}_x = \sin(\theta_\omega)\cdot\cos(\phi_\omega)\cdot\boldsymbol{T} + \sin(\theta_\omega)\cdot\sin(\phi_\omega)\cdot\boldsymbol{B} + \cos(\theta_\omega)\cdot\boldsymbol{N} \tag{10.38}$$

To recapitulate, given a set of general directions $\mathcal{D} \subseteq \mathcal{S}^2$ (distributed about the direction vector $\boldsymbol{\omega}_x$), the solid angle occupied by $\mathcal{D}$ is denoted by $\sigma(\mathcal{D})$. Similarly, at a point $\boldsymbol{x}$ the solid angle subtended by a surface patch $\mathcal{P}$ is computed by projecting the surface patch $\mathcal{P}$ onto a unit sphere (centred at point $\boldsymbol{x}$) and determining the area of the sphere corresponding to the resulting set of projected directions.

Another useful concept is that of the *projected solid angle*, introduced by Nicodemus et al. [Nicodemus 77]. Given a point $x \in \mathcal{M}$, given that $\boldsymbol{N}$ is the unit normal at point $\boldsymbol{x}$ and given a set of general directions $\mathcal{D} \subseteq \mathcal{S}^2$ (distributed about the direction vector $\boldsymbol{\omega}_x$), then the projected solid angle is defined as

$$\sigma^{\perp}(\mathcal{D}) = \int_{\mathcal{D}} |\boldsymbol{\omega}_x \odot \boldsymbol{N}| \cdot \partial\sigma(\boldsymbol{\omega}_x) = \int_{\mathcal{D}} \cos(\theta_\omega) \cdot \partial\sigma(\omega_x) \tag{10.39}$$

where θ_ω is the angle between vectors $\boldsymbol{\omega}_x$ and $\boldsymbol{N}$.

In order to describe the illumination events taking place above or below the interface surface between two participating media we introduce the concept of the *illuminating hemisphere*, which for computational convenience is assumed to be a unit hemisphere. The plane defined by the tangent vector $\boldsymbol{T}$ and the bitangent vector $\boldsymbol{B}$ is known as the *tangent plane* and divides the set of all directions $\mathcal{S}^2$ into two hemispheres. The *upper illuminating hemisphere* $\mathcal{H}_+^2$ and the *lower illuminating hemisphere* $\mathcal{H}_-^2$. The upper illuminating hemisphere, which lies above the tangent plane, is defined as

$$\mathcal{H}_+^2(x) = \left\{\boldsymbol{\omega}_x \in \mathcal{S}^2 | (\boldsymbol{\omega}_x \odot \boldsymbol{N}) > 0\right\} \tag{10.40}$$

The lower illuminating hemisphere, which lies below the tangent plane, is defined as

$$\mathcal{H}_-^2(x) = \left\{\boldsymbol{\omega}_x \in \mathcal{S}^2 | (\boldsymbol{\omega}_x \odot \boldsymbol{N}) < 0\right\} \tag{10.41}$$

The upper illuminating hemisphere is used in the computation of the reflection of light and the lower illuminating hemisphere is used in the computation of the transmission of light.

Illumination events, such as light area sources or reflecting surfaces, are projected onto the illuminating hemispheres and for each of these we compute a solid angle $\mathcal{D}\boldsymbol{\omega}_x$. See Fig. 10.19.

The solid angle of an entire illuminating hemisphere (denoted by the name of the hemisphere $\mathcal{H}_+^2$ or $\mathcal{H}_-^2$, or sometimes by Ω – the capital Greek letter *omega*) is half the surface area of the unit sphere. Thus,

$$\sigma\left(\mathcal{H}_+^2\right) = 2\pi \tag{10.42}$$

Given a set of directions $\mathcal{D}$ contained in a single hemisphere, the projected solid angle can be found by first projecting orthographically the set of directions $\mathcal{D}$ onto the tangent plane and then computing the area of this projected region. For instance if $\mathcal{D}$ is equal to the entire upper illuminating hemisphere $\mathcal{H}_+^2$, then the corresponding projection region is a disc with area

$$\sigma^{\perp}(\mathcal{D}) = \pi \tag{10.43}$$

See also Fig. 10.24 and Eq. (10.64) in the next section.

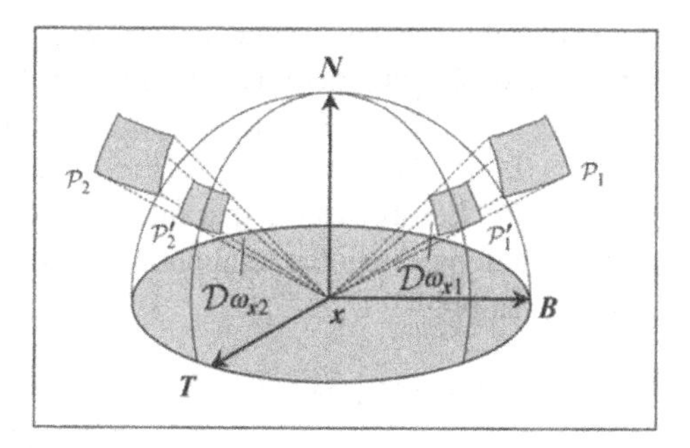

FIGURE 10.19. The definition of a number of solid angles.

10.5 Some Basic Terminology of Lighting

As we have seen before, light sources radiate energy in the form of photons. The energy of a photon, whose wavelength is λ, is denoted by e_λ and is given as

$$e_\lambda = \frac{h \cdot c}{\lambda} \tag{10.44}$$

where h is Planck's constant and c is the speed of light.

The energy of a beam of n photons of wavelength λ is known as the *spectral radiant energy* Q_λ and is defined as

$$Q_\lambda = n \cdot e_\lambda = n \cdot \frac{h \cdot c}{\lambda} \tag{10.45}$$

The energy of a collection of photons over the entire visible spectrum is known as the *radiant energy* Q and is defined as the sum or integral of spectral radiant energy over the entire wavelength spectrum. Thus,

$$Q = \int_0^\infty Q_\lambda \partial\lambda \tag{10.46}$$

The radiant energies Q_λ and Q are measured in Joules per nano-meter ($J \cdot nm^{-1}$) and Joules (J), respectively.

The rate of flow of the radiant energy Q over time t is known as the *radiant power*, *radiant flux* or simply *flux* and is given by

$$\Phi = \frac{\partial Q}{\partial t} \tag{10.47}$$

Flux is the radiant energy flowing through a surface per unit time and is measured in Watts (W), where watts are joules per second.

Similarly, the rate of flow of spectral radiant energy Q_λ over time t is called the *spectral radiant flux* or simply *spectral flux* and is given by

$$\Phi_\lambda = \frac{\partial Q_\lambda}{\partial t} \tag{10.48}$$

The propagation (flow) of light in an environment is best explained using *transport theory*. Transport theory deals with the transport of abstract particles devoid of physical meaning [Duderstadt 79]. In transport theory, the flow of light is a differential quantity and the best way to visualise it is to think of it in terms of particles of light (i.e. photons). Let us assume that in a unit volume V, centred at a point $\boldsymbol{x}$, there are $p(\boldsymbol{x})$ photons. Then the total number of photons in a small differential volume ∂V is given by:

$$P(\boldsymbol{x}) = p(\boldsymbol{x}) \cdot \partial V \tag{10.49}$$

Next, let us assume that all photons (in this volume) travel in the same direction with a velocity vector $\boldsymbol{v}$ and that we wish to count the total number of photons that flow through a small differential surface patch $\mathcal{P}$ with area ∂A. To count these photons we first construct a prism, with differential volume ∂V, by sweeping the *differential area* ∂A in the direction from which the photons are coming (i.e. in the direction $-\boldsymbol{v}$) by a small distance $|\boldsymbol{v} \cdot \partial t|$, where ∂t is a small time interval). Now, the photons that lie inside the differential volume ∂V, between times t and $t + \partial t$, will flow through the differential area ∂A. Thus,

$$\begin{aligned} P(\boldsymbol{x}) &= p(\boldsymbol{x}) \cdot \partial V \\ &= p(\boldsymbol{x}) \cdot \cos\theta \cdot |\boldsymbol{v} \cdot \partial t| \cdot \partial A \\ \therefore \quad \partial V &= \cos\theta \cdot |\boldsymbol{v} \cdot \partial t| \cdot \partial A \end{aligned} \tag{10.50}$$

As we can see from Fig. 10.20, the vector $\boldsymbol{N}$ represents the unit normal to the small differential surface patch $\mathcal{P}$ with area ∂A, θ is the angle between the patch normal and the direction of the sweep $-\boldsymbol{v} \cdot \partial t$ and $\cos\theta \cdot |\boldsymbol{v} \cdot \partial t|$ is the length of the projection of the vector $-\boldsymbol{v} \cdot \partial t$ onto vector $\boldsymbol{N}$.

Observe that the maximum flow of photons through the differential surface patch occurs when the patch is perpendicular to the direction of flow of the photons.

In the more general case, not all the photons passing through the small deferential area ∂A_x, at point $\boldsymbol{x}$, will be flowing in the same direction and the length

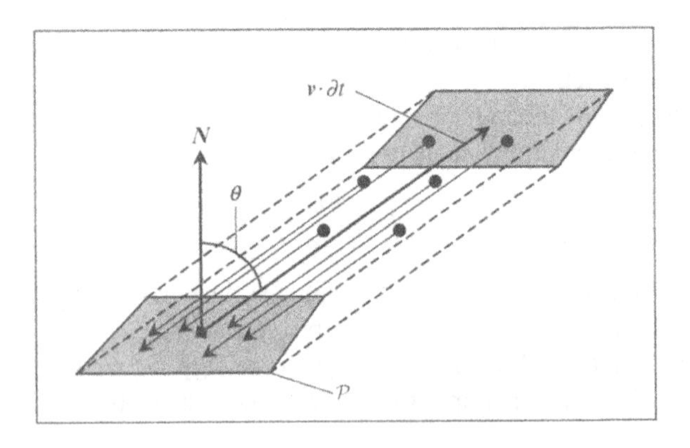

FIGURE 10.20. Sweeping a differential volume ∂V.

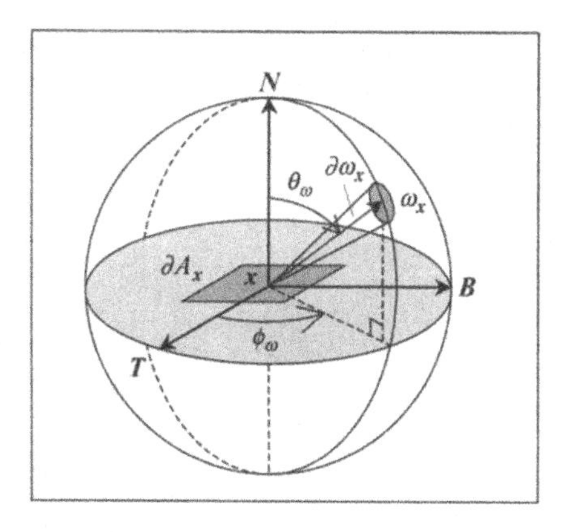

FIGURE 10.21. The differential solid angle $\partial\omega_x$.

of the side of the above prism will form a small *differential solid angle* $\partial\omega_x$ of different directions about the direction vector ω_x. See Fig. 10.21.

In this case, the number of photons flowing through the differential surface area ∂A_x is given by

$$P(\boldsymbol{x}, \omega_x) = p(\boldsymbol{x}, \omega_x) \cdot \cos(\theta_\omega) \cdot \partial\omega_x \cdot \partial A_x \tag{10.51}$$

and the volume of the prism is $\cos(\theta_\omega) \cdot \partial\omega_x \cdot \partial A_x$. As the direction vector ω_x is a unit vector, it can be represented by a point on a unit sphere with spherical coordinates $(\theta_\omega, \phi_\omega)$, where θ_ω is the *altitude* or *zenith* angle and ϕ_ω is the *azimuth* angle.

As we have seen in Section 10.4.4, the differential solid angle $\partial\omega_x$ is given by

$$\partial\omega_x = sin(\theta_\omega) \cdot \partial\theta_\omega \cdot \partial\phi_\omega \tag{10.52}$$

The differential solid angle represents both the size and the direction of a beam of photons; i.e. it represents both a direction and an infinitesimal area on the unit sphere. Thus θ_ω is the angle between the unit normal $\boldsymbol{N}$ of the differential surface and the general direction ω_x of the beam of photons and ϕ_ω is the angle between the projection of vector ω_x onto the tangent plane and the $\boldsymbol{T}$ axis of the local surface Cartesian frame $\{\boldsymbol{T}, \boldsymbol{B}, \boldsymbol{N}, \boldsymbol{x}\}$.

The *differential radiant flux* or *differential radiant power* $\partial\Phi(\boldsymbol{x})$ arriving at a differential area ∂A_x (at a point $\boldsymbol{x}$ on a surface) is known as the *irradiance* $E(\boldsymbol{x})$, which is defined as

$$E(\boldsymbol{x}) = \frac{\partial\Phi(\boldsymbol{x})}{\partial A_x} \tag{10.53}$$

This quantity, which is the unit energy falling on a unit area surface, is measured in Watts per square meter ($W \cdot m^{-2}$) and corresponds to the photometric quantity of *illuminance* (see Fig. 10.22a).

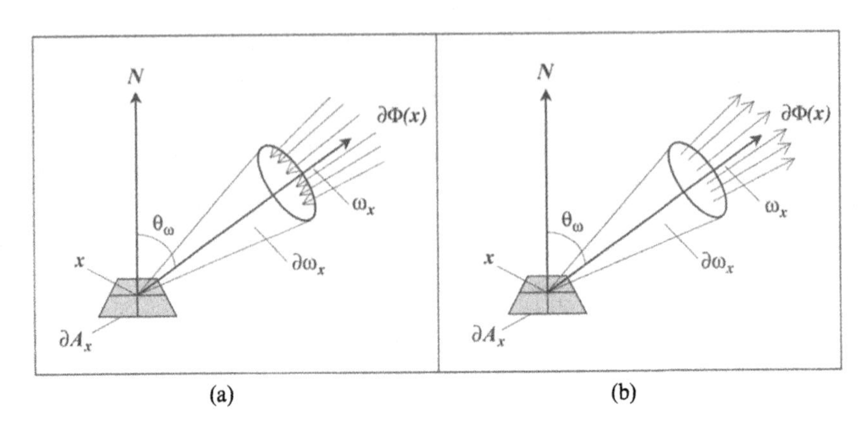

FIGURE 10.22. (a) Irradiance (Illuminance) arriving at ∂A_x. (b) Radiosity (Luminosity) leaving from ∂A_x.

The differential radiant power $\partial\Phi(\boldsymbol{x})$ departing from a differential area ∂A_x (at a point $\boldsymbol{x}$ on a surface) is known as the *radiant exitance* $M(\boldsymbol{x})$ or the *radiosity* $B(\boldsymbol{x})$, which are defined as

$$M(\boldsymbol{x}) = B(\boldsymbol{x}) = \frac{\partial\Phi(\boldsymbol{x})}{\partial A_x} \tag{10.54}$$

This quantity is also measured in Watts per square meter ($\mathrm{W}{\cdot}\mathrm{m}^{-2}$). Its photometric equivalent is *luminosity* (see Fig. 10.22b).

The differential radiant power $\partial\Phi(\boldsymbol{x})$ arriving from a differential solid angle $\partial\omega_x$ at a given point $\boldsymbol{x}$ is known as the *radiant intensity* or *luminous intensity*, which is given as

$$I(\omega_x) = \frac{\partial\Phi(\boldsymbol{x})}{\partial\omega_x} \tag{10.55}$$

This quantity, which is the unit power per unit solid angle, is measured in Watts per steradians ($\mathrm{W}{\cdot}\mathrm{sr}^{-1}$).

Finally, the *radiance* $L(\boldsymbol{x}, \omega_x)$ for a given point $\boldsymbol{x}$ and direction ω_x is defined as the differential radiant power $\partial\Phi(\boldsymbol{x})$ radiated per unit projected area and per unit time from point $\boldsymbol{x}$ in the direction ω_x. The radiance is given by

$$L(\boldsymbol{x}, \omega_x) = \frac{\partial E(\boldsymbol{x})}{\partial\omega_x} = \frac{\partial^2\Phi(\boldsymbol{x})}{\partial A_x^{\perp}{\cdot}\partial\omega_x} = \frac{\partial^2\Phi(\boldsymbol{x})}{\cos(\theta_\omega){\cdot}\partial A_x{\cdot}\partial\omega_x} = \frac{\partial \mathrm{I}(\boldsymbol{x})}{\cos(\theta_\omega){\cdot}\partial A_x} \tag{10.56}$$

where $\partial A_x^{\perp}$ is the *differential projected area* (i.e. the differential area around a point $\boldsymbol{x}$ which is perpendicular to the direction ω_x), $\partial\omega_x$ is the differential solid angle around the direction ω_x and $\partial^2\Phi(\boldsymbol{x})$ is the differential radiant power through the differential surface area $\partial A_x^{\perp}$ and solid angle $\partial\omega_x$. In other words, to measure the radiance at (x, ω_x), we determine how many photons pass through a differential area $\partial A_x^{\perp}$ that is perpendicular to the direction ω_x in each unit of time. These

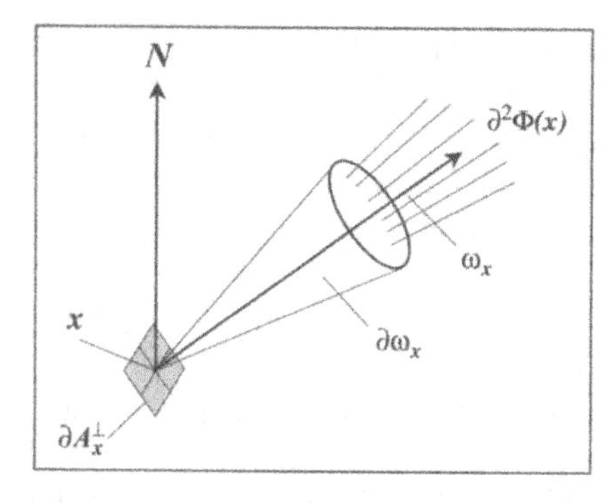

FIGURE 10.23. The differential projected area $\partial A_x^{\perp}$.

photons have trajectories that are contained in a differential solid angle $\partial\omega_x$ surrounding the direction ω_x. Thus radiance is defined to be the limiting ratio of the differential radiant power $\partial\Phi(\boldsymbol{x})$ represented by these photons, arriving at a differential projected area $\partial A_x^{\perp}$ from a solid angle $\partial\omega_x$. See Fig. 10.23. Radiance is measured in Watts per meters squared per steradians ($\mathrm{W{\cdot}m^{-2}{\cdot}sr^{-1}}$).

As radiance is defined for all points of 3D space and for all direction of a 2D directional space, we say that radiance is a 5D function. The radiance just before light arrives at a surface is called the *incoming radiance* or *surface field radiance* $L_i(\boldsymbol{x}, \omega_x)$ and the radiance just after light leaves a surface is called the *outgoing radiance* or *surface radiance* $L_o(\boldsymbol{x}, \omega_x)$. When measuring the radiance leaving the surface patch $\mathcal{P}$, a more convenient form of the equation is

$$L_o(\boldsymbol{x}, \omega_x) = \frac{\partial^2\Phi(\boldsymbol{x})}{|\omega_x \odot \boldsymbol{N}| \cdot \partial A_x \cdot \partial\omega_x} = \frac{\partial^2\Phi(\boldsymbol{x})}{\cos(\theta_\omega) \cdot \partial A_x \cdot \partial\omega_x} \tag{10.57}$$

where A_x is the area measured on the surface patch $\mathcal{P}$. Thus, the relationship between the differential projected area $\partial A_x^{\perp}$ and the ordinary differential area ∂A_x is

$$\partial A_x^{\perp} = |\omega_x \odot \boldsymbol{N}| \cdot \partial A_x = \cos(\theta_\omega) \cdot \partial A_x \tag{10.58}$$

Alternatively, the radiant energy per unit volume is the product of the *photon volume density* by the power of a single photon. Thus,

$$L(\boldsymbol{x}, \omega_x) = \int_0^{\infty} P(\boldsymbol{x}, \omega_x, \lambda) \cdot \frac{h \cdot c}{\lambda} \partial\lambda \tag{10.59}$$

where $P(\boldsymbol{x}, \omega_x, \lambda)$ is the photon volume density at a given wavelength λ.

To recapitulate, incoming radiance or field radiance is a measure of energy per unit time arriving at a small area (centred about a point) from a given direction defined by small solid angle and outgoing radiance or surface radiance is a measure of energy per unit time radiating from a small area (centred about a point) towards a given direction defined by small solid angle. Thus we can surmise that radiance is most important radiometric quantity in physically based image synthesis, as its distribution completely determines the distribution of light in the scene and all other radiometric quantities can be derived from it. For instance, the rate of radiant power (flux) per unit time $\partial\Phi(\boldsymbol{x})$ arriving at a differential area ∂A_x from a differential solid angle $\partial\omega_x$ is given as

$$\partial^2\Phi(\boldsymbol{x}) = L_i(\boldsymbol{x}, \omega_x)\cdot\cos(\theta_\omega)\cdot\partial A_x\cdot\partial\omega_x \tag{10.60}$$

Now if $L_i\,(\boldsymbol{x}, \omega_x)$ is the radiance incident onto a surface with a fixed orientation, we can compute the total energy per unit area arriving at this surface by integrating the incoming radiance over the entire illuminating hemisphere at a point $\boldsymbol{x}$.

$$\partial\Phi(\boldsymbol{x}) = \left(\int_\Omega L_i(\boldsymbol{x}, \omega_x)\cdot\cos(\theta_\omega)\cdot\partial\omega_x\right)\cdot\partial A_x \tag{10.61}$$

As we have seen above the irradiance is given by $E(\boldsymbol{x}) = \dfrac{\partial\Phi(\boldsymbol{x})}{\partial A_x}$, thus:

$$\begin{aligned} E(\boldsymbol{x}) &= \int_\Omega L_i(\boldsymbol{x}, \omega_x)\cdot\cos\,(\theta_\omega)\cdot\partial\omega_x \\ &= \int_0^{2\pi}\int_0^{\pi} L_i(\boldsymbol{x}, \omega_x)\cdot\cos(\theta_\omega)\cdot sin\,(\theta_\omega)\cdot\partial\theta_x\cdot\partial\phi_x \end{aligned} \tag{10.62}$$

where the quantity $\cos(\theta_\omega)\cdot\partial\omega_x$ is known as the *projected solid angle* and can be thought of as the projection of a differential surface which is defined on the illuminating hemisphere and projected onto the base of this hemisphere. See Fig. 10.24. The projected solid angle is often denoted by $\partial\omega_x^\perp$, thus

$$\partial\omega_x^\perp = \cos(\theta_\omega)\cdot\partial\omega_x \tag{10.63}$$

It turns out that the integral of the projected solid angle over the entire illuminating hemisphere reduces to π, since

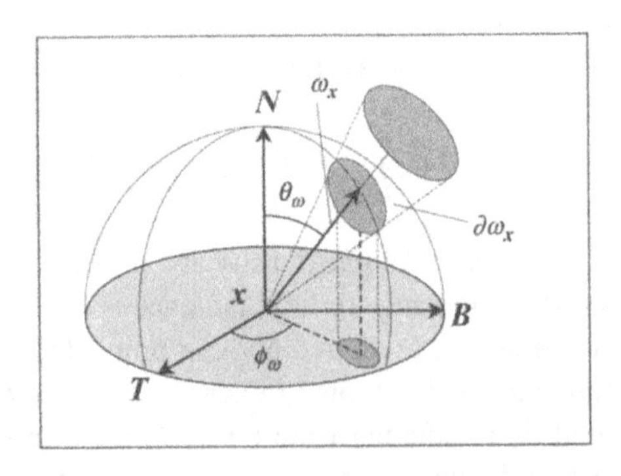

FIGURE 10.24. The projected solid angle $\cos\,(\theta_\omega)\cdot\partial\omega_x$.

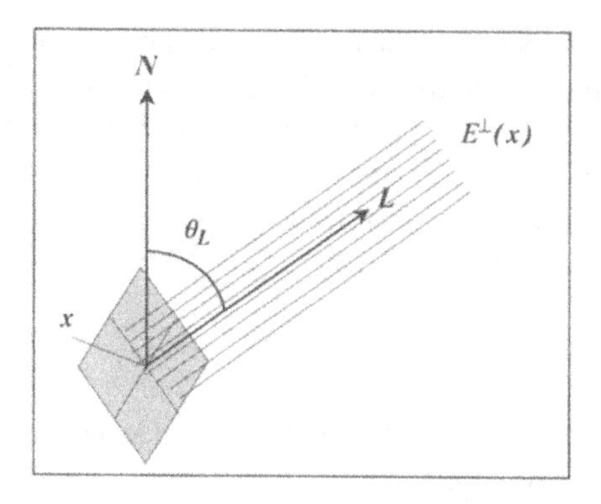

FIGURE 10.25. The projected irradiance $E^{\perp}(\boldsymbol{x})$.

$$\begin{aligned}\int_{\Omega} \cos(\theta_\omega) \cdot \partial\omega_x &= \int_0^{2\pi} \int_0^{\pi} \cos(\theta_\omega) \cdot \sin(\theta_\omega) \cdot \partial\theta_x \cdot \partial\phi_x \\ &= -\int_0^{2\pi} \int_0^{\pi} \cos(\theta_\omega) \cdot \partial(\cos(\theta_\omega)) \cdot \partial\phi_x \\ &= -2\pi \left.\frac{\cos^2(\theta_\omega)}{2}\right|_0^{\pi/2} \\ &= \pi\end{aligned} \tag{10.64}$$

Similarly, the radiosity is given by

$$\begin{aligned}B(\boldsymbol{x}) &= \int_{\Omega} L_o(\boldsymbol{x}, \omega_x) \cdot \cos(\theta_\omega) \cdot \partial\omega_x \\ &= \int_0^{2\pi} \int_0^{\pi} L_o(\boldsymbol{x}, \omega_x) \cdot \cos(\theta_\omega) \cdot sin(\theta_\omega) \cdot \partial\theta_x \cdot \partial\phi_x\end{aligned} \tag{10.65}$$

where $L_o(\boldsymbol{x}, \omega_x)$ is the outgoing radiance or surface radiance.

If a single distant light source irradiates the surface, then all the rays from this source will be parallel and the integral in Eq. (10.62) reduces to a simpler form

$$E(\boldsymbol{x}) = E^{\perp}(\boldsymbol{x}) \cdot \cos(\theta_L) \tag{10.66}$$

where $E^{\perp}(\boldsymbol{x})$ is the *projected irradiance* (i.e. the energy arriving from the light source onto a unit surface which is perpendicular to the rays from the light source) and θ_L is the incidence angle. See Fig. 10.25.

10.6 Light Emission

Although the use of radiance is convenient for characterising the light transport between surface elements, it is not appropriate for the description of the energy emanating from a point light source. It turns out that the radiant intensity I is best suited for this purpose.

Isotropic point light sources emanate energy in the form of photons equally in all directions and their energy distribution expands outwards from the centre.

As we have seen above, radiant power (energy flux) Φ_s is energy per unit time, irradiance (energy density) E_s is energy per unit time per unit area and radiant intensity (brightness) I_s is energy flux per solid angle ω_s through which the source radiates energy. Thus the radiant intensity is given by

$$I_s = \frac{\Phi_s}{\omega_s} \tag{10.67}$$

For instance, consider a point light source situated at a point $\boldsymbol{x}_s$ radiating a total energy of Φ_s (per unit time) equally in all directions. The solid angle through which the source radiates is 4π (i. e. the solid angle of the entire sphere). Now the radiant intensity of the source in any direction is given by

$$I_s = \frac{\Phi_s}{4\pi} \tag{10.68}$$

The intensity (brightness) of the light source is independent of the distance of the viewer from the source, but the irradiance (energy density) reaching the viewer or a surface is dependant on this distance. A sphere of radius r, containing the point source, has a surface area $4\pi r^2$. The energy flux Φ_s (of the light source) radiates equally through all the points of the surface of this sphere and the projected energy density $E_s^{\perp}$ at any point $\boldsymbol{x}$ on this sphere is given by dividing the energy flux by the surface area of the sphere, thus

$$E_s^{\perp}(\boldsymbol{x}) = \frac{\Phi_s}{4\pi r^2} \tag{10.69}$$

The projected incident irradiance $E_s^{\perp}(\boldsymbol{x})$ is the energy received by the surface element of any point $\boldsymbol{x}$ that lies on the surface of a sphere of radius r that is centred at the point light source. This surface element is assumed to lie on a plane that is tangential to this sphere (at point $\boldsymbol{x}$) and to have a normal that points towards the centre of the sphere from where the light radiates. See Fig. 10.26a.

If a point $\boldsymbol{x}$ is situated at a distance r from the light source point $\boldsymbol{x}_s$ and its surface element lies on a surface with a unit normal vector $\boldsymbol{N}$ pointing in a different direction, then the incident irradiance must be scaled by the cosine of the incidence angle of the incoming rays. Thus, the incident irradiance $E_s(\boldsymbol{x})$ is

$$E_s(\boldsymbol{x}) = E_s^{\perp}(\boldsymbol{x}) \cdot (\boldsymbol{N} \odot \boldsymbol{L}) = \frac{\Phi_s}{4\pi r^2} \cdot \cos(\theta_x) \tag{10.70}$$

where $r = |\boldsymbol{x}_s - \boldsymbol{x}|$ is the distance of surface point from the light source point, $\boldsymbol{L}$ is the vector pointing in the direction of the light source and θ_x is the incidence angle of the light rays. See Fig. 10.26b.

Now combining Eqs. (10.68) and (10.70), we can express the incident irradiance $E_s(\boldsymbol{x})$ in terms of the incident intensity I_s as

$$E_s(\boldsymbol{x}) = \frac{I_s}{r^2} \cdot (\boldsymbol{N} \odot \boldsymbol{L}) = \frac{I_s}{r^2} \cdot \cos(\theta_x) \tag{10.71}$$

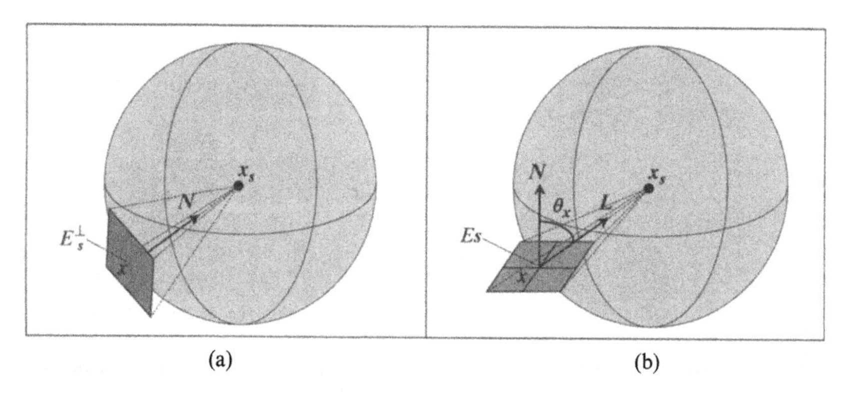

FIGURE 10.26. (a) The projected incident irradiance $E_s^{\perp}(\boldsymbol{x})$. (b) The incident irradiance $E_s(\boldsymbol{x})$.

10.7 The Scattering and Reflection Functions

As we have seen previously, when a light beam strikes an interface surface between two participating media it can undergo reflection, refraction followed by transmission, or subsurface scattering followed by reflection. In this section we will examine the scattering and reflection functions that model these phenomena.

Let us start by examining what happens to a single photon when it strikes an interface surface. Figure 10.27a depicts a photon striking the interface surface and undergoing reflection, while Fig. 10.27b depicts a photon striking the interface surface and undergoing subsurface scattering followed by reflection.

As we can see from the latter diagram, a light beam may strike the surface at one point, undergo subsurface scattering and leave the surface from another point. For this reason, we momentarily drop any reference to points from our notation. Thus, the incoming (field) radiance will be denoted by $L_i(\omega_i)$ and the outgoing (surface) radiance will be denoted by $L_r(\omega_r)$. In general the outgoing radiance L_r depends on the radiance arriving at a given point from all directions. To simplify things, we will fix a particular direction ω_i and consider only the incident light arriving at a point from a solid angle $\partial\omega_i$. Let us assume that this light strikes the surface at a point and generates an irradiance $E_i(\omega_i)$ which given by

$$\partial E_i(\omega_i) = L_i(\omega_i) \cdot \partial\omega_i^{\perp} \qquad (10.72)$$

where $\partial\omega_i^{\perp}$ is the projected solid angle in the direction ω_i. This light is subsequently scattered by the surface in all directions.

Let $\partial L_r(\omega_r)$ represent the contribution made to the outgoing radiance in the direction ω_r with a solid angle $\partial\omega_r$. Experiments have shown that the incident light energy $\partial E_i(\omega_i)$ arriving onto the surface is proportional to the reflected radiance $L_r(\omega_r)$ [Clarke 85]. Thus,

$$\partial L_r(\omega_r) \propto \partial E_i\,(\omega_i) \qquad (10.73)$$

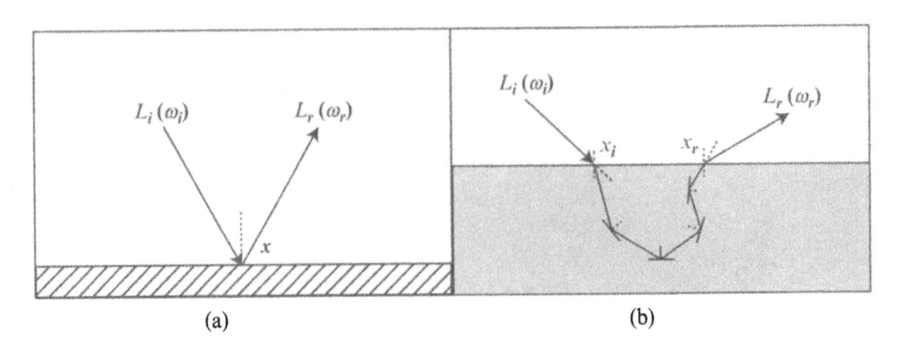

FIGURE 10.27. (a) The reflection of a photon. (b) The scattering and reflection of a photon.

This relationship means that an increase in incident irradiance results in a corresponding increase in the reflected radiance. An increase in the incident irradiance can be achieved either by increasing the solid angle subtended by the light source (i.e. by bringing the source closer to the surface) or by increasing the energy destiny of the light beam (i.e. by increasing the intensity of the light source).

It should be equally apparent that the reflected radiance is proportional to the incident radiant flux, thus

$$\partial L_r(\omega_r) \propto \partial \Phi_i(\omega_i) \tag{10.74}$$

10.7.1 Bi-directional Scattering Surface Reflectance Distribution Function (BSSRDF)

When subsurface scattering followed by reflection takes place, a photon that strikes an interface surface enters the second participating medium is scattered around by various obstacles and finally re-emerges from the surface. The scattered photon can even re-emerge from a different surface of the object being lit, as shown in Fig. 10.28. This phenomenon is most noticeable in translucent materials such as skin, leafs or marble and is examined in detail in a monograph by Nicodemus et al. [Nicodemus 77].

This form of scattering and reflection can be modelled using a BSSRDF function f_s which represents the constant of proportionality of the differential reflected radiance $\partial L_r(\omega_r)$ (leaving from a point $\boldsymbol{x}_r$ in a direction ω_r) and the differential incident energy flux $\partial \Phi_i(\omega_i)$ (arriving at a point $\boldsymbol{x}_i$ from a direction ω_i). Thus this function, which is the most comprehensive model of light transport, is given by

$$f_s(\boldsymbol{x}_i, \boldsymbol{x}_r, \omega_i, \omega_r) = \frac{\partial L_r(\boldsymbol{x}_r, \omega_r)}{\partial \Phi_i(\boldsymbol{x}_i, \omega_i)} \tag{10.75}$$

where $\boldsymbol{x}_i$ and $\boldsymbol{x}_r$ are the points on the surface where the incident and the reflected rays enter and leave the surface respectively, ω_i and ω_r are the direction vectors of these rays, $\partial \Phi_i(\omega_i)$ is the incident differential energy flux and $\partial L_r(\omega_r)$ is the

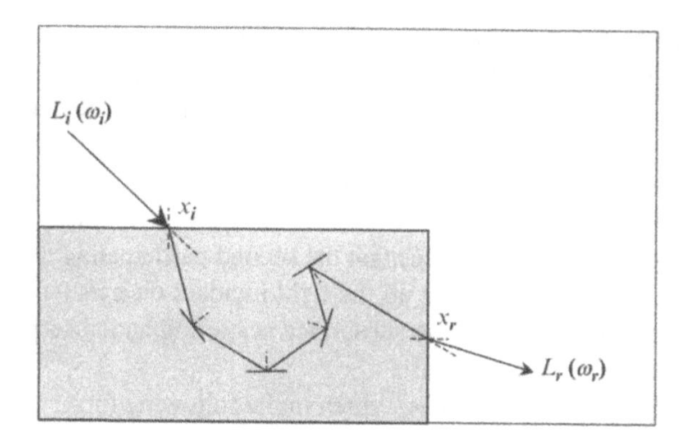

FIGURE 10.28. Subsurface scattering followed by reflection.

reflected differential radiance. The BSSRDF is measured in dimensionless units per meters squared per steradians ($\mathrm{m}^{-2} \cdot \mathrm{sr}^{-1}$).

This model of reflectance assumes that the second participating medium is non-homogenous. Here it is assumed that some of the light incident on a material is directly reflected off its surface, some is transmitted and scattered inside the material before re-emerging from a different point of its surface and some is absorbed by the material and stored as heat.

The BSSRDF is a very expensive function to evaluate and has only been used by very few researchers in computer graphics, mainly to compute subsurface scattering followed by reflection, [Dorsey 99], [Hanrahan 93], [Jensen 99], [Jensen 01a], [Pharr 00].

As we shall see in the next section a BSSRDF can be expressed as a collection of two BRDFs and two BTDFs.

10.7.2 *Bi-directional Reflectance Distribution Function (BRDF)*

In the simpler case of reflection where the point of incidence and point of reflection coincide, a much simpler function (known as a *Bi-directional Reflectance Distribution Function*) was introduced by Nicodemus et al. [Nicodemus 77] to represent this form of reflection. The BRDF function f_r represents the constant of proportionality of the differential reflected radiance $\partial L_r(\boldsymbol{x}, \omega_r)$ (leaving from a point $\boldsymbol{x}$ in a direction ω_r) and the differential incident irradiance $\partial E_i(\boldsymbol{x}, \omega_i)$ (arriving at a point $\boldsymbol{x}$ from a direction ω_i). Thus this function is given by

$$\begin{aligned} f_r(\boldsymbol{x}, \omega_i, \omega_r) &= \frac{\partial L_r(\boldsymbol{x}, \omega_r)}{\partial E_i(\boldsymbol{x}, \omega_i)} = \frac{\partial L_r(\boldsymbol{x}, \omega_r)}{L_i(\boldsymbol{x}, \omega_i) \cdot \partial \omega_i^{\perp}} = \frac{\partial L_r(\boldsymbol{x}, \omega_r)}{L_i(\boldsymbol{x}, \omega_i) \cdot \cos(\theta_i) \cdot \partial \omega_i} \\ &= \frac{\partial L_r(\boldsymbol{x}, \omega_r)}{L_i(\boldsymbol{x}, \omega_i) \cdot (\boldsymbol{L} \odot \boldsymbol{N}) \cdot \partial \omega_i} \end{aligned} \tag{10.76}$$

where $\boldsymbol{x}$ is the point on the surface, ω_i and ω_r are the direction vectors of the incident and the reflected rays, $\partial\omega_i$ and $\partial\omega_r$ are the solid angles in the directions ω_i and ω_r, θ_i is the incidence angle, $\partial E_i(\boldsymbol{x}, \omega_i)$ is the differential incident irradiance, $L_i(\boldsymbol{x}, \omega_i)$ is the incident radiance and $\partial L_r(\boldsymbol{x}, \omega_r)$ is the differential reflected radiance. See Fig. 10.29. The BRDF function is measured in dimensionless units per steradians (sr^{-1}).

This model of reflectance assumes that the second participating medium is homogenous. Here it is assumed that all the light incident on a material is directly reflected off its surface and that the subsurface scattering and absorption characteristics of the material are ignored.

The BRDF function is sometimes written in the following form:

$$\begin{aligned} f_r(\boldsymbol{x}, \omega_i \to \omega_r) &= \frac{\partial L(\boldsymbol{x} \to \omega_r)}{\partial E(\boldsymbol{x} \leftarrow \omega_i)} = \frac{\partial L(\boldsymbol{x} \to \omega_r)}{L(\boldsymbol{x} \leftarrow \omega_i) \cdot \cos(\theta_i) \cdot \partial\omega_i} \\ &= \frac{\partial L(\boldsymbol{x} \to \omega_r)}{L(\boldsymbol{x} \leftarrow \omega_i) \cdot (\boldsymbol{L} \odot \boldsymbol{N}) \cdot \partial\omega_i} \end{aligned} \tag{10.77}$$

where the arrows indicate the direction in which the light travels (.i.e. $\to$ meaning outgoing and $\leftarrow$ meaning incoming light).

The BRDF can also be expressed in terms of the zenith and azimuth angles of the incident and the reflected ray directions.

$$\begin{aligned} f_r(\boldsymbol{x}, \theta_i, \phi_i, \theta_r, \phi_r) &= \frac{\partial L_r(\boldsymbol{x}, \theta_r, \phi_r)}{\partial E_i(\boldsymbol{x}, \theta_i, \phi_i)} = \frac{\partial L_r(\boldsymbol{x}, \theta_r, \phi_r)}{L_i(\boldsymbol{x}, \theta_i, \phi_i) \cdot \cos(\theta_i) \cdot \partial\omega_i} \\ &= \frac{\partial L_r(\boldsymbol{x}, \theta_r, \phi_r)}{L_i(\boldsymbol{x}, \theta_i, \phi_i) \cdot (\boldsymbol{L} \odot \boldsymbol{N}) \cdot \partial\omega_i} \end{aligned} \tag{10.78}$$

As we have seen in Section 10.4, surfaces can be classified as being isotropic or anisotropic. With isotropic surfaces, the reflection remains unchanged when the surface is rotated about its normal while the incoming irradiance and outgoing radiance are left unchanged. Anisotropic surfaces, on the other hand, reflect light differently when rotated about their normal. Such surfaces have micro-facets that are strongly oriented. This orientation of their micro-facets causes light to reflect differently in different directions, with some directions being preferred. The observer does not see these micro-facets but only perceives their effect on the reflected light. Anisotropic surfaces include brushed and brandished metal surfaces, cloth, feathers, fur, and hair. Most natural and man-made materials have isotropic surfaces with randomly distributed micro-facets.

When dealing with isotropic surfaces a simplified version of the BRDF, known as an *isotropic BRDF*, can be used.

$$\begin{aligned} f_{i,r}(\boldsymbol{x}, \theta_i, \theta_r, \phi) &= \frac{\partial L_r(\boldsymbol{x}, \theta_r, \phi)}{\partial E_i(\boldsymbol{x}, \theta_i, \phi)} = \frac{\partial L_r(\boldsymbol{x}, \theta_r, \phi)}{L_i(\boldsymbol{x}, \theta_i, \phi) \cdot \cos(\theta_i) \cdot \partial\omega_i} \\ &= \frac{\partial L_r(\boldsymbol{x}, \theta_r, \phi)}{L_i(\boldsymbol{x}, \theta_i, \phi) \cdot (\boldsymbol{L} \odot \boldsymbol{N}) \cdot \partial\omega_i} \end{aligned} \tag{10.79}$$

where ϕ is the difference of the azimuth angles of the reflected and the incident rays (i.e. $\phi = \phi_r - \phi_i$).

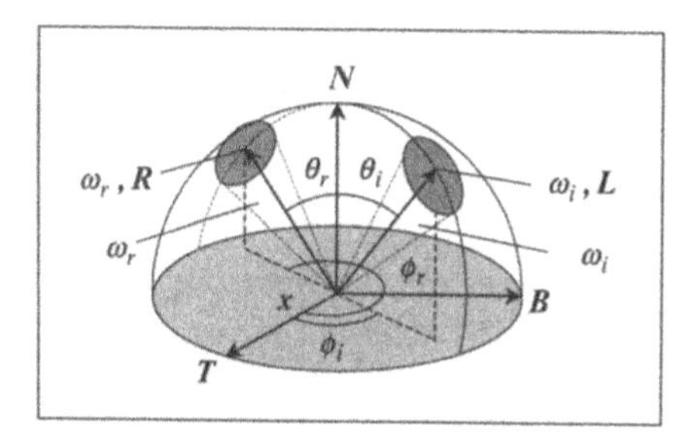

FIGURE 10.29. The BRDF function.

A corresponding function f_t, known as the *Bi-directional Transmission Distribution Function* (BTDF), can be constructed to deal with the refraction and transmission of light passing through the interface surface between two participating media with differing refraction coefficients.

The BTDF function f_t represents the constant of proportionality of the differential transmitted radiance $\partial L_t(\boldsymbol{x}, \omega_t)$ (leaving from a point $\boldsymbol{x}$ in a direction ω_t) and the differential incident irradiance $\partial E_i(\boldsymbol{x}, \omega_i)$ (arriving at a point $\boldsymbol{x}$ from a direction ω_i). Thus this function is given by

$$\begin{aligned} f_t(\boldsymbol{x}, \omega_i, \omega_t) &= \frac{\partial L_t(\boldsymbol{x}, \omega_t)}{\partial E_i(\boldsymbol{x}, \omega_i)} = \frac{\partial L_t(\boldsymbol{x}, \omega_t)}{L_i(\boldsymbol{x}, \omega_i) \cdot \partial \omega_i^{\perp}} = \frac{\partial L_t(\boldsymbol{x}, \omega_t)}{L_i(\boldsymbol{x}, \omega_i) \cdot \cos(\theta_i) \cdot \partial \omega_i} \\ &= \frac{\partial L_t(\boldsymbol{x}, \omega_t)}{L_i(\boldsymbol{x}, \omega_i) \cdot (\boldsymbol{L} \odot \boldsymbol{N}) \cdot \partial \omega_i} \end{aligned} \tag{10.80}$$

where $\boldsymbol{x}$ is the point on the surface, ω_i and ω_t are the direction vectors of the incident and the transmitted rays, $\partial\omega_i$ and $\partial\omega_t$ are the solid angles in the directions ω_i and ω_t, θ_i is the incidence angle, $\partial E_i(\boldsymbol{x}, \omega_i)$ is the differential incident irradiance, $L_i(\boldsymbol{x}, \omega_i)$ is the incident radiance and $\partial L_t(\boldsymbol{x}, \omega_t)$ is the differential transmitted radiance.

In its most general form a BRDF can be seen as a function whose domain is the Cartesian product of the *incident illuminating hemisphere* $\mathcal{H}_i^2$ and the *reflected illuminating hemisphere* $\mathcal{H}_r^2$ and whose co-domain is the set real numbers $\mathbb{R}$, thus

$$f_r : \mathcal{H}_i^2 \times \mathcal{H}_r^2 \to \mathbb{R} \tag{10.81}$$

In this case, both illuminating hemispheres refer to the same set of directions, i.e. $\mathcal{H}_i^2 = \mathcal{H}_r^2 = \mathcal{H}_+^2$.

Similarly, a BTDF is defined as

$$f_t : \mathcal{H}_i^2 \times \mathcal{H}_t^2 \to \mathbb{R} \tag{10.82}$$

In this case, the *transmitted illuminating hemisphere* $\mathcal{H}_t^2$ is the complement of the incident illuminating hemisphere $\mathcal{H}_i^2$, i.e. $\mathcal{H}_t^2 = -\mathcal{H}_i^2 = \mathcal{H}_-^2$.

The combination of a BRDF function (defined over $\mathcal{H}_+^2$ at $\boldsymbol{x}_i$), a BTDF function (defined over $\mathcal{H}_-^2$ at $\boldsymbol{x}_i$), a BTDF function (defined using $\mathcal{H}_-^2$ at $\boldsymbol{x}_r$) and a BRDF

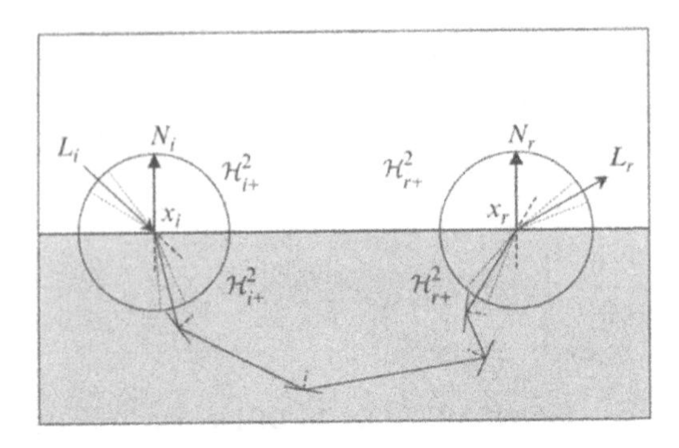

FIGURE 10.30. A Bi-directional scattering distribution function.

function (defined using $\mathcal{H}_+^2$ at $\boldsymbol{x}_r$), is known as a *Bi-directional Scattering Distribution Function* (BSDF). This function is an alternative form of the BSSRDF function examined in the pervious section (see Fig. 10.30).

The BSDF function f_s represents the constant of proportionality of the differential reflected radiance $\partial L_r(\boldsymbol{x}_r, \omega_r)$ (leaving from a point $\boldsymbol{x}_r$ in a direction ω_r) and the differential incident irradiance $\partial E_i(\boldsymbol{x}_i, \omega_i)$ (arriving at a point $\boldsymbol{x}_i$ from a direction ω_i). Thus this function is given by

$$\begin{aligned} f_s(\boldsymbol{x}_i, \boldsymbol{x}_r, \omega_i, \omega_r) &= \frac{\partial L_r(\boldsymbol{x}_r, \omega_r)}{\partial E_i(\boldsymbol{x}_i, \omega_i)} = \frac{\partial L_r(\boldsymbol{x}_r, \omega_r)}{L_i(\boldsymbol{x}_i, \omega_i) \cdot \partial \omega_i^{\perp}} \\ &= \frac{\partial L_r(\boldsymbol{x}_r, \omega_r)}{L_i(\boldsymbol{x}_i, \omega_i) \cdot \cos(\theta_i) \cdot \partial \omega_i} \end{aligned} \tag{10.83}$$

where $\boldsymbol{x}_i$ and $\boldsymbol{x}_r$ are the points on the surface where the incident and the reflected rays enter and leave the surface respectively, ω_i and ω_r are the direction vectors of the incident and the reflected rays, $\partial \omega_i$ and $\partial \omega_r$ are the solid angles in the directions ω_i and ω_r, θ_i is the incidence angle, $\partial E_i(\boldsymbol{x}_i, \omega_i)$ is the differential incident irradiance, $L_i(\boldsymbol{x}_i, \omega_i)$ is the incident radiance and $\partial L_r(\boldsymbol{x}_r, \omega_r)$ is the differential reflected radiance (see Fig. 10.30).

A BSDF can be seen as a function whose domain is the Cartesian product of four illuminating hemispheres and whose co-domain is the set of real numbers $\mathbb{R}$, thus

$$f_s : \mathcal{H}_i^2(\boldsymbol{x}_i) \times \left[-\mathcal{H}_i^2(\boldsymbol{x}_i)\right] \times \left[-\mathcal{H}_r^2(\boldsymbol{x}_r)\right] \times \mathcal{H}_r^2(\boldsymbol{x}_r) \rightarrow \mathbb{R} \tag{10.84}$$

10.7.3 Reflectance, Transmittance and Scattering Equations

Rearranging Eq. (10.76) we get

$$\partial L_r(\boldsymbol{x}, \omega_r) = f_r(\boldsymbol{x}, \omega_i, \omega_r) \cdot L_i(\boldsymbol{x}, \omega_i) \cdot \partial \omega_i^{\perp} \tag{10.85}$$

This is the differential reflected radiance (leaving point $\boldsymbol{x}$ in the direction ω_r) expressed in terms of the incident radiance (arriving at point $\boldsymbol{x}$ from the direction ω_i). To get the total reflected radiance (leaving point $\boldsymbol{x}$ in the direction ω_r) we must sum the incident radiance arriving at point $\boldsymbol{x}$ from all possible directions of the incident illuminating hemisphere $\mathcal{H}_i^2$ (or equivalently Ω_i). Thus,

$$L_r(\boldsymbol{x}, \omega_r) = \int_{\mathcal{H}_i^2} f_r(\boldsymbol{x}, \omega_i, \omega_r) \cdot L_i(\boldsymbol{x}, \omega_i) \cdot \partial\omega_i^{\perp} \tag{10.86}$$

This equation is known as the *surface reflectance equation* and is used to predict the appearance of a surface given the incident illumination. Observe that here we integrate the incoming radiance over the incident illuminating hemisphere.

Similarly, starting from Eqs.(10.80) and (10.83), we can develop the *surface transmittance equation* and the *surface scattering equation* respectively.

Thus, the surface transmittance equation is given by

$$L_t(\boldsymbol{x}, \omega_t) = \int_{\mathcal{H}_i^2} f_t(\boldsymbol{x}, \omega_i, \omega_t) \cdot L_i(\boldsymbol{x}, \omega_i) \cdot \partial\omega_i^{\perp} \tag{10.87}$$

10.7.4 Properties of the BRDFs

Despite the fact that the co-domain of BRDFs is the real number set, BRDFs are not arbitrary functions and have to satisfy a number of constraints in order to be physically correct or at least physically plausible.

10.7.4.1 Non-Negativity Property

BRDFs can only assume non-negative values in the half-open interval $[0, \infty)$. From Eqn. (10.76) it is self-evident that the BRDF $f_r(\boldsymbol{x}, \omega_i, \omega_r) = \partial L_r(\boldsymbol{x}, \omega_r) / \partial E_i(\boldsymbol{x}, \omega_i)$ is non-negative, as both the numerator of this fraction (i.e. the differential reflected radiance) and its denominator (i.e. the differential incident irradiance) are non-negative. Thus,

$$f_r(\boldsymbol{x}, \omega_i, \omega_r) \geq 0 \quad \forall\omega_i \in \mathcal{H}_i^2 \wedge \forall\omega_r \in \mathcal{H}_r^2 \tag{10.88}$$

10.7.4.2 Symmetry Property or the Helmholtz Reciprocity Property

Helmholtz's law of reciprocity states that the reflective properties of a surface do not depend on the direction in which the light travels. This means that reversing the direction of the incident and the reflected light has no effect on the value of the BRDF.

Thus,

$$f_r(\boldsymbol{x}, \omega_i, \omega_r) = f_r(\boldsymbol{x}, \omega_r, \omega_i) \quad \forall\omega_i \in \mathcal{H}_i^2 \wedge \forall\omega_r \in \mathcal{H}_r^2 \tag{10.89}$$

Figure 10.31 illustrates the reciprocity property. By convention all directions are defined as pointing away from point $\boldsymbol{x}$. Observe that on the left-hand diagram of the figure, light travels in the directions $-\omega_i$ and $+\omega_r$ with solid angles

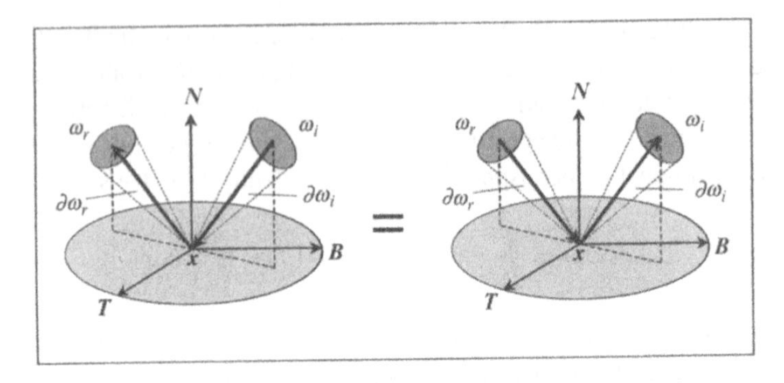

FIGURE 10.31. Helmholtz's law of reciprocity.

$-\partial\omega_i$ and $+\partial\omega_r$, respectively, while on the right-hand diagram light travels in the directions $-\omega_r$ and $+\omega_i$ with solid angles $-\partial\omega_r$ and $+\partial\omega_i$, respectively.

A more detailed discussion of the reciprocity property can be found in [Clarke 85] and [Veach 97a].

10.7.4.3 Energy Conservation Property

Physically correct BRDFs must be energy conserving, which means that for any incident direction the power reflected over the entire reflected illuminating hemisphere cannot exceed the incident power. Similarly, the power reflected from a given point, in any particular direction, does not exceed the total power arriving at that point from all directions of the illuminating hemisphere. Any power that is not reflected is absorbed by the surface and transformed into heat. The underlying assumption here is that surface does not emit any energy by itself, which assumes that the surface in not fluorescent.

More precisely, the total amount of power reflected over the entire illuminating hemisphere $\mathcal{H}_r^2$ (or equivalently Ω_r) must be less than or equal to the power incident from the direction ω_i. In this case, the fraction of the outgoing to the incoming power is known as the *directional-hemispherical reflectance* and is given by

$$\rho(\boldsymbol{x}, \omega_i) = \int_{\mathcal{H}_r^2} f_r(\boldsymbol{x}, \omega_i, \omega_r) \cdot \partial\omega_r^{\perp} \leq 1 \quad \forall \omega_i \in \mathcal{H}_i^2 \qquad (10.90)$$

Observe that here we integrate over the outgoing direction.

In some literature the directional-hemispherical reflectance is denoted as $\rho(\boldsymbol{x}, \omega_i, 2\pi)$ or as $\rho(\boldsymbol{x}, \omega_i \rightarrow 2\pi)$.

10.8 Reflectance Function of a Surface

As we have noted in Section 10.7.2, the BRDF is a function that represents the ratio of the differential reflected radiance $\partial L_r(\boldsymbol{x}, \omega_r)$ (leaving from a point $\boldsymbol{x}$ in a direction ω_r) and the differential incident irradiance $\partial E_i(\boldsymbol{x}, \omega_i)$ (arriving at a point $\boldsymbol{x}$ from a direction ω_i). Although this function gives us the ratio of the reflected to the incident light energy for a surface, it is not very intuitive, as its range is the infinite interval $[0, \infty)$. To overcome this difficulty we define another function, which represents the fraction of the incident light reflected off a surface. This function is known as the *reflectance* of the surface and has a range $[0, 1]$. To determine this function we proceed as follows. We start by considering the ratio of the reflected differential flux to the incident differential flux at a point $\boldsymbol{x}$. Using Eqs. (10.61) and (10.63) we get

$$\frac{\partial \Phi_r(\boldsymbol{x})}{\partial \Phi_i(\boldsymbol{x})} = \frac{\left(\int_{\mathcal{H}_r^2} L_r(\boldsymbol{x}, \omega_r) \cdot \partial \omega_r^{\perp}\right) \cdot \partial A_x}{\left(\int_{\mathcal{H}_i^2} L_i(\boldsymbol{x}, \omega_i) \cdot \partial \omega_i^{\perp}\right) \cdot \partial A_x} = \frac{\int_{\mathcal{H}_r^2} L_r(\boldsymbol{x}, \omega_r) \cdot \partial \omega_r^{\perp}}{\int_{\mathcal{H}_i^2} L_i(\boldsymbol{x}, \omega_i) \cdot \partial \omega_i^{\perp}}$$

Using Eq. (10.86) we can expand this result to get

$$\frac{\partial \Phi_r(\boldsymbol{x})}{\partial \Phi_i(\boldsymbol{x})} = \frac{\int_{\mathcal{H}_r^2} \int_{\mathcal{H}_i^2} f_r(\boldsymbol{x}, \omega_i, \omega_r) \cdot L_i(\boldsymbol{x}, \omega_i) \cdot \partial \omega_i^{\perp} \cdot \partial \omega_r^{\perp}}{\int_{\mathcal{H}_i^2} L_i(\boldsymbol{x}, \omega_i) \cdot \partial \omega_i^{\perp}}$$

If we assume that the incident radiance $L_i(\boldsymbol{x}, \omega_i)$ is both uniform and isotropic, then we can remove it from both integrals of the numerator and the denominator of this fraction. The simplified fraction defines the *reflectance* function of the surface.

$$\rho(\boldsymbol{x}, \omega_i, \omega_r) = \frac{\int_{\mathcal{H}_r^2} \int_{\mathcal{H}_i^2} f_r(\boldsymbol{x}, \omega_i, \omega_r) \cdot \partial \omega_i^{\perp} \cdot \partial \omega_r^{\perp}}{\int_{\mathcal{H}_i^2} \partial \omega_i^{\perp}} \tag{10.91a}$$

or equivalently

$$\rho(\boldsymbol{x}, \omega_i, \omega_r) = \frac{\int_{\mathcal{H}_r^2} \int_{\mathcal{H}_i^2} f_r(\boldsymbol{x}, \omega_i, \omega_r) \cdot \cos(\theta_i) \cdot \partial \omega_i \cdot \cos(\theta_r) \cdot \partial \omega_r}{\int_{\mathcal{H}_i^2} \cos(\theta_i) \cdot \partial \omega_i} \tag{10.91b}$$

Observe that the reflectance of the surface changes as the incidence angle θ_i changes.

In the arrow notation this function is denoted by $\rho(\boldsymbol{x}, \omega_i \to \omega_r)$.

The reflectance function involves the evaluation of a double integral over the incident and reflected directions. The limits of these integrals can be set to a differential solid angle range $[\partial \omega_a, \partial \omega_b]$, a finite solid angle range $[\mathcal{D}\omega_a, \mathcal{D}\omega_b]$ or the entire illuminating hemisphere $\mathcal{H}_i^2$ or $\mathcal{H}_r^2$.

Depending on the chosen integration limits for the incident and reflected directions we can use a qualifier that characterises the reflectance function. This qualifier is constructed by juxtaposing the type of the integration limits used in the incident and the reflected directions. The integration limit types are referred to by the terms *directional* (for a differential solid angle range), *conical* (for a finite solid angle range) and *hemispherical* (for an entire illuminating hemisphere). By permutating these three integration limit types we can generate nine such reflectance qualifiers. Thus, the qualifiers for the following reflectance functions are:

$\rho(\boldsymbol{x}, \omega_i \to \omega_r)$ is a *bi-directional reflectance* function,
$\rho(\boldsymbol{x}, \omega_i \to \mathcal{D}\omega_r)$ is a *directional-conical reflectance* function,
$\rho\left(\boldsymbol{x}, \omega_i \to \mathcal{H}_r^2\right)$ is a *directional-hemispherical reflectance* function,
$\rho(\boldsymbol{x}, \mathcal{D}\omega_i \to \omega_r)$ is a *conical-directional reflectance* function,
$\rho(\boldsymbol{x}, \mathcal{D}\omega_i \to \mathcal{D}\omega_r)$ is a *bi-conical reflectance* function,
$\rho\left(\boldsymbol{x}, \mathcal{D}\omega_i \to \mathcal{H}_r^2\right)$ is a *conical-hemispherical reflectance* function,
$\rho\left(\boldsymbol{x}, \mathcal{H}_i^2 \to \omega_r\right)$ is a *hemispherical-directional reflectance* function,
$\rho\left(\boldsymbol{x}, \mathcal{H}_i^2 \to \mathcal{D}\omega_r\right)$ is a *hemispherical-conical reflectance* function,
$\rho\left(\boldsymbol{x}, \mathcal{H}_i^2 \to \mathcal{H}_r^2\right)$ is a *bi-hemispherical reflectance* function.

10.9 Transmittance Function of a Surface

Following an analogous argument for the transmission of light, to that we used in the previous section for the reflection of light, we can derive the *transmittance* function of a surface. As before, we start by considering the ratio of the transmitted differential flux to the incident differential flux at a point $\boldsymbol{x}$:

$$\frac{\partial\Phi_t(\boldsymbol{x})}{\partial\Phi_i(\boldsymbol{x})} = \frac{\left(\int_{\mathcal{H}_t^2} L_t(\boldsymbol{x}, \omega_t) \cdot \partial\omega_t^{\perp}\right) \cdot \partial A_x}{\left(\int_{\mathcal{H}_i^2} L_i(\boldsymbol{x}, \omega_i) \cdot \partial\omega_i^{\perp}\right) \cdot \partial A_x} = \frac{\int_{\mathcal{H}_t^2} L_t(\boldsymbol{x}, \omega_t) \cdot \partial\omega_t^{\perp}}{\int_{\mathcal{H}_i^2} L_i(\boldsymbol{x}, \omega_i) \cdot \partial\omega_i^{\perp}}$$

Expanding this result we get

$$\frac{\partial\Phi_t(x)}{\partial\Phi_i(x)} = \frac{\int_{\mathcal{H}_t^2}\int_{\mathcal{H}_i^2} f_t(x, \omega_i, \omega_t) \cdot L_i(x, \omega_i) \cdot \partial\omega_i^{\perp} \cdot \partial\omega_t^{\perp}}{\int_{\mathcal{H}_i^2} L_i(x, \omega_i) \cdot \partial\omega_i^{\perp}}$$

Once again, assuming that the incident radiance $L_i(\boldsymbol{x}, \omega_i)$ is both uniform and isotropic, we can remove it from both integrals of the numerator and the denominator of this fraction. The simplified fraction defines the transmittance function of the surface.

$$\tau(\boldsymbol{x}, \omega_i, \omega_t) = \frac{\int_{\mathcal{H}_t^2}\int_{\mathcal{H}_i^2} f_t(\boldsymbol{x}, \omega_i, \omega_t) \cdot \partial\omega_i^{\perp} \cdot \partial\omega_t^{\perp}}{\int_{\mathcal{H}_i^2} \partial\omega_i^{\perp}} \tag{10.92a}$$

or equivalently

$$\tau(\boldsymbol{x}, \omega_i, \omega_t) = \frac{\int_{\mathcal{H}_t^2}\int_{\mathcal{H}_i^2} f_t(\boldsymbol{x}, \omega_i, \omega_t) \cdot \cos(\theta_i) \cdot \partial\omega_i \cdot \cos(\theta_t) \cdot \partial\omega_t}{\int_{\mathcal{H}_i^2} \cos(\theta_i) \cdot \partial\omega_i} \tag{10.92b}$$

Again, observe that the transmittance of the surface changes as the incidence angle θ_i changes and that in the arrow notation this function is denoted by $\tau(\boldsymbol{x}, \omega_i \to \omega_t)$.

10.10 Reflection and Transmission Models

Determining the reflection model of a surface is a difficult task and so, in practice, it is often convenient to think of the general BRDF as the sum of three more basic reflection models, namely: the *ideal diffuse* (or *Lambertian diffuse*) reflection model, the *ideal specular* (*or perfect mirror specular*) reflection model and the *glossy* reflection model. See Fig. 10.32.

The Lambertian diffuse and the perfect mirror specular reflection models are mathematical abstractions and can not deal with the complexity of most surfaces found in nature. Such surfaces are rough (i.e. composed of micro-facets) and exhibit surface and sub-surface scattering phenomena.

An alternative way of characterising the reflection of light from an interface surface and for that matter the transmission of light through an interface surface is to think of the type of image generated by the reflected or the transmitted light. This image can be characterised either as being *coherent* or *incoherent*.

A *coherent reflection* is a mirror type reflection, where the reflected rays of light are not scattered in any way and reproduce a perfectly coherent image. Similarly, *coherent transmission* of light occurs when it passes through a flat, transparent (clear) pane of glass.

On the other hand, when light is reflected off a rough or grooved (scratched) interface surface an *incoherent reflection* occurs, as the micro-facet characteristics of the interface surface cause the light rays to be scattered giving rise to a defocused reflection. Most surfaces that we observe produce incoherent reflections that are difficult to predict and understand. The degree of incoherence of the reflection is related to the degree of roughness of the surface. The rougher the surface is the more incoherent the reflected image becomes. At the extreme of this type of reflection we have a Lambertian diffuse reflection, which is perfectly diffusing and produces a totally incoherent reflection. In this case we characterise the interface surface as being matt.

Similarly, *incoherent transmission* occurs when light passes through a rough interface surface or a semitransparent (translucent) material. For instance, the image seen through a frosted pane of glass or through a thin sheet of paper is incoherent. The degree of incoherence is related to both the roughness of the

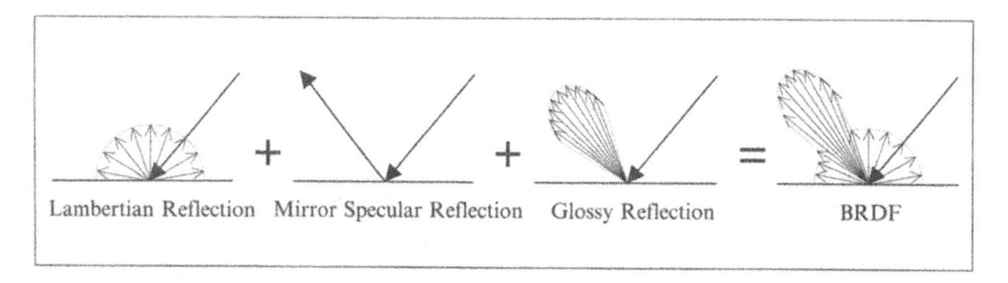

FIGURE 10.32. A BRDF is composed of the sum of Lambertian, mirror specular and glossy reflections.

interface surface and transparency/translucency of the material through which the light has to travel before it reaches us.

Thus one can see that light reflection and transmission can be characterised as being perfectly diffuse (i.e. *totally incoherent*), perfectly specular (i.e. *totally coherent*) and something in-between these two extremes (i.e. *semi-coherent*). In order to simulate these three types of reflection and transmission, we can use correspondingly the *diffuse reflection model*, the *specular reflection model* and a mixture of these two idealised models.

10.10.1 Diffuse Reflection Model

A surface that exhibits diffuse reflection characteristics scatters all incoming energy in all possible directions of the reflected illuminating hemisphere $\mathcal{H}_r^2$. In the ideal case of Lambertian diffuse reflection, it is assumed that the micro-facets of the surface are perfectly evenly distributed in all directions and thus they scatter light evenly in all directions of $\mathcal{H}_r^2$. In this case, the reflected radiance is constant in all directions regardless of the direction of the incoming radiance. Which means that the BRDF is constant and does not depend on the incident and the reflected directions ω_i and ω_r. Thus, Eq. (10.86) becomes

$$
\begin{aligned}
L_{r,d}(\boldsymbol{x}, \omega_r) &= \int_{\mathcal{H}_i^2} f_{r,d}(\boldsymbol{x}) \cdot L_i(\boldsymbol{x}, \omega_i) \cdot \partial\omega_i^{\perp} \\
&= f_{r,d}(\boldsymbol{x}) \cdot \int_{\mathcal{H}_i^2} L_i(\boldsymbol{x}, \omega_i) \cdot \partial\omega_i^{\perp} \\
&= f_{r,d}(\boldsymbol{x}) \cdot E_i(\boldsymbol{x}) \qquad (10.93)
\end{aligned}
$$

From the above equation we can see that the reflected radiance is proportional to the incident irradiance and that it is constant and has the same value in all directions, since neither the BRDF $f_{r,d}(\boldsymbol{x})$ nor the irradiance $E_i(\boldsymbol{x})$ depend on the incident or reflected directions.

By forcing the bi-hemispherical reflectance $\rho_d\left(\boldsymbol{x}, \mathcal{H}_i^2 \rightarrow \mathcal{H}_r^2\right)$ to be less than or equal to 1, we make sure that the energy conservation property is adhered to. Thus,

$$
\begin{aligned}
\rho_d\left(\boldsymbol{x}, \mathcal{H}_i^2, \mathcal{H}_r^2\right) &= \frac{\Phi_{r,d}(\boldsymbol{x})}{\Phi_i(\boldsymbol{x})} \\
&= \frac{\int_{\mathcal{H}_r^2} L_{r,d}(\boldsymbol{x}, \omega_r) \cdot \partial\omega_r^{\perp}}{\int_{\mathcal{H}_i^2} L_i(\boldsymbol{x}, \omega_i) \cdot \partial\omega_i^{\perp}} \\
&= \frac{L_{r,d}(\boldsymbol{x}, \omega_r) \cdot \int_{\mathcal{H}_r^2} \partial\omega_r^{\perp}}{E_i(\boldsymbol{x})} \\
&= \frac{L_{r,d}(\boldsymbol{x}, \omega_r) \cdot \pi}{E_i(\boldsymbol{x})} \\
\therefore \quad \rho_d\left(\boldsymbol{x}, \mathcal{H}_i^2, \mathcal{H}_r^2\right) &= f_{r,d}(\boldsymbol{x}) \cdot \pi \qquad (10.94)
\end{aligned}
$$

since $L_{r,d}(\boldsymbol{x}, \omega_r)$ is constant and has the same value in all directions, and $\int_{\mathcal{H}_r^2} \partial \omega_r^{\perp} = \pi$.

But if the Lambertian BRDF is constant, since it does not depend on the incident direction ω_i, then the Lambertian reflectance ρ_d must also be constant. Now we can express the Lambertian BRDF in terms of the Lambertian reflectance as

$$f_{r,d}(\boldsymbol{x}) = \frac{\rho_d(\boldsymbol{x})}{\pi} \tag{10.95}$$

A physically plausible *Lambertian reflectance* is usually denoted by a constant $k_d \leq 1$, known as the *diffuse coefficient of reflectance.*

10.10.2 Specular Reflection Model

In a perfect mirror reflection the light reaching a perfectly smooth (polished) surface is reflected in only one direction and no surface scattering occurs. In this case, the angle of incidence θ_i is equal to the angle of reflection θ_r and, since a perfect mirror surface is isotropic, the reflected ray vector $\boldsymbol{R_L}$ lies on the plane defined by the light ray vector $\boldsymbol{L}$ and the surface normal vector $\boldsymbol{N}$. Recall that in Section 10.4.1 we have shown how the reflected ray vector could be computed.

Thus, for perfect mirror reflections we have

$$\begin{aligned} \theta_r &= \theta_i \\ \phi_r &= \phi_i \pm \pi \end{aligned} \tag{10.96}$$

As we have also seen in Section 10.4.1, the reflection occurs in the direction

$$\omega_r = 2 \cdot (\boldsymbol{N} \odot \omega_i) \cdot \boldsymbol{N} - \omega_i \tag{10.97}$$

In some literature the direction of the reflection vector is given as

$$\omega_r = \boldsymbol{R}_{(\boldsymbol{N})} \cdot \omega_i \tag{10.98}$$

where $\boldsymbol{R}_{(N)}$ is a matrix known as the *Householder transformation matrix.* This matrix is defined as follows:

$$\boldsymbol{R}_{(\boldsymbol{N})} = 2 \cdot \boldsymbol{N} \cdot \boldsymbol{N}^T - \boldsymbol{I} \tag{10.99}$$

i.e. as twice the Cartesian product of the unit surface normal by its transposed minus the identity matrix.

Expanding this product we get

$$\begin{aligned} \boldsymbol{R}_{(\boldsymbol{N})} &= 2 \cdot \begin{bmatrix} N_x \\ N_y \\ N_z \end{bmatrix} \cdot \begin{bmatrix} N_x & N_y & N_z \end{bmatrix} - \begin{bmatrix} 1 & 0 & 0 \\ 0 & 1 & 0 \\ 0 & 0 & 1 \end{bmatrix} \\ \therefore \quad \boldsymbol{R}_{(\boldsymbol{N})} &= \begin{bmatrix} 2N_x^2 - 1 & 2N_xN_y & 2N_xN_z \\ 2N_yN_x & 2N_y^2 - 1 & 2N_yN_z \\ 2N_zN_x & 2N_zN_y & 2N_z^2 - 1 \end{bmatrix} \end{aligned} \tag{10.100}$$

As this matrix is symmetric, it is equal to its transpose (i.e. $\boldsymbol{R}_{(N)} = \boldsymbol{R}^T_{(N)}$).

With a perfect mirror surface, the incoming light is only reflected in the mirror direction and we have a perfect specular reflection. For such reflections, the reflected radiance is related to the incident radiance by

$$L_r(\boldsymbol{x}, \omega_r) = \rho_s(\boldsymbol{x}, \omega_i, \omega_r) \cdot L_i(\boldsymbol{x}, \omega_i) \tag{10.101}$$

For perfect mirror specular reflections, the value of the specular reflectance $\rho_s(\boldsymbol{x}, \omega_i, \omega_r)$ must be equal to 1 in the direction ω_r and equal to 0 in all other directions. The corresponding BRDF is usually defined in terms of two *Dirac delta* functions.

Dirac's delta function $\delta(x)$ can informally be thought of as a function that has the value

$$\delta(x) = \begin{cases} \infty, & x = 0 \\ 0, & x \neq 0 \end{cases} \tag{10.102a}$$

and has a total integral equal to 1, i.e.

$$\int_{-\infty}^{+\infty} \delta(x) \cdot \partial x = 1 \tag{10.102b}$$

The delta function obeys the *sifting property*, i.e.

$$\int_{-\infty}^{+\infty} f(x) \cdot \delta(x - x_0) \cdot \partial x = f(x_0) \tag{10.102c}$$

Cohen and Wallace [Cohen 93] have developed a perfect mirror surface BRDF that uses two Dirac delta functions to achieve a non-zero result in only one particular direction. In their definition of the BRDF they use the spherical coordinates of the incident and the reflected directions. Thus, they define their BRDF as follows :

$$f_{r,s}(\boldsymbol{x}, \omega_i, \omega_r) = \frac{\delta(\cos(\theta_i) - \cos(\theta_r))}{\cos(\theta_i)} \delta(\phi_i - (\phi_r \pm \pi)) \tag{10.103}$$

In this function the first delta function has the value zero unless $\theta_r = \pm\theta_i$ and the second delta function has the value zero unless $\phi_r = \phi_i \pm \pi$, which guarantees the BRDF only has a non-zero value in the direction ω_r.

Cohen and Wallace, using the above BRDF for a perfect mirror reflection, showed that the radiance incident (at a surface point $\boldsymbol{x}$) is equal to the reflected radiance (from that point). They did so by expressing the incident radiance $L_i(\theta_i, \phi_i)$ and the reflected radiance $L_r(\theta_r, \phi_r)$ in terms of spherical coordinates.

Thus, they showed that

$$L_r(\theta_r, \phi_r) = L_i(\theta_i, \phi_i) = L_i(\theta_r, (\phi_r \pm \pi)) \tag{10.104}$$

They proved this by evaluating the integral over the entire incident illuminating hemisphere

$$L_r(\theta_r, \phi_r) = \int_{\mathcal{H}_i^2} \frac{\delta(\cos(\theta_i) - \cos(\theta_r))}{\cos(\theta_i)} \cdot \delta(\phi_i - (\phi_r \pm \pi)) \cdot L_i(\theta_i, \phi_i) \cdot \cos(\theta_i) \cdot \partial\theta_i \cdot \partial\phi_i$$
$$= L_i(\theta_r, (\phi_r \pm \pi))$$

10.10.3 Fresnel Effect

As we have observed in Section 10.3.2, when light strikes a smooth interface surface between two participating media with different refractive indices $\eta_i \neq \eta_t$ that are transparent, some of the incident light is reflected and some is refracted and transmitted into the second participating medium. The portion of the light energy that is reflected, is reflected in a specular fashion. Both the reflection and transmission of light is known as the *Fresnel effect*. This effect is responsible for the stronger specular reflections at larger incidence angles and for the stronger transmission at lower incidence angles.

The Fresnel effect is a direct consequence of the electromagnetic nature of light. As we have explained in Section 10.2, light is a combination of an electrical field $\boldsymbol{E}$ and magnetic field $\boldsymbol{B}$, which are perpendicular to each other and to the direction $\boldsymbol{k}$ of the propagation of the wave. The Fresnel effect both depends on and influences the *polarisation* of light during specular reflection and transmission. The directional nature of both the electrical and magnetic fields of light explains the phenomenon of polarisation.

Polarisation represents the orientation of the electrical field of the light with respect to the *plane of incidence*, which contains the surface normal, the incident, the reflected and the transmitted rays. If the electric field is parallel to the plane of incidence, then we say that the light is *P-polarised* and if the electric field is perpendicular to the plane of incidence, then we say that the light is *S-polarised.* The letters "S" and "P" in the terms S-polarised and P-polarised are derived from the German words "Senkrecht" and "Parallele" meaning perpendicular and parallel respectively (see Figs. 10.33 and 10.34). These figures show two different depictions of the incident, reflected and transmitted rays of light at the interface surface between two participating media with refractive indices η_i and η_t, respectively. In these figures, the vectors $\boldsymbol{e}_P^{(i)}$ and $\boldsymbol{e}_S^{(i)}$ represent the orientation of the electrical fields of the P-polarised and S-polarised incident light, respectively and vector $\boldsymbol{k}^{(i)}$ represents the direction of the propagation of the light wave. Similarly, the $\boldsymbol{e}_P^{(r)}, \boldsymbol{e}_S^{(r)}, \boldsymbol{k}^{(r)}$ and $\boldsymbol{e}_P^{(t)}, \boldsymbol{e}_S^{(t)}, \boldsymbol{k}^{(t)}$ represent these directions for the reflected and transmitted light respectively.

Recall that the energy (power) of an electromagnetic wave is proportional to its amplitude. To determine the ratios of the reflected to the incident and the transmitted to incident electrical field magnitudes we use the Fresnel equations. There are four such equations that define the four Fresnel factors: one pair of *amplitude reflection and amplitude transmission coefficients* for S-polarised light (r_s and t_s) and one pair of amplitude reflection and amplitude transmission coefficients for P-polarised light (r_P and t_P).

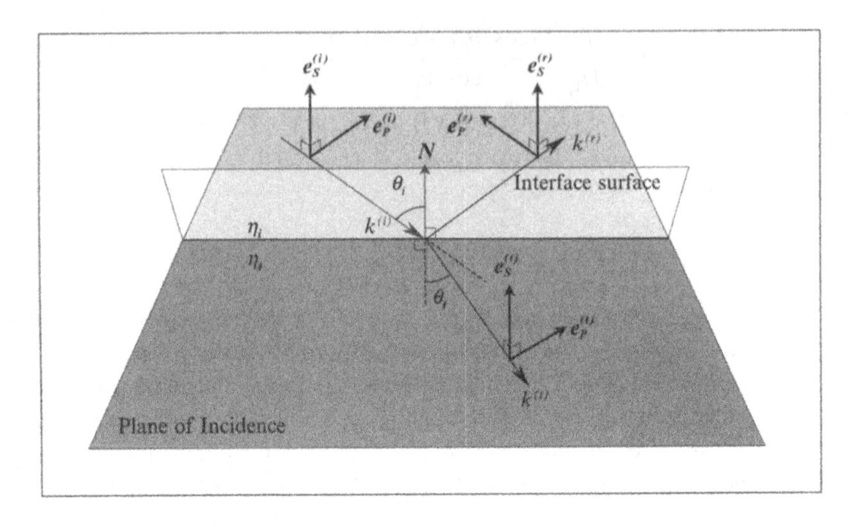

FIGURE 10.33. The reflection and transmission of S-polarised and P-polarised incident light.

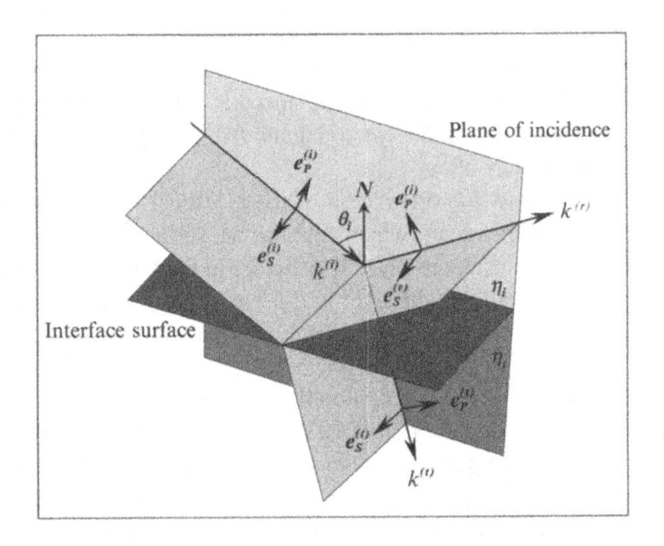

FIGURE 10.34. The reflection and transmission of S-polarised and P-polarised incident light.

The amplitude reflection and transmission coefficients for S-polarised light are given by

$$r_S = \frac{E_S^{(r)}}{E_S^{(i)}} = \frac{\eta_i \cdot \cos(\theta_i) - \eta_t \cdot \cos(\theta_t)}{\eta_i \cdot \cos(\theta_i) + \eta_t \cdot \cos(\theta_t)} \tag{10.105}$$

$$t_S = \frac{E_S^{(t)}}{E_S^{(i)}} = \frac{2 \cdot \eta_i \cdot \cos(\theta_i)}{\eta_i \cdot \cos(\theta_i) + \eta_t \cdot \cos(\theta_t)} \tag{10.106}$$

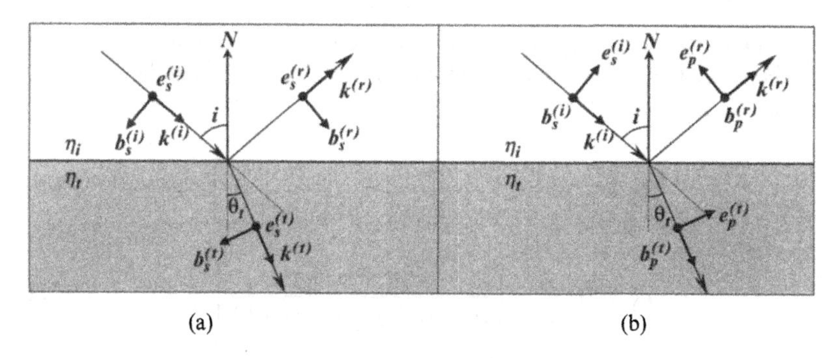

FIGURE 10.35. (a) Refraction and transmission of S-polarised incident light. (b) Refraction and transmission of P-polarised incident light.

where $E_S^{(i)}$, $E_S^{(r)}$ and $E_S^{(t)}$ are the amplitudes of the electrical fields of the incident, reflected and transmitted S-polarised light rays (pointing in the directions $\boldsymbol{e}_S^{(i)}$, $\boldsymbol{e}_S^{(r)}$ and $\boldsymbol{e}_S^{(t)}$, respectively), η_i and η_t are the refractive indices of the media on either side of the interface surface, and θ_i and $\theta_t$are the incidence and transmission angles (see Fig. 10.35a). In this figure the vectors $\boldsymbol{e}_S^{(i)}$, $\boldsymbol{e}_S^{(r)}$ and $\boldsymbol{e}_S^{(t)}$ point out of the page and vectors $\boldsymbol{b}_S^{(i)}$, $\boldsymbol{b}_S^{(r)}$ and $\boldsymbol{b}_S^{(t)}$ represent the orientations of the magnetic fields of the incident, reflected and transmitted S-polarised light. Here the magnetic field direction vectors lie on the plane of incidence.

If $\eta_i \geq \eta_t$, then the amplitude reflection coefficient $r_S > 0$ otherwise $r_S < 0$. This means that the electric field of the light is subjected to a *phase shift* upon reflection, i.e.

$$\Delta(\phi) = \begin{cases} 0, & \eta_i \geq \eta_t \\ \pi, & \eta_i < \eta_t \end{cases}$$

where ϕ is the *phase angle* of the electrical field.

Similarly, the amplitude reflection and transmission coefficients for P-polarised light are given by

$$r_P = \frac{E_P^{(r)}}{E_P^{(i)}} = \frac{\eta_t \cdot \cos(\theta_i) - \eta_i \cdot \cos(\theta_t)}{\eta_t \cdot \cos(\theta_i) + \eta_i \cdot \cos(\theta_t)} \tag{10.107}$$

$$t_P = \frac{E_P^{(t)}}{E_P^{(i)}} = \frac{2 \cdot \eta_i \cdot \cos(\theta_i)}{\eta_t \cdot \cos(\theta_i) + \eta_i \cdot \cos(\theta_t)} \tag{10.108}$$

where $E_P^{(i)}$, $E_P^{(r)}$ and $E_P^{(t)}$ are the amplitudes of the electrical fields of the incident, reflected and transmitted P-polarised light rays (pointing in the directions $\boldsymbol{e}_P^{(i)}$, $\boldsymbol{e}_P^{(r)}$ and $\boldsymbol{e}_P^{(t)}$, respectively), η_i and η_t are the refractive indices of the media on either side of the interface surface, and θ_i and $\theta_t$are the incidence and transmission angles (see Fig. 10.35b). In this figure the vectors $\boldsymbol{b}_P^{(i)}$, $\boldsymbol{b}_P^{(r)}$ and $\boldsymbol{b}_P^{(t)}$, which represent the orientations of the magnetic fields of the incident, reflected and transmitted P-polarised light, point out of the page. Here the electrical field direction vectors lie on the plane of incidence.

The squares of these four amplitude coefficients are known as the *reflectance coefficients* and the *transmittance coefficients*, which can be thought of as power or intensity coefficients.

$$R_S = (r_S)^2 = \left[\frac{\eta_i \cdot \cos(\theta_i) - \eta_t \cdot \cos(\theta_t)}{\eta_i \cdot \cos(\theta_i) + \eta_t \cdot \cos(\theta_t)}\right]^2 \tag{10.109}$$

$$T_S = (t_S)^2 = \left[\frac{2 \cdot \eta_i \cdot \cos(\theta_i)}{\eta_i \cdot \cos(\theta_i) + \eta_t \cdot \cos(\theta_t)}\right]^2 \tag{10.110}$$

$$R_P = (r_P)^2 = \left[\frac{\eta_t \cdot \cos(\theta_i) - \eta_i \cdot \cos(\theta_t)}{\eta_t \cdot \cos(\theta_i) + \eta_i \cdot \cos(\theta_t)}\right]^2 \tag{10.111}$$

$$T_P = (t_P)^2 = \left[\frac{2 \cdot \eta_i \cdot \cos(\theta_i)}{\eta_t \cdot \cos(\theta_i) + \eta_i \cdot \cos(\theta_t)}\right]^2 \tag{10.112}$$

The reflectance coefficients R_S and R_P represent the directional-hemispherical spectral reflectivities of the interface surface.

The power reflectance and transmittance coefficients are sometimes given as

$$R_S = \left[-\frac{\sin(\theta_i - \theta_t)}{\sin(\theta_i + \theta_t)}\right]^2 \tag{10.113}$$

$$T_S = \left[-\frac{2 \cdot \cos(\theta_i) \cdot \sin(\theta_t)}{\sin(\theta_i + \theta_t)}\right]^2 \tag{10.114}$$

$$\begin{aligned} R_P &= \left[\frac{\tan(\theta_i - \theta_t)}{\tan(\theta_i + \theta_t)}\right]^2 \\ &= \left[\frac{\sin(\theta_t) \cdot \cos(\theta_t) - \sin(\theta_i) \cdot \cos(\theta_i)}{\sin(\theta_t) \cdot \cos(\theta_t) + \sin(\theta_i) \cdot \cos(\theta_i)}\right]^2 \end{aligned} \tag{10.115}$$

$$T_P = \left[\frac{2 \cdot \cos(\theta_i) \cdot \sin(\theta_t)}{\sin(\theta_i + \theta_t) \cdot \cos(\theta_i - \theta_t)}\right]^2 \tag{10.116}$$

Equations (10.109) – (10.112), however, are more frequently used in computer graphics, as they only involve the cosines of the incidence and the transmission angles which are easy to compute from the surface normal and, the incidence and transmission directions (i.e. $\cos(\theta_i) = \boldsymbol{N} \odot \boldsymbol{L}$ and $\cos(\theta_t) = -\boldsymbol{N} \odot \boldsymbol{T}_L$).

Because the amplitude reflection and transmission coefficients are different for S-polarised and P-polarised light both the reflected and transmitted light are polarised. In computer graphics we usually assume that the incident light is unpolarised, thus the unpolarised reflectance and transmittance coefficients are given as

$$R = \frac{(r_S)^2 + (r_P)^2}{2} \tag{10.117}$$

$$T = \frac{(t_S)^2 + (t_P)^2}{2} \tag{10.118}$$

Despite the assumption that the incoming light is unpolarised, the outgoing reflected and transmitted light will be polarised.

The Fresnel effect does not take into account light absorption, thus

$$R + T = 1$$

When the incident light direction is near-normal to the interface surface (i.e. when $\theta_i \approx \theta_r \approx \theta_t \approx 0^0$), then the amplitude reflection and transmission coefficients of the S-polarised and the P-polarised light coincide, giving

$$r_S = r_P = \frac{\eta_i - \eta_t}{\eta_i + \eta_t} \tag{10.119}$$

$$t_S = t_P = \frac{2\eta_i}{\eta_i + \eta_t} \tag{10.120}$$

Thus,

$$R = \left[\frac{\eta_i - \eta_t}{\eta_i + \eta_t}\right]^2 \tag{10.121}$$

$$T = \left[\frac{2\eta_i}{\eta_i + \eta_t}\right]^2 \tag{10.122}$$

In all the equations presented above, the transmission angle θ_t is computed using Snell's law (that we have examined in detail in Section 10.3.2). Recall that

$$\theta_i = \arccos\left(\boldsymbol{L} \odot \boldsymbol{N}\right)$$

where $\boldsymbol{N}$ and $\boldsymbol{L}$ are the unit surface normal and the unit vector pointing in the direction of the of the light source. Using Snell's law, we have

$$\theta_t = \arcsin\left(\frac{\eta_i}{\eta_t} \cdot \sin\left(\theta_i\right)\right)$$

The formulae for the computation of the reflectance and transmittance coefficients given above are valid whenever the transmission angle θ_t has a real value. This is always the case for *external reflections* where $\eta_i < \eta_t$ and for *internal reflections* with $\eta_i > \eta_t$ but where the incidence angle is smaller than the *critical angle* (i.e. when $\theta_i < \theta_c$). See Fig. 10.36.

The latter case was examined in Section 10.3.3. Figure 10.37 depicts external reflections. It shows a graph of the *reflectivity* (reflectance coefficient) for S-polarised, P-polarised and unpolarised light at the interface surface between air (with $\eta_i = 1.0$) and glass (with $\eta_t = 1.5$). From this figure we can see that as θ_i increases above zero the P-polarised reflectivity R_P first decreases, reaching a zero value at approximately 56^0, and then increases reaching a value of one at 90^0. The angle at which the reflectivity reaches the value zero is known as *Brewster's angle*.

As we have seen in Section 10.3.3, internal reflections occur at the interface of a dense and a less dense material (i.e. when $\eta_i > \eta_t$). When $\theta_i < \theta_c$, then

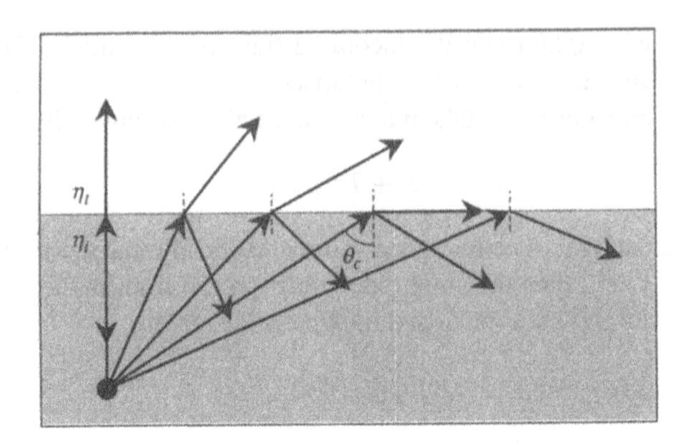

FIGURE 10.36. Partial and total internal reflections.

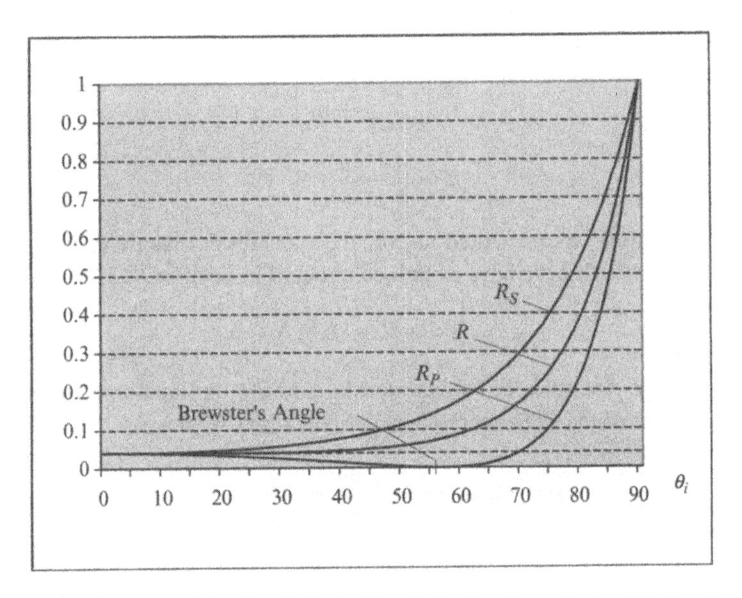

FIGURE 10.37. A plot of the S-polarised , P-polarised and unpolarised light reflectance against the angle of incidence.

partial internal reflection and diffraction followed by transmission occurs. But when $\theta_i > \theta_c$, then only *total internal reflection* takes place (see Fig. 10.36). From Eq. (10.17) we know that the critical angle is given by

$$\theta_c = \arcsin\left(\frac{\eta_i}{\eta_t}\right)$$

For light striking the interface surface with $\theta_i < \theta_c$ we can use the Fresnel equations presented above for the computation of the reflectance and transmittance coefficients. Above the critical angle, however, for $\theta_i > \theta_c$ and $\eta_i > \eta_t$ the following equations must be used to compute the amplitude reflection and transmission coefficients.

$$r_S = \frac{\cos(\theta_i) - i \cdot \sqrt{\sin^2(\theta_i) - \left(\frac{\eta_t}{\eta_i}\right)^2}}{\cos(\theta_i) + i \cdot \sqrt{\sin^2(\theta_i) - \left(\frac{\eta_t}{\eta_i}\right)^2}} \tag{10.123}$$

$$t_S = 0 \tag{10.124}$$

$$r_P = \frac{\left(\frac{\eta_t}{\eta_i}\right)^2 \cdot \cos(\theta_i) - i \cdot \sqrt{\sin^2(\theta_i) - \left(\frac{\eta_t}{\eta_i}\right)^2}}{\left(\frac{\eta_t}{\eta_i}\right)^2 \cdot \cos(\theta_i) - i \cdot \sqrt{\sin^2(\theta_i) - \left(\frac{\eta_t}{\eta_i}\right)^2}} \tag{10.125}$$

$$t_P = 0 \tag{10.126}$$

where $i = \sqrt{-1}$ is the imaginary unit.

In practice however, above the critical angle, we only need to compute the power reflectance and transmittance coefficients, which can be taken to be:

$$R_S = (r_S)^2 = 1 \tag{10.127}$$

$$T_S = 0 \tag{10.128}$$

$$R_P = (r_P)^2 = 1 \tag{10.129}$$

$$T_P = 0 \tag{10.130}$$

Above the critical angle, both r_S and r_P have an absolute value of one and are pure phase shifts. In this case, the phase shift of the electrical field is given by

$$\Delta(\phi) = 2 \cdot \arctan\left(\cos(\theta_i) \cdot \sqrt{\sin^2(\theta_i) - \left(\frac{\eta_t}{\eta_i}\right)^2} \Big/ \sin^2(\theta_i)\right)$$

where ϕ is the phase angle of the electrical field.

Figure 10.38 depicts this case. It shows a graph of the reflectivity for S-polarised, P-polarised and unpolarised light at the interface surface between glass (with $\eta_i = 1.5$) and air (with $\eta_t = 1.0$). In this figure Brewster's angle is approximately $34^0$and the critical angle is approximately 41.8^0.

The electrical properties of a material have the greatest influence on how light travels through it. As we have seen in Section 10.2, the magnetic field of a wave effects the electrons of the material and the freedom of these electrons to respond to this field determines the optical characteristics of the material.

Dielectric materials are largely unaffected by the passage of light through them, as their electrons have very stable orbits that are not disturbed by the passing electromagnetic wave.

Conductors, on the other hand, have free electrons (i.e. electrons that are free to move inside the material) which are made to oscillate under the influence of the magnetic field of the wave. The oscillations of the free electrons match the

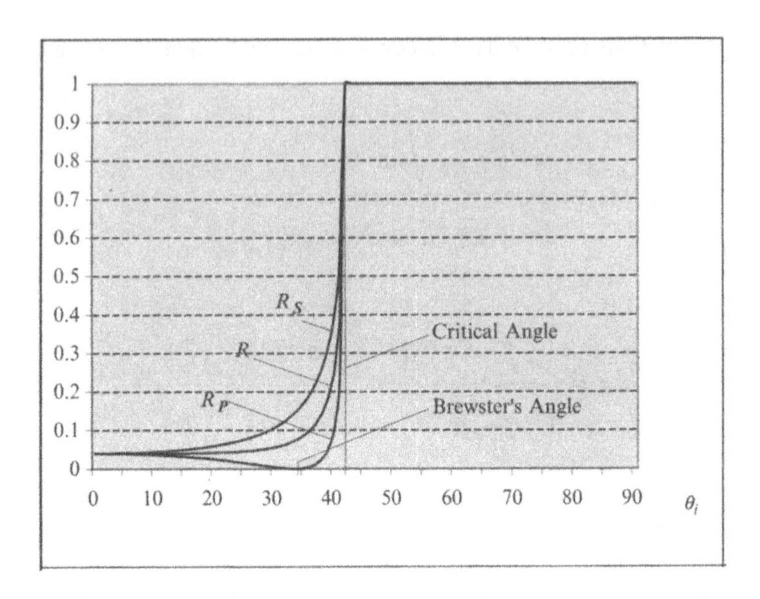

FIGURE 10.38. A plot of the S-polarised, P-polarised and unpolarised light reflectance against the angle of incidence.

frequency of the wave. The oscillating free electrons generate electrical and magnetic fields that cause the material to reflect the light (by re-emitting it). In conductors, there still exist forces that resist the motion of free electrons, resulting in the resistance of the material to the passage of an electrical current. This resistance slows down the motion of the free electrons, absorbing the lost energy in the form of heat. As a result of this absorption the reflected energy of light is less than the incident energy on a conducting surface. The ratio of the incident energy to the reflected energy is known as the *absorption coefficient* of the material which is denoted by k. The *absorption index* k′ is a measure of the decrease of the wave energy that is caused by the absorption of this energy per unit distance travelled that occurs in an electromagnetic wave (of a given wavelength λ) propagating through a material (with a given refractive index η).

The absorption index is defined as

$$k' = \frac{k \cdot \lambda}{4\pi \cdot \eta} \tag{10.131}$$

The absorption index of a material can vary from 0.001 to 70 approximately.

A reasonable approximation of the Fresnel amplitude reflection coefficients for S-polarised and P-polarised light can be found in [Ditchburn 76], when the sum of the squares of the refractive index and absorption coefficient of a conductive material is much greater than one (i.e. when $\eta_t^2 + k_t^2 >> 1$)

$$r_S^2 = \frac{\left(\eta_t^2 + k_t^2\right) - 2\eta_t \cdot \cos(\theta_i) + \cos^2(\theta_i)}{\left(\eta_t^2 + k_t^2\right) - 2\eta_t \cdot \cos(\theta_i) + \cos^2(\theta_i)} \tag{10.132}$$

$$r_P^2 = \frac{\left(\eta_t^2 + k_t^2\right) \cdot \cos^2(\theta_i) - 2\eta_t \cdot \cos(\theta_i) + 1}{\left(\eta_t^2 + k_t^2\right) \cdot \cos^2(\theta_i) + 2\eta_t \cdot \cos(\theta_i) + 1} \tag{10.133}$$

In these formulae, the interface surface is assumed to be between air (with a refractive index $\eta_i = 1.0$) and a conducting material with refractive index η_t and absorption coefficient k_t. In the case where the first participating medium is not air and has a refractive index $\eta_i \neq 1.0$, then we assume that the interface surface is composed of a pseudo-material that has a relative refractive index $\eta_a = \frac{\eta_t}{\eta_i}$ and we use this relative index in the above formulae.

Also, the power reflectance coefficient of the conducting surface for S-polarised and P-polarised light is given by

$$R_S = (r_S)^2 \tag{10.134}$$

$$R_P = (r_P)^2 \tag{10.135}$$

$$\text{and} \quad R = \frac{(r_S)^2 + (r_P)^2}{2} \tag{10.136}$$

Figure 10.39 shows the plots for the S-polarised, P-polarised and unpolarised reflectance coefficients for the interface between air and glass R_S^G, R_P^G and R^G, as well as, the plots for the S-polarised, P-polarised and unpolarised reflectance coefficients for the interface between air and copper R_S^C, R_P^C and R^C. The refractive index of air is taken to be $\eta_i = 1.0$, the refractive index of glass is taken to be $\eta_t = 1.5$, the refractive index of copper is taken to be $\eta_t = 0.617$ and its absorption coefficient is taken to be $k_t = 2.63$. To compute the reflectance coefficients of the air-glass interface we used Eqs. (10.109), (10.111) and (10.117), and

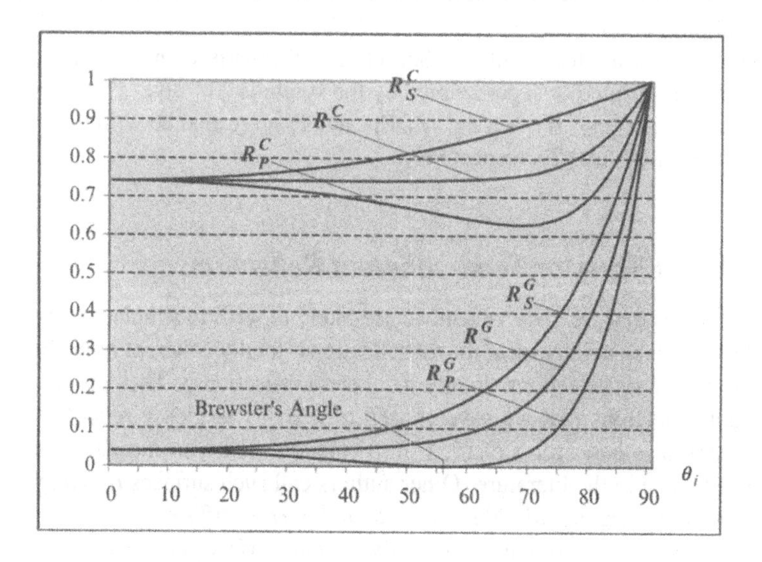

FIGURE 10.39. A plot of the S-polarised, P-polarised and unpolarised light reflectance against the incidence angle for the air-glass and air-copper interface surfaces.

for the reflectance coefficients of the air-copper interface we used Eqs. (10.132), (10.133) and (10.136).

The low reflectance of glass is characteristic of all dielectric materials and the high reflectance of copper is characteristic of all conductors. As can also be seen from this diagram, all materials become 100% reflective at grazing angles (i.e. when $\theta_i \approx 90^0$). This is why when we observe a surface at an acute angle we see glare from the surface.

Although the Fresnel relationship provides a valuable insight as to how light behaves at an interface surface, its usefulness is often limited in practice as there is very little reliable data on the refractive indices and absorption coefficients of materials. This is exacerbated by the fact that the refractive index and absorption coefficient of a given material is wavelength dependent. Frequently, these values only exist for a single wavelength value of light and this information is not sufficient for our purposes in computer graphics. Useful source of such data are [Jenkins 76] and [Palik 85]. Also the website [WWW 1] is a source of very useful lighting data, listing indices of refraction and providing a photonics calculator.

Where there is missing or insufficient data for the refractive indices of participating materials and the absorption coefficients of conducting materials, we can use a set of techniques, developed by Cook and Torrance, to approximate this data [Cook 82].

For unpolarised light, a good approximation of the average reflectance coefficient of a given material is given in [Schlick 93]

$$R(\theta_i) \approx R_0 + (1 - R_0) \cdot (1 - \cos(\theta_i))^5 \tag{10.137}$$

where $R(\theta_i)$ is the average reflectance at incidence angle θ_i and R_0 is the measured value of the reflectance at normal incidence (i.e. measured in the direction of the normal of the surface.

In some literature the quantities that relate to S-polarised and P-polarised light are either subscripted or superscripted by the symbols "$\perp$" and "$\|$", respectively. Thus the terms r_S, t_S, R_S, T_S, r_P, t_P, R_P and T_P may also be written as $r_\perp$, $t_\perp$, $R_\perp$, $T_\perp$, $r_\|$, $t_\|$, $R_\|$ and $T_\|$, respectively.

10.10.4 Glossy or Semi-coherent Reflections

Most surfaces that we observe and we are likely to wish to simulate in computer graphics produce neither ideal mirror reflections (totally coherent reflections) nor ideal Lambertian reflections (totally incoherent reflections). They produce semi-coherent reflections that we have chosen to refer to as glossy reflections. The name chosen is descriptive of their properties, but is by no means the only established name in the literature. Other authors call such surfaces *rough specular*, *semi-coherent*, *directional diffuse* or *wide and narrow diffuse*.

As we have observed before, perfectly smooth surfaces produce totally coherent reflections, rough surfaces produce semi-coherent reflections and perfectly rough surfaces produce totally incoherent reflections. The degree of smoothness or roughness of a surface is explained in terms of the micro-facets that make up the surface.

Perfectly smooth surfaces are composed of micro-facets that are perfectly aligned and have normals that point in the same direction. Thus, they reflect and transmit light in the directions and proportions predicted by the Fresnel equations and Snell's law.

Perfectly rough surfaces (matt surfaces) have micro-facets that are totally randomly oriented and have normals that are evenly distributed over the entire illuminating hemisphere. Thus, they reflect and transmit light evenly in all directions and obey Lambert's law.

Rough surfaces are composed of micro-facets that are unevenly distributed and have normals that are statistically distributed around the main (average) normal of the surface. This micro-facet normal distribution can be biased in certain directions to account for grooves, scratches, cracks or some other surface finish. Thus, such surfaces reflect and transmit light in a more complex and unpredictable way and give rise to semi-coherent (defocused) reflections. In this case, the amount of light from a given light source being reflected in the direction of the viewer is proportional to the number of micro-facets that have their normals aligned halfway between the vectors pointing, out of the surface, towards the light source and the viewer. This direction is represented by the $\boldsymbol{H}$ vector in Fig. 10.10. This unit vector is computed as

$$\boldsymbol{H} = \frac{\boldsymbol{L}+\boldsymbol{E}}{|\boldsymbol{L}+\boldsymbol{E}|} \tag{10.138}$$

where $\boldsymbol{L}$ is the unit vector pointing in the direction of the light source and $\boldsymbol{E}$ is the unit vector pointing in the direction of the viewer (eye).

Various models with varying degrees of sophistication have been developed by computer graphics researchers over the years. Some of these models are purely empirical in nature, while others are physically plausible or even more physically based. A good survey of the early models can be found in [Hall 89] and a more up to date treatment in [Lewis 93].

For most surfaces that occur in nature, getting the precise physical model of the interaction of light with the surface is very difficult if not impossible. Most researchers in the field have concentrated on developing models that are good approximations of physical phenomena while being both visually accurate and easy to compute.

For simple renderers, visual believability and ease of computation are the primary criteria for selecting a particular model. With more sophisticated renderers, using radiosity computations which are energy conserving, the selected model must also be physically plausible. This of course does not necessarily mean that the model is physically accurate, but only that it does not violate the laws of physics and it is thus a believable model of a physical process.

10.11 Some Classical and Physically Plausible Shading Models

In the ensuing discussion we will assume the following. An infinitesimal (differential) surface element with area ∂A is situated at a point $\boldsymbol{x}$ on a surface. This surface element is both opaque and non-emissive and is being lit by incident radiance $L_i(\boldsymbol{x}, \omega_i)$ arriving from an infinitesimal (differential) solid angle $\partial\omega_i$ surrounding the direction unit vector ω_i (i.e. $\boldsymbol{L}$). $\boldsymbol{N}$ is the unit normal of the surface element at $\boldsymbol{x}$. $\boldsymbol{E}$ is the unit vector in the direction of the viewer. The reflected radiance $L_r(\boldsymbol{x}, \omega_r)$ leaving the patch with an infinitesimal (differential) solid angle $\partial\omega_r$ surrounding the direction unit vector ω_r (i.e. $\boldsymbol{E}$). $\boldsymbol{R}_L$ is the unit vector in the direction of the perfect mirror reflection (i.e. the reflection of vector $\boldsymbol{L}$ with respect to $\boldsymbol{N}$). See figure 10.40.

We further assume that all reflected light passes unobstructed through the illuminating hemisphere $\mathcal{H}_+^2$ of point $\boldsymbol{x}$.

In computer graphics a shading model is often referred to as a *shader*.

To consider the shader from the standpoint of energy conservation, we start from the equation of energy balance [i.e. Eq. (10.56)]:

$$\partial E_i(\boldsymbol{x}) = L_i(\boldsymbol{x}, \omega_i) \cdot \partial\omega_i \tag{10.139}$$

where $\partial E_i(\boldsymbol{x})$ is the change in the irradiance resulting from the illumination of the patch from the solid angle $\partial\omega_i$.

Also from the definition of the BRDF [Eq. (10.76)], we have

$$\partial L_r(\boldsymbol{x}, \omega_r) = f_r(\boldsymbol{x}, \omega_i, \omega_r) \cdot \partial E_i(\boldsymbol{x}, \omega_i) \tag{10.140}$$

The BRDF may also vary over the surface that we wish to shade but we treat this as part of the texturing rather than shading, so we will ignore this effect.

As we have assumed that our surface is opaque and non-emissive, the only contribution to the reflected radiance $L_r(\boldsymbol{x}, \omega_r)$ can only come from the incident

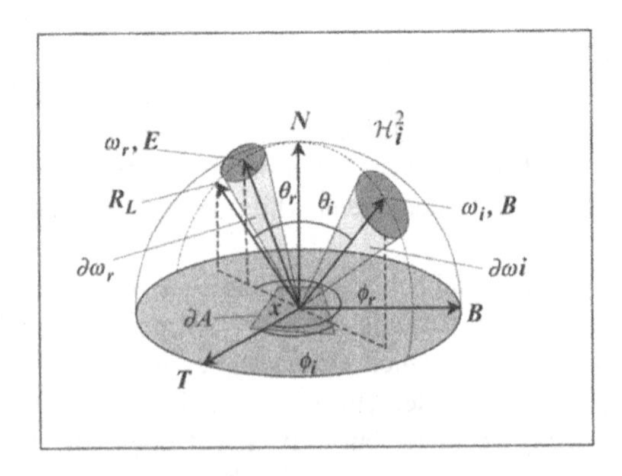

FIGURE 10.40. Physically plausible shading models.

illuminating hemisphere $\mathcal{H}_i^2$, thus

$$\begin{aligned} L_r(\boldsymbol{x}, \omega_t) &= \int_{\mathcal{H}_i^2} f_r(\boldsymbol{x}, \omega_i, \omega_r) \cdot L_i(\boldsymbol{x}, \omega_i) \cdot \partial \omega_i^{\perp} \\ &= \int_{\mathcal{H}_i^2} f_r(\boldsymbol{x}, \omega_i, \omega_r) \cdot L_i(\boldsymbol{x}, \omega_i) \cdot (\boldsymbol{N} \odot \omega_i) \cdot \partial \omega_i \\ &= \int_{\mathcal{H}_i^2} f_r(\boldsymbol{x}, \omega_i, \omega_r) \cdot L_i(\boldsymbol{x}, \omega_i) \cdot \cos(\theta_i) \cdot \partial \omega_i \end{aligned} \quad (10.141)$$

Next we will consider some of the most common shaders used in computer graphics expressed as BRDFs.

10.11.1 Phong Shader

As we have seen in Chapter 9, the Phong shader is given by the equation

$$L_r = k_a \cdot L_a + (k_d \cdot (\boldsymbol{N} \odot \boldsymbol{L}) + k_s \cdot F_s(\boldsymbol{N}, \boldsymbol{L}, \boldsymbol{E}, n_s)) \cdot L_d \quad (10.142)$$

where k_a is the ambient reflection coefficient, L_a is the ambient radiance evenly distributed over the incident illuminating hemisphere $\mathcal{H}_i^2$, k_d is the diffuse reflection coefficient, k_s is the specular reflection coefficient, F_s is the specular reflection function, n_s is the specular sharpness and L_d is the radiance coming from the direction $\boldsymbol{L}$ of the light source.

As we have seen in Section 9.3.3, the specular function can be defined using Horn's method:

$$\mathrm{F}_s(\boldsymbol{N}, \boldsymbol{L}, \boldsymbol{E}, n_s) = (2 \cdot (\boldsymbol{N} \odot \boldsymbol{E}) \cdot (\boldsymbol{N} \odot \boldsymbol{L}) - (\boldsymbol{E} \odot \boldsymbol{L}))^{n_s} \quad (10.143)$$

or Blinn's method:

$$\mathrm{F}_s(\boldsymbol{N}, \boldsymbol{L}, \boldsymbol{E}, n_s) = \left(\boldsymbol{N} \odot \frac{\boldsymbol{L} + \boldsymbol{E}}{|\boldsymbol{L} + \boldsymbol{E}|}\right)^{n_s} \quad (10.144)$$

Let us examine how the Phong shader of Eq. (10.142) corresponds to Eq. (10.141). Our surface patch is illuminated by the directed radiance L_d coming from the direction of the light source and by the ambient radiance L_a which is constant over $\mathcal{H}_i^2$. From the directional illumination we have $\theta_i = \theta_l$ and $\phi_i = \phi_l$. We can now express the resulting incident radiance as

$$L_i = L_a + L_d \cdot \delta(\cos(\theta_i) - \cos(\theta_l)) \cdot \delta(\phi_i - \phi_l) \quad (10.145)$$

Equation (10.141) can be rewritten as

$$L_r = \int_{\mathcal{H}_i^2} f_r(\theta_i, \phi_i, \boldsymbol{E}) \cdot L_i \cdot \cos(\theta_i) \cdot \partial \theta_i \cdot \partial \phi_i \quad (10.146)$$

Substituting equation (10.145) into equation (10.146) we get

$$L_r = L_a \cdot \int_{\mathcal{H}_i^2} f_r(\theta_l, \phi_l, \boldsymbol{E}) \cdot \cos(\theta_l) \cdot \partial \theta_l \cdot \partial \phi_l + L_d \cdot f_r(\theta_i, \phi_i, \boldsymbol{E}) \cdot \cos(\theta_i)$$

which can be rewritten as

$$L_r = L_a \cdot \int_{\mathcal{H}_i^2} f_r(\boldsymbol{L}, \boldsymbol{E}) \cdot (\boldsymbol{N} \odot \boldsymbol{L}) \cdot \partial\omega_i + L_d \cdot f_r(\boldsymbol{L}, \boldsymbol{E}) \cdot (\boldsymbol{N} \odot \boldsymbol{L}) \tag{10.147}$$

By combining Eqs. (10.142) and (10.147) we get

$$\begin{aligned} &k_a \cdot L_a + (k_d \cdot (\boldsymbol{N} \odot \boldsymbol{L}) + \mathrm{k}_s \cdot F_s(\boldsymbol{N}, \boldsymbol{L}, \boldsymbol{E}, n_s)) \cdot L_d \\ &\quad = L_a \cdot \int_{\mathcal{H}_i^2} f_r(\boldsymbol{L}, \boldsymbol{E}) \cdot (\boldsymbol{N} \odot \boldsymbol{L}) \cdot \partial\omega_i + L_d \cdot f_r(\boldsymbol{L}, \boldsymbol{E}) \cdot (\boldsymbol{N} \odot \boldsymbol{L}) \end{aligned}$$

This equality must be valid for all values of L_a and L_d. Equating the components of this equation that deal with the directed radiance L_d, we get

$$\begin{aligned} &L_d \cdot f_r(\boldsymbol{L}, \boldsymbol{E}) \cdot (\boldsymbol{N} \odot \boldsymbol{L}) = (k_d \cdot (\boldsymbol{N} \odot \boldsymbol{L}) + k_s \cdot F_s(\boldsymbol{N}, \boldsymbol{L}, \boldsymbol{E}, n_s)) \cdot L_d \\ \therefore\ \ &f_r(\boldsymbol{L}, \boldsymbol{E}) \cdot (\boldsymbol{N} \odot \boldsymbol{L}) = k_d \cdot (\boldsymbol{N} \odot \boldsymbol{L}) + k_s \cdot F_s(\boldsymbol{N}, \boldsymbol{L}, \boldsymbol{E}, n_s) \\ \therefore\ \ &f_r(\boldsymbol{L}, \boldsymbol{E}) = \frac{k_d \cdot (\boldsymbol{N} \odot \boldsymbol{L}) + k_s \cdot \mathrm{F}_s(\boldsymbol{N}, \boldsymbol{L}, \boldsymbol{E}, n_s)}{(\boldsymbol{N} \odot \boldsymbol{L})} \end{aligned}$$

Thus, the Phong shader BRDF is given by

$$f_r(\boldsymbol{x}, \boldsymbol{L}, \boldsymbol{E}) = k_d + k_s \cdot \frac{F_s(\boldsymbol{N}, \boldsymbol{L}, \boldsymbol{E}, n_s)}{(\boldsymbol{N} \odot \boldsymbol{L})} \tag{10.148}$$

Note that the specular function in Eq. (10.143) can also be written as

$$F_s(\boldsymbol{N}, \boldsymbol{L}, \boldsymbol{E}, n_s) = (\boldsymbol{E} \odot \boldsymbol{R}_L)^{n_s} \tag{10.149a}$$

where $\boldsymbol{R}_L$ is the unit vector representing the reflection of $\boldsymbol{L}$ with respect to $\boldsymbol{N}$, which as we have seen in Section 10.4.2 can be computed as

$$\boldsymbol{R}_L = 2 \cdot (\boldsymbol{N} \odot \boldsymbol{L}) \cdot \boldsymbol{N} - \boldsymbol{L}$$

or alternatively using the Householder matrix as

$$\boldsymbol{R}_L = R_{(N)} \cdot \boldsymbol{L}$$

Equally, the specular function can be rewritten as

$$F_s(\boldsymbol{N}, \boldsymbol{L}, \boldsymbol{E}, n_s) = (\boldsymbol{L} \odot \boldsymbol{R}_E)^{n_s} \tag{10.149b}$$

where

$$\begin{aligned} &\boldsymbol{R}_E = 2 \cdot (\boldsymbol{N} \odot \boldsymbol{E}) \cdot \boldsymbol{N} - \boldsymbol{E} \\ \text{or}\ \ &\boldsymbol{R}_E = \boldsymbol{R}_{(N)} \cdot \boldsymbol{E} \end{aligned}$$

The original Phong model is not physically plausible, since it obeys neither the energy conservation property nor the reciprocity property.

10.11.2 Modified Phong Shader

Lafortune and Willems proposed a modified version of the Phong model, which deals with the lack of reciprocity and energy conservation of the original model [Lafortune 94]. Their BRDF is given as the sum of a diffuse BRDF and a specular BRDF.

$$f_r(\boldsymbol{x}, \omega_i, \omega_r) = f_{r,d}(\boldsymbol{x}, \omega_i, \omega_r) + f_{r,s}(\boldsymbol{x}, \omega_i, \omega_r)$$
$$\text{i.e.} \quad f_r(\boldsymbol{x}, \boldsymbol{L}, \boldsymbol{E}) = k_d \cdot \frac{1}{\pi} + k_s \cdot \frac{n_s + 2}{2\pi} \cdot \cos^{n_s} \alpha \tag{10.150}$$

where k_d is the diffuse reflectivity (i.e. the fraction of incoming energy that is reflected diffusely), k_s is the specular reflectivity (i.e. the fraction of incoming energy that is reflected in a specular manner), n_s is the specular sharpness (i.e. the higher this exponent is the sharper the specular reflections) and α is the highlight angle (i.e. the angle between vectors $\boldsymbol{E}$ and $\boldsymbol{R}_L$ or between $\boldsymbol{L}$ and $\boldsymbol{E}_R$). This angle must be clamped to $\pi/2$, to avoid negative cosine values. The cosine of this angle is given as

$$\cos \alpha = (\boldsymbol{E} \odot \boldsymbol{R}_L) = (\boldsymbol{L} \odot \boldsymbol{R}_E) \tag{10.151}$$

Lafortune and Willems show that the modified Phong model obeys the Helmholtz reciprocity property and that the total hemispherical reflectivity of this model becomes

$$\begin{aligned}
\rho(\boldsymbol{x}, \omega_i, \omega_r) &= \int_{\mathcal{H}_r^2} f_r(\boldsymbol{x}, \omega_i, \omega_r) \cdot \cos(\theta_r) \cdot \partial\omega_r \\
&= \int_{\mathcal{H}_r^2} \left(k_d \cdot \frac{1}{\pi} + k_s \cdot \frac{n_s + 2}{2\pi} \cdot \cos^{n_s} \alpha \right) \cdot \cos(\theta_r) \cdot \partial\omega_r \\
&= k_d + k_s \cdot \frac{n_s + 2}{2\pi} \int_{\mathcal{H}_r^2} \cos^{n_s} \alpha \cdot \cos(\theta_r) \cdot \partial\omega_r \\
&= \rho_d + \rho_s(\boldsymbol{x}, \omega_i, \omega_r)
\end{aligned}$$

As the integral $\int_{\mathcal{H}_r^2} \cos^{n_s} \alpha \cdot \cos(\theta_r) \cdot \partial\omega_r$ reaches its maximum value of $2\pi/(n_s + 2)$ at a normal incident direction (i.e. when $\alpha = \theta_r$), we have

$$\rho_{\max} = k_d + k_s$$

In order for the BRDF to obey the energy conservation property we must have

$$k_d + k_s \leq 1 \tag{10.152}$$

Thus, the Lafortune and Willems modifications of the Phong model make it physically plausible.

10.11.3 The Cook-Torrance Shader

A model for the off-specular reflections from a rough surface was first developed by Torrance and Sparrow [Torrance 67], first used in computer graphics by Blinn [Blinn 77] and later perfected by Cook and Torrance [Cook 81 & 82].

The Cook-Torrance shader model is based on the micro-facet nature of surfaces and accounts for the probability distribution of the micro-facet normals and the shadowing and masking of micro-facets, but does not account for multiple scattering which could be significant with certain types of rough surfaces.

The Cook-Torrance BRDF is a combination of a diffuse and a specular BRDF, thus

$$f_r(\boldsymbol{x}, \boldsymbol{L}, \boldsymbol{E}) = k_d \cdot f_{r,d} + k_s \cdot f_{r,s}(\boldsymbol{x}, \boldsymbol{L}, \boldsymbol{E}) \tag{10.153}$$

where k_d is the diffuse reflectivity (i.e. the fraction of incoming energy that is reflected diffusely) and k_s is the specular reflectivity (i.e. the fraction of incoming energy that is reflected in a specular manner). To enforce energy conservation it is required that

$$k_d + k_s \leq 1$$

The diffuse BRDF is given as

$$f_{r,d} = \frac{1}{\pi} \tag{10.154}$$

and the specular BRDF is given as

$$f_{r,s}(\boldsymbol{x}, \boldsymbol{L}, \boldsymbol{E}) = \frac{F_r \cdot G \cdot D}{(\boldsymbol{N} \odot \boldsymbol{L}) \cdot (\boldsymbol{N} \odot \boldsymbol{E})} \tag{10.155}$$

where F_r is the Fresnel reflectance coefficient for unpolarised light (i.e. $F_r = R$), G is the *geometric attenuation factor* (which represents the shadowing and masking of the micro-facets) and D is the *micro-facet slope distribution function* (which accounts for the distribution of the orientation of the micro-facets).

Torrance and Sparrow derived a function for the geometric attenuation data G by making a number of simplifying assumptions. They assumed that:

- Each micro-facet represents one side (bank) of a symmetrical V-shaped groove.
- The longitudinal axis of each groove is parallel to the average surface plane (i.e. that a given groove does not have a varying depth).
- The lips (top edges) of all grooves are situated at the surface plane.
- The grooves are randomly distributed and do not have a biased orientation (i.e. that grooves do not have a preferred direction).

See Fig. 10.41(a) and (b).

They defined the geometric attenuation function as

$$G_C = \min\left(1, \frac{2 \cdot (\boldsymbol{N} \odot \boldsymbol{H}) \cdot (\boldsymbol{N} \odot \boldsymbol{E})}{(\boldsymbol{E} \odot \boldsymbol{H})}, \frac{2 \cdot (\boldsymbol{N} \odot \boldsymbol{H}) \cdot (\boldsymbol{N} \odot \boldsymbol{L})}{(\boldsymbol{E} \odot \boldsymbol{H})}\right) \tag{10.156a}$$

or equivalently

$$G_C = \min\left(1, \frac{2 \cdot (\boldsymbol{N} \odot \boldsymbol{H}) \cdot (\boldsymbol{N} \odot \boldsymbol{E})}{(\boldsymbol{E} \odot \boldsymbol{H})}, \frac{2 \cdot (\boldsymbol{N} \odot \boldsymbol{H}) \cdot (\boldsymbol{N} \odot \boldsymbol{L})}{(\boldsymbol{L} \odot \boldsymbol{H})}\right) \tag{10.156b}$$

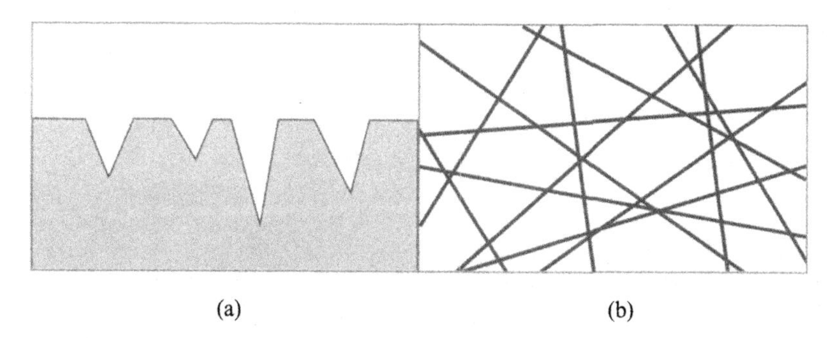

(a) (b)

FIGURE 10.41. (a) The V-shaped grooves of the Torrance-Sparrow rough surface model (side view). (b) The V-shaped Grooves of the Torrance-Sparrow rough surface model (top view).

This second formulation shows that the two fractions are symmetrical.

This function does not account for surface roughness, exhibits sharp discontinuities, and is asymmetric and centred around the $\boldsymbol{R}_L$ direction.

An attenuation function introduced by Sancer [Sancer 69] accounts for the self-shadowing and masking of rough surfaces, although it is more expensive to compute. This function is given as

$$G_S = \frac{1}{G_i + G_r + 1} \tag{10.157}$$

where

$$G_i = \frac{e^{-g_1}}{2\sqrt{\pi \cdot g_1}} - \frac{1}{2} \cdot \operatorname{erfc}\left(\sqrt{g_1}\right) \tag{10.158a}$$

$$G_r = \frac{e^{-g_2}}{2\sqrt{\pi \cdot g_2}} - \frac{1}{2} \cdot \operatorname{erfc}\left(\sqrt{g_2}\right) \tag{10.158b}$$

$$g_1 = \frac{(\boldsymbol{N} \odot \boldsymbol{L})^2}{2\overline{m}^2 \cdot \left[1 - (\boldsymbol{N} \odot \boldsymbol{L})^2\right]} \tag{10.159a}$$

$$g_2 = \frac{(\boldsymbol{N} \odot \boldsymbol{E})^2}{2\overline{m}^2 \cdot \left[1 - (\boldsymbol{N} \odot \boldsymbol{E})^2\right]} \tag{10.159b}$$

The G_i function relates to incident energy and controls the amount of masking that is caused by the surface roughness, while the G_r function relates to reflected energy and controls the amount of blocking that is caused by the surface roughness. See Fig. 10.42.

In the definition of the functions G_i and G_r, $erfc(\boldsymbol{x})$ is the *conjugate error function* or *complimentary error function* (i.e. the integral of the residual error under the *Gauss function* curve), which is defined as

$$\operatorname{erfc}(x) = \frac{2}{\sqrt{\pi}} \int_x^{\infty} e^{-t^2} \partial t = 1 - \operatorname{erf}(x)$$

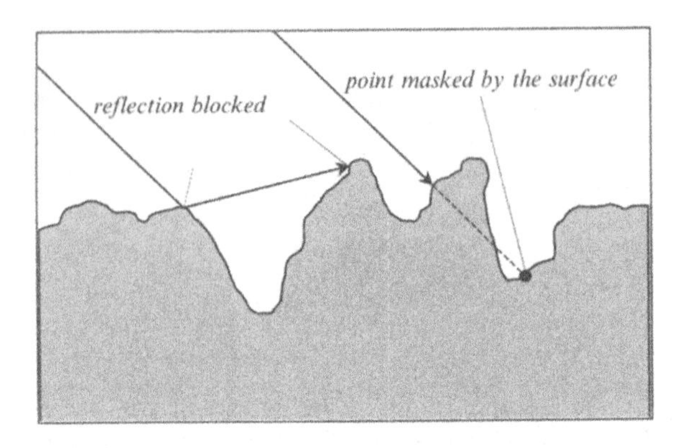

FIGURE 10.42. Self-shadowing of a rough surface.

where erf (x), known as the *error function*, is defined to be the integral of the Gauss function e^{-ax^2} and is given by

$$\begin{aligned}
\operatorname{erf}(x) &= \frac{2}{\sqrt{\pi}} \int_0^x e^{-t^2} \partial t \\
&= \frac{2}{\sqrt{\pi}} \cdot \sum_{n=0}^{\infty} \frac{(-1)^n \cdot x^{2n+1}}{n!\,(2n+1)} \\
&= \frac{2}{\sqrt{\pi}} \cdot \left(x - \frac{x^3}{3} + \frac{x^5}{10} - \frac{x^7}{42} + \cdots \right)
\end{aligned}$$

In the definition of the functions g_1 and g_2, $\overline{m}$ represents the *mean slope* of the surface grooves and it is defined by Bennett and Porteus [Bennett 61] as

$$\overline{m} = \frac{\sigma\sqrt{2}}{\tau} \tag{10.160}$$

where τ is known as the *correlation distance* and represents the average peak-to-valley distance between surface grooves, while σ is known as the *RMS (root mean square) roughness* and represents the average variation of the top or bottom of a groove from the average surface height. See Fig. 10.43.

In the Sancer geometric attenuation function G_s, increasing $\overline{m}$ either by decreasing τ or increasing σ causes self-shadowing (i.e. blocking and masking) to begin at a shallower incidence angle. This function is symmetrical about the normal and only depends on the elevation angle of the incident direction.

There are several choices for the micro-facet slope distribution function D. Blinn suggested three alternatives [Blinn 77]. The first alternative corresponds to the Phong shader and is given as

$$D_1 = b_1 \cdot \cos^{c_1}(\alpha) \tag{10.161a}$$

where α is the *gloss angle* (i.e. the angle between the surface normal $\boldsymbol{N}$ and the halfway vector $\boldsymbol{H}$). Thus,

$$\cos\alpha = (\boldsymbol{N} \odot \boldsymbol{H}) \tag{10.161b}$$

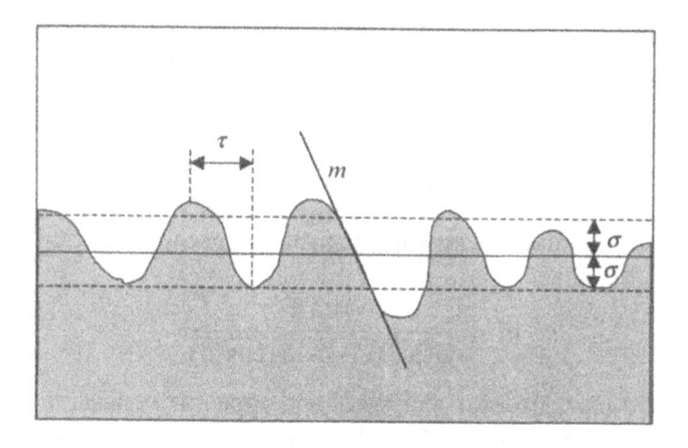

FIGURE 10.43. The geometry of a rough surface.

Here b_1 is an empirical constant (corresponding to the k constants of the Phong model) that determines the sharpness of the micro-facet slopes and c_1 is an empirical constant that represents the width of the spectral lobe and is given as

$$c_1 = -\frac{\ln(2)}{\ln(\cos\beta)} \tag{10.161c}$$

where β is the value of angle α at which the distribution drops at half its peak value.

The second micro-facet slope distribution function is derived from [Torrance 67], is Gaussian in nature and is given as

$$D_2 = b_2 \cdot e^{-(c_2 \cdot \alpha)^2} \tag{10.162a}$$

where, as before, b_2 is an empirical constant that determines the sharpness of the micro-facet slopes and c_2 is an empirical constant that represents the width of the spectral lobe and is given as

$$c_2 = \frac{\sqrt{\ln(2)}}{\beta} \tag{10.162b}$$

The third micro-facet slope distribution function is also derived from [Torrance 67] and is given as

$$D_3 = b_3 \cdot \left(\frac{c_3^2}{\cos^2\alpha \cdot \left(c_3^2 - 1\right) + 1} \right)^2 \tag{10.163a}$$

where, as before, b_3 is an empirical constant that determines the sharpness of the micro-facet slopes and c_3 is an empirical constant that represents the width of the spectral lobe and is given as

$$c_3 = \sqrt{\frac{\cos^2\beta - 1}{\cos^2\beta - \sqrt{2}}} \tag{10.163b}$$

Cook and Torrance introduced an alternative micro-facet slope distribution function [Cook 82], which is derived from [Beckmann 63]. This function is given as

$$D_4 = \frac{1}{4\overline{m}^2 \cdot \cos^4 \alpha} \cdot e^{-\left(\frac{1-\cos^2 \alpha}{\overline{m}^2 \cdot \cos^2 \alpha}\right)} \tag{10.164a}$$

where $\overline{m}$ is the RMS micro-facet slope distribution of the rough surface, which is give as

$$\overline{m} = \frac{\tan \beta}{\sqrt{\ln(2) - 4 \cdot \ln(\cos \beta)}} \tag{10.164b}$$

The Cook-Torrance shader is reciprocal and appears to conserve energy, but it fails to account for the blocked and masked energy. The shader does not treat secondary reflections and instead it treats blocked or masked light as being absorbed by the surface.

10.11.4 The Ashikmin-Shirley Shader

Ashikmin and Shirley developed a shading model based on the Blinn variation of the Phong shading model [Ashikmin 00a, 00b & 00c]. Their version of the Phong model incorporates and refines some of the improvements first introduced to this model by Schlick [Schlick 93], Newmann et al. [Neumann 99a & 99b] and Lafortune et al. [Lafortune 97]. The Ashikmin and Shirley model is a modern version of the Phong model which is energy preserving, reciprocal, has a sophisticated non-Lambertian non-constant diffuse term (which decreases as the incidence angle increases), takes into account the Fresnel effect, is capable of describing both isotropic and anisotropic surface finishes and uses an intuitive set of parameters. In their publications, they state that their BRDF is the sum of a diffuse and a specular component, i.e.

$$f_r(\boldsymbol{x}, \boldsymbol{L}, \boldsymbol{E}) = f_{r,d}(\boldsymbol{x}, \boldsymbol{L}, \boldsymbol{E}) + f_{r,s}(\boldsymbol{x}, \boldsymbol{L}, \boldsymbol{E}) \tag{10.165}$$

If we examine their model more carefully however, we realise that it is not just the sum of a diffuse and a specular component but a linear combination of these two components. This relationship can be expressed as follows:

$$f_r(\boldsymbol{x}, \boldsymbol{L}, \boldsymbol{E}) = [1 - k_s(\alpha)] \cdot k_d \cdot f'_{r,d}(\boldsymbol{x}, \boldsymbol{L}, \boldsymbol{E}) + k_s(\alpha) \cdot f'_{r,s}(\boldsymbol{x}, \boldsymbol{L}, \boldsymbol{E}) \tag{10.166}$$

where $f'_{r,d}$ and $f'_{r,s}$ are the diffuse and specular BRDF components before scaling, k_d is the diffuse reflection coefficient (a constant), $k_s(\alpha)$ is the specular reflection coefficient function and α is the angle whose cosine is used to shape the specular lobe.

This type of shading model was designed to simulate the reflection of light from polished surfaces and an earlier version of it is described in Shirley's Ph.D thesis [Shirley 91]. A polished surface reflector can be thought of as a diffuse reflector covered by a transparent specular coating. A simplified depiction of such a polished surface is shown in Fig. 10.44.

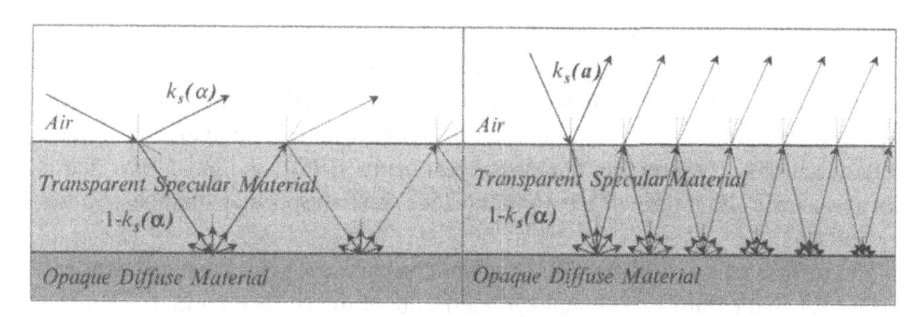

FIGURE 10.44. (a) Large incidence angle reflection off a polished surface. (b) Small incidence angle reflection off a polished surface.

In this figure $k_s\,[\alpha]$ and $[1 - k_s\,(\alpha)]$ are the fractions of the light incident on the first interface surface (air to transparent coating) that are reflected by or transmitted through this surface (i.e. $k_s\,(\alpha) + [1 - k_s\,(\alpha)] = 1$). As the incidence angle increases $k_s\,(\alpha)$ increases and $[1 - k_s\,(\alpha)]$ decreases. Near the grazing angle $k_s\,[\alpha]$ approaches 1 and $[1 - k_s\,(\alpha)]$ approaches 0. When the transmitted light reaches the second interface surface (transparent coating to opaque reflector), it is scattered and some of it will be absorbed though most of it will be reflected in various directions. When the light reflected in this way reaches the first interface surface again (from the interior this time), some of it will undergo an internal reflection and some of it will be transmitted. This process is repeated until all the incident energy is dissipated either by being absorbed or by being scattered/reflected. In this situation the light initially reflected from the first interface surface is represented by the specular reflection component of the model and all the light subsequently transmitted through the first interface surface (having been diffusely reflected off the second interface) is represented by the diffuse reflection component of the model. Thus, in this model it is assumed that the diffuse reflection is the result of subsurface scattering.

The Ashikmin and Shirley model emulates this reflection and subsurface scattering process by using four parameters to control it. These parameters are:

R_s which controls the colour of the specular lobe and specifies the reflectance coefficient of the surface measured at normal incidence (i.e. when $\theta_i = 0^0$). As the specular reflectance coefficient of the surface is wavelength dependant, there will be three such coefficients – one for each primary colour.

R_d which controls the colour of the diffuse lobe and specifies the diffuse reflectance coefficient of the opaque substrate surface under the specular coating. This coefficient should be measured away from the diffuse lobe. As the diffuse reflectance coefficient of the surface is wavelength dependant, there will be three such coefficients – one for each primary colour.

n_T which is the Phong-like specular sharpness in the direction of the unit tangent vector $\boldsymbol{T}$ in the local frame of the surface.

n_B which is the Phong-like specular sharpness in the direction of the unit bitangent vector $\boldsymbol{B}$ in the local frame of the surface.

As we have seen above, the specular part $f_{r,s}$ of the BRDF does not represent a pure specular reflection and cannot be modelled with a combination of Dirac's delta functions. It represents the glossy part of the BRDF and has a lobe that is not symmetric about the surface normal (as we use two specular sharpness values n_T and n_B).

The diffuse part $f_{r,d}$ of the BRDF is not constant, as it is view dependent and has a lobe that exhibits rotational symmetry about the surface normal. This part of the BRDF models the phenomenon by which the diffuse colour of a surface disappears near the grazing angle, as most of the light is reflected in a specular fashion at that angle. Thus the fraction of energy that is scattered diffusely is dependent on the fraction of energy that is reflected in a specular fashion, which in turn depends on the incidence angle.

With this model, to represent metal surfaces $f_{r,d}$ is forced to zero by setting the R_d parameter to zero. To represent brushed metal surfaces we can vary the values of the specular sharpens in the $\boldsymbol{T}$ and $\boldsymbol{B}$ directions in a non-uniform fashion. To represent polished surfaces, such as smooth plastics, we must set R_d and R_s to non-zero values. For plastics typically $R_s = 0.05$. Finally, to represent diffuse surfaces we must set the R_s coefficient to a very small value.

The Ashikmin and Shirley specular BRDF is given as

$$f_{r,s}(\boldsymbol{x}, \boldsymbol{E}, \boldsymbol{L}) = \frac{\sqrt{(n_T+1)\cdot(n_B+1)}}{8\pi} \cdot \frac{(\boldsymbol{N}\odot\boldsymbol{H})^{[n_T(\boldsymbol{T}\odot\boldsymbol{H})^2+n_B(\boldsymbol{B}\odot\boldsymbol{H})^2]/[1-(\boldsymbol{N}\odot\boldsymbol{H})^2]}}{\cos\alpha\cdot\max((\boldsymbol{N}\odot\boldsymbol{L}),(\boldsymbol{N}\odot\boldsymbol{E}))} \cdot F_r(\cos\alpha) \tag{10.167}$$

where $\boldsymbol{H}$ is the halfway vector $\boldsymbol{H} = (\boldsymbol{L}+\boldsymbol{E})/|\boldsymbol{L}+\boldsymbol{E}|$, $\boldsymbol{T}$ and $\boldsymbol{B}$ are the tangent and bitangent vectors, respectively. $F_r()$ is the Fresnel reflectance coefficient computed using Schlick's approximation [Schlick 93], i.e.

$$F_r(\cos\alpha) = R_s + (1-R_s)\cdot(1-\cos\alpha)^5 \tag{10.168}$$

and where α is the angle that is used to shape the lobe of the specular BRDF. Normally $\cos\alpha = (\boldsymbol{L}\odot\boldsymbol{H})$ or $\cos\alpha = (\boldsymbol{E}\odot\boldsymbol{H})$.

The Ashikmin and Shirley diffuse BRDF is given as

$$f_{r,d}(\boldsymbol{x}, \boldsymbol{L}, \boldsymbol{E}) = R_d\cdot(1-R_s)\cdot\left(\frac{28}{23\pi}\right)\cdot\left(1-\left(1-\frac{(\boldsymbol{N}\odot\boldsymbol{L})}{2}\right)^5\right) \cdot\left(1-\left(1-\frac{(\boldsymbol{N}\odot\boldsymbol{E})}{2}\right)^5\right) \tag{10.169}$$

The constant $28/23\pi$ was selected to enforce energy conservation. Observe that the diffuse component of the BRDF is isotropic, as it does not depend on n_T and n_B.

To represent isotropic surfaces we can set $n_T = n_B = n_S$, in which case the specular component BRDF can be simplified to

$$f_{r,s}(\boldsymbol{x}, \boldsymbol{E}, L) = \frac{(n_S + 1)}{8\pi} \cdot \frac{(\boldsymbol{N} \odot \boldsymbol{H})^{ns}}{\cos\alpha \cdot \max\left((\boldsymbol{N} \odot \boldsymbol{L}), (\boldsymbol{N} \odot \boldsymbol{E})\right)} \cdot F_r(\cos\alpha) \tag{10.170}$$

The Ashikmin and Shirley BRDF although based on an empirical model, simulates a number of physical effects and it is physically plausible, thus it is considered to be a physically based BRDF.

10.12 Illumination Models and the Rendering Equation

The application of our knowledge of optics and physics to computer generated rendering is an evolutionary process which at any given time is dependant on the following three factors: the sophistication of the currently available hidden-surface removal algorithms, the degree of realism that is currently deemed acceptable and the computational cost per computer generated frame currently considered affordable. The illumination models used in the computer graphics and animation production process differ significantly from the theoretical physically based models. Computer generated rendering has always been and is likely to continue to be a feasibility versus quality versus cost compromise.

In Section 10.7.3, we have seen how the surface reflectance, transmittance and scattering equations can be used to compute the reflected, transmitted and scattered light distribution from the incident light distribution and the BRDF, BTDF and BSDF of the interface surface. All these equations are of the form

$$L_o(\boldsymbol{x}, \omega_o) = \int_{\mathcal{H}_i^2} f_T(\boldsymbol{x}, \omega_i, \omega_o) \cdot L_i(\boldsymbol{x}, \omega_i) \cdot \cos(\theta_i) \cdot \partial\omega_i \tag{10.171}$$

where the subscripts i and o indicate the *incoming* and *outgoing* radiance and angles, respectively, and the subscript $\boldsymbol{T}$ can be one of r, t or s indicating the bi-directional distribution function of the surface for reflection, transmission or scattering, respectively.

In Section 10.11, we have seen how some empirical (classical) and some physically plausible shading models can be defined.

The one remaining task left to do is to determine how we compute the *incoming radiance* $L_i(\boldsymbol{x}, \omega_i)$. The simulation of the incoming light distribution is often referred to as the *illumination model.* We can distinguish two different types of illumination model, namely: the *local illumination model*, also known as the *direct illumination model*, and the *global illumination model*, also known as the *indirect illumination model.*

The local or direct illumination model is the simplest to explain and understand, as it only deals with the light, from simple light sources, arriving on the surface unobstructed by other surfaces and objects in the scene. This model can only deal with a small number of point or parallel light sources, which can be shaped, in some fashion, to simulate spotlights and spots with barn doors, but can not deal with shadows. Thus this model is considered to be *local*, as it only deals with the local geometry of the surface being lit, and it is considered to be *direct*, as only light arriving directly from a light source is accounted for.

The global or indirect illumination model is considerably more complex, as it has to deal with the indirect illumination that arrives at a given surface, emanating from other surfaces and objects in the scene and accounts for shadows and colour bleeding. Broadly speaking we can distinguish two types of indirect illumination, namely: *specular inter-object illumination* and *diffuse inter-object illumination.*

Specular inter-object illumination is the *coherent illumination* that results from the coherent reflection, transmission or emission of light from or through smooth surfaces. These phenomena are referred to as *specular inter-object reflection, specular inter-object transmission* and *specular emission.*

Diffuse inter-object illumination, on the other hand, is the *incoherent illumination* that results from the incoherent reflection, transmission or emission of light from or through rough surfaces. These phenomena are referred to as *diffuse inter-object reflection, diffuse inter-object transmission* and *diffuse emission.*

Roughly speaking, specular inter-object illumination is best handled by the Monte Carlo ray-tracing family of rendering algorithms and diffuse inter-object illumination is best handled by the finite element radiosity family of rendering algorithms. More recent physically-based algorithms have managed to combine these two types of inter-object illumination using various hybrid techniques.

Both local and global illumination can be represented by a *rendering equation,* which allows us to compute the *outgoing radiance* from any given point on a surface.

10.12.1 Local or Direct Illumination Model

The rendering equation for the local or direct illumination model is relatively easy to explain. To develop this equation we reason as follows.

As we have seen in Section 10.6, if a point $\boldsymbol{x}$ is situated at a distance r from a point light source (located at point $\boldsymbol{x}_s$) and its surface normal subtends an angle θ_s with the direction unit vector, ω_i, pointing towards the light source, then the incident irradiance arriving at a point $\boldsymbol{x}$ is given as

$$E_s(\boldsymbol{x}) = \frac{\Phi_s}{4\pi r^2} \cdot \cos(\theta_s) \tag{10.172}$$

where Φ_s is the radiant flux of the point source and r is distance between the point and the light source (i.e. $r = |\boldsymbol{x}_s - x|$) (see Fig. 10.45).

Now, the incoming radiance from this point source can be expressed in terms of two Dirac delta functions as

$$L_i(\boldsymbol{x}, \omega_i) = \frac{\Phi_s}{4\pi r^2} \cdot \delta(\cos(\theta_i) - \cos(\theta_s)) \cdot \delta(\phi_i - \phi_s) \tag{10.173}$$

The two delta functions ensure that only light arriving from the direction ω_s is taken into account.

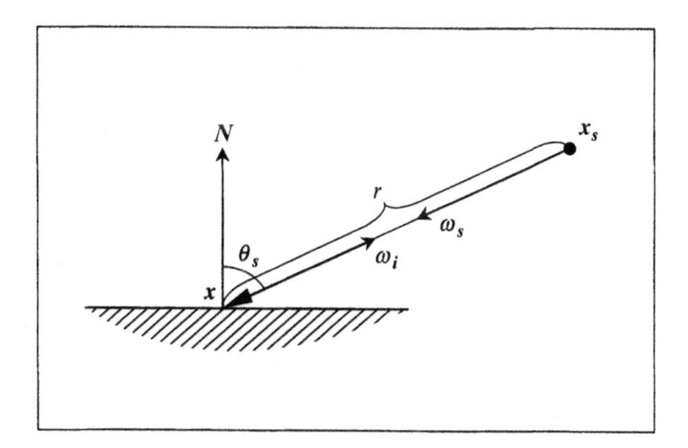

FIGURE 10.45. A point $\boldsymbol{x}$ situated at a distance r from a point light source.

Substituting Eq. (10.173) into Eq. (10.171) we get

$$\begin{aligned} L_o(\boldsymbol{x}, \omega_o) &= \int_{\mathcal{H}_i^2} f_r(\boldsymbol{x}, \omega_i, \omega_o) \cdot L_i(\boldsymbol{x}, \omega_i) \cdot \cos(\theta_i) \cdot \partial\omega_i \\ &= \frac{\Phi_s}{4\pi r^2} \cdot f_r(\boldsymbol{x}, \omega_s, \omega_o) \cdot \cos(\theta_s) \end{aligned} \qquad (10.174)$$

If the virtual scene contains n point sources, then the hemispherical integral simplifies to a sum over the n light sources. This is the equivalent of the traditional Phong illumination model.

The direct illumination model can be easily extended to arbitrary directional light sources, as well as, linear and area sources. See [Amanatides 84] and [Nishita 85].

10.12.2 Global or Indirect Illumination Model

The rendering equation for the global or indirect illumination model is slightly more complex. To develop this equation we reason as follows.

For any point $\boldsymbol{x}$ (on a surface) the radiance leaving the surface, known as the *outgoing radiance* L_o, is equal to the sum of the radiance reflected off the surface (L_r) and the radiance emitted by the surface (L_e). Thus,

$$L_o(\boldsymbol{x}, \omega_o) = L_e(\boldsymbol{x}, \omega_o) + L_r(\boldsymbol{x}, \omega_o) \qquad (10.175)$$

In the case of the *emitted radiance* L_e, the surface acts as a light source and emits light that is not the direct result of incident light being reflected off the surface. A surface may emit light, if it becomes *incandescent*, due to a rise in its temperature or as a result of molecular activity, or if it becomes *phosphorescent*, due to the release of light energy absorbed at an earlier instance. In computer graphics we often use lighting textures to represent features such as window pains or complex lighting fixtures, which can be implemented as *emitter surfaces*.

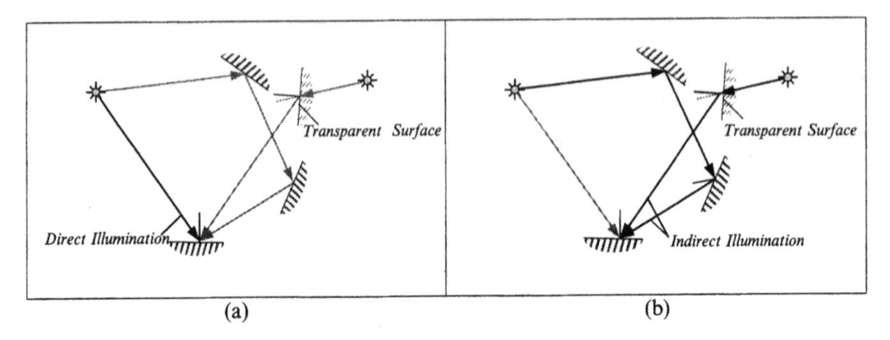

FIGURE 10.46. (a) Radiance due to direct illumination. (b) Radiance due to indirect illumination.

This illumination model can be adjusted to simulate both coherent and incoherent emitted illumination.

The emitted radiance L_e is usually given as part of the scene description, in some pre-computed form, and it does not need to be computed by the rendering equation.

The *reflected radiance* L_r is due to the incident *incoming radiance* L_i that is subsequently reflected by the surface in question. The incoming radiance is due to either direct illumination, arriving on the surface directly from a light source, or due to indirect illumination, arriving at the surface indirectly having been reflected from one or more other surfaces in the scene. See Fig. 10.46.

As we have seen above, there are two main families of rendering algorithms: the Monte Carlo ray-tracing family of rendering algorithms and the finite element radiosity family of rendering algorithms. Each family of algorithms requires a distinct formulation of the rendering equation.

Let us examine first the rendering equation used by the Monte Carlo ray-tracing family of rendering algorithms, which includes the *Photon Mapping* technique [Jensen 96] that we will examine in detail later. For these algorithms, we can construct the rendering equation by combining Eqs. (10.175) and (10.86) to get

$$\begin{aligned} L_o(\boldsymbol{x}, \omega_o) &= L_e(\boldsymbol{x}, \omega_o) + \int_{\mathcal{H}_i^2} f_r(\boldsymbol{x}, \omega_i, \omega_o) \cdot L_i(\boldsymbol{x}, \omega_i) \cdot \partial\omega_i^{\perp} \\ &= L_e(\boldsymbol{x}, \omega_o) \\ &\quad + \int_{\mathcal{H}_i^2} f_r(\boldsymbol{x}, \omega_i, \omega_o) \cdot L_i(\boldsymbol{x}, \omega_i) \\ &\quad \cdot (\boldsymbol{N} \odot \omega_i) \cdot \partial\omega_i \end{aligned} \tag{10.176}$$

As the indirect illumination model is a global illumination model, we must relate the incident incoming illumination of one surface to the outgoing illumination reflected from another surface. Consider the scene geometry shown in Fig. 10.47a. To account for possible occlusions we introduce a *visibility function*

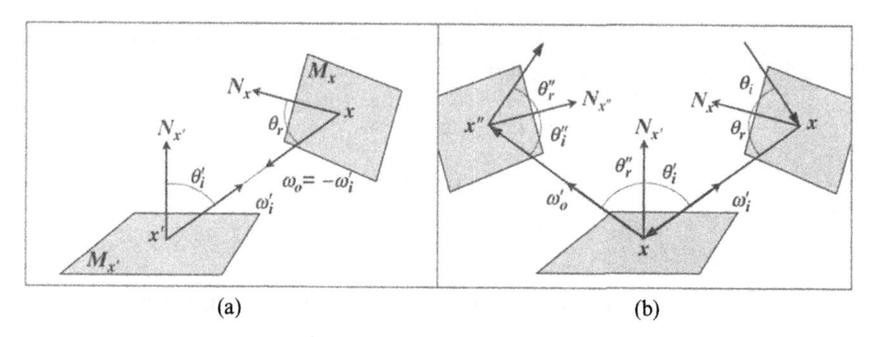

FIGURE 10.47. (a) Two mutually visible points $\boldsymbol{x}$ and $\boldsymbol{x}'$. (b) The three-point form geometry of the light transport equation.

$V\left(\boldsymbol{x}, \boldsymbol{x}'\right)$, which is defined as

$$\forall \boldsymbol{x}, \boldsymbol{x}' \in \mathcal{M} : V\left(\boldsymbol{x}, \boldsymbol{x}'\right) = \begin{cases} 1, & \boldsymbol{x} \text{ and } \boldsymbol{x}' \text{ are mutually visible} \\ 0, & \text{otherwise} \end{cases} \tag{10.177}$$

where $\boldsymbol{x}$ and $\boldsymbol{x}'$ are points on the manifolds $\mathcal{M}_x$ and $\mathcal{M}_{x'}$, which are members of the set of all manifolds in the scene $\boldsymbol{\mathcal{M}}$.

With the ray-tracing family of rendering algorithms, the visibility function is part of the hidden surface elimination algorithms and is normally not included in the rendering equation.

Now, we can express the incoming radiance at point $\boldsymbol{x}'$, on surface $\mathcal{M}_{x'}$, in terms of the outgoing radiance at a point $\boldsymbol{x}$, on a surface $\mathcal{M}_x$. Thus,

$$L_i\left(\boldsymbol{x}', \omega_i'\right) = V\left(\boldsymbol{x}, \boldsymbol{x}'\right) \cdot L_o\left(\boldsymbol{x}, \omega_o\right) \tag{10.178}$$

where ω_i' is the direction unit vector from point $\boldsymbol{x}'$ to point $\boldsymbol{x}$ and ω_o is the opposite direction vector. Thus,

$$\omega_i' = -\omega_o = \frac{\boldsymbol{x} - \boldsymbol{x}'}{\left|\boldsymbol{x} - \boldsymbol{x}'\right|} \tag{10.179}$$

See Fig. 10.47b.

Similarly, ω_o' is the direction unit vector from point $\boldsymbol{x}'$ to point $\boldsymbol{x}''$ and is given as

$$\omega_o' = -\omega_i'' = \frac{\boldsymbol{x}'' - \boldsymbol{x}'}{\left|\boldsymbol{x}'' - \boldsymbol{x}'\right|} \tag{10.180}$$

If we now switch from the hemispherical integral over all the incident directions $\mathcal{H}_i^2$, of Eq. (10.176), to the integral over all the other surfaces in the scene $\boldsymbol{\mathcal{M}}$, we get the rendering equation

$$\begin{aligned} L_o\left(\boldsymbol{x}', \omega_o'\right) = {} & L_e\left(\boldsymbol{x}', \omega_o'\right) \\ & + \int_{\mathcal{M}} f_r\left(\boldsymbol{x}', \omega_i', \omega_o'\right) \cdot L_i\left(\boldsymbol{x}', \omega_i'\right) \cdot \\ & \cdot \boldsymbol{V}\left(\boldsymbol{x}, \boldsymbol{x}'\right) \cdot \boldsymbol{G}\left(\boldsymbol{x}, \boldsymbol{x}'\right) \cdot \partial A_{x'} \end{aligned} \tag{10.181}$$

This equation is sometimes written in arrow notation, in which case it is known as the *three-point form* of the *light transport equation* and is given as

$$
\begin{aligned}
L\left(\boldsymbol{x}' \rightarrow \boldsymbol{x}''\right) = L_e\left(\boldsymbol{x}' \rightarrow \boldsymbol{x}''\right) \\
+ \int_{\mathcal{M}} f_r\left(\boldsymbol{x} \rightarrow \boldsymbol{x}' \rightarrow \boldsymbol{x}''\right) \cdot L\left(\boldsymbol{x} \rightarrow \boldsymbol{x}'\right) \\
\cdot V\left(\boldsymbol{x} \leftrightarrow \boldsymbol{x}'\right) \cdot G\left(\boldsymbol{x} \leftrightarrow \boldsymbol{x}'\right) \cdot \partial A_{\boldsymbol{x}'}
\end{aligned}
\tag{10.182}
$$

In this notation the arrows indicate the flow of light energy from point $\boldsymbol{x}$, through point $\boldsymbol{x}'$, to point $\boldsymbol{x}''$.

In the above equations, $\mathcal{M}$ is the union of all the surfaces in the scene, A_x is the area measure on $\mathcal{M}$, the function G, known as the *geometry function*, represents the change of variables from the original integration variable (step) $\partial\omega_i^{\perp}$ to the new integration variable $\partial A_{\boldsymbol{x}'}$. This relationship is given as

$$\partial\omega_i^{\perp} = G\left(\boldsymbol{x}, \boldsymbol{x}'\right) \cdot \partial A_{\boldsymbol{x}'} \tag{10.183a}$$

$$\text{or} \quad \partial\omega_i^{\perp} = G\left(\boldsymbol{x} \leftrightarrow \boldsymbol{x}'\right) \cdot \partial A_{\boldsymbol{x}'} \tag{10.183b}$$

where

$$G\left(\boldsymbol{x}, \boldsymbol{x}'\right) = \frac{\left|\cos\left(\theta_o\right) \cdot \cos\left(\theta_i'\right)\right|}{\left|\boldsymbol{x} - \boldsymbol{x}'\right|^2} \tag{10.184a}$$

$$\text{or} \quad G\left(\boldsymbol{x} \leftrightarrow \boldsymbol{x}'\right) = \frac{\left|\cos\left(\theta_o\right) \cdot \cos\left(\theta_i'\right)\right|}{\left|\boldsymbol{x} - \boldsymbol{x}'\right|^2} \tag{10.184b}$$

For the radiosity family of rendering algorithms we can simplify the rendering Eq. (10.181), which we have developed for the Monte Carlo ray-tracing family of algorithms, and replace it by a much simpler expression of the outgoing radiosity.

With the radiosity family of rendering algorithms we assume that all the surfaces in the scene are perfect Lambertian reflectors and restrict ourselves to dealing with only diffuse inter-object reflections and diffuse emittance. Under these conditions the outgoing radiance from any point $\boldsymbol{x}'$ is constant in all directions, as the BRDF of a perfectly diffusing surface is independent of the incoming and outgoing directions of light. Thus, we can replace the complex computation of the outgoing radiance $L_o\left(\boldsymbol{x}', \omega_o'\right)$ of Eq. (10.181) by a much simpler expression for the outgoing radiosity $B\left(\boldsymbol{x}'\right)$. To explain how this simplification is justified, we rewrite Eq. (10.65) to get

$$B\left(\boldsymbol{x}'\right) = \int_{\mathcal{H}_i^2} L_o\left(\boldsymbol{x}', \omega_o'\right) \cdot \cos\left(\theta_r'\right) \cdot \partial\omega_o \tag{10.185}$$

But since we are dealing with a perfectly diffuse surface, the outgoing radiance $L_o\left(\boldsymbol{x}'\right)$ is the same in all directions and can be moved outside the integral. Thus,

$$B\left(\boldsymbol{x}'\right) = L_o\left(\boldsymbol{x}'\right) \cdot \int_{\mathcal{H}_i^2} \cos\left(\theta_r'\right) \cdot \partial\omega_o \tag{10.186}$$

Using Equ. (10.64), this reduces to

$$B\left(x'\right) = L_o\left(x'\right) \cdot \pi$$
$$\therefore \quad L_o\left(x'\right) = \frac{B\left(x'\right)}{\pi} \tag{10.187}$$

It should now be apparent that Eq. (10.181) can be simplified to

$$B\left(x'\right) = B_e\left(x'\right) + \int_{\mathcal{M}} f_{r,d}\left(x'\right) \cdot B(x) \cdot V\left(x, x'\right) \cdot G\left(x, x'\right) \cdot \partial A_{x'} \tag{10.188}$$

where $B\left(x'\right)$ is the emitted radiosity from a differential patch at point x'.

Since the BRDF $f_{r,d}\left(x'\right)$ is independent of the incoming and outgoing directions, it can be moved outside the integral. Thus,

$$B\left(x'\right) = B_e\left(x'\right) + f_{r,d}\left(x'\right) \cdot \int_{\mathcal{M}} B(x) \cdot V\left(x, x'\right) \cdot G\left(x, x'\right) \cdot \partial A_{x'} \tag{10.189}$$

Using Eq. (10.95) we can rewrite the above equation as

$$B\left(x'\right) = B_e\left(x'\right) + \frac{\rho_d\left(x'\right)}{\pi} \cdot \int_{\mathcal{M}} B(x) \cdot V\left(x, x'\right) \cdot G\left(x, x'\right) \cdot \partial A_{x'} \tag{10.190}$$

This equation is known as the *radiosity equation*.

10.13 Monte Carlo Method and Monte Carlo Integration

Before we proceed with examining the implementation of physically based renderers we must introduce a mathematical technique that underlies almost all state of the art physically based rendering algorithms. This technique is known as the *Monte Carlo method*. We will use this method to compute accurate estimates of integrals which are required for the evaluation of the components of the rendering equation. We refer to this technique as *Monte Carlo integration*.

The concept of the Monte Carlo method has existed for a long time, but was first formalised by Metropolis and Ulam [Metropolis 49]. Monte Carlo methods allow us to solve problems by estimating the value of an equation using random numbers. Historically they have been applied to the solution of problems of a probabilistic nature.

Let us now examine how the Monte Carlo method can be used to perform numerical integration (also known as *quadrature*). Given the definite integral of a function $f(x)$ over the interval $[a, b]$ its value, known as the *estimand*, is given by

$$I = \int_a^b f(x) \cdot \partial x \tag{10.191}$$

The value of the integral, representing the area under the graph of the function (see Fig. 10.48), can be estimated by computing the mean value of the

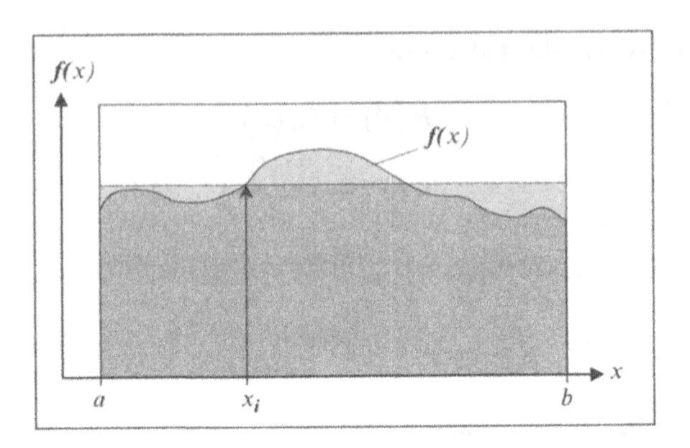

FIGURE 10.48. Monte Carlo Integration.

function $f(x)$ over the interval $[a, b]$ and then multiplying this mean value by the length of the interval $(b - a)$.

The value of this integral can be estimated by picking a uniform random number χ_i from the interval $[a, b]$ and evaluating $f(\chi_i)$. Here the value of the integral of $f(\chi_i)$ is called the *primary estimator* of the integral and is denoted as

$$\langle I \rangle_{prim} = \int_a^b f(\chi_i) \cdot \partial x = (b - a) \cdot f(\chi_i) \tag{10.192}$$

The evaluation of the primary estimator for a specific sample χ_i is known as an *estimate*. Thus, $(b - a) \cdot f(\chi_i)$ gives us an *estimate* of the area under the graph of $f(\boldsymbol{x})$ in Fig. 10.48. If we use N uniformly distributed random sample points $\chi_1, \chi_2, \ldots, \chi_N$ from the range $[a, b]$ to compute N estimates of the integral and average these, then we get a more accurate estimate of the integral $\int_a^b f(\boldsymbol{x}) \cdot \partial x$.

$$\langle I \rangle_{sec} = \frac{1}{N} \cdot \sum_{i=1}^{N} \langle I \rangle_{prim} = (b - a) \cdot \frac{1}{N} \cdot \sum_{i=1}^{N} f(\chi_i) \tag{10.193}$$

This is known as the *secondary estimate* or the *Monte Carlo estimate* of the integral. As we increase the number of samples in the computation of the secondary estimate of the integral, this estimate becomes more accurate and at the limit it becomes equal to the estimand, i.e.

$$\lim_{N \to \infty} \langle I \rangle_{sec} = I \tag{10.194}$$

The standard deviation σ of the secondary estimate $\langle I \rangle_{sec}$ from the true value of the integral I is proportional to the square root of the sample size, i.e.

$$\sigma \propto \frac{1}{\sqrt{N}} \tag{10.195}$$

which means that in order to half the estimation error, we must quadruple the sample size. See [Dutré 94].

10.14 Physically-Based Rendering Algorithms

The eventual output of any computer visualisation process is the production of a rendered image of the input scene. The geometry of a virtual scene can be described by a combination of geometric primitives which can either take the form of a *solid representation* or a *boundary representation*.

A solid representation is usually based on an *implicit functional representation* that describes the shape (i.e. the volume and boundary) of the object in question. A representation of this type is often referred to as an F-rep.

Complex object representations of this form are composed by blending and combining the functional representations of primitive solid objects, such as cubes, cylinders, cones, spheres and super-ellipsoids. Each of these primitives can be represented by a unique implicit function that allows us to test if a given point in space lies inside of, on the surface of or outside the solid object. These simple solid primitives can be combined through blending operations or Boolean algebra operations (such as union, intersection and difference) to form more complex solid objects. Most manufactured objects can easily be composed in this fashion. The F-rep is also ideally suited for the representation of soft objects and ethereal phenomena, such as steam, smoke and fire. It is common to use ray-tracing rendering techniques to render such arbitrary geometries.

A boundary representation, on the other hand, is composed of a set of boundary surfaces that represent the *skin* of the solid object. Such boundary surfaces are represented in the form of parametric equations that describe curved surfaces delimited by their boundary curves or polygons delimited by their outlines. A representation of this type is often referred to as a B-rep.

Boolean algebra operations can also be applied to B-reps to generate more complex forms. With a boundary representation it is much more difficult to determine if a given point in space lies inside, on the surface or outside a solid object. This makes the implementation of Boolean algebra operations on B-reps more difficult. B-reps are ideally suited, however, for the representation of free-form surfaces, which are best suited for the description of bodies of cars, ships and aeroplanes.

Other information that is required by the renderer, in order to be able to render a virtual scene, includes a description of the light scattering properties of the surfaces in the scene, a description of the light sources in the scene and a description of the virtual camera.

The description of the light scattering properties of the various surfaces is provided through the specification of their respective BRDFs, BTDFs and BSDFs, as we have seen earlier in this chapter.

The description of each light source in the scene should include its type (i.e. point, parallel, linear, area or solid shape light source), its position and direction of emission, and its emittance characteristics (i.e. its power or intensity and its colour).

Additionally, for each surface in the scene that is a light emitter we must describe its *emittance distribution function*.

Finally, the description of the virtual camera should include, its position, its direction of view, its up direction and its horizontal and vertical fields of view (i.e. the aspect ratio of the frame), and its resolution.

Having got all this information the renderer evaluates the rendering equation in order to generate an appropriate representation of the illumination in the virtual scene.

Ideally, physically-based renderers should be able to produce a rendering of the virtual scene which is indistinguishable from a photograph of its physical counterpart. Thus, achieving the illusive holy-grail of photorealism. Photorealistic results are best achieved by the accurate evaluation of global illumination effects, such as glossy reflections and caustics (achieved through *specular inter-object illumination*) and soft shadows, indirect illumination and *colour bleeding* (achieved through *diffuse inter-object illumination*). In physically-based rendering correctness of the resulting image is paramount, immaterial of the way in which it is achieved.

There are two main approaches for achieving the production of photo-realistic images. We can produce such images by using either an *object-space approach* or an *image space approach.*

10.14.1 Object-Space Rendering Algorithms

Object-space approaches compute and store a representation of the outgoing radiance function for each surface in the scene, in a pre-processing step. Then, in order to generate an image from a given viewpoint, for each pixel of the image we determine the visible surface or surfaces, using a scan-line, a depth-buffer or a ray-casting visibility algorithm. The average radiance for this pixel is determined by computing the average radiance radiated towards the camera from all the surfaces that are visible through the pixel using the stored radiance values computed in the pre-processing step.

Some examples of object-space rendering algorithms include the basic diffuse radiosity algorithm and other non-diffuse radiosity-style algorithms.

10.14.1.1 The Radiosity Algorithm

The radiosity method was first introduced by Goral et. al [Goral 84]. See also [Cohen 85] and [Nishita 85]. Radiosity is an object-space physically based rendering technique that deals exclusively with perfectly diffuse inter-object reflections (i.e. totally incoherent inter-object reflections). The surfaces of objects in the scene can only be Lambertian reflectors or emitters. Transmitted light and light reflected in a specular fashion cannot be handled by the basic radiosity technique.

In static diffuse scenes, of this sort, the emittance distribution function (EDF) and the BRDFs of the surfaces are direction independent and are only determined by their spatial location and the wavelength of the light incident on them. This simplification implies that the outgoing reflected radiance from a given point on a surface will be perceived with the same intensity and colour regardless of the viewing position. By reducing the dimension of the radiance, employing these simplifications, it becomes feasible to store accurate object-space representations of the radiance of complex scenes. This of course means that the output produced by the pre-processing step can be used to render images from different viewpoints in the scene.

With such diffuse environments it is more appropriate to use the radiosity (i.e. radiant exitance) rather than the radiance to quantify the illumination at a given location and wavelength.

Some non-diffuse radiosity-style algorithms compute the average radiance emitted by each surface in the scene towards all other surfaces in the scene. See [Aupperle 93] and [Stamminger 98]. Some other implementations use an angular rather than a spatial parameterisation to represent the directional dependence of the radiance. See [Immel 86] and [Sillion 91].

The major advantage of representing the scene illumination in object space is that we can reuse the results of the radiance computation, output from the pre-processing step of the these algorithms, to render images of the scene from different viewpoints. Thus these algorithms exhibit *frame-to-frame coherence.* 3D graphics hardware can be used to accelerate the rendering stage of these algorithms.

With these algorithms, a significant proportion of the pre-computed radiance can often be reused even if the geometry of the light emission or the surface scattering properties of the virtual scene are altered. See [George 90], [Chen 90] and [Drettakis 97].

The main disadvantage of the object-space rendering algorithms is the excessive amount of storage required to accurately represent the highly direction dependant radiance functions required to represent specular inter-object illumination. Even for the diffuse inter-object illumination of large virtual scenes, the storage requirements of these algorithms become prohibitive.

10.14.2 Image-Space Rendering Algorithms

Image-space rendering algorithms compute the average incoming radiance at each pixel on the fly, without relying on a pre-processing step to compute an object-space representation of the radiance reflected from each surface of the virtual scene. This family of algorithms solves the rendering equation for each pixel or group of pixels in the rendered image. Although the underlying global illumination method used by all the image-space algorithms is basically the same, the algorithms themselves are quite different. Examples of such algorithms are the *ray-tracing algorithm* [Whitted 80], the *distributed ray-tracing algorithm* [Cook 84], the *path-tracing algorithm* [Kajiya 86][Dutré 94], the *bi-directional path-tracing algorithm* [Lafortune 93][Veach 94], the *Metropolis light transport algorithm* [Veach 97b] and the *photon-mapping algorithm* [Jensen 95a].

The major advantage of the image-space algorithms is that they require very little storage, as they compute the visible surface solution on a per-pixel basis. In general these algorithms are more suitable than the object-space algorithms for more complex virtual scenes and can handle more complex illumination models.

The more sophisticated algorithms of this type can handle both specular inter-object and diffuse inter-object illumination, as well as, deal with volume rendering for participating media.

Let us now examine the family of image-space rendering algorithms in more detail.

10.14.2.1 The Recursive Ray-Tracing Algorithm

The concept of ray tracing first appeared in a 1971 paper by Goldstein and Nagel [Goldstein 71]. In the early eighties, Whitted introduced the concept of recursive ray tracing [Whitted 80].

Ray tracing is a simple and elegant algorithm that is able to handle specular inter-object illumination. Thus, it deals with specular reflection and transmission and generates sharp shadows.

For each pixel of the image, we spawn one *primary ray* that starts from the viewing point and passes through the centre of the pixel. If this ray intersects none of the surfaces in the scene, then the pixel is painted with the background colour and we proceed to the next pixel. If the ray, however, intersects some of the surfaces in the scene, we pick the surface whose intersection point is closest to the observer. Once we have identified the closest surface, we compute its unit normal at the point of intersection. This vector is then used in the shading calculations.

How the light is reflected from a surface depends on the type of surface that we are dealing with. If the surface is diffuse (rough), the reflected light will depend only on the illumination arriving directly from the light sources in the scene. Alternatively, if the surface is specular (smooth), the reflected light will depend on the illumination arriving directly from the light sources and on the indirect illumination arriving on this surface after having been reflected off other surfaces in the scene or been transmitted through the surface (if it is transparent). To discover if the point is directly illuminated, we have to trace a *shadow ray* to each of the light sources. If the surface point is visible from a light source, then it will receive direct illumination from this source, otherwise it will not.

If the surface in question is smooth, then we recursively spawn a ray in the reflection direction and if the surface is transparent, a second ray in the transmission direction. We follow these rays repeating the process recursively until we satisfy one of the recursion termination conditions. The recursion will terminate when one of the three following conditions is met. When the ray hits a rough surface, in which case the light reflected from this surface is returned by the ray since this will be the incoming light at the point of origin of the ray. When the ray misses all surfaces in the scene, in which case the ray returns the background colour. Finally, when a user defined maximum level of recursion has been reached, no new rays are generated. Upon returning from a recursive call we accumulate all the calculated radiance contributions.

An outline of the recursive ray-tracing algorithm is presented in Algorithm 10.1. In this algorithm, the function **emitter_shader()** computes the light emitted by an emitter surface and the function **direct_shader()** computes the direct illumination component of the light reflected from the surface.

```
function recursive_ray_tracing_renderer(scene, image)
{
 for each pixel in the image do
  {
   level = 0;

   generate a primary_ray from the eye to the centre of the
   pixel;

   pixel_colour = trace_ray(primary_ray);
  }
} /* recursive_ray_tracing_renderer */

function trace_ray(ray)
{
 find the closest point of intersection of the ray with a
 surface of the scene;

 if there are no intersections then return(background_colour);

 compute the unit normal of the closest surface at the point
 of intersection;
 if surface is emitter
        then colour = emitter_shader(ray, surface, point);
        else colour = 0;

 for each light source do
  {
   generate a shadow_ray to the light source;

   if light source is visible then
    colour = colour
    + direct_shader(point, surface, normal, light) / n_lights;
  }

 if surface is specular then
  {
   level = level + 1;

   if level > max_level then return(colour);
    {
     generate the reflected_ray;
     colour = colour + trace_ray(reflected_ray);
    }

   if surface is transparent then
    {
     generate the transmitted_ray;
     colour = colour + trace_ray(transmitted_ray);
    }
  }
 return(colour);
} /* trace_ray */
```

Algorithm 10.1 The outline of the recursive ray-tracing rendering algorithm.

10.14.2.2 The Distributed Ray-Tracing Algorithm

In 1984, Cook et. al introduced the distributed ray-tracing algorithm [Cook 84]. This algorithm is a refinement of the recursive ray-tracing algorithm that provided correct and easy solutions to a number of previously unresolved problems, including semi-coherent reflections and transmissions, shadows with penumbras, depth of field and motion blur.

With distributed ray-tracing we spawn a number of primary rays for each pixel using an appropriately selected probability. The precise probability distribution of the spawned rays, as well as, their individual direction and origin depend on the effect that we are attempting to simulate. For instance, to get a semi-coherent reflection (i.e. a fuzzy reflection) at a ray intersection point with a diffuse surface instead of spawning a single reflected ray we spawn a number of rays stochastically distributed around the mirror reflection direction. Similarly, to simulate a semi-coherent transmission (i.e. translucency) instead of spawning a single transmitted ray we spawn a number of rays stochastically distributed around the transmittance direction.

To generate penumbras (i.e. soft shadows), which result from the illumination produced by area light sources, instead of spawning a single shadow-ray towards the light source we spawn a number of shadow-rays stochastically distributed on the surface of the area light source. The illumination received from this light source is then made proportional to the number of shadow rays that are unobstructed by other surfaces in the scene.

An outline of the distributed ray-tracing algorithm is presented in Algorithm 10.2. In this algorithm, the function **emitter_shader()** computes the light emitted by an emitter surface and the function **direct_shader()** computes the direct illumination component of the light reflected from the surface.

```
function distributed_ray_tracing_renderer(scene, image)
{
  for each pixel in the image do
   {
    colour = 0;

    for each primary_ray sample do
     {
      generate a primary_ray from the eye through a random
      point in the pixel;

      /* Extension for depth of field */
      perturb the ray to account for the lens position;

      /* Extension for motion blur */
      pick a random time within the inter-frame interval to
      trace the ray;

      colour = colour + trace_ray(primary_ray);
     }
```

```
      pixel_colour = colour / n_ray_samples;
    }
} /* distributed_ray_tracing_renderer */

  function trace_ray(ray)
  {
   find the closest point of intersection of the ray with a
   surface of the scene;

  if there are no intersections then return(background_colour);

  compute the unit normal of the closest surface at the point
  of intersection;

  if surface is emitter then
   {
    if surface is diffuse then
     {
      colour = 0;

      for each ray sample do
       {
        stochastically perturb the ray direction;
        colour = colour +
                emitter_shader(perturbed_ray, surface, point)
                / n_ray_samples;
       }
     }
    else
     colour = emitter_shader(ray, surface, point);
   }
  else
   colour = 0;

  for each light source do
   {
    direct_colour = 0;

    for every shadow_ray do
     {
      generate a shadow_ray to a stochastically selected point
      on the light source;

      if the light source is visible then
       direct_colour = direct_colour +
                   direct_shader(point, surface, normal,
                              light) / n_shadow_rays;
     }

    colour = colour + direct_colour / n_lights;
   }

  determine if the ray is reflected or absorbed using a Russian
  roulette procedure;

  if ray is absorbed then return(colour);

  generate the reflected_ray;

  if surface is diffuse then
```

```
    {
      for each ray sampledo
        {
          stochastically perturb the reflected_ray direction;
          colour = colour + trace_ray(perturbed_reflected_ray)
                                   / n_ray_samples;
        }
    }
  else
    colour = colour + trace_ray(reflected_ray);

  if surface is transparent then
    {
      generate the transmitted_ray;

      if surface is diffuse then
        {
          for each ray sample do
            {
              stochastically perturb the transmitted_ray direction;
              colour = colour + trace_ray(perturbed_transmitted_ray)
                                     / n_ray_samples;
            }
        }
      else
        colour = colour + trace_ray(transmitted_ray);
    }

  return(colour);

} /* trace_ray */
```

Algorithm 10.2 The outline of the distributed ray-tracing rendering algorithm.

This algorithm represents the first effort in an attempt to introduce a certain degree of physical plausibility in the way the rendering equation is computed. The physically based image-space renderers that we will examine next attempt to be physically correct in the way they evaluate the rendering equation.

10.14.2.3 The Path-Tracing Algorithm

Path tracing was introduced by Kajiya in the mid-eighties to provide an efficient way to solve the rendering equation for both local and global illumination [Kajiya 86].

Path tracing is concerned with solving the integration of light energy resulting from the direct illumination arriving directly from area light sources and the indirect illumination arriving from other surfaces in the scene. These integration problems are solved using the Monte Carlo integration method, which as we have seen in Section 10.13, produces an estimate of the value of an integral by averaging a number of random primary estimates of the value of the integral. In the context of Monte Carlo ray-tracing algorithms, this means that we need to stochastically spawn a large number of rays within the integration domain in order to estimate the value of the integral representing the incoming light energy. As with all Monte Carlo methods, by spawning more stochastically scattered rays (i.e. by evaluating more randomly selected primary estimates) we improve the accuracy of our result. Thus path-tracing can be seen as a progressive refinement of the distributed ray-tracing algorithm examined above.

In distributed ray-tracing we spawn a number of stochastically positioned primary rays through the pixel. When one of these rays is reflected from or transmitted through a diffuse (rough) surface it spawns a number of secondary rays, which are stochastically distributed around the specular reflection or transmission directions. This process is repeated recursively until it satisfies one of the recursion termination criteria. The recursion termination criteria are one of the following. A reflected or transmitted ray does not intersect any of the surfaces of the scene. Alternatively, a stochastic test, known as *Russian roulette*, is used to determine if a ray is reflected/transmitted or absorbed. When a ray is absorbed, no further rays are spawn. This recursive approach can very quickly lead to a combinatorial explosion of secondary rays. For instance, starting with 100 primary rays, in a perfectly diffuse scene, after the first reflective bounce we get $100^2 = 10{,}000$ rays, after the second reflective bounce we get $100^3 = 1{,}000{,}000$ rays and so on.

To avoid this type of combinatorial explosion the number of rays spawn at each bounce had to be kept to a minimum. Kajiya noticed that it was better to focus the bulk of our computing effort on the first few lighting events that have undergone the smallest number of reflection or transmission bounces. He decided that it was a lot more cost effective to spawn a large number of primary rays though the pixel and only spawn one ray for every secondary bounce stochastically distributed around the reflection or transmission directions. With this approach we could afford to spawn thousands of primary rays through each pixel. It is not uncommon to spawn between 1,000 and 10,000 rays per pixel.

The only problem with this approach is that we need a large number of primary rays to avoid noise in the generated image. This noise is related to the error in estimating the incoming radiance integral due to the use of Monte Carlo integration method. The better behaved the incoming radiance function is, the fewer primary rays we need to estimate its value. A well-balanced radiance function means that that there are few bright highlights coming from specific directions in the scene. Thus, in a perfectly diffuse environment we can get away with using as few as 100 primary rays per pixel. Trying to reduce the variance/standard deviation of the Monte Carlo integration method, thus allowing us to reduce the number of primary rays that are required to produce an accurate picture, remains an active research topic.

An outline of the path-tracing algorithm is presented in Algorithm 10.3. In this algorithm, the function **emitter_shader()** computes the light emitted by an emitter surface and the function **direct_shader()** computes the direct illumination component of the light reflected from the surface.

```
function path_tracing_renderer(scene, image)
{
  for each pixel in the image do
   {
     colour = 0;
```

```
    for each primary_ray sample do
     {
      generate a primary_ray from the eye through a random
      point in the pixel;

      /* Extension for depth of field */
      perturb the ray to account for the lens position;

      /* Extension for motion blur */
      pick a random time within the inter-frame interval to
      trace the ray;

      colour = colour + trace_ray(primary_ray);
     }

    pixel_colour = colour / n_ray_samples;
   }
} /* path_tracing_renderer */

function trace_ray(ray)
{
 find the closest point of intersection with a surface of the
 scene;

 if there are no intersections then return(background_colour);

 compute the unit normal of the closest surface at the point
 of intersection;

 if surface is emitter then
  {
   if surface is diffuse then
    {
     stochastically perturb the ray direction;
     colour = emitter_shader(perturbed_ray, surface, point);
    }
   else
    colour = emitter_shader(ray, surface, point);
  }
 else
  colour = 0;

 for each light source do
  {
   generate a shadow_ray to a stochastically selected point on
   the light source;

   if light source is visible then
   colour = colour + direct_shader(point, surface, normal,
                                   light) / n_lights;
  }

determine if the ray is reflected, transmitted or absorbed
using a Russian roulette procedure;

 if ray is absorbed then return(colour);

if surface is transparent and ray is transmitted then
  {
   compute the transmitted_ray;

   if surface is diffuse then
```

```
      {
        stochastically perturb transmitted_ray;
        colour = colour + trace_ray(perturbed_transmitted_ray);
      }
    else
      colour = colour + trace_ray(transmitted_ray);
  }
else
  {
    compute the reflected_ray;

    if surface is diffuse then
      {
        stochastically perturb reflected_ray;
        colour = colour + trace_ray(perturbed_reflected_ray);
      }
    else
      colour = colour + trace_ray(reflected_ray);
  }

return(colour);
} /* trace_ray */
```

Algorithm 10.3 The outline of the path-tracing rendering algorithm.

10.14.2.4 The Bi-directional Path-Tracing Algorithm

The bi-directional path-tracing algorithm was first developed by Lafortune and Willems [Lafortune 93] and a year later it was independently developed by Veach and Guibas [Veach 94]. Although both algorithms achieve very similar results, their underlying theoretical framework is quite different.

As we have seen above, with path tracing the primary estimator of the radiance for a given pixel is calculated by tracing a primary ray from the viewing point through the pixel being considered. At the intersection point of this ray with the surface of the scene, closest to the eye, one or more shadow rays are traced towards each of the light sources to determine the direct illumination contribution of each source. This contribution is only accumulated if the light source and the intersection point are mutually visible. Then, we use the Russian roulette stochastic test to determine if the ray (incident on this surface) is absorbed or if it continues its random walk, being reflected/transmitted from surface to surface. This process is repeated recursively until the ray misses all the surfaces in the scene or is absorbed by a surface. The primary estimator determined in this way by a single random walk is likely to have a large variance (i.e. it is a poor estimate of the true value of the radiance). This large variance is mainly due to the way indirect illumination is sampled. The path-tracing algorithm attempts to remedy this by computing a more accurate secondary estimate that is the average of a large number of primary estimates.

In a scene that is primarily illuminated by indirect illumination very few shadow rays are likely to reach any given light source. This will cause a large variance in both the primary and secondary estimates of the radiance due to direct illumination, resulting in high frequency noise in the rendered image. Bi-directional path tracing attempts to resolve this problem by the following technique. Instead of

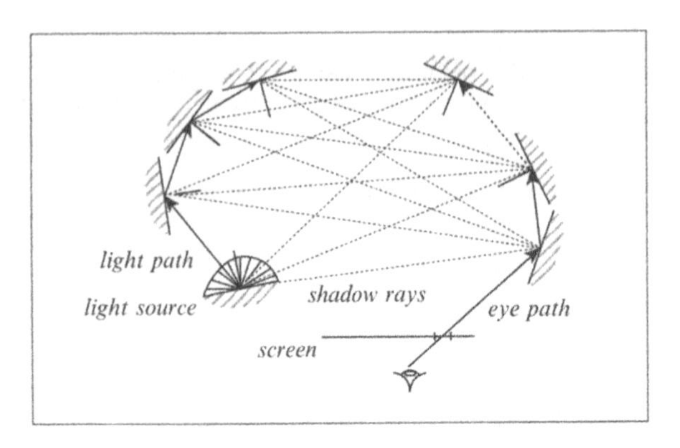

FIGURE 10.49. The geometry of bi-directional path tracing.

just tracing a random walk from the eye into the scene, known as the *eye-path*, in parallel we also trace a random walk from a randomly selected light source into the scene, known as the *light-path*. See Fig. 10.49.

In constructing the light-path, a light source is selected probabilistically depending on its power (brightness). Thus, brighter light sources are more frequently selected than dimmer sources. The starting point on the light source and the starting direction of the light-path are also selected probabilistically depending on the direction of the power distribution of the source. Thus, directions towards which the light source is brighter (i.e. emits more light) are selected more frequently than directions in which the source is dimmer.

In constructing the eye-path, the starting direction of the primary ray from the eye through the pixel is also selected probabilistically according to some random distribution.

The construction of both the eye and light-paths proceeds as follows. When a ray hits a surface it is reflected, transmitted or absorbed, depending on a probability computed by a Russian roulette procedure. This process is repeated recursively until the ray is absorbed or it misses all the surfaces in the scene.

When both the eye and light random walk paths have been constructed, we can proceed with the computation of the primary estimate of the incoming radiance at the pixel by following the eye-path and accumulating the radiance arriving at each hit point (intersection point) on this path. To compute the radiance arriving at a particular hit point, we connect this hit point with every hit point on the light-path (including the first point, which is a point on the light source). This allows us to accurately compute the incoming radiance contributions from both direct and indirect illumination. Each shadow ray that is unobstructed by another surface contributes to this computation. Having computed the incoming radiance at this point, we can, in turn, compute the outgoing radiance from this point. Once we have reached the end of the eye-path, the algorithm returns a primary estimate of the incoming radiance at the pixel.

Once again, as this is a probabilistic estimate, a more accurate secondary estimate of the radiance, incoming at the pixel, can be computed by averaging a large number of primary estimates, which is achieved by spawning a large number of such random walks per pixel.

An outline of the bi-directional path-tracing algorithm is presented in Algorithm 10.4. This algorithm uses three primitive functions that are not shown, namely: **store_light_hit_point(), store_eye_hit_point()** and **accumulate_shade()**.

```
function bidirectional_path_tracing_renderer(scene, image)
{
  for each pixel in the image do
   {
    colour = 0;

    for each ray sample do
     {
      light_path = generate_light_path(scene);

      eye_path = generate_eye_path(scene, pixel);

      colour = colour + combine_paths(light_path, eye_path);
     }

    pixel_colour = colour / n_ray_samples;
   }

} /* bidirectional_path_tracing_renderer */

function generate_light_path(scene);
{
  empty light_path;

  stochastically select a light source, a point on the source
  and an initial direction for the light_path;

  trace_light_path(scene, light_path, light_ray, light_energy);

  return(light_path);

} /* generate_light_path */

function trace_light_path(scene, path, ray, incoming_energy);
{
  compute intersection of the light ray with all the surfaces
  in the scene and select the closest intersection_point;

  if no intersections exist then return();

  determine if the ray is to be reflected, transmitted or
  absorbed using a Russian roulette procedure;

  if ray is absorbed then return();

  if ray is transmitted and surface is transparent then
   {
    compute the transmitted_ray and the outgoing_energy from
    the incoming_energy and the surface BTDF;
```

```
    if surface is diffuse then
     {
      perturb transmitted_ray;
      trace_light_path(scene, path, perturbed_transmitted_ray,
                       outgoing_energy);
     }
    else
     trace_light_path(scene, path, transmitted_ray,
                      outgoing_energy);
  }
 else
  {
    store_light_hit_point(path, intersection_point,
                          incoming_energy);

    compute the reflected_ray and the outgoing_energy from the
    incoming_energy and the surface BRDF;

    if surface is diffuse then
     {
      perturb reflected_ray;
      trace_light_path(scene, path, perturbed_reflected_ray,
                       outgoing_energy);
      }
    else
     trace_light_path(scene, path, reflected_ray,
                      outgoing_energy);
  }

 return();

} /* trace_light_path */

function generate_eye_path(scene, pixel);
{
 empty eye_path;

 generate a primary eye_ray from the eye through a random
 point in the pixel area;

 /* Extension for depth of field */
 perturb the eye_ray to account for the lens position;

 /* Extension for motion blur */
 pick a random time within the inter-frame interval to trace
 the eye_ray;

 transmission_factor = 1;

 trace_eye_path(scene, eye_path, eye_ray, transmission_factor);

 return(eye_path);

} /* generate_eye_path */

function trace_eye_path(scene, path, ray, transmission_factor)
{
 compute intersection of the eye ray with all the surfaces in
 the scene and select the closest intersection_point;
```

```
 if no intersections exist then return();

 determine if the ray is to be reflected, transmitted or
 absorbed using a Russian roulette procedure;
 if ray is absorbed then return();

 if ray is transmitted and surface is transparent then
  {
   compute the transmitted_ray and scale the
    transmission_factor using the surface BTDF;

   if surface is diffuse then
    {
     perturb transmitted_ray;
     trace_eye_path(scene, path, perturbed_transmitted_ray,
                    transmission_factor);
    }
   else
    trace_eye_path (scene, path, transmitted_ray,
                    transmission_factor);
  }
 else
  {
   store_eye_hit_point(path, intersection_point,
                       transmission_factor);

   compute the reflected_ray;

   if surface is diffuse then
    {
     perturb reflected_ray;
     trace_eye_path(scene, path, perturbed_reflected_ray,
                    transmission_factor);
    }
   else
    trace_eye_path(scene, path, reflected_ray,
                   transmission_factor);
  }

 return();

} /* trace_eye_path */

function combine_paths(light_path, eye_path)
{
 path_colour = 0;

 trace_path(light_path, eye_path, path_colour);

 return(path_colour);

} /* combine_paths */

function trace_path(light_path, eye_path, colour)
{
 if eye_path.next_node is not empty then
  {
   trace_path(light_path, eye_path.next_node, colour);
```

```
    for each node in the light path do
     {
      generate a shadow ray from the eye_path_node.point to the
      light_path_node.point;

      if these two points are mutually visible then
       accumulate_shade(eye_path_node, eye_path_node, colour);
     }
  }

 return();

} /* trace_path */
```

Algorithm 10.4 The outline of the bi-directional path-tracing rendering algorithm.

The **store_light_hit_point()** function progressively builds the light-path by creating a linked list of light-path hit points, while the **store_eye_hit_point()** function progressively builds the eye-path by creating a linked list of eye-path hit points.

As can be seen from the function **trace_light_path()**, when the light-ray is reflected from an opaque surface, then a hit point is entered in the light-path and the incoming light energy is recorded, and the energy of the reflected ray is attenuated to account for the reflectance characteristics of the surface. But, when the light-ray is transmitted through a transparent surface, then a hit point is not entered in the light-path and the energy of the transmitted ray is attenuated to account for the transmittance characteristics of the surface. See the left-hand side of Fig. 10.50. Figure 10.51a shows that after the light path has been constructed we now have one direct light source and three indirect light sources representing the four hit points on the light-path.

Analogously, as can be seen from the function **trace_eye_path()**, when the eye-ray is reflected from an opaque surface, then a hit point is entered in the eye-path and the cumulative transmission factor is recorded. But, when the eye-ray is transmitted through a transparent surface, then a hit point is not entered in the eye-path and the cumulative transmission factor is attenuated to account for

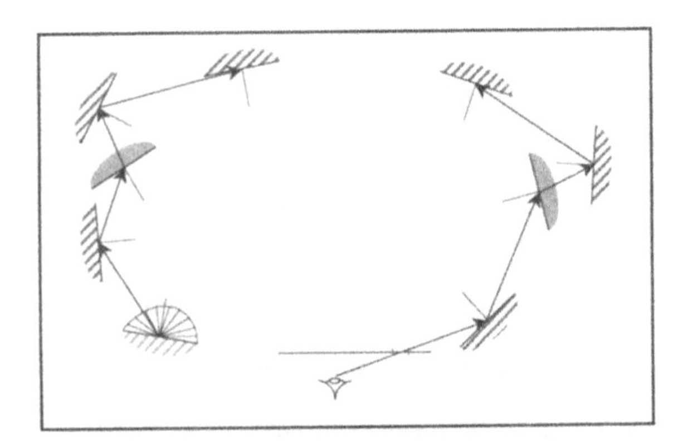

FIGURE 10.50. Tracing the eye-path and the light-path.

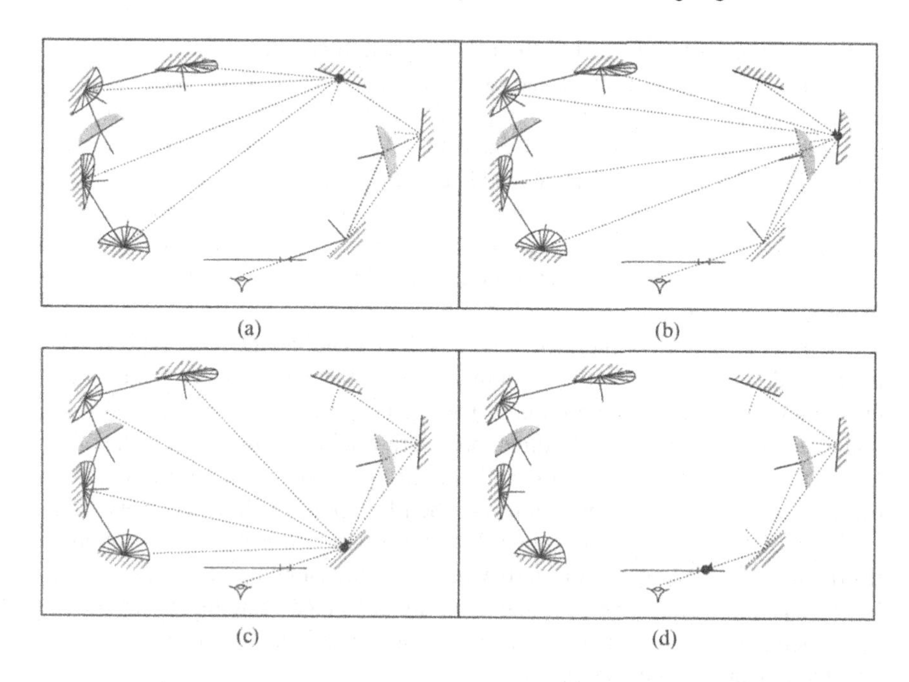

FIGURE 10.51. Tracing the eye-path from the deepest to the shallowest hit point.

the transmittance characteristics of the surface (thus, accounting for the amount of radiance being absorbed by this surface). See the right-hand side of Fig. 10.50.

The **accumulate_shade()** function computes the reflected radiance outgoing from a given eye-path hit point and arriving at the hit point that is positioned immediately before it on the eye-path (i.e. the hit point that is one ray-bounce closer to the eye than the given hit point). Thus, the computation of the reflected radiance proceeds from the deepest eye-path hit point (i.e. the hit point which is the largest number of bounces away from the eye) to the shallowest eye-path hit point (i.e. the hit point which is the smallest number of bounces away from the eye) and finally the eye. As this recursive computation proceeds, indirect illumination arriving from deeper hit points of the eye-path, indirect illumination arriving from light-path hit points and direct illumination arriving from the light source are taken into account. See Fig. 10.51(a)–(d).

A more detailed description of this algorithm and the expressions for the calculation of the illumination contributions can be found in [Lafortune 93].

10.14.2.5 The Metropolis Light Transport Algorithm

The Metropolis Light Transport (MLT) algorithm was introduced by Veach and Guibas [Veach 97b]. This algorithm, which is used to solve the light transport problem, is based on a Monte Carlo statistical simulation approach inspired by the Metropolis sampling technique. The Metropolis sampling technique was first used in computational physics by Metropolis and Ulam while they were working on the Manhattan Project [Metropolis 49], [Metropolis 53].

The general approach used in the MLT algorithm can be outlined as follows. To render an image of a 3D scene, we start with a single light transport path, known as the *current path*, and we progressively generate a set of alternative paths by stochastically mutating the current path. We use an appropriately selected probability, to accept or reject a mutated path, ensuring that the retained paths are sampled with an order that reflects their statistical contribution to the ideal image. Then, we estimate this ideal image by sampling a large number of the mutated paths and by recording their positions on the image plane that is represented by a 2D array in memory.

The MLT algorithm is unbiased, it is capable of handling the most general geometric and BSDF models, it is economical in storage and it can be significantly faster than other unbiased algorithms. The MLT algorithm is very different from both the path-tracing and the bi-directional path-tracing algorithms. Unlike other Monte Carlo methods, instead of randomly sampling the value of a function in order to estimate the value of its integral, the Metropolis method generates a distribution of samples to the unknown function value. To achieve this sampling distribution the MLT algorithm starts with a random sampling of the space of all light paths in the scene. These initial paths are generated using the bi-directional path-tracing algorithm and are subsequently cloned and mutated in order to compute the radiance of the final image.

The most important advantage of the MLP algorithm is that it explores the path space locally, selecting more frequently mutations arrived at by applying minor incremental modifications to the current path. Using this progressive refinement approach has the following beneficial consequences. The average cost for each sample path is relatively small, as very few rays are used. Once an important path is identified, nearby paths are employed as well, thus spreading the cost of determining such a "good" path over many neighbouring paths. The set of mutation operations applied, by the MLT algorithm, to a "good" path is easy to extend. By selecting mutations that retain certain of the properties of a given "good" path while changing other properties, we can take advantage of any type of coherence that is present in the scene. In this way, it is often possible to deal with different types of lighting problems more effectively by designing specialised mutations to handle these particular situations.

The propensity of the MLT algorithm to concentrate on incremental changes to a path, once it has found a "good" path, also leads to one of the main weaknesses of this algorithm. With scenes that do not exhibit space coherence, the algorithm may be caught by one particular feature and be prevented from converging quickly. Consider, for instance, a scene containing a surface with a grid of holes lit from behind. The MLT algorithm can be "trapped" by one of the holes and will fail to investigate properly the illumination from neighbouring holes.

A more detailed explanation of the algorithm can be found in [Veach 97a] and [Veach 97b].

10.14.2.6 The Photon-Mapping Technique

The photon-mapping technique was developed by Jensen and Christensen as an efficient alternative to pure Monte Carlo ray-tracing techniques [Jensen 95a]. This technique de-couples the representation of the illumination from the representation of the geometry of the scene. Thus, allowing us to handle arbitrarily complex geometric models and BRDFs.

To best visualise the *photon map* we may think of it as the cache of all the light paths in the bi-directional path-tracing algorithm. The photon map could indeed be used for this purpose. It is however used to estimate the illumination in the scene based on an estimation of the light energy density. The estimation error resulting from the use of the photon map, to estimate the illumination of the scene, results in low frequency noise, as opposed to the high frequency noise resulting from using the traditional Monte Carlo techniques. The density estimation method that uses a photon map is much faster than the pure Monte Carlo techniques. The main disadvantage, however, of this estimation method is that it is biased.

The algorithm that generates, stores and uses illumination as points on the surfaces of objects in the scene is known as *photon mapping* and the data structure that is used to store these illumination points is known as the photon map. The technique that is used to generate the illumination points is known as the *photon-tracing algorithm*. Thus, a renderer that uses the photon-mapping technique has two distinct passes. The *photon-mapping pass*, which builds the photon map data structure by spawning *photon rays* from the light sources and tracing them through the objects in the scene and the *rendering pass*, which renders the scene using the illumination information stored in the photon map (thus speeding up the rendering process).

As this has proved to be a very influential algorithm, we will examine it in some detail.

10.14.2.6.1 The Photon-Mapping Pass

The photon-mapping pass is an essential pre-processing step of any rendering algorithm that uses the photon-mapping technique. During this pass photons are emitted from the light sources, their paths are traced through the scene and when they hit a diffuse surface their location and power are recorded in the photon map.

10.14.2.6.2 Emission of Photons

A large number of photons are emitted by each light source in the scene. The power (i.e. the wattage) of a light source is divided equally among all the photons that it emits. Thus, each emitted photon transports a fraction of the power of the light source. The Jensen-Christensen model supports many different types of light source.

Diffuse point light sources emit photons uniformly in all directions using one of two Monte Carlo sampling techniques. *Explicit sampling*, which randomly selects two spherical coordinate angles, and *rejection sampling*, which randomly generates points inside a unit cube and selects the first such point that lies inside the unit sphere.

Spherical light sources emit photons in all directions. First a random point is selected on the surface of the light source sphere and then a random direction is selected on the hemisphere above this point. A similar procedure is followed for polygonal *square light sources*.

For *directed light sources*, which are used to simulate very distant light sources, we enclose the scene in a bounding sphere which when projected onto the ground plane produces a circle. Random points in this circle can be used as the terminating points of incoming photon beams from the direction of the parallel light source.

Complex three-dimensional shapes can also be used as light sources. In this case, the photon ray emission points and directions are selected using a rejection-sampling scheme.

10.14.2.6.3 Scattering and Tracing of Photons

After a photon is emitted by a light source, it is traced through the scene by the photon-tracing algorithm. The photon tracing algorithm works in a very similar fashion to a ray tracing algorithm, except that photons distribute flux while rays accumulate radiance. This is significant in the case of refraction, where radiance changes according to the relative refractive index of the interface surface while flux does not.

When a photon arrives at an interface surface it can be reflected, transmitted or absorbed. This determination is made using a Russian roulette procedure, which acts as an importance-sampling technique. Here, a probability distribution function serves to eliminate the statistically insignificant parts of the domain of our problem.

10.14.2.6.4 Storage of Photons

As we have seen above, when a photon hits a specular surface it can be reflected, transmitted or absorbed. When it hits a non-specular surface, however, it is stored in the photon map. See Fig. 10.52. Photons represent incoming illumination (flux) at a given point on the surface. Thus, we can use the photons, stored in the photon map, to approximate the reflected illumination at several points on the surface.

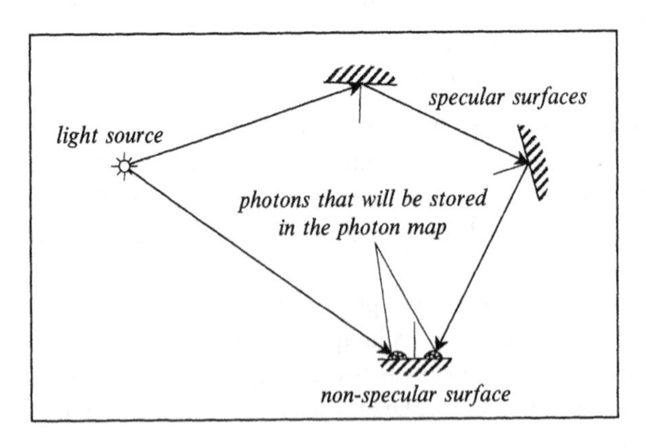

FIGURE 10.52. Photon tracing.

The photon map is stored as a *left-balanced kd-tree* data structure, which is very efficient to traverse [Bentley 79]. Essentially, the kd-tree is an axis-aligned *binary space partition tree* (*BSP-tree*). Each node of this tree stores a photon. Each photon is represented by the x, y, z coordinates of the point of incidence of the photon ray on a surface (which is stored as three floats), by its power (which is stored as four bytes), by its direction vector (which is given by the θ and ϕ angles and stored in compressed form as two bytes) and by a kd-tree flag (which is stored as a short).

Once the photon-tracing algorithm is completed the kd-tree is balanced to speed up the random access of its nodes.

10.14.2.6.5 Photon Density Estimation

The photon map represents the incoming flux on the surfaces of the scene. Each photon can be thought of as transporting a package of energy that represents a fraction of the power of the light source that emitted it. Thus, the photon map contains information indicating that a given region of the scene has received some direct or indirect illumination from a light source.

Looking at a single photon we can not tell how much light a given region has received and we must compute the photon density $\partial\Phi/\partial A$ and to estimate the irradiance for a small region surrounding a given point on a surface of the scene. We can approximate the incoming flux $\Phi_i(\boldsymbol{x})$ at a point $\boldsymbol{x}$ on a surface of the scene by finding the n photons, stored in the photon map, which are the closest neighbours of this point. All these photons will be enclosed in a sphere of radius r_x and each photon will have power $\Delta\Phi_p(\omega_p)$. See Fig. 10.53.

Now, the outgoing radiance $L_r\,(\boldsymbol{x}, \omega_x)$ reflected from this point can be approximated as

$$L_r\,(\boldsymbol{x}, \omega_x) \approx \sum_{p=1}^{n} f_r\left(\boldsymbol{x}, \omega_p, \omega_x\right) \cdot \frac{\Delta\Phi_p\left(\omega_p\right)}{\Delta A} \tag{10.196}$$

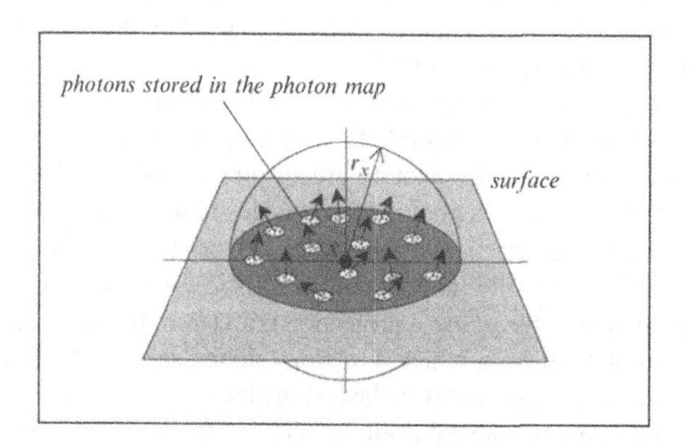

FIGURE 10.53. Photons in the neighbourhood of point $\boldsymbol{x}$.

If we assume that the region around point $\boldsymbol{x}$ is flat, then the intersection of the surface and the sphere containing the photons can be taken as

$$\Delta A = \pi \cdot r_x^2 \tag{10.197}$$

Substituting Eq. (10.197) into Eq. (10.196) we get an estimate of the outgoing reflected radiance from point $\boldsymbol{x} \cdot \nabla$ Thus,

$$L_r(\boldsymbol{x}, \omega_x) \approx \frac{1}{\pi \cdot r_x^2} \sum_{p=1}^{n} f_r(\boldsymbol{x}, \omega_p, \omega_x) \cdot \Delta \Phi_p(\omega_p) \tag{10.198}$$

The accuracy of this estimate depends on the number of photons in the photon map. The larger this number the better the estimate.

10.14.2.6.6 The Rendering Pass

The photon map created during the photon-tracing pass can now be used to render an image of the scene. The renderer is composed of a simple ray tracer that uses the radiance estimate to determine the reflected radiance component due to all the diffuse reflections and a recursive ray tracer that determines the reflected radiance component due to all the specular reflections and transmissions.

Unlike other Monte Carlo ray-tracing techniques, photon mapping is ideally suited for rendering *caustics*. Caustics occur when light that has been reflected from or transmitted through one or more smooth (specular) surfaces reaches a rough (diffuse) surface.

To improve the quality of the rendered diffuse inter-object reflections, often perceived as colour bleeding, we need to increase the number of photons that are stored in the photon map and the number of photons that are included in the neighbourhood of a given point when computing the diffuse component of the reflected radiance estimate at this point.

For scenes that exhibit an even balance of specular and diffuse reflections we can use two photon maps. One map, known as the *global photon map*, to store the indirect illumination, and a second map, known as the *caustics photon map*, to store caustics. The caustics photon map can be used by the recursive ray tracer that computes the direct illumination.

The caustics photon map contains photons that have undergone at least one specular reflection or transmission before arriving at a diffuse surface. After a collision with a diffuse surface photons are absorbed. In the photon-tracing pass, while populating the caustics photon map it is desirable to concentrate the emission of photons in the directions of specular surfaces in the scene. These could be identified either manually (allowing more artistic control) or automatically by the renderer. It is possible to use a projection map (from the point of view of the light source) to determine in which directions shinny surfaces lie so that we can concentrate the photon emissions in these directions.

The global photon map contains all the photons that reached a diffuse surface in the scene. These photons represent direct and indirect illumination, as well as, caustics. The rendering algorithm must ensure that the caustics term is not added more than once in the rendering equation. The global photon map is populated by tracing photons towards all the surfaces in the scene and storing them when they reach a diffuse surface. Such photons may be absorbed or further reflected or refracted to reach other surfaces.

The final image is rendered using a distributed ray-tracing algorithm. Here, the incoming radiance at each pixel is computed by averaging a large number of sample estimates. As we have seen earlier in this chapter, a BRDF, f_r, is often composed of a specular term, $f_{r,s}$, and a diffuse term, $f_{r,d}$. Thus,

$$f_r(\boldsymbol{x}, \omega_i, \omega_o) = f_{r,s}(\boldsymbol{x}, \omega_i, \omega_o) + f_{r,d}(\boldsymbol{x}, \omega_i, \omega_o) \tag{10.199}$$

Similarly, the incoming radiance can be thought of as the sum of three incoming radiances. The *incoming direct illumination radiance* $L_{i,l}(\boldsymbol{x}, \omega_i)$, which is illumination arriving on the surface directly from the light sources. The *incoming caustics illumination radiance* $L_{i,c}(\boldsymbol{x}, \omega_i)$, which is indirect illumination arriving on the surface as a result of one or more specular reflections or transmissions. The *incoming indirect illumination radiance* $L_{i,d}(\boldsymbol{x}, \omega_i)$, which is indirect illumination arriving on the surface as a result of one or more diffuse inter-object reflections or transmissions. Thus, the incoming radiance can be written as

$$L_i(\boldsymbol{x}, \omega_i) = L_{i,l}(\boldsymbol{x}, \omega_i) + L_{i,c}(\boldsymbol{x}, \omega_i) + L_{i,d}(\boldsymbol{x}, \omega_i) \tag{10.200}$$

Recall, from Section 10.7.3, that the outgoing reflected radiance from a point $\boldsymbol{x}$ on a surface is given as

$$L_o(\boldsymbol{x}, \omega_o) = \int_{\mathcal{H}_i^2} f_r(\boldsymbol{x}, \omega_i, \omega_o) \cdot L_i(\boldsymbol{x}, \omega_i) \cdot (\omega_i \odot N_x) \cdot \partial\omega_i \tag{10.201}$$

where $\boldsymbol{N_x}$ is the unit normal vector of the surface at point $\boldsymbol{x}$.

Substituting Eqs. (10.199) and (10.200) into Eq. (10.201), we get

$$\begin{aligned} L_o(\boldsymbol{x}, \omega_o) = &\int_{\mathcal{H}_i^2} f_r(\boldsymbol{x}, \omega_i, \omega_o) \cdot L_{i,l}(\boldsymbol{x}, \omega_i) \cdot (\omega_i \odot N_x) \cdot \partial\omega_i \\ &+ \int_{\mathcal{H}_i^2} f_{r,s}(\boldsymbol{x}, \omega_i, \omega_o) \cdot \left[L_{i,c}(\boldsymbol{x}, \omega_i) + L_{i,d}(\boldsymbol{x}, \omega_i)\right] \cdot (\omega_i \odot N_x) \cdot \partial\omega_i \\ &+ \int_{\mathcal{H}_i^2} f_{r,d}(\boldsymbol{x}, \omega_i, \omega_o) \cdot L_{i,c}(\boldsymbol{x}, \omega_i) \cdot (\omega_i \odot N_x) \cdot \partial\omega_i \\ &+ \int_{\mathcal{H}_i^2} f_{r,d}(\boldsymbol{x}, \omega_i, \omega_o) \cdot L_{i,d}(\boldsymbol{x}, \omega_i) \cdot (\omega_i \odot N_x) \cdot \partial\omega_i \end{aligned} \tag{10.202}$$

This equation or some approximation of it is used by the ray tracer to compute each sample estimate for the radiance incoming at a given point. At the closest intersection of the primary ray emanating from the eye, we evaluate Eq. (10.202). Ideally we would use the same equation at all subsequent ray bounces (ray hits), but this would be computationally too expensive and in any case would not be the most appropriate use of computing time. Thus, we try to use the accurate computation as infrequently as possible and use an approximate computation in all other cases. An accurate computation is only used on the first bounce of the primary ray and on any subsequent bounces where the ray-surface intersection point is closer to the ray-origin than a given threshold. This latter condition is necessary to enhance the likelihood of accurate colour-bleeding occurring at convex corners of the scene, where the distance between two ray bounces is short.

Next, let us consider the individual components of the outgoing radiance from Eq. (10.202).

The *direct illumination reflected radiance* term, $L_{o,l}(\boldsymbol{x}, \omega_o)$, is given by

$$L_{o,l}(\boldsymbol{x}, \omega_o) = \int_{\mathcal{H}_i^2} f_r(\boldsymbol{x}, \omega_i, \omega_o) \cdot L_{i,l}(\boldsymbol{x}, \omega_i) \cdot (\omega_i \odot N_x) \cdot \partial\omega_i \qquad (10.203)$$

This term is frequently the most important part of the outgoing reflected radiance, since it is responsible for depicting shadows (to which the eye is most sensitive). So it must be computed accurately.

From every intersection point $\boldsymbol{x}$ shadow rays are cast towards each light source in the scene. With area light sources, more than one shadow rays are required per light source so as to generate convincing penumbra areas. This is an expensive process. In an attempt to improve its efficiency the algorithm can be modified to emit shadow photons (anti-photons) with negative light energy [Jensen 95b]. In the photon-tracing pass, when the photon map is being populated, shadow photons are emitted by each light source. When a shadow-photon ray is cast, starting from the second closest ray-surface intersection and including all subsequent intersections we deposit a negative photon if the intersected surface is facing the light source. In the rendering pass, if all the photons in the region of a surface point $\boldsymbol{x}$ are positive, then the point is deemed to be visible by the light source, if all the neighbouring photons are negative, then it is deemed to be hidden from this light source, otherwise it is deemed to be in the penumbra region of this light source. In the latter case we have to use a number of rays to discover the fraction of lighting that the point receives from this light source. Thus, we are only forced to perform the expensive shadow-ray casting operation for the regions of the scene that fall within the penumbra areas associated with this particular light source. This implementation of the algorithm requires a modification of the photon data structure to identify the light source that emitted the photon and to indicate whether the photon is positive or negative.

The *specular illumination reflected radiance* term, $L_{o,s}(\boldsymbol{x}, \omega_o)$, of the outgoing radiance of Eq. (10.202) is given by

$$L_{o,s}(\boldsymbol{x}, \omega_o) = \int_{\mathcal{H}_i^2} f_{r,s}(\boldsymbol{x}, \omega_i, \omega_o) \cdot \left[L_{i,c}(\boldsymbol{x}, \omega_i) + L_{i,d}(\boldsymbol{x}, \omega_i)\right] \cdot (\omega_i \odot N_x) \cdot \partial\omega_i \qquad (10.204)$$

This integral is evaluated using a Monte Carlo ray-tracing algorithm with an *importance sampling* optimisation, which is based on the specular BRDF $f_{r,s}$. Importance sampling is an optimisation technique employed to improve the performance of the Monte Carlo method [Jensen 95c].

The *caustic illumination reflected radiance* term, $L_{o,c}(\boldsymbol{x}, \omega_o)$, of the outgoing radiance of Equ. (10.202) is given by

$$L_{o,c}(\boldsymbol{x}, \omega_o) = \int_{\mathcal{H}_i^2} f_{r,d}(\boldsymbol{x}, \omega_i, \omega_o) \cdot L_{i,c}(\boldsymbol{x}, \omega_i) \cdot (\omega_i \odot \boldsymbol{N}_x) \cdot \partial\omega_i \qquad (10.205)$$

When an accurate value for $L_{o,c}$ is required, then we evaluate this integral using Monte Carlo integration and the contents of the caustics photon map. When only an approximate value for $L_{o,c}$ is required, then we do not evaluate this equation at all but we rely on the caustics contribution included in the specular radiance estimate from the global photon map.

Finally, the *indirect illumination reflected radiance* term, $L_{o,d}(\boldsymbol{x}, \omega_o)$, of the outgoing radiance of Equ. (10.202) is given by

$$L_{o,d}(\boldsymbol{x}, \omega_o) = \int_{\mathcal{H}_i^2} f_{r,d}(\boldsymbol{x}, \omega_i, \omega_o) \cdot L_{i,d}(\boldsymbol{x}, \omega_i) \cdot (\omega_i \odot \boldsymbol{N}_x) \cdot \partial\omega_i \qquad (10.206)$$

This outgoing radiance term represents light that since leaving the light source has been reflected, at least once, from a diffuse surface, resulting in incoherent (soft) illumination. When an accurate value for $L_{o,d}$ is required, then we evaluate this integral using Monte Carlo ray tracing. This is done by spawning a large number of rays, stochastically distributed around the reflection/transmission direction, and averaging the computed radiance from all the primary estimates. When only an approximate value for $L_{o,d}$ is required, then we compute it using a radiance estimate from the global photon map, which contains the direct, indirect and caustic illumination contributions.

10.14.2.6.7 Observations

Photon mapping is a very elegant technique that allows us to handle many different lighting phenomena and to generate photo-realistic images that are physically based or at least physically plausible. In conclusion we note that:

- Photon mapping is an elegant rendering technique that provides a complete global illumination solution for large scene geometries with complex material properties.
- The photon mapping technique separates the storage of the photons from the storage of the geometric representation of the scene. With complex scenes this represents a clear advantage over object-space finite element methods of rendering.
- Any rendering algorithm that uses photon maps must perform a final gather (aggregation) of light flux which requires a number of expensive near neighbourhood operations.
- Several optimisations exist that speed up the performance of this algorithm significantly.

A more detailed description of this technique can be found in Jensen's book [Jensen 01b] on which the above discussion is based.

10.14.3 Hybrid Multi-Pass Rendering Algorithms

A promising approach is to use a hybrid algorithm for rendering, which involves multiple passes. The first pass of such an algorithm adopts an object-space approach and deals with the diffuse inter-object illumination problem, by computing and storing the outgoing radiance diffusely reflected from the surfaces of the scene. The second pass of such an algorithm adopts an image-space approach and deals with specular inter-object illumination, by accessing the pre-stored object-space solution and by computing (on the fly) for each pixel the radiance contributed by the specular inter-object illumination, which would be impossible to compute and store in the first pass.

Such algorithms can be designed to take advantage of the strengths of both object-space and image-space techniques.

Examples of such multi-pass algorithms are found in [Wallace 87], [Sillion 89], [Chen 91], [Jensen 96] and [Suykens 99].

References

[Amanatides 84] Amanatides, J. Ray tracing with cones. *Proceedings of SIGGRAPH 84. Computer Graphics*, Vol. 18, No. 3, p.p. 129–135, 1984.

[Ashikmin 00a] Ashikmin, M. and Shirley, P. An anisotropic Phong light reflection model. Technical Report UUCS-00-014, Computer Science Department, University of Utah, p.p. 1–11, June 2000.

[Ashikmin 00b] Ashikmin, M. and Shirley, P. An anisotropic phong BRDF model. *Journal of Graphics Tools*, Vol. 5, No. 2, p.p. 25–32, 2000.

[Ashikmin 00c] Ashikmin, M., Premoze, S., and Shirley, P. A microfacet-based BRDF generator, *Proceedings of SIGGRAPH 2000, Computer Graphics, ACM Press/ACM SIGGRAPH/Addison Wesley Longman*, p.p. 65–74, 2000.

[Aupperle 93] Aupperle, L. and Hanrahan, P. A hierarchical illumination algorithm for surfaces with glossy reflection". *Proceedings of SIGGRAPH 93, Computer Graphics*, Vol. 27, No. 4, p.p. 155–162 1993.

[Beckmann 63] Beckmann, P. and Spizzichino, A. *The Scattering of Electromagnetic Waves from Rough Surfaces*. Pergamin Press, Oxford, England (1963).

[Bennett 61] Bennett, R. A. and Porteus, J. O. Relation between surface roughness and specular reflectance at normal incidence, *Journal of the Optical Society of America*, Vol. 51, p.p. 123–129, 1961.

[Bentley 79] Bentley, J. L. and Friedman, J. H. Data structures for range searching, *Computing Surveys*, Vol. 11, No. 4, p.p. 397–409, 1979.

[Blinn 77] Blinn, J. F. Models of light reflection for computer synthesized pictures. *Proceedings of SIGGRAPH 77, Computer Graphics*, Vol. 11, No. 2, p.p. 192–198, 1977.

[Chen 90] Chen, S. E. Incremental radiosity: An extension of progressive radiosity to an interactive image synthesis system. *Proceedings of SIGGRAPH 90, Computer Graphics*, Vol. 24, No. 4, p.p. 135–144, (1990).

[Chen 91] Chen, S. E., Rushmeier, H. E., Miller, G., and Turner, D. A progressive multi-path method for global illumination, *Proceedings of SIGGRAPH 91, Computer Graphics*, Vol. 25, No. 4, p.p. 165–174, 1991.

[Clarke 85] Clarke, F. J. J. and Parry, D. J. Helmholz reciprocity: Its VIvidity and application to reflectometry, *Lighting Research and Technology*, Vol. 17, No. 1, p.p. 1–11, 1985.

[Cohen 93] Cohen, M. F. and Wallace, J. R. *Radiosity and Realistic Image Synthesis*, Academic Press, San Diego, CA (1993).

[Cohen 85] Cohen, M. F. and Greenberg, D. P. The hemi-cube: A Rdriosity solution for complex environments. *Proceedings of SIGGRAPH 85, Computer Graphics,* Vol. 19, No. 3, p.p. 31–40, 1985.

[Cook 81] Cook, R. L. and Torrance, K. E. A reflection model for computer graphics. *Proceedings SIGGRAPH 81, Computer Graphics*, Vol. 15, No. 4, p.p. 307–316, 1981.

[Cook 82] Cook, R. L. and Torrance, K. E. A reflection model for computer graphics. *ACM Transactions on Graphics*, Vol. 1, No. 1, p.p. 7–24, 1982.

[Cook 84] Cook, R., L., Porter, T. and Carpenter, L. Distributed ray tracing. *Proceedings of SIGGRAPH 84, Computer Graphics*, Vol. 18, No. 3, p.p. 137–145, 1984.

[Ditchburn 76] Ditchburn, R. W. *Light*, Vols. 1 and 2. Academic Press, London (1976).

[Dorsey 99] Dorsey, J., Edelman, A., Jensen, H. W., Legakis, J., and Pedersen, H. K. Modelling and rendering of weathered stone. *Proceedings of SIGGRAPH 99*, pp. 225–234. Addison-Wesley, Reading, MA (1999).

[Drettakis 97] Drettakis, G. and Sillion, F., X. Interactive update of global illumination using a line-space hierarchy. *Proceedings of SIGGRAPH 97. Computer Graphics*, Vol. 31, No. 4, p.p. 57–64, 1997.

[Duderstadt 79] Duderstadt, J. J. and Martin, W. R. *Transport Theory*. John Wiley & Sons, New York (1979).

[Dutré 94] Dutré, Ph. and Willems, Y. D., Importance-Driven Monte Carlo light tracing. *Proceedings of the Fifth Eurographics Workshop on Rendering*, Darmstadt, Germany, Eurographics Association, p.p. 185–194 (1994).

[Dutré 98] Dutré, Ph. Mathematical Framework and Monte Carlo Algorithms for Global Illumination in Computer Graphics. Ph.D Thesis, University of Leuven (1998).

[George 90] George, D. W., Sillion, F. X., and Greenberg, D., P. Radiosity Redistribution for Dynamic Environments, *IEEE, Computer Graphics and Applications*, Vol. 10, No. 4, p.p. 26–34, July 1990.

[Goldstein 71] Goldstein, R. A. and Nagel, R. 3-D Visual Simulation. *Simulation*, p.p. 25–31 Jan 1971.

[Goral 84] Goral, C. M., Torrance, K. E., Greenberg, D. P., and Battaile, B. Modeling the interaction of light between diffuse surfaces. *Proceedings of SIGGRAPH 84, Computer Graphics*, Vol. 18, No. 3, p.p. 213–222, 1984.

[Hall 89] Hall, R. *Illumination and Color in Computer Generated Imagery*. Springer-Verlag, New York (1989).

[Hanrahan 93] Hanrahan, P. and Krueger, W. Reflection from layered surfaces due to subsurface scattering. *Proceedings of SIGGRAPH 93*, 165–174, ACM Press, New York (1993).

[Immel 86] Immel, D. S., Cohen, M. F. and Greenberg, D. P. A radiosity method for non-diffuse environments. *Proceedings of SIGGRAPH 86, Computer Graphics*, Vol. 20, No. 4, p.p. 133–142, 1986.

[Jenkins 76] Jenkins, F. A. and White, H. E. *Fundamentals of Optics*. McGraw-Hill, New York (1976).

[Jensen 95a] Jensen, H. W. and Christensen, N. J. Photon maps in bi-directional Monte Carlo ray tracing of complex objects. *Computers & Graphics*, Vol. 19, No. 2, p.p. 215–224, 1995.

[Jensen 95b] Jensen, H. W. and Christensen, N. J. Efficiently rendering shadows using the photon map. *Proceedings of CompuGraphics 95*, p.p. 285–291 Dec 1995.

[Jensen 95c] Jensen, H. W. Importance driven path tracing using the photon map. *Proceedings of the Eurographics Rendering Workshop 95*, Eurographics Association, p.p. 326– 335, June 1995.

[Jensen 96] Jensen, H. W. Global illumination using photon maps. *Proceedings of the Eurographics Rendering Workshop 1996*. Springer-Verlag, Vienna, p.p. 21–30, 1996.

[Jensen 99] Jensen, H. W., Legakis, J., and Dorsey, J. Rendering of wet materials, In: C. Lischinski and G. W. Larson (Eds.), *Rendering Techniques'99*. Springer-Verlag, Vienna (1999).

[Jensen 01a] Jensen, H. W., Marschner, S., Levoy, M., and Hanrahan, P. A practical model for subsurface light transport. *Proceedings of SIGGRAPH 2001*, p.p. 399–408. Addison-Wesley, Reading MA (2001).

[Jensen 01b] Jensen, H. W. *Realistic Image Synthesis using Photon Mapping*. A. K. Peters Ltd., Natick, Massachusetts (2001).

[Kajiya 86] Kajiya, J. T. The rendering equation. *Proceedings of SIGGRAPH 86, Computer Graphics*, Vol. 20, No. 4, p.p. 143–150, 1986.

[Lafortune 93] Lafortune, E. P. and Willems, Y. D. Bi-directional path tracing. *Proceedings of the International Computer Graphics and Visualisation Techniques, CompuGraphics 93*. Avlor, Portugal, p.p. 145–153 1993.

[Lafortune 94] Lafortune, E. and Willems, Y. Using the modified Phong reflectance model for physically based rendering, Technical Report CW 194, Department of Computer Science, K.U., Leuven, p.p. 1–18, Nov 1994.

[Lafortune 97] Lafortune, E. P. F., Foo, S. C., Torrance, K. E., and Greenberg D. P. Non-linear approximation of reflectance functions. *Proceedings of SIGGRAPH 97, Computer Graphics*, No. 31, p.p. 117–126, 1997.

[Lewis 93] Lewis, R. R. Making shaders more physically plausible. *Proceedings of the Fourth Eurographics Workshop on Rendering*, Paris, France. Eurographics Series EG 93 RW, pp. 47–62 June 1993.

[Metropolis 49] Metropolis, N. and Ulam, S. The Monte Carlo method, *Journal of the American Statistical Association*, Vol. 44, No. 247, p.p. 335–341, 1949.

[Metropolis 53] Metropolis, N., Rosenbluth, A. W., Rosenbluth, M. N., Teller, A. H., and Teller, E. Equations of state calculations by fast computing machines, *Journal of Chemical Physics 21*, p.p. 1087–1091, 1953.

[Neumann 99a] Neumann, L., Neumann, A., and Szirmay-Kalos L. Compact metallic reflectance models, *Computer Graphics Forum (Eurographics '99)*. The Eurographics Association and Blackwell Publishers, Vol. 18, No. 3, p.p. 161–172, 1999.

[Neumann 99b] Neumann, L., Neumann, A., and Szirmay-Kalos L. Reflectance models with fast importance sampling, *Computer Graphics Forum*, Vol. 18, No. 4, p.p. 249–265, 1999.

[Nicodemus 77] Nicodemus, F. E., Richmond, J. C., Hsia, J. J., Ginburg, I. W., and Limperis, T. Geometric considerations and nomenclature for reflectance, *Monograph 161*, National Bureau of Standards (US) Oct 1977.

[Nishita 85] Nishita, T. and Nakamae, E. Continuous tone representation of three-dimensional objects taking account of shadows and inter-reflection. *Proceedings of SIGGRAPH 85, Computer Graphics*, Vol. 19, No. 3, p.p. 23–30, 1985.

[Palik 85] Palik, E. D. *Handbook of Optical Constants of Solids*. Academic Press, New York, NY (1985).

[Pharr 00] Pharr, M. and Hanrahan, P. Monte Carlo evaluation of non-linear scattering equations for subsurface reflection. *Proceedings of SIGGRAPH 2000*, pp. 75–84. Addison-Wesley, Reading, MA (2000).

[Sancer 69] Sancer, M. I. Shadow-corrected electromagnetic scattering from a randomly rough surface, *IEEE Transactions on Antennas and Propagation*, Vol. 17, No. 5, p.p. 577–585, 1969.

[Schlick 93] Schlick, C. A customizable reflectance model for everyday rendering. *Proceedings of the Fourth Eurographics Workshop on Rendering*. Series EG 93 RW. Paris, France, p.p. 73–84 June 1993.

[Shirley 91] Shirley, P. Physically Based Lighting Calculations for Computer Graphics, Ph.D Thesis, University of Illinois at Urbana Champaign Jan 1991.

[Sillion 89] Sillion, F. and Puech, C. A general two-pass method integrating specular and diffuse reflection. *Proceedings of SIGGRAPH 89, Computer Graphics*, Vol. 23, No. 4, p.p. 335–344, 1989.

[Sillion 91] Sillion, F., Avro, J., Westin, S., and Greenberg, D. P. A global illumination solution for general reflectance distributions. *Proceedings of SIGGRAPH 91, Computer Graphics*, Vol. 25, No. 4, p.p. 187–196 1991.

[Stamminger 98] Stamminger, M., Slusallek, Ph., and Seidel, P. H. Three point clustering for radiance computations. *Proceedings of the Eurographics Rendering Workshop 98*. Eurographics Association, p.p. 211–222, 1998.

[Suykens 99] Suykens, F. and Willems, Y. D. Weighted multi-pass method for global illumination. *Proceedings of Eurographics 99, Computer Graphics Forum*, Vol. 18, No. 3, p.p. 209–220, 1999.

[Torrance 67] Torrance, K. E. and Sparrow, E. M. Theory of off-specular reflection from roughened surfaces, *Journal of the Optical Society of America*, Vol. 57, No. 9, p.p. 1105–1114, 1967.

[Veach 94] Veach, E. and Guibas, L. J. Bi-directional estimates for light transport. *Proceedings of the Fifth Eurographics Workshop on Rendering*, Darmstadt, Germany. Eurographics Association, p.p. 147–162 1994.

[Veach 97a] Veach, E. Robust Monte Carlo Methods for Light Transport. Ph.D Thesis, Department of Computer Science, Stanford University (1997).

[Veach 97b] Veach, E. and Guibas, L. J. Metropolis light transport. *Proceedings of SIGGRAPH 97, Computer Graphics*, Vol. 31, No. 4, p.p. 65–76, 1997.

[Wallace 87] Wallace, J. R., Cohen, M. F., and Greenberg, D. P. A two-pass solution to the rendering equation: A synthesis of ray tracing and radiosity methods. *Proceedings of SIGGRAPH 97, Computer Graphics*, Vol. 21, No. 4, p.p. 311–320, 1987.

[Whitted 80] Whitted, T. An improved illumination model for shaded display. *Communications of the ACM*, Vol. 23, No. 6, p.p. 343–349, 1980.

[WWW 1] http://www.luxpop.com

Appendix 1

A Simple Vector Algebra C Library

In this Appendix we examine a simple **C** library that implements all the vector algebra operations we have defined in Chapter 2.

```
#include <stdio.h>
#include <math.h>

/*---------------------------------------------------------------*/

/*
 * Three-dimensional vectors are stored as three doubles and a
 * single byte header which is a vector type indicator.
 *
 * Temporary vectors have their type indicator set to 1 and
 * Permanent vectors have their type indicator set to 2.
 *
 * All vector operations return temporary vectors so they can be
 * nested without memory leakage. Each vector function checks its
 * vector parameters and destroys them if they are found to be
 * temporary.
 *
 * In its current implementation this library deals with 3D vectors
 * and all vector components are stored as doubles.
 */

#define temporary_vector 1
#define permanent_vector 2

typedef double        *vector_t;
typedef unsigned char *header_t;

/*
 * Key to global constants:
 *
 * vector_length_bytes       - Length of the vector components in
 *                             bytes.
 *
 * vt_length_bytes           - Length of the header excluding the
 *                             filler bytes.
```

```
 * filler_length_bytes        - Length of the filler in bytes.
 *
 * vt_displacement_bytes      - Header displacement relative to
 *                              first component of the vector.
 * vector_displacement_bytes - Displacement of first vector
 *                              component relative to the start of
 *                              memory chunk allocated for storing
 *                              the vector.
 */


#define vector_length_bytes sizeof(double)*3
#define vt_length_bytes     sizeof(char)


#if (0)
/*
 * Here the header is placed after the elements of the vector.
 * This is the most memory efficient solution, as it requires
 * only 25 bytes and no padding bytes.
 */

#define filler_length_bytes       (0)
#define vt_displacement_bytes     (sizeof(double)*3)
#define vector_displacement_bytes (0)
#endif


#if (1)
/*
 * Here the header is placed before the elements of the vector.
 * This method necessitates the introduction of some padding bytes.
 * The filler bytes are need for the alignment of doubles by the
 * HP C compiler and other compilers. Such compilers assume that a
 * double starts at an address that is a multiple of 8 bytes. Thus,
 * as our header is 1 byte long we must have a filler of 7 bytes.
 * This method of storage requires 32 bytes to store a vector.
 */

#define filler_length_bytes       (7)
#define vt_displacement_bytes     (-1)
#define vector_displacement_bytes (8)
#endif


vector_t Vnew(char *function_name, int vector_type)
{
  /*
   * Create a new vector and return the address of its first element.
   * For internal use only!
   * function_name is the name the function that called this function.
   */
 vector_t v;   /* pointer to vector components */
 header_t vt;  /* pointer to vector type       */
```

```
 char      *a;

 a = (char *) malloc(vector_length_bytes + vt_length_bytes
                     + filler_length_bytes);

 if (!a)
  {
   printf("%s: can not malloc vector!\n", function_name);
   return(NULL);
  }

 a = a + vector_displacement_bytes;
 v = (vector_t) a;

 v[0] = 0.0;
 v[1] = 0.0;
 v[2] = 0.0;

 vt  = ((header_t) v) + vt_displacement_bytes;
 *vt = (unsigned char) vector_type;

return(v);

}  /* Vnew */


/*------------------------------------------------------------------*/


vector_t Vtemp(double x, double y, double z)
{
 /*
  * Create a temporary vector with the given components and return
  * the address of its first element.
  */

 vector_t v;  /* pointer to vector components */

 v = Vnew("Vtemp", temporary_vector);

 if (v)
  {
   v[0] = x;
   v[1] = y;
   v[2] = z;
  }

return(v);

}  /* Vtemp */

/*------------------------------------------------------------------*/
vector_t Vperm(double x, double y, double z)
{
```

```
 /*
  * Create a permanent vector with the given components and return
  * the address of its first element.
  */

 vector_t v;  /* pointer to vector components */

 v = Vnew("Vperm", permanent_vector);

 if (v)
  {
   v[0] = x;
   v[1] = y;
   v[2] = z;
  }

 return(v);

}  /* Vperm */

void Vfree(vector_t v)
{
 /*
  * Destroy a vector.
  *
  * Expects the address of the first element of the vector.
  */

 if (!v)
  {
   printf("Vfree: invalid vector parameter!\n");

   return;
  }

 free(((char *) v) - vector_displacement_bytes);

}  /* Vfree */

/*------------------------------------------------------------------*/

void Vfree_if_temp(vector_t v)
{
 /*
  * Destroy a temporary vector.
  *
  * Expects the address of the first element of the vector.
  */

 header_t vt;  /* pointer to vector type */

if (!v)
  {
   printf("Vfree_if_temp: invalid vector parameter!\n");
```

```
   return;
  }

 vt = ((header_t) v) + vt_displacement_bytes;

 if (*vt == temporary_vector) free(((char *) v)
         - vector_displacement_bytes);

}  /* Vfree_if_temp */

/*------------------------------------------------------------*/

vector_t Vsave(vector_t v)
{
 /*
  * Convert a temporary vector into a permanent vector.
  */

 header_t vt;  /* pointer to vector type */

if (!v)
  {
   printf("Vsave: invalid vector parameter!\n");
   return(NULL);
  }

 vt  = ((header_t) v) + vt_displacement_bytes;
 *vt = (unsigned char) permanent_vector;

 return(v);

}  /* Vsave */

/*------------------------------------------------------------*/

double Vmagnitude(vector_t v)
{
 /*
  * Compute the magnitude of a vector.
  */

 double m;

 if (!v)
  {
   printf("Vmagnitude: Invalid vector parameter!\n");
   return(0.0);
  }

 m = sqrt(v[0]*v[0] + v[1]*v[1] + v[2]*v[2]);

 Vfree_if_temp(v);

 return(m);
```

```
} /* Vmagnitude */

/*-----------------------------------------------------------------*/

vector_t Vnormalise(vector_t vi)
{
 /*
  * Return the normalised version of the input vector.
  */

 vector_t v;  /* pointer to output vector components */
 double   m;  /* magnitude of the input vector       */

 if (!vi)
  {
   printf("Vnormalise: invalid vector parameter!\n");
   return(NULL);
  }

 v = Vnew("Vnormalise", temporary_vector);

 if (v)
  {
   m = sqrt(vi[0]*vi[0] + vi[1]*vi[1] + vi[2]*vi[2]);

   if (m <= 0) m = 1;

   v[0] = vi[0] / m;
   v[1] = vi[1] / m;
   v[2] = vi[2] / m;
  }

 Vfree_if_temp(vi);

 return(v);

} /* Vnormalise */

/*-----------------------------------------------------------------*/

vector_t Vnegate(vector_t vi)
{
 /*
  * Return the negated version of the input vector.
  */
 vector_t v;  /* pointer to output vector components */

 if (!vi)
  {
   printf("Vnegate: invalid vector parameter!\n");
   return(NULL);
  }
 v = Vnew("Vnegate", temporary_vector);
```

```
 if (v)
  {
   v[0] = -vi[0];
   v[1] = -vi[1];
   v[2] = -vi[2];
  }

 Vfree_if_temp(vi);

 return(v);

}  /* Vnegate */

/*---------------------------------------------------------------*/

 vector_t Vadd(vector_t v1, vector_t v2)
{
 /*
  * Add two vectors and return their sum (v1 + v2)
  */

 vector_t v;  /* pointer to vector components */

 if ((!v1) || (!v2))
  {
   printf("Vadd: invalid vector parameter!\n");
   return(NULL);
  }

 v = Vnew("Vadd", temporary_vector);

 if (v)
  {
   v[0] = v1[0] + v2[0];
   v[1] = v1[1] + v2[1];
   v[2] = v1[2] + v2[2];
  }

 Vfree_if_temp(v1);
 Vfree_if_temp(v2);

 return(v);

}  /* Vadd */

/*---------------------------------------------------------------*/

vector_t Vsubtract(vector_t v1, vector_t v2)
{
 /*
  * Subtract one vector from another and return their difference
  * (v1 - v2).
  */
```

```
 vector_t v;  /* pointer to vector components */

 if ((!v1) || (!v2))
  {
   printf("Vsubtract: invalid vector parameter!\n");
   return(NULL);
  }

 v = Vnew("Vadd", temporary_vector);

 if (v)
  {
   v[0] = v1[0] - v2[0];
   v[1] = v1[1] - v2[1];
   v[2] = v1[2] - v2[2];
  }

 Vfree_if_temp(v1);
 Vfree_if_temp(v2);

 return(v);

} /* Vsubtract */

/*-----------------------------------------------------------------*/

vector_t Vproduct_by_scalar(vector_t vi, double s)
{
 /*
  * Return the product of the input vector by a scalar.
  */

 vector_t v;  /* pointer to output vector components */

 if (!vi)
  {
   printf("Vproduct_by_scalar: invalid vector parameter!\n");
   return(NULL);
  }

 v = Vnew("Vproduct_by_scalar", temporary_vector);

 if (v)
  {
   v[0] = vi[0] * s;
   v[1] = vi[1] * s;
   v[2] = vi[2] * s;
  }

 Vfree_if_temp(vi) ;
 return(v);

} /* Vproduct_by_scalar */
```

```
/*-------------------------------------------------------------*/

double Vdot_product(vector_t v1, vector_t v2)
{
 /*
  * Compute the dot product of the two input vectors.
  */

 double dp;  /* dot product */

 if ((!v1) || (!v2))
  {
   printf("Vdot_product: invalid vector parameter!\n");
   return(0.0);
  }

 dp = v1[0] * v2[0] + v1[1] * v2[1] + v1[2] * v2[2];

 Vfree_if_temp(v1);
 Vfree_if_temp(v2);

 return(dp);

}  /* Vdot_product */

/*-------------------------------------------------------------*/

vector_t Vcross_product(vector_t v1, vector_t v2)
{
 /*
  * Compute the cross product of the two input vectors.
  */

 vector_t v;  /* pointer to vector components */

 if ((!v1) || (!v2))
  {
   printf("Vcross_product: invalid vector parameter!\n");
   return(NULL);
  }

 v = Vnew("Vcross_product", temporary_vector);

 if (v)
  {
   v[0] =v1[1] * v2[2] - v1[2] * v2[1];
   v[1] =v1[2] * v2[0] - v1[0] * v2[2];
   v[2] =v1[0] * v2[1] - v1[1] * v2[0];
  }
 Vfree_if_temp(v1);
 Vfree_if_temp(v2);

 return (v);
```

```
}  /* Vcross_product */

/*-----------------------------------------------------------------*/

double Vtriple_scalar_product(vector_t v1, vector_t v2, vector_t v3)
{
 /*
  * Compute the triple scalar product of the three input vectors.
  */

 double tsp;  /* triple scalar product */

 if ((!v1) || (!v2) || (!v3))
  {
   printf("Vtriple_scalar_product: invalid vector parameter!\n");
   return(0.0);
  }

 tsp =v1[0] * (v2[1] * v3[2] - v2[2] * v3[1]) +
      v1[1] * (v2[2] * v3[0] - v2[0] * v3[2]) +
      v1[2] * (v2[0] * v3[1] - v2[1] * v3[0]);

 Vfree_if_temp(v1);
 Vfree_if_temp(v2);
 Vfree_if_temp(v3);

 return(tsp);

}  /* Vtriple_scalar_product */

/*-----------------------------------------------------------------*/

vector_t Vtriple_vector_product_1(vector_t v1, vector_t v2,
        vector_t v3)
{
 /*
  * Compute the triple vector product (v1 x v2) x v3 of the three
  * input vectors.
  */

 vector_t v;  /* pointer to components of the resulting vector */

 if ((!v1) || (!v2) || (!v3))
  {
   printf("Vtriple_vector_product_1: invalid vector parameter!\n");
   return(NULL);
  }

 v = Vcross_product(v1, v2);
 v = Vcross_product(v,  v3);

 return(v);
}  /* Vtriple_vector_product_1 */
```

```
/*--------------------------------------------------------------*/

vector_t Vtriple_vector_product_2(vector_t v1, vector_t v2,
        vector_t v3)
{
 /*
  * Compute the triple vector product v1 x (v2 x v3) of the three
  * input vectors.
  */

 vector_t v;  /* pointer to components of the resulting vector */

 if ((!v1) || (!v2) || (!v3))
  {
   printf("Vtriple_vector_product_2: invalid vector parameter!\n");
   return(NULL);
  }

 v = Vcross_product(v2, v3);
 v = Vcross_product(v1, v);

 return(v);

}  /* Vtriple_vector_product_2 */

/*--------------------------------------------------------------*/

double Vscalar_product_of_4_vectors(vector_t v1,
                                    vector_t v2,
                                    vector_t v3,
                                    vector_t v4
                                   )

{
 /*
  * Compute the scalar product of the four input vectors
  * (v1 x v2) . (v3 x v4).
  */

 vector_t va, vb;  /* pointers to components of the temp cross
          product vectors */

 if ((!v1) || (!v2) || (!v3) || (!v4))
  {
   printf("Vscalar_product_of_4_vectors: invalid vector parameter!\n");
   return(0.0);
  }

 va = Vcross_product(v1, v2);
 vb = Vcross_product(v3, v4);

 return(Vdot_product(va, vb));
}  /* Vscalar_product_of_4_vectors */
```

```
/*--------------------------------------------------------------*/

vector_t Vvector_product_of_4_vectors(vector_t v1,
                                      vector_t v2,
                                      vector_t v3,
                                      vector_t v4
                                     )

{
 /*
  * Compute the vector product of the four input vectors
  * (v1 x v2) x (v3 x v4).
  */

 vector_t va, vb;  /* pointers to components of the temp cross
                      product vectors */

 if ((!v1) || (!v2) || (!v3) || (!v4))
  {
   printf("Vvector_product_of_4_vectors: invalid vector parameter!\n");
   return(NULL);
  }

 va = Vcross_product(v1, v2);
 vb = Vcross_product(v3, v4);

 return(Vcross_product(va, vb));

}  /* Vvector_product_of_4_vectors */

/*--------------------------------------------------------------*/

void Vprint_debug(char *name, vector_t v)
{
 /*
  * Print the vector with out disposing of temporary vectors.
  */

 header_t vt;

 if (!v)
  {
   printf("Vprint_debug: invalid vector parameter!\n");
   return;
  }

 printf("%s=[%f, %f, %f]", name, v[0], v[1], v[2]);

 vt = ((header_t) v) + vt_displacement_bytes;
 if (*vt == temporary_vector) printf(" temporary vector\n");
 else                         printf(" permanent vector\n");
```

```
}  /* Vprint_debug */

/*----------------------------------------------------------------*/

void Vprint(char * name , vector_t v)
{
 /*
  * Print the vector and dispose of it if it is temporary.
  */

 if (!v)
  {
   printf("Vprint: invalid vector parameter!\n");
   return;
  }

 printf("%s=[%f, %f, %f]\n", name, v[0], v[1], v[2]);

 Vfree_if_temp(v);

} /* Vprint */

/*----------------------------------------------------------------*/

/*
 * Test main program.
 */

main()
{
 vector_t v1, v2, v3;

 v1 = Vtemp(1.5, 2.5, 3.5);
 Vprint_debug("v1" , v1);

 v1 = Vsave(v1);
 Vprint_debug("v1" , v1);

 v2 = Vperm(3.5, 2.5, 1.5);
 Vprint_debug("v2" , v2);

 Vprint("v1+v2 ", Vadd(v1, v2));

 Vprint("v1-v2", Vsubtract(v1, v2));

 v3 = Vsave(Vnegate(v2));
 Vprint("-v3", v3);

 v3 = Vsave(Vnormalise(v3));
 Vprint("normalise(v3)", v3);
 printf("|v3|=%f\n", Vmagnitude(v3));
 printf("[4,0,-1].[2,-1,3]=%f\n", Vdot_product(Vtemp(4,0,-1),
        Vtemp(2,-1,3)));
```

```
 Vprint("[2,1,0]x[2,-1,1]", Vcross_product(Vtemp(2,1,0),
        Vtemp(2,-1, 1)));

 printf("([2,-3,4]x[1,3,-1]).[3,-1,2]=%f\n",
        Vtriple_scalar_product(Vtemp(2,-3,4),Vtemp(1,3,-1),
         Vtemp(3,-1,2))
       );

 Vprint("([3,-2,1]x[-1,3,4])x[2,1,-3]",
        Vtriple_vector_product_1(Vtemp(3,-2,1),Vtemp(-1,3,4),
         Vtemp(2,1,-3))
       );

 Vprint("[3,-2,1]x([-1,3,4]x[2,1, -3])",
        Vtriple_vector_product_2(Vtemp(3,-2,1),Vtemp(-1,3,4),
         Vtemp(2,1,-3))
       );

 printf("([3,-2,1]x[-1,3,4]).([2,1,-3]x[-2,1,4])]=%f\n",
        Vscalar_product_of_4_vectors(Vtemp(3,-2,1), Vtemp(-1,3,4),
         Vtemp(-1,3,4), Vtemp(-2,-1,2))
       );

 Vprint("([3,-2,1]x[-1,3,4])x([2,1,-3]x[-2,1,4])",
        Vvector_product_of_4_vectors(Vtemp(3,-2,1), Vtemp(-1,3,4),
         Vtemp(2,1,-3), Vtemp(-2,-1,2))
       );

 Vfree(v1);
 Vfree(v2);
 Vfree(v3);
}
```

The above test main program produces the following results:

```
v1=[1.500000, 2.500000, 3.500000] temporary vector
v1=[1.500000, 2.500000, 3.500000] permanent vector
v2=[3.500000, 2.500000, 1.500000] permanent vector
v1+v2=[5.000000, 5.000000, 5.000000]
v1-v2=[-2.000000, 0.000000, 2.000000]
-v3=[-3.500000, -2.500000, -1.500000]
normalise(v3) = [-0.768350, -0.548821, -0.329293]
|v3|=1.000000
[4,0,-1].[2,-1,3]=5.000000
[2,1,0]x[2,-1,1]=[1.000000, -2.000000, -4.000000]
([2,-3,4]x[1,3,-1]).[3,-1,2]=-15.000000
([3,-2,1]x[-1,3,4])x[2,1,-3]=[32.000000, -19.000000, 15.000000]
[3,-2,1]x([-1,3,4]x[2,1,-3])=[9.000000, 8.000000, -11.000000]
([3,-2,1]x[-1,3,4]).([2,1,-3]x[-2,1,4])]=-15.000000
([3,-2,1]x[-1,3,4])x([2,1,-3]x[-2,1,4])=[-14.000000, -7.000000,
                                          -35.000000]
```

Appendix 2

A Simple Matrix Algebra C Library

In this appendix we examine a simple **C** library that implements all the matrix algebra operations we have defined in Chapter 3.

```
#include <stdio.h>
#include <math.h>
#include <string.h>


/*---------------------------------------------------------------*/

#define MOrder 3  /* Matrix Order */

typedef double matrix_t[MOrder][MOrder];

typedef struct {
               matrix_t dummy_matrix;
              } matrix_structure_t, *matrix_ptr_t;

#define M_copy_matrix(src, dst)\
 *((matrix_ptr_t)(dst)) = *((matrix_ptr_t)(src))

typedef double vector_t[MOrder];

typedef struct {
               vector_t dummy_vector;
              } vector_structure_t, *vector_ptr_t;

#define M_copy_vector(src, dst)\
 *((vector_ptr_t)(dst)) = *((vector_ptr_t)(src))

#define M_round_to_zero(x) ((fabs(x) > le-33) ? (x) : 0.0)

/*---------------------------------------------------------------*/

extern void M_print_matrix(char *name, matrix_t m);
extern void M_print_sub_matrix(char *name, matrix_t m, int order);
```

```
/*-----------------------------------------------------------------*/

void M_set_identity_matrix(matrix_t m)
{
 int r, c;

 for (r = 0; r < MOrder; r++)
 for (c = 0; c < MOrder; c++)
  if (r == c) m[r][c] = 1.0;
  else        m[r][c] = 0.0;

} /* M_set_identity_matrix */

/*-----------------------------------------------------------------*/

void M_set_rc_reversal_matrix(matrix_t m)
{
 int r, c;

 for (r = 0; r < MOrder; r++)
 for (c = 0; c < MOrder; c++)
  if (r == (MOrder - (c + 1))) m[r][c] = 1.0;
  else                         m[r][c] = 0.0;

} /* M_set_rc_reversal_matrix */

/*-----------------------------------------------------------------*/

void M_set_scalar_matrix(matrix_t m, double s)
{
 int r, c;

 for (r = 0; r < MOrder; r++)
 for (c = 0; c < MOrder; c++)
  if (r == c) m[r][c] = s;
  else        m[r][c] = 0.0;

} /* M_set_scalar_matrix */

/*-----------------------------------------------------------------*/

void M_set_zero_matrix(matrix_t m)
{
 int r, c;

 for (r = 0; r < MOrder; r++)
 for (c = 0; c < MOrder; c++)
  m[r][c] = 0.0;

} /* M_set_zero_matrix */

/*-----------------------------------------------------------------*/
```

```
void M_transpose_matrix(matrix_t m, matrix_t tm)
{
 int r, c;

 for (r = 0; r < MOrder; r++)
 for (c = 0; c < MOrder; c++)
  tm[c][r] = m[r][c];

} /* M_transpose_matrix */

/*------------------------------------------------------------------*/


void M_matrix_add(matrix_t m1, matrix_t m2, matrix_t m3)
{
 int r, c;

 for (r = 0; r < MOrder; r++)
 for (c = 0; c < MOrder; c++)
  m3[r][c] = m1[r][c] + m2[r][c];

} /* M_matrix_add */

/*------------------------------------------------------------------*/

void M_matrix_subtract(matrix_t m1, matrix_t m2, matrix_t m3)
{
 int r, c;

 for (r = 0; r < MOrder; r++)
 for (c = 0; c < MOrder; c++)
  m3[r][c] = m1[r][c] - m2[r][c];

} /* M_matrix_subtract */

/*------------------------------------------------------------------*/

void M_matrix_by_scalar(matrix_t m1, double s, matrix_t m2)
{
 int r, c;

 for (r = 0; r < MOrder; r++)
 for (c = 0; c < MOrder; c++)
  m2[r][c] = m1[r][c] * s;

}  /* M_matrix_by_scalar */

/*------------------------------------------------------------------*/

double M_dot_product(vector_t v1, vector_t v2)
{
 double dp;
 int    e;
```

```
 dp = 0;
 for (e = 0; e < MOrder; e++)
  dp += v1[e] * v2[e];

 return(dp);

}  /* M_dot_product */

/*-------------------------------------------------------------*/

void M_tensor_product(vector_t v1, vector_t v2, matrix_t m)
{
 int r, c;

 for (r = 0; r < MOrder; r++)
 for (c = 0; c < MOrder; c++)
  m[r][c] = v1[r] * v2[c];

}  /* M_tensor_product */

/*-------------------------------------------------------------*/

void M_matrix_by_vector(matrix_t m, vector_t v1, vector_t v2)
{
 int r, c;

 for (r = 0; r < MOrder; r++)
  {
   v2[r] = 0;

   for (c = 0; c < MOrder; c++)
    v2[r] += m[r][c] * v1[c];
  }

}  /* M_matrix_by_vector */

/*-------------------------------------------------------------*/

void M_vector_by_matrix(vector_t v1, matrix_t m, vector_t v2)
{
 int r, c;

 for (c = 0; c < MOrder; c++)
  {
   v2[c] = 0;

   for (r =0; r < MOrder; r++)
    v2[r] += v1[c] * m[r][c];
  }

}  /* M_vector_by_matrix */

/*-------------------------------------------------------------*/
```

```
void M_matrix_multiply(matrix_t m1, matrix_t m2, matrix_t m3)
{
 int r1, c1, c2;

 for (r1 = 0; r1 < MOrder; r1++)
 for (c2 = 0; c2 < MOrder; c2++)
  {
   m3[r1][c2] = 0.0;

   for (c1 =0; c1 < MOrder; c1++)
    m3[r1][c2] += m1[r1][c1] * m2[c1][c2];
  }

} /* M_matrix_multiply */

/*------------------------------------------------------------------*/

void M_minor_matrix(matrix_t m, int order, int i, int j,
                    matrix_t mm)
{
 int r, c, mr, mc;

 mr = -1;

 for (r = 0; r < order; r++)
  if (r != i)
   {
    mr++;
    mc = -1;

    for (c = 0; c < order; c++)
     if (c != j)
      {
       mc++;
       mm[mr][mc] = m[r][c];
      }
   }
} /* M_minor_matrix */

/*------------------------------------------------------------------*/

double M_determinant(matrix_t m, int order)
{
 matrix_t mm;
 double   d;
 int      r;

 if (order == 1) d = m[0][0];
 else
  {
   d = 0;

   for (r = 0; r < order; r++)
```

```
    {
     M_minor_matrix(m, order, r, 0, mm);

     d = d + pow(-1.0, (double) r) * m[r][0] *
               M_determinant(mm, order-1);
    }
  }

 return(d);

}  /* M_determilnant */

/*-----------------------------------------------------------------*/

void M_cofactor_matrix(matrix_t m, matrix_t cm)
{
 matrix_t mm;
 int      r, c;
 for (r = 0; r < MOrder; r++)
 for (c = 0; c < MOrder; c++)
  {
   M_minor_matrix(m, MOrder, r, c, mm);

   cm[r][c] = pow(-1.0, (double) (r+c)) * M_determinant(mm, MOrder-1);
  }
}  /* M_cofactor_matrix */

/*-----------------------------------------------------------------*/

void M_adjugate_matrix(matrix_t m, matrix_t am)
{
 matrix_t cm;

 M_cofactor_matrix(m, cm);
 M_transpose_matrix(cm, am);

}  /* M_adjugate_matrix */

/*-----------------------------------------------------------------*/

void M_inverse_matrix(matrix_t m, matrix_t im)
{
 matrix_t am;
 double   d;
 int      r, c;

 d = M_determinant(m, MOrder);

 if (fabs(d) > 0.0)
  {
   M_adjugate_matrix(m, am);

    for (r = 0; r < MOrder; r++)
    for (c = 0; c < MOrder; c++)
```

```
      im[r][c] = am[r][c] / d;
    }

  else
   {
    printf("M_inverse_matrix: Input matrix is singular!\n");
   }
}  /* M_inverse_matrix */

/*---------------------------------------------------------------*/

void M_solve_system(matrix_t A, vector_t C, vector_t X)
{
 matrix_t im;
 int      r, c;

 M_set_identity_matrix(im);
 M_inverse_matrix(A, im);

 for (r = 0; r < MOrder; r++)
  {
   X[r] = 0;

   for (c = 0; c < MOrder; c++)
    X[r] += im[r][c] * C[c];
  }
}  /* M_solve_system */

/*---------------------------------------------------------------*/

void M_print_matrix(char *name, matrix_t m)
{
 char fs[512];
 int  r, c, l, ln;

 l = strlen(name);
 ln = (MOrder - 1) / 2;
 sprintf(fs, "%%%ds", l);

 for (r =0; r < MOrder; r++)
  {
   if (r != ln)
    {
     printf(fs, " ");
     printf(" |");
    }
   else
    {
     printf(fs, name);
     printf("=|");
    }
   for (c = 0; c < MOrder; c++)
   printf(" % e", M_round_to_zero(m[r][c]));
```

```
    printf("  |\n");
   }

  printf("\n");

} /* M_print_matrix */

/*------------------------------------------------------------------*/

void M_print_sub_matrix(char *name, matrix_t m, int order)
{
 char fs[512];
 int  r, c, l, ln;

 l = strlen(name);
 ln = (order - 1) / 2;
 sprintf(fs, "%%%ds", l);

 for (r =0; r < order; r++)
  {
   if (r != ln)
    {
     printf(fs, " ");
     printf("  |");
    }
   else
    {
     printf(fs, name);
     printf("=|");
    }
   for (c = 0; c < order; c++)
   printf(" % e", M_round_to_zero(m[r][c]));

   printf("  |\n");
  }

 printf("\n");

} /* M_print_sub_matrix */

/*------------------------------------------------------------------*/

void M_print_vector(char *name, vector_t v)
{
 int e;

 printf("%s=[", name);

 for (e = 0; e < MOrder; e++)
  printf(" %e", M_round_to_zero(v[e]));

 printf("]\n\n");

} /* M_print_vector */
```

```
/*-------------------------------------------------------------*/

void M_print_sub_vector(char *name, vector_t v, int order)
{
 int e;

 printf("%s=[", name);

 for (e = 0; e < order; e++)
  printf(" %e", M_round_to_zero(v[e]));

 printf("]\n\n");

}  /* M_print_sub_vector */

/*-------------------------------------------------------------*/

/*
 * Test main program.
 */

main ()
{
 int       i, j;
 char      s[25];
 matrix_t a, b, c;

 matrix_t test = {
                  {1.0, 2.0, 3.0},
                  {4.0, 5.0, 6.0),
                  {7.0, 8.0, 9.0}
                 };
 matrix_t test_2 = {
                    {1.0, 2.0, 3.0},
                    {1.0, 0.0, 1.0},
                    {1.0, 1.0, 1.0}
                   };
 matrix_t A = {
               {3.0, 2.0, 1.0},
               {2.0, 3.0, 1.0),
               {1.0, 2.0, 3.0}
              };
 vector_t C = {39.0, 34.0, 26.0};
 vector_t X;

 M_set_identity_matrix(a);
 M_set_zero_matrix(b);

 M_copy_matrix(a, b);
 M_print_matrix("b", b);

 M_set_rc_reversal_matrix(a);
 M_print_matrix("a", a);
```

```
 M_print_matrix("original matrix", test);
 M_set_rc_reversal_matrix(b);

 M_matrix_multiply(test, b, c);
 M_print_matrix("column reversed matrix", c);

 M_matrix_multiply(b, test, c);
 M_print_matrix("row reversed matrix", c);

 M_transpose_matrix(test, c);
 M_print_matrix("transposed matrix", c);

 M_print_sub_matrix("sub matrix", test, 2);

 M_print_matrix("original matrix", test);
 for (i = 0; i < MOrder; i++)
 for (j = 0; j < MOrder; j++)
  {
   M_minor_matrix(test, MOrder, i, j, c);
   sprintf(s, "minor[d][d] matrix", i, j);
   M_print_sub_matrix(s, c, MOrder-1);
  }

 M_set_identity_matrix(a);
 M_print_matrix("a", a);
 printf("|a|=%f\n", M_determinant(a, MOrder));

 M_print_matrix("test_2", test_2);
 printf("|test_2|=%f\n\n", M_determinant(test_2, MOrder));

 M_set_identity_matrix(a);
 M_print_matrix("a", a);
 M_inverse_matrix(a, b);
 M_print_matrix("a^(-1)", b);
 M_inverse_matrix(b, c);
 M_print_matrix("(a"(-1))"(-!)", c);

 M_print_matrix("A", A);
 M_print_vector("C", C);
 M_solve_system(A, C, X);
 M_print_vector("A^(-1)*C", X);
}
```

The above test main program produces the following results:

```
   | 1.000000e+00  0.000000e+00  0.000000e+00 |
b= | 0.000000e+00  1.000000e+00  0.000000e+00 |
   | 0.000000e+00  0.000000e+00  1.000000e+00 |

   | 0.000000e+00  0.000000e+00  1.000000e+00 |
a= | 0.000000e+00  1.000000e+00  0.000000e+00 |
   | 1.000000e+00  0.000000e+00  0.000000e+00 |
```

```
                 | 1.000000e+00  2.000000e+00  3.000000e+00 |
original matrix= | 4.000000e+00  5.000000e+00  6.000000e+00 |
                 | 7.000000e+00  8.000000e+00  9.000000e+00 |

                        | 3.000000e+00  2.000000e+00  1.000000e+00 |
column reversed matrix= | 6.000000e+00  5.000000e+00  4.000000e+00 |
                        | 9.000000e+00  8.000000e+00  7.000000e+00 |

                     | 7.000000e+00  8.000000e+00  9.000000e+00 |
 row reversed matrix=| 4.000000e+00  5.000000e+00  6.000000e+00 |
                     | 1.000000e+00  2.0000Q0e+00  3.000000e+00 |

                   | 1.000000e+00  4.000000e+00  7.000000e+00 |
transposed matrix= | 2.000000e+00  5.000000e+00  8.000000e+00 |
                   | 3.000000e+00  6.000000e+00  9.000000e+00 |

sub matrix= | 1.000000e+00  2.000000e+00 |
            | 4.000000e+00  5.000000e+00 |

                 | 1.000000e+00  2.000000e+00  3.000000e+00 |
original matrix= | 4.000000e+00  5.000000e+00  6.000000e+00 |
                 | 7.000000e+00  8.000000e+00  9.000000e+00 |

minor[0][0] matrix= | 5.000000e+00  6.000000e+00 |
                    | 8.000000e+00  9.000000e+00 |

minor[0][1] matrix= | 4.000000e+00  6.000000e+00 |
                    | 7.000000e+00  9.000000e+00 |

minor[0][2] matrix= | 4.000000e+00  5.000000e+00 |
                    | 7.000000e+00  8.000000e+00 |

minor[1][0] matrix= | 2.000000e+00  3.000000e+00 |
                    | 8.000000e+00  9.000000e+00 |

minor[1][1] matrix= | 1.000000e+00  3.000000e+00 |
                    | 7.000000e+00  9.000000e+00 |


minor[1][2] matrix= | 1.000000e+00  2.000000e+00 |
                    | 7.000000e+00  8.000000e+00 |

minor[2][0] matrix= | 2.000000e+00  3.000000e+00 |
                    | 5.000000e+00  6.000000e+00 |

minor[2][1] matrix= | 1.000000e+00  3.000000e+00 |
                    | 4.000000e+00  6.000000e+00 |

minor[2][2] matrix= | 1.000000e+00  2.000000e+00 |
                    | 4.000000e+00  5.000000e+00 |

   | 1.000000e+00  0.000000e+00  0.000000e+00 |
a= | 0.000000e+00  1.000000e+00  0.000000e+00 |
   | 0.000000e+00  0.000000e+00  1.000000e+00 |
```

```
|a|=1.000000


        | 1.000000e+00  2.000000e+00  3.000000e+00 |
test_2= | 1.000000e+00  0.000000e+00  1.000000e+00 |
        | 1.000000e+00  1.000000e+00  1.000000e+00 |

|test_2|=2.000000

  | 1.000000e+00  0.000000e+00  0.000000e+00 |
a=| 0.000000e+00  1.000000e+00  0.000000e+00 |
  | 0.000000e+00  0.000000e+00  1.000000e+00 |

        | 1.000000e+00  0.000000e+00  0.000000e+00 |
a^(-1)= | 0.000000e+00  1.000000e+00  0.000000e+00 |
        | 0.000000e+00  0.000000e+00  1.000000e+00 |


              | 1.000000e+00  0.000000e+00  0.000000e+00 |
(a^(-1))^(-1)=| 0.000000e+00  1.000000e+00  0.000000e+00 |
              | 0.000000e+00  0.000000e+00  1.000000e+00 |

  | 3.000000e+00  2.000000e+00  1.000000e+00 |
A=| 2.000000e+00  3.000000e+00  1.000000e+00 |
  | 1.000000e+00  2.000000e+00  3.000000e+00 |

C=[ 3.900000e+01  3.400000e+01  2.600000e+01]

A^(-1)*C=[ 9.250000e+00  4.250000e+00  2.750000e+00]
```

Appendix 3

A Simple C Library for 2D Transformations

In this appendix we examine a simple **C** library that implements all the 2D transformations we have examined in Chapter 5.

```
#include <stdio.h>
#include <math.h>

/*------------------------------------------------------------*/

/*
 * Constants and typedefs for 2D transformation routines.
 */

typedef double g2d_matrix_t[3][3];

#define g2d_Pi                   3.1415927     /* Pi                 */
#define g2d_DtoR                 0.0174532925  /* Degrees to Radians */
#define g2d_RtoD                 57.295778     /* Radians to Degrees */
#define g2d_ieee_small_single 3.4e-45          /* Small float near
                                                  zero               */


/*------------------------------------------------------------*/
/*!!!!!!!!!!!!!!!!!!!!!!!!!!!!!!!!!!!!!!!!!!!!!!!!!!!!!!!!!!!!*/
/*!                  General Matrix Routines                  !*/
/*!!!!!!!!!!!!!!!!!!!!!!!!!!!!!!!!!!!!!!!!!!!!!!!!!!!!!!!!!!!!*/
/*------------------------------------------------------------*/


#define g2d_copy_matrix(src, dst) \
{                                 \
 int r, c;                        \
                                  \
 for (r = 0; r < 3; r++)          \
 for (c = 0; c < 3; c++)          \
  (dst)[r][c] = (src)[r][c];      \
}
```

```
/*-------------------------------------------------------------------*/

#define g2d_set_unit_matrix(t) \
{                              \
 int r, c;                     \
                               \
 for (r = 0; r < 3; r++)       \
 for (c = 0; c < 3; c++)       \
 if (r == c) (t)[r][c] = 1.0;  \
 else        (t)[r][c] = 0.0;  \
}

/*-------------------------------------------------------------------*/

#define g2d_set_zero_matrix(t) \
{                              \
 int r, c;                     \
                               \
 for (r = 0; r < 3; r++)       \
 for (c = 0; c < 3; c++)       \
  (t)[r][c] = 0.0;             \
}

/*-------------------------------------------------------------------*/

#define g2d_matrix_mutiply(m1, m2, m3)                 \
{                                                      \
 /*                                                    \
  * Computes the product: m3 : = m1 * m2.              \
  */                                                   \
                                                       \
 int r1, c1, c2;                                       \
                                                       \
 for (r1 = 0; r1 < 3; r1++)                            \
 for (c2 = 0; c2 < 3; c2++)                            \
  {                                                    \
   (m3)[r1][c2] = 0.0;                                 \
                                                       \
   for (c1 = 0; c1 < 3; c1++)                          \
    (m3)[r1][c2] += (m1)[r1][c1] * (m2)[c1][c2];      \
  }                                                    \
}

/*-------------------------------------------------------------------*/
/*!!!!!!!!!!!!!!!!!!!!!!!!!!!!!!!!!!!!!!!!!!!!!!!!!!!!!!!!!!!!!!!!!!!*/
/*!              2D Initialise Transformation Routine              !*/
/*!!!!!!!!!!!!!!!!!!!!!!!!!!!!!!!!!!!!!!!!!!!!!!!!!!!!!!!!!!!!!!!!!!!*/
/*-------------------------------------------------------------------*/

void g2d_initialise_tm(g2d_matrix_t t)
{
 /*
  * Initialises the transformation matrix (t) to the Identity matrix.
  */
```

```
  int r, c;

  for (r = 0; r < 3; r++)
  for (c = 0; c < 3; c++)
   if (r == c) t[r][c] = 1.0;
   else        t[r][c] = 0.0;

} /* g2d_initialise_tm */

/*------------------------------------------------------------*/
/*!!!!!!!!!!!!!!!!!!!!!!!!!!!!!!!!!!!!!!!!!!!!!!!!!!!!!!!!!!!!*/
/*!              2D Global Transformation Routines           !*/
/*!!!!!!!!!!!!!!!!!!!!!!!!!!!!!!!!!!!!!!!!!!!!!!!!!!!!!!!!!!!!*/
/*------------------------------------------------------------*/

void g2d_concatenate(g2d_matrix_t ct, g2d_matrix_t t)
{
 /*
  * Computes the product; ct = ct * t.
  */

 g2d_matrix_t h;   /* Temporary matrix */
 int           r1, c1, c2;

 for (r1 = 0; r1 < 3; r1++)
 for (c2 = 0; c2 < 3; c2++)
  {
   h[r1][c2] = 0.0;

   for (c1 = 0; c1 < 3; c1++)
    h[r1][c2] += ct[r1][c1] * t[c1][c2];
  }

 g2d_copy_matrix(ct, h);

} /* g2d_concatenate */

/*------------------------------------------------------------*/

void g2d_translate(g2d_matrix_t t, double dx, double dy)
{
 /*
  * Concatenates a global translation transformation into the
  * transformation matrix (t).
  *
  * | t00 t01 0 |   | 1  0  0 |   | t00    t01    0 |
  * | t10 t11 0 | * | 0  1  0 | = | t10    t11    0 |
  * | t20 t21 1 |   | dx dy 1 |   | t20+dx t21+dy 1 |
  */

 t[2][0] += dx;
 t[2][1] += dy;

} /* g2d_translate */
```

```
/*--------------------------------------------------------------*/

void g2d_scale(g2d_matrix_t t, double sx, double sy)
{
 /*
  * Concatenates a global scale relative the origin transformation
  * into the transformation matrix (t).
  *
  * | t00 t01 0 |     | sx 0  0 |     | t00*sx t01*sy 0 |
  * | t10 t11 0 |  *  | 0  sy 0 |  =  | t10*sx t11*sy 0 |
  * | t20 t21 1 |     | 0  0  1 |     | t20*sx t21*sy 1 |
  */

 int r;

 for (r = 0; r < 3; r++)
  {
   t[r][0] *= sx;
   t[r][1] *= sy;
  }

}  /* g2d_scale */

/*--------------------------------------------------------------*/

void g2d_rotate(g2d_matrix_t t, double a)
(
 /*
  * Concatenates a global rotation about the origin transformation
  * into the transformation matrix (t).
  *
  * | t00 t01 0 |     |  cos(a) sin(a) 0 |
  * | t10 t11 0 |  *  | -sin(a) cos(a) 0 |  =
  * | t20 t21 1 |     |  0      0      1 |
  *
  * | t00*cos(a)-t01*sin(a)  t00*sin(a)+t01*cos(a)  0 |
  * | t10*cos(a)-t11*sin(a)  t10*sin(a)+t11*cos(a)  0 |
  * | t20*cos(a)-t21*sin(a)  t20*sin(a)+t21*cos(a)  1 |
  *
  * The rotation angle (a) is given in degrees and must be
  * converted to radians.
  */

 g2d_matrix_t h;    /* Temporary matrix */
 int          r;
 double       c, s;

 c = cos(a * g2d_DtoR);
 s = sin(a * g2d_DtoR);

 /* Copy matrix (t) into matrix (h) */

 g2d_copy_matrix(t, h);
```

```
for (r = 0; r < 3; r++)
  {
   t[r][0] = h[r][0] * c - h[r][1] * s;
   t[r][1] = h[r][0] * s + h[r][1] * c;
  }

} /* g2d_rotate */

/*-------------------------------------------------------------------*/

void g2d_shear_x(g2d_matrix_t t, double a)
(
 /*
  * Concatenates a global shear transformation of the X-axis
  * parallel to the Y-axis into the transformation matrix (t).
  *
  * | t00 t01 0 |   | 1 tan(a) 0 |   | t00 t00*tan(a)+t01 0 |
  * | t10 t11 0 | * | 0 1      0 | = | t10 t10*tan(a)+t11 0 |
  * | t20 t21 1 |   | 0 0      1 |   | t20 t20*tan(a)+t21 1 |
  *
  * The shearing angle (a) is given in degrees and must be converted
  * to radians.
  */

 double tangent;
 int    r;

 if (!((a > -90.0) && (a < 90.0)))
  {
   printf("g2d_shear_x: the searing angle must be -90 < a <",
          "90 degrees!\n");
   return;
  }

 tangent = tan(a * g2d_DtoR);

 for (r = 0; r < 3; r++)
  t[r][1] += t[r][0] * tangent;

} /* g2d_shear_x */

/*-------------------------------------------------------------------*/

void g2d_shear_y (g2d_matrix_t t, double a)
{
 /*
  * Concatenates a global shear transformation of the Y-axis
  * parallel to the X-axis into the transformation matrix (t).
  *
  * | t00 t01 0 |   | 1       0 0 |   | t00-t01*tan(a) t01 0 |
  * | t10 t11 0 | * | -tan(a) 1 0 | = | t10-t11*tan(a) t11 0 |
  * | t20 t21 1 |   | 0       0 1 |   | t20-t21*tan(a) t21 1 |
  *
  * The shearing angle (a) is given in degrees and must be converted
```

```
 * to radians.
 */
double tangent;
int    r;

if (!((a > -90.0) && (a < 90.0)))
 {
  printf("g2d_shear_y: the searing angle must be -90 < a <",
         "90 degrees!\n");
  return;
 }

tangent = tan(a * g2d_DtoR);

for (r = 0; r < 3; r++)
 t[r][0] -= t[r][1] * tangent;

} /* g2d_shear_y */

/*-------------------------------------------------------------*/

void g2d_scale_arbitrary(g2d_matrix_t t, double xc, double yc,
       double sx, double sy)
{
 /*
  * Concatenates a global scale transformation about an arbitrary
  * point <xc,yc> into the Transformation Matrix (t).
  *
  * Done in three steps:
  *
  *  1. Translate the centre to the origin.
  *  2. Scale about the origin.
  *  3. Translate the centre to its original position.
  */

  g2d_translate(t, -xc, -yc);
  g2d_scale(t, sx, sy);
  g2d_translate(t, xc, yc);

} /* g2d_scale_arbitrary */

/*-------------------------------------------------------------*/

void g2d_rotate_arbitrary(g2d_matrix_t t, double xc, double yc,
       double a)
{
 /*
  * Concatenates a global rotation transformation about an
  * arbitrary point <xc,yc> into the transformation matrix (t).
  *
  * Done in three steps:
  *
  *  1. Translate the centre to the origin.
  *  2. Rotate about the origin.
```

```
     *  3. Translate the centre to its original position.
     */

    g2d_translate(t, -xc, -yc);
    g2d_rotate(t, a);
    g2d_translate(t, xc, yc);

} /* g2d_rotate_arbitrary */

/*------------------------------------------------------------------*/

void g2d_reflect_arbitrary(g2d_matrix_t t, double x1,double y1,
        double x2,double y2)
 {
  /*
   * Concatenates a global reflection transformation about an
   * arbitrary axis <<x1,y1>,<x2,y2>> into the transformation
   * matrix (t).
   *
   * If the two points coincide, then this is a reflection about an
   * arbitrary point and is done in three steps:
   *
   *  1. Translate the arbitrary point to the origin.
   *  2. Reflect about the origin.
   *  3. Translate the arbitrary point to its original position.
   *
   * If the two points are distinct, then this is a reflection about
   * an arbitrary axis and is done in five steps:
   *
   *  1. Translate the point <x1,y1> to the origin.
   *  2. Rotate the arbitrary axis until it coincides with the x-axis.
   *  3. Reflect about the x-axis.
   *  4. Undo the rotation of step 2.
   *  5. Undo the translation of step 1.
   */

  g2d_matrix_t m;   /* Temporary matrix */
  double       h, dx, dy, s, c;

  dx = x2 - x1;
  dy = y2 - y1;
  h  = sqrt(dx * dx + dy * dy);

  if (fabs(h) < g2d_ieee_small_single)
    {
     /*
      * Reflection about an arbitrary point.
      */

     g2d_translate(t, -x1, -y1);
     g2d_scale(t, -1.0, -1.0);
     g2d_translate(t, x1, y1);
   }
  else
```

```
  {
   /*
    * Reflection about an arbitrary axis.
    */

   g2d_translate(t, -x1, -y1);

   /* Set-up the rotation matrix */

   c = dx / h;
   s = dy / h;

   m[0][0] = c;    m[0][1] = -s;   m[0][2] = 0.0;
   m[1][0] = s;    m[1][1] = c;    m[1][2] = 0.0;
   m[2][0] = 0.0;  m[2][1] = 0.0;  m[2][2] = 1.0;

   g2d_concatenate(t, m);

   g2d_scale(t, 1.0, -1.0);

   /* Set-up the inverse rotation matrix */

   m[0][1] = s;
   m[1][0] = -s;

   g2d_concatenate(t, m);
   g2d_translate(t, x1, y1);
  }

} /* g2d_reflect_arbitrary */

/*-----------------------------------------------------------------*/
/*!!!!!!!!!!!!!!!!!!!!!!!!!!!!!!!!!!!!!!!!!!!!!!!!!!!!!!!!!!!!!!!!!*/
/*!               2D Local Transformation Routines               !*/
/*!!!!!!!!!!!!!!!!!!!!!!!!!!!!!!!!!!!!!!!!!!!!!!!!!!!!!!!!!!!!!!!!!*/
/*-----------------------------------------------------------------*/

void g2d_local_concatenate(g2d_matrix_t ct, g2d_matrix_t t)
{
 /*
  * Computes the products ct = t * ct.
  */

 g2d_matrix_t h;    /* Temporary matrix */
 int            r1, c1, c2 ;

 for (r1 = 0; r1 < 3; r1++)
 for (c2 = 0; c2 < 3; c2++)
  {
   h[r1][c2] = 0.0;

   for (c1 = 0; c1 < 3; c1++)
    h[r1][c2] += t[r1][c1] * ct[c1][c2];
  }
```

```
 g2d_copy_matrix(h, ct);

}  /* g2d_local_concatenate */

/*-------------------------------------------------------------------*/

void g2d_local_translate(g2d_matrix_t t, double dx, double dy)
{
 /*
  * Concatenates a local translation transformation into the
  * transformation matrix (t).
  *
  * | 1  0  0 |   | t00 t01 0 |   | t00            t01          0 |
  * | 0  1  0 | * | t10 t11 0 | = | t10            t11          0 |
  * | dx dy 1 |   | t20 t21 1 |   | t00*dx+        t01*dx+        |
  */                              |   t10*dy+t20     t11*dy+t21 1 |

 t[2][0] += (t[0][0] * dx + t[1][0] * dy);
 t[2][1] += (t[0][1] * dx + t[1][1] * dy);

}  /* g2d_local_translate */

/*-------------------------------------------------------------------*/

void g2d_.local_scale(g2d_matrix_t t, double sx, double sy)
{
 /*
  * Concatenates a local scale relative to the local origin
  * transformation into the transformation matrix (t).
  *
  * | sx 0  0 |   | t00 t01 0 |   | t00*sx t01*sx 0 |
  * | 0  sy 0 | * | t10 t11 0 | = | t10*sy t11*sy 0 |
  * | 0  0  1 |   | t20 t21 1 |   | t20    t21    1 |
  */

 t[0][0] *= sx;  t[0][1] *= sx;
 t[1][0] *= sy;  t[1][1] *= sy;

}  /* g2d_local_scale */

/*-------------------------------------------------------------------*/

void g2d_local_rotate(g2d_matrix_t t, double a)
{
 /*
  * Concatenates a local rotation about the local origin
  * transformation into the transformation matrix (t).
  *
  * |  cos(a)  sin(a) 0 |   | t00 t01 0 |
  * | -sin(a)  cos(a) 0 | * | t10 t11 0 | =
  * |  0       0      1 |   | t20 t21 1 |
  *
  * |  t00*cos(a)+ t10*sin(a)  t01*cos(a)+t11*sin(a) 0 |
```

```
 * | -t00*sin(a)+ t10*cos(a)  -t01*sin(a)+t11*cos(a)  0 |
 * |  t20                      t21                     1 |
 *
 * The rotation angle (a) is given in degrees and must be
 * converted to radians.
 */

g2d_matrix_t h;    /* Temporary matrix */
double        c, s;

c = cos(a * g2d_DtoR);
s = sin(a * g2d_DtoR);

/* Copy matrix (t) into matrix (h) */

g2d_copy_matrix(t, h);

t[0][0] =  h[0][0] * c + h[1][0] * s;
t[1][0] = -h[0][0] * s + h[1][0] * c;

t[0][1] =  h[0][1] * c + h[1][1] * s;
t[1][1] = -h[0][1] * s + h[1][1] * c;

} /* g2d_local_rotate */

/*-----------------------------------------------------------------*/

void g2d_local_shear_x(g2d_matrix_t t, double a)
{
 /*
  * Concatenates a local shear transformation of the X-axis parallel
  * to the Y-axis into the transformation matrix (t).
  *
  * | 1 tan(a) 0 |  | t00 t01 0 |  | t00+t10*tan(a) t01+t11*tan(a) 0 |
  * | 0 1      0 |*| t10 t11 0 |=| t10            t11            0 |
  * | 0 0      1 |  | t20 t21 1 |  | t20            t21            1 |
  *
  * The shearing angle (a) is given in degrees and must be converted
  * to radians.
  */

 double tangent;

 if (!((a > -90.0) && (a < 90.0)))
  {
   printf("g2d_local_shear_x: the searing angle must be -90 < a <",
          "90 degrees!\n");
   return;
  }

 tangent = tan(a * g2d_DtoR);

 t[0][0] += t[1][0] * tangent;
 t[0][1] += t[1][1] * tangent;
```

```
} /* g2d_local_shear_x */

/*--------------------------------------------------------------*/

void g2d_local_shear_y(g2d_matrix_t t, double a)
{
 /*
  * Concatenates a local shear transformation of the Y-axis parallel
  * to the X-axis into the transformation matrix (t).
  *
  *  | 1       0 0 |  | t00 t01 0 |  |  t00           t01          0 |
  *  | -tan(a) 1 0 |*| t10 t11 0 |=| -t00*tan(a)+ -t01*tan (a)+    |
  *  | 0       0 1 |  | t20 t21 1 |  |   t10            t11         0 |
  *                                  |  t20           t21          1 |
  *
  * The shearing angle (a) is given in degrees and must be converted
  * to radians.
  */

 double tangent;

 if (!((a > -90.0) && (a < 90.0)))
  {
   printf("g2d_shear_ys the searing angle must be -90 < a <",
          "90 degrees!\n");
   return;
  }

 tangent = tan(a * g2d_DtoR);

 t[1][0] -= t[0][0] * tangent;
 t[1][1] -= t[0][1] * tangent;

} /* g2d_local_shear_y */

/*--------------------------------------------------------------*/

void g2d_local_scale_arbitrary(g2d_matrix_t t,
                               double       xc,
                               double       yc,
                               double       sx,
                               double       sy
                              )

{
 /*
  * Concatenates a local scale transformation about an arbitrary
  * point <xc,yc> into the Transformation Matrix (t).
  *
  * Done in three steps:
  *
  * 1. Translate the centre to the local origin.
  * 2. Scale about the local origin.
```

```
   * 3. Translate the centre to its original position.
   */

  g2d_local_translate(t, -xc, -yc);
  g2d_local_scale(t, sx, sy);
  g2d_local_translate(t, xc, yc);

} /* g2d_local_scale_arbitrary */

/*-----------------------------------------------------------*/
void g2d_local_rotate_arbitrary(g2d_matrix_t t,
                                double       xc,
                                double       yc,
                                double       a
                               )
{
  /*
   * Concatenates a local rotation transformation about an arbitrary
   * point <xc,yc> into the transformation matrix (t).
   *
   * Done in three steps:
   *
   * 1. Translate the centre to the local origin.
   * 2. Rotate about the local origin.
   * 3. Translate the centre to its original position.
   */

  g2d_local_translate(t, -xc, -yc);
  g2d_local_rotate(t, a);
  g2d_local_translate(t, xc, yc);

} /* g2d_local_rotate_arbitrary */

/*-----------------------------------------------------------*/

void g2d_local_reflect_arbitrary(g2d_matrix_t t,
                                 double       x1,
                                 double       y1,
                                 double       x2,
                                 double       y2
                                )
{
  /*
   * Concatenates a local reflection transformation about a local
   * arbitrary axis <<x1,y1>,<x2 ,y2>> into the transformation
   * matrix (t).
   *
   * If the two points coincide, then this is a reflection about an
   * arbitrary point and is done in three steps:
   *
   * 1. Translate the arbitrary point to the local origin,
   * 2. Reflect about the local origin.
   * 3. Translate the arbitrary point to its original position.
   *
```

```
 * If the two points are distinct, then this is a Reflection about
 * an arbitrary axis and is done in five steps:
 *
 * 1. Translate the point <x1,y1> to the local origin.
 * 2. Rotate the arbitrary axis until it coincides with the local
 *    x-axis.
 * 3. Reflect about local the x-axis.
 * 4. Undo the rotation of step 2.
 * 5. Undo the translation of step 1.
 */


g2d_matrix_t m;    /* Temporary matrix */
double       h, dx, dy, s, c;

dx = x2 - x1;
dy = y2 - y1;
h = sqrt(dx * dx + dy * dy);

if (fabs(h) < g2d_ieee_small_single)
 {
  /*
   * Reflection about an arbitrary point.
   */

  g2d_local_translate(t, -x1, -y1);
  g2d_local_scale(t, -1.0, -1.0);
  g2d_local_translate(t, x1, y1);
 }
else
 {
  /*
   * Reflection about an arbitrary axis.
   */

  g2d_local_translate(t, -x1, -y1);

  /* Set-up the rotation matrix */

  c = dx / h;
  s = dy / h;

  m[0][0] = c;   m[0][1] = -s;  m[0][2] = 0.0;
  m[1][0] = s;   m[1][1] = c;   m[1][2] = 0.0;
  m[2][0] = 0.0; m[2][1] = 0.0; m[2][2] = 1.0;

  g2d_local_concatenate(t, m);

  g2d_local_scale(t, 1.0, -1.0);

  /* Set-up the inverse rotation matrix */

  m[0][1] = s;
  m[1][0] = -s;
```

```
    g2d_local_concatenate(t, m);

    g2d_local_translate(t, x1, y1);
  }

} /* g2d_local_reflect_arbitrary */


/*---------------------------------------------------------------*/
/*!!!!!!!!!!!!!!!!!!!!!!!!!!!!!!!!!!!!!!!!!!!!!!!!!!!!!!!!!!!!!!!*/
/*!              2D Viewing Transformation Routines            !*/
/*!!!!!!!!!!!!!!!!!!!!!!!!!!!!!!!!!!!!!!!!!!!!!!!!!!!!!!!!!!!!!!!*/
/*---------------------------------------------------------------*/

void g2d_viewing_transformation
(
 g2d_matrix_t t,    /* Transformation Matrix (In/Out) */
 double       wl,   /* Window Limits (In)             */
 double       wb,
 double       wr,
 double       wt,
 int          vl,   /* Viewport Limits (In)           */
 int          vb,
 int          vr,
 int          vt
)
{
  /*
   * Concatenates a viewing transformation into the transformation
   * matrix (t).
   *
   * This version of the viewing transformation is defined by
   * specifying the coordinates of the bottom-left and top-right
   * corners of the window and the viewport.
   */

  double sx, sy, dx, dy;


  /*
   * Compute the coefficients of the viewing transformation.
   *
   * Xs = sx * Xw + dx
   * Ys = sy * Yw + dy
   */

  sx = (vr - vl) / (wr - wl);
  sy = (vt - vb) / (wt - wb);
  dx = vl - sx * wl;
  dy = vb - sy * wb;

  /*
   * Perform the matrix multiplication:
```

```
   *
   *  | t00 t01 0 |   | sx 0  0 |   | t00*sx     t01*sy    0 |
   *  | t10 t11 0 | * | 0  sy 0 | = | t10*sx     t11*sy    0 |
   *  | t20 t21 1 |   | dx dy 1 |   | t20*sx+dx t21*sy+dy 1 |
   */

  t[0][0] *= sx;
  t[1][0] *= sx;
  t[2][0] = t[2][0] * sx + dx;

  t[0][1] *= sy;
  t[1][1] *= sy;
  t[2][1] = t[2][1] * sy + dy;

} /* g2d_viewing_transformation */

/*------------------------------------------------------------------*/

void g2d_viewing_transformation_sc
(
 g2d_matrix_t t,     /* Transformation Matrix (In/Out) */
 double      wsx,   /* Window Size (In)               */
 double      wsy,
 double      wcx,   /* Window Centre (In)             */
 double      wcy,
 int         vsx,   /* Viewport Size (In)             */
 int         vsy,
 int         vcx,   /* Viewport Centre (In)           */
 int         vcy
)
{
 /*
  * Concatenates a viewing transformation into the transformation
  * matrix (t).
  *
  * This version of the viewing transformation is defined by
  * specifying the size and the coordinates of the centre of the
  * window and the viewport.
  */

 double wl, wr, wb, wt;
 int    vl, vr, vb, vt;

 /*
  * Compute the bottom-left and top-right corners of the Window and
  * the Viewport.
  */

 wl = wcx - wsx / 2;
 wr = wl  + wsx;
 wb = wcy - wsy / 2;
 wt = wb  + wsy;

 vl = vcx - vsx / 2;
```

```
 vr = vl  + vsx;
 vb = vcy - vsy / 2;
 vt = vb  + vsy;

 g2d_viewing_transformation(t, wl, wb, wr, wt, vl, vb, vr, vt);

} /* g2d_viewing_transformation_sc */

/*-----------------------------------------------------------*/
void g2d_viewing_transformation_sd
(
 g2d_matrix_t t,     /* Transformation Matrix (In/Out) */
 double       wsx,   /* Window Size (In)               */
 double       wsy,
 double       wdx,   /* Window Displacement (In)       */
 double       wdy,
 int          vsx,   /* Viewport Size (In)             */
 int          vsy,
 int          vdx,   /* Viewport Displacement (In)     */
 int          vdy
)

{
 /*
  * Concatenates a viewing transformation into the transformation
  * matrix (t).
  *
  * This version of the viewing transformation is defined by
  * specifying the size and the displacement of the window and the
  * viewport.
  */

 double wl, wr, wb, wt;
 int    vl, vr, vb, vt;


 /*
  * Compute the bottom-left and top-right corners of the window
  * and the viewport.
  */
 wl = wdx;
 wr = wl + wsx;
 wb = wdy;
 wt = wb + wsy;

 vl = vdx;
 vr = vl + vsx;
 vb = vdy;
 vt = vb + vsy;

 g2d_viewing_transformation(t, wl, wb, wr, wt, vl, vb, vr, vt);

} /* g2d_viewing_transformation_sd */
```

```
/*-------------------------------------------------------------------*/
/*!!!!!!!!!!!!!!!!!!!!!!!!!!!!!!!!!!!!!!!!!!!!!!!!!!!!!!!!!!!!!!!!!!!*/
/*!       Apply 2D Viewing Transformation to Point Routines        !*/
/*!!!!!!!!!!!!!!!!!!!!!!!!!!!!!!!!!!!!!!!!!!!!!!!!!!!!!!!!!!!!!!!!!!!*/
/*-------------------------------------------------------------------*/

void g2d_transform_point(g2d_matrix_t t, double *x, double *y)
{
 /*
  * Apply the transformation stored in matrix (t) to the point <x,y>,
  * i.e. perform perform the vector by matrix multiplication:
  *
  *                | t00 t01 0 |
  *   [x, y, 1] *  | t10 t11 0 |=[x*t00+y*t10+t20, x*t01+y*t11+t21, 1]
  *                | t20 t21 1 |
  *
  */

 double ox, oy;


 ox = *x;
 oy = *y;

 *x = ox * t[0][0] + oy * t[1][0] + t[2][0] ;
 *y = ox * t[0][1] + oy * t[1][1] + t[2][1];

}  /* g2d_transform_point */
```

Appendix 4

A Simple C Library for 3D Transformations

In this appendix we examine a simple C library that implements all the 3D transformations we have examined in Chapters 7 and 8.

```
#include <stdio.h>
#include <math.h>
#include <string.h>

/*------------------------------------------------------------------*/

/*
 * Common Constants and typedefs associated with Transformations.
 */

typedef unsigned char boolean_t;

#define False (boolean_t) 0
#define True  (boolean_t) 1

#define g3d_Pi              3.1415927
#define g3d_DtoR            0.0174532925  /* Degrees to Radians */
#define g3d_RtoD            57.295778     /* Radians to Degrees */

#define g3d_IEEE_small_single 3.4e-45     /* single precision real
                                             near 0     */
#define g3d_small_e             0.0000005

typedef double g3d_matrix_t[4][4];
typedef double g3d_vector_t[3];
typedef double g4d_vector_t[4];

/*
 * Constants and typedef for Choen and Sutherland 3D line clipping
 * routine.
 */

#define g3d_small_t          0.000000005
```

```
#define g3d_z_displacement 0.0005

typedef unsigned char g3d_region_code_t;

#define g3d_left_plane   ((g3d_region_code_t) 1)
#define g3d_right_plane  ((g3d_region_code_t) 2)
#define g3d_bottom_plane ((g3d_region_code_t) 4)
#define g3d_top_plane    ((g3d_region_code_t) 8)

/*------------------------------------------------------------------*/
/*!!!!!!!!!!!!!!!!!!!!!!!!!!!!!!!!!!!!!!!!!!!!!!!!!!!!!!!!!!!!!!!!!!*/
/*!                     General Vector Macros                      !*/
/*!!!!!!!!!!!!!!!!!!!!!!!!!!!!!!!!!!!!!!!!!!!!!!!!!!!!!!!!!!!!!!!!!!*/
/*------------------------------------------------------------------*/

#define Mg3d_vector_set(v, x, y, z) \
{                                   \
 (v)[0] = x;                        \
 (v)[1] = y;                        \
 (v)[2] = z;                        \
}

/*------------------------------------------------------------------*/

#define Mg3d_vector_copy(src, dst) \
{                                  \
 (dst)[0] = (src)[0];              \
 (dst)[1] = (src)[1];              \
 (dst)[2] = (src)[2];              \
}

/*------------------------------------------------------------------*/

#define Mg3d_vector_magnitude(v) \
 (sqrt((v)[0] * (v)[0] + (v)[1] * (v)[1] + (v)[2] * (v)[2]))

/*------------------------------------------------------------------*/

#define Mg3d_vector_normalise(v)                                      \
{                                                                     \
 double m;                                                            \
                                                                      \
 m = sqrt((v)[0] * (v)[0] + (v)[1] * (v)[1] + (v)[2] * (v)[2]);      \
                                                                      \
 if (m <= 0.0) m = 1.0;                                               \
                                                                      \
 (v)[0] /= m;                                                         \
 (v)[1] /= m;                                                         \
 (v)[2] /= m;                                                         \
}

/*------------------------------------------------------------------*/
```

```
#define Mg3d_vector_dot(v1, v2) \
 ((v1)[0] * (v2)[0] + (v1)[1] * (v2)[1] + (v1)[2] * (v2)[2])

/*-------------------------------------------------------------*/

#define Mg3d_vector_cross(v1, v2, v3)                        \
{                                                            \
 (v3)[0] = (v1)[1] * (v2)[2] - (v1)[2] * (v2)[1];  \
 (v3)[1] = (v1)[2] * (v2)[0] - (v1)[0] * (v2)[2];  \
 (v3)[2] = (v1)[0] * (v2)[1] - (v1)[1] * (v2)[0];  \
}

/*-------------------------------------------------------------*/
/*!!!!!!!!!!!!!!!!!!!!!!!!!!!!!!!!!!!!!!!!!!!!!!!!!!!!!!!!!!!!!*/
/*!                   General Matrix Routines                  !*/
/*!!!!!!!!!!!!!!!!!!!!!!!!!!!!!!!!!!!!!!!!!!!!!!!!!!!!!!!!!!!!!*/
/*-------------------------------------------------------------*/

void g3d_initialize_matrix(g3d_matrix_t tm)  /* Transformation
                                                Matrix (Out)  */
{
 /*
  * Initialises the Matrix (tm) to the Identity Matrix.
  */

 int r, c;

 for (r = 0; r <= 3; r++)
 for (c = 0; c <= 3; c++)
  if (r == c) tm[r][c] = 1.0;
  else        tm[r][c] = 0.0;

}  /* g3d_initialize_matrix */

/*-------------------------------------------------------------*/

void g3d_matrix_multiply(g3d_matrix_t m1,  /* (In)  */
                         g3d_matrix_t m2,  /* (In)  */
                         g3d_matrix_t m3   /* (Out) */
                        )
{
 /*
  * Computes the product: m3 := m1 * m2.
  */

 int r1, c1, c2;

 for (r1 = 0; r1 <= 3; r1++)
 for (c2 = 0; c2 <= 3; c2++)
  {
   m3[r1][c2] = 0.0;
   for (c1 = 0; c1 <= 3; c1++)
    m3[r1][c2] += m1[r1][c1] * m2[c1][c2];
  }
```

```
} /* g3d_matrix_multiply */

/*-------------------------------------------------------------------*/

void g3d_concatenate(g3d_matrix_t tm,  /* Transformation Matrix
                                          (In/Out)             */
                     g3d_matrix_t t    /* Transformation Matrix
                                          (In)                 */
                    )
{
 /*
  * Computes the product: tm := tm * t.
  */

 g3d_matrix_t h;
 int          r1, c1, c2;

 for (r1 = 0; r1 <= 3; r1++)
 for (c2 = 0; c2 <= 3; c2++)
  {
   h[r1][c2] = 0.0;

   for (c1 = 0; c1 <= 3; c1++)
    h[r1][c2] += tm[r1][c1] * t[c1][c2];
  }

 memcpy(tm, h, sizeof(g3d_matrix_t));

} /* g3d_concatenate */

/*-------------------------------------------------------------------*/

void g3d_invert_matrix(g3d_matrix_t tm) /* Transformation Matrix
                                           (In/Out)            */
{
 g3d_matrix_t cm;
 double       a, b, c, d, e, f, det;
 int          ri, ci;

/* Compute the Cofactor Matrix of the 4*4 Transformation Matrix */

 a = tm[2][2] * tm[3][3] - tm[3][2] * tm[2][3];
 b = tm[1][2] * tm[3][3] - tm[3][2] * tm[1][3];
 c = tm[1][2] * tm[2][3] - tm[2][2] * tm[1][3];
 d = tm[0][2] * tm[3][3] - tm[3][2] * tm[0][3];
 e = tm[0][2] * tm[2][3] - tm[2][2] * tm[0][3];
 f = tm[0][2] * tm[1][3] - tm[1][2] * tm[0][3];

 cm[0][0] = tm[1][1] * a - tm[2][1] * b + tm[3][1] * c;
 cm[1][0] = tm[2][1] * d - tm[0][1] * a - tm[3][1] * e;
 cm[2][0] = tm[0][1] * b - tm[1][1] * d + tm[3][1] * f;
 cm[3][0] = tm[1][1] * e - tm[0][1] * c - tm[2][1] * f;

 cm[0][1] = tm[2][0] * b - tm[1][0] * a - tm[3][0] * c;
```

```
 cm[1][1] = tm[0][0] * a - tm[2][0] * d + tm[3][0] * e;
 cm[2][1] = tm[1][0] * d - tm[0][0] * b - tm[3][0] * f;
 cm[3][1] = tm[0][0] * c - tm[1][0] * e + tm[2][0] * f;

 a = tm[2][1] * tm[3][3] - tm[3][1] * tm[2][3];
 b = tm[1][1] * tm[3][3] - tm[3][1] * tm[1][3];
 c = tm[1][1] * tm[2][3] - tm[2][1] * tm[1][3];
 d = tm[0][1] * tm[3][3] - tm[3][1] * tm[0][3];
 e = tm[0][1] * tm[2][3] - tm[2][1] * tm[0][3];
 f = tm[0][1] * tm[1][3] - tm[1][1] * tm[0][3];

 cm[0][2] = tm[1][0] * a - tm[2][0] * b + tm[3][0] * c;
 cm[1][2] = tm[2][0] * d - tm[0][0] * a - tm[3][0] * e;
 cm[2][2] = tm[0][0] * b - tm[1][0] * d + tm[3][0] * f;
 cm[3][2] = tm[1][0] * e - tm[0][0] * c - tm[2][0] * f;

 a = tm[2][1] * tm[3][2] - tm[3][1] * tm[2][2];
 b = tm[1][1] * tm[3][2] - tm[3][1] * tm[1][2];
 c = tm[1][1] * tm[2][2] - tm[2][1] * tm[1][2];
 d = tm[0][1] * tm[3][2] - tm[3][1] * tm[0][2];
 e = tm[0][1] * tm[2][2] - tm[2][1] * tm[0][2];
 f = tm[0][1] * tm[1][2] - tm[1][1] * tm[0][2];

 cm[0][3] = tm[2][0] * b - tm[1][0] * a - tm[3][0] * c;
 cm[1][3] = tm[0][0] * a - tm[2][0] * d + tm[3][0] * e;
 cm[2][3] = tm[1][0] * d - tm[0][0] * b - tm[3][0] * f;
 cm[3][3] = tm[0][0] * c - tm[1][0] * e + tm[2][0] * f;

 /* Now Compute the Determinant of the Matrix */

 det = tm[0][0]*cm[0][0] + tm[1][0]*cm[1][0] + tm[2][0]*cm[2][0]
       + tm[3][0]*cm[3][0];


 if (fabs(det) <= g3d_IEEE_small_single)
  {
   printf("g3d_invert_matrix: The matrix is Singular!\n");
   return;
  }

 for (ri = 0; ri <= 3; ri++)
 for (ci = 0; ci <= 3; ci++)
  tm[ri][ci] = cm[ci][ri] / det;

} /* g3d_invert_matrix */

/*-----------------------------------------------------------*/
boolean_t g3d_compute_inverse_matrix
(
 g3d_matrix_t tm,  /* Transformation Matrix (In)         */
 g3d_matrix_t itm  /* Inverse Transformation Matrix (Out) */
)
{
 g3d_matrix_t cm;
```

```
double        a, b, c, d, e, f, det;
int           ri, ci;
boolean_t     result;

/* Compute the Cofactor Matrix of the 4*4 Transformation Matrix */

a = tm[2][2] * tm[3][3] - tm[3][2] * tm[2][3];
b = tm[1][2] * tm[3][3] - tm[3][2] * tm[1][3];
c = tm[1][2] * tm[2][3] - tm[2][2] * tm[1][3];
d = tm[0][2] * tm[3][3] - tm[3][2] * tm[0][3];
e = tm[0][2] * tm[2][3] - tm[2][2] * tm[0][3];
f = tm[0][2] * tm[1][3] - tm[1][2] * tm[0][3];

cm[0][0] = tm[1][1] * a - tm[2][1] * b + tm[3][1] * c;
cm[1][0] = tm[2][1] * d - tm[0][1] * a - tm[3][1] * e;
cm[2][0] = tm[0][1] * b - tm[1][1] * d + tm[3][1] * f;
cm[3][0] = tm[1][1] * e - tm[0][1] * c - tm[2][1] * f;

cm[0][1] = tm[2][0] * b - tm[1][0] * a - tm[3][0] * c;
cm[1][1] = tm[0][0] * a - tm[2][0] * d + tm[3][0] * e;
cm[2][1] = tm[1][0] * d - tm[0][0] * b - tm[3][0] * f;
cm[3][1] = tm[0][0] * c - tm[1][0] * e + tm[2][0] * f;

a = tm[2][1] * tm[3][3] - tm[3][1] * tm[2][3];
b = tm[1][1] * tm[3][3] - tm[3][1] * tm[1][3];
c = tm[1][1] * tm[2][3] - tm[2][1] * tm[1][3];
d = tm[0][1] * tm[3][3] - tm[3][1] * tm[0][3];
e = tm[0][1] * tm[2][3] - tm[2][1] * tm[0][3];
f = tm[0][1] * tm[1][3] - tm[1][1] * tm[0][3];

cm[0][2] = tm[1][0] * a - tm[2][0] * b + tm[3][0] * c;
cm[1][2] = tm[2][0] * d - tm[0][0] * a - tm[3][0] * e;
cm[2][2] = tm[0][0] * b - tm[1][0] * d + tm[3][0] * f;
cm[3]12] = tm[1][0] * e - tm[0][0] * c - tm[2][0] * f;

a = tm[2][1] * tm[3][2] - tm[3][1] * tm[2][2];
b = tm[1][1] * tm[3][2] - tm[3][1] * tm[1][2];
c = tm[1][1] * tm[2][2] - tm[2][1] * tm[1][2];
d = tm[0][1] * tm[3][2] - tm[3][1] * tm[0][2];
e = tm[0][1] * tm[2][2] - tm[2][1] * tm[0][2];
f = tm[0][1] * tm[1][2] - tm[1][1] * tm[0][2];

cm[0][3] = tm[2][0] * b - tm[1][0] * a - tm[3][0] * c;
cm[1][3] = tm[0][0] * a - tm[2][0] * d + tm[3][0] * e;
cm[2][3] = tm[1][0] * d - tm[0][0] * b - tm[3][0] * f;

/* Now Compute the Determinant of the Matrix */
det = tm[0][0]*cm[0][0] + tm[1][0]*cm[1][0] + tm[2][0]*cm[2][0]
      + tm[3][0]*cm[3][0];

result = (fabs(det) > g3d_IEEE_small_single);

if (!result) return(result);
```

```
 for (ri = 0; ri <= 3; ri++)
 for (ci = 0; ci <= 3; ci++)
  itm[ri][ci] = cm[ci][ri] / det;

 return(result);

} /* g3d_compute_inverse_matrix */
void g3d_concatenate_inverse(g3d_matrix_t tm, /* Transformation
                                                   Matrix (In/Out) */
                             g3d_matrix_t t   /* Transformation
                                                   Matrix (In)     */
                            )

{
 /*
  * Computes the product: tm := tm * t^(-1).
  */

 g3d_matrix_t it, h;
 int          r1, c1, c2;

 memcpy(it, t, sizeof(g3d_matrix_t));

 g3d_invert_matrix(it);

 for (r1 = 0; r1 <= 3; r1++)
 for (c2 = 0; c2 <= 3; c2++)
  {
   h[r1][c2] = 0.0;

  for (c1 = 0; c1 <= 3; c1++)
   h[r1][c2] += tm[r1][c1] * it[c1][c2];
 }

 memcpy(tm, h, sizeof(g3d_matrix_t));

} /* g3d_concatenate_inverse */

/*-------------------------------------------------------------------*/

void g3d_concatenate_transpose
(
 g3d_matrix_t tm,  /* Transformation Matrix (In/Out) */
 g3d_matrix_t t    /* Transformation Matrix (In)     */
)
{
 /*
  * Computes the product: tm := tm * transpose(t).
  */

 g3d_matrix_t h;
 int          r1, c1, c2;

 for (r1 = 0; r1 <= 3; r1++)
```

```
 for (c2 = 0; c2 <= 3; c2++)
  {
   h[r1][c2] = 0.0;

   for (c1 = 0; c1 <= 3; c1++)
    h[r1][c2] += tm[r1][c1] * t[c2][c1];
  }

 memcpy(tm, h, sizeof(g3d_matrix_t));

} /* g3d_concatenate_transpose */

/*-------------------------------------------------------------*/

void g3d_local_concatenate(g3d_matrix_t tm,  /* Transformation
                                                 Matrix (In/Out) */
                           g3d_matrix_t t    /* Transformation
                                                 Matrix (In)     */
                          )
{
 /* Computes the product: tm := t * tm */

 g3d_matrix_t m;
 int          r1, c1, c2;

 for (r1 = 0; r1 <= 3; r1++)
 for (c2 = 0; c2 <= 3; c2++)
  {
   m[r1][c2] = 0.0;

   for (c1 = 0; c1 <= 3; c1++)
    m[r1][c2] += t[r1][c1] * tm[c1][c2];
  }

 memcpy(tm, m, sizeof(g3d_matrix_t));

} /* g3d_local_concatenate */

/*-------------------------------------------------------------*/

void g3d_local_concatenate_inverse
(
 g3d_matrix_t tm,  /* Transformation Matrix (In/Out) */
 g3d_matrix_t t    /* Transformation Matrix (In)     */
)
{
 /* Computes the product: tm := t^(-1) * tm */

 g3d_matrix_t it, m;
 int          r1, c1, c2;

 memcpy(it, t, sizeof(g3d_matrix_t));

 g3d_invert_matrix(it);
```

```
  for (r1 = 0; r1 <= 3; r1++)
  for (c2 = 0; c2 <= 3; c2++)
   {
    m[r1][c2] = 0.0;

    for (c1 = 0; c1 <= 3; c1++)
     m[r1][c2] += it[r1][c1] * tm[c1][c2];
   }

  memcpy(tm, m, sizeof(g3d_matrix_t));

} /* g3d_local_concatenate_inverse */

/*------------------------------------------------------------*/

void g3d_local_concatenate_transpose
(
 g3d_matrix_t tm,  /* Transformation Matrix (In/Out) */
 g3d_matrix_t t    /* Transformation Matrix (In)     */
)
{
 /* Computes the product: tm := transpose(t) * tm */

 g3d_matrix_t m;
 int          r1, c1, c2;

 for (r1 = 0; r1 <= 3; r1++)
 for (c2 = 0; c2 <= 3; c2++)
  {
   m[r1][c2] = 0.0;

   for (c1 = 0; c1 <= 3; c1++)
    m[r1][c2] += t[c1][r1] * tm[c1][c2];
  }

 memcpy(tm, m, sizeof(g3d_matrix_t));

} /* g3d_local_concatenate_transpose */

/*------------------------------------------------------------*/
/*!!!!!!!!!!!!!!!!!!!!!!!!!!!!!!!!!!!!!!!!!!!!!!!!!!!!!!!!!!!!*/
/*!           3D Global Transformation Routines             !*/
/*!!!!!!!!!!!!!!!!!!!!!!!!!!!!!!!!!!!!!!!!!!!!!!!!!!!!!!!!!!!!*/
/*------------------------------------------------------------*/

void g3d_translate(g3d_matrix_t tm,  /* Transformation Matrix
                                          (In/Out)        */
                   double       dx,  /* Displacements (In) */
                   double       dy,
                   double       dz
                  )
{
 /* Concatenates a Translation Transformation into the
```

```
  * Transformation Matrix(tm)
  */
 g3d_matrix_t m; /* Temporary Matrix */
 int          r;

 memcpy(m, tm, sizeof(g3d_matrix_t));

 for (r = 0; r <= 3; r++)
  {
   tm[r][0] += m[r][3] * dx;
   tm[r][1] += m[r][3] * dy;
   tm[r][2] += m[r][3] * dz;
  }
}  /* g3d_translate */

void g3d_scale(g3d_matrix_t tm, /* Transformation Matrix (In/Out) */
               double       sx, /* Scale Factors (In)             */
               double       sy,
               double       sz
              )
{
 /*
  * Concatenates a Scale relative the Origin Transformation into the
  * Transformation Matrix (tm).
  */

 int r;

 for (r = 0; r <= 3; r++)
  {
   tm[r][0] *= sx;
   tm[r][1] *= sy;
   tm[r][2] *= sz;
  }
}  /* g3d_scale */

/*------------------------------------------------------------------*/

void g3d_rotate_x(g3d_matrix_t tm,    /* Transformation Matrix
                                         (In/Out) */
                  double       angle  /* Rotation Angle in Degrees
                                         (In) */
                 )
{
 /*
  * Concatenates a Rotation about the X-Axis Transformation into the
  * Transformation Matrix (tm).
  */

 g3d_matrix_t m;  /* Temporary Matrix */
 double       cosine, sine;
 int          r;

 memcpy(m, tm, sizeof(g3d_matrix_t));
```

```
 cosine = cos(angle * g3d_DtoR);
 sine   = sin(angle * g3d_DtoR);

 for (r = 0; r <= 3; r++)
  {
   tm[r][1] = m[r][1] * cosine + m[r][2] * sine;
   tm[r][2] = m[r][2] * cosine - m[r][1] * sine;
  }
}  /* g3d_rotate_x */

/*----------------------------------------------------------*/

void g3d_rotate_y(g3d_matrix_t tm,     /* Transformation Matrix
                                          (In/Out) */
                  double       angle   /* Rotation Angle in Degrees
                                          (In) */
                 )
{
 /*
  * Concatenates a Rotation about the Y-Axis Transformation into the
  * Transformation Matrix (tm).
  */

 g3d_matrix_t m;  /* Temporary Matrix */
 double       cosine, sine;
 int          r;

 memcpy(m, tm, sizeof(g3d_matrix_t));

 cosine = cos(angle * g3d_DtoR);
 sine   = sin(angle * g3d_DtoR);

 for (r = 0; r <= 3; r++)
 {
  tm[r][0] = m[r][0] * cosine - m[r][2] * sine;
  tm[r][2] = m[r][2] * cosine + m[r][0] * sine;
 }
}  /* g3d_rotate_y */

/*----------------------------------------------------------*/

void g3d_rotate_z(g3d_matrix_t tm,     /* Transformation Matrix
                                          (In/Out) */
                  double       angle   /* Rotation Angle in Degrees
                                          (In) */
                 )
{
 /*
  * Concatenates a Rotation about the Z-Axis Transformation into the
  * Transformation Matrix (tm).
  */

 g3d_matrix_t m;  /* Temporary Matrix */
```

```
 double        cosine, sine;
 int           r;

 memcpy(m, tm, sizeof(g3d_matrix_t));

 cosine = cos(angle * g3d_DtoR);
 sine   = sin(angle * g3d_DtoR);

 for (r = 0; r <= 3; r++)
  {
   tm[r][0] = m[r][0] * cosine + m[r][1] * sine;
   tm[r][1] s m[r][1] * cosine - m[r][0] * sine;
  }
}  /* g3d_rotate_z */

/*-------------------------------------------------------------------*/

void g3d_shear_x_parallel_to_y
(
 g3d_matrix_t tm,     /* Transformation Matrix (In/Out) */
 double       angle   /* Shear Angle in Degrees (In)    */
)
{
 /*
 * Concatenates a Shear Transformation of the X-Axis parallel to the
 * Y-Axis into the Transformation Matrix (tm).
 */

 double tangent;
 int    r;

 tangent = tan(angle * g3d_DtoR);

 for (r = 0; r <= 3; r++)
  tm[r][1] -= tm[r][0] * tangent;

}  /* g3d_shear_x_parallel_to_y */

/*-------------------------------------------------------------------*/

void g3d_shear_x_parallel_to_z
(
 g3d_matrix_t tm,     /* Transformation Matrix (In/Out) */
 double       angle   /* Shear Angle in Degrees (In)    */
)
{
 /*
 * Concatenates a Shear Transformation of the X-Axis parallel to the
 * Z-Axis into the Transformation Matrix (tm).
 */

 double tangent;
 int    r;
```

```
 tangent = tan(angle * g3d_DtoR);

 for (r = 0; r <= 3; r++)
  tm[r][2] += tm[r][0] * tangent;

} /* g3d_shear_x_parallel_to_z */

/*-----------------------------------------------------------------*/

void g3d_shear_y_parallel_to_x
(
 g3d_matrix_t tm,     /* Transformation Matrix (In/Out) */
 double       angle   /* Shear Angle in Degrees (In)    */
)
{
 /*
  * Concatenates a Shear Transformation of the Y-Axis parallel to
  * the X-Axis into the Transformation Matrix (tm).
  */

 double tangent;
 int    r;

 tangent = tan(angle * g3d_DtoR);

 for (r = 0; r <= 3; r++)
  tm[r][0] += tm[r][1] * tangent;

} /* g3d_shear_y_parallel_to_x */

/*-----------------------------------------------------------------*/

void g3d_shear_y_parallel_to_z
(
 g3d_matrix_t tm,     /* Transformation Matrix (In/Out) */
 double       angle   /* Shear Angle in Degrees (In)    */
)
{
 /*
  * Concatenates a Shear Transformation of the Y-Axis parallel to
  * the Z-Axis into the Transformation Matrix (tm).
  */

 double tangent;
 int    r;

 tangent = tan(angle * g3d_DtoR);

 for (r = 0; r <= 3; r++)
  tm[r][2] -= tm[r][1] * tangent;

} /* g3d_shear_y_parallel_to_z */

/*-----------------------------------------------------------------*/
```

```
void g3d_shear_z_parallel_to_x
(
 g3d_matrix_t tm,    /* Transformation Matrix (In/Out) */
 double       angle /* Shear Angle in Degrees (In)    */
)
{
 /*
  * Concatenates a Shear Transformation of the Z-Axis parallel to
  * the X-Axis into the Transformation Matrix (tm).
  */

 double tangent;
 int    r;

 tangent = tan(angle * g3d_DtoR);

 for (r = 0; r <= 3; r++)
  tm[r][0] -= tm[r][2] * tangent;

} /* g3d_shear_z_parallel_to_x */

/*-----------------------------------------------------------------*/

void g3d_shear_z_parallel_to_y
(
 g3d_matrix_t tm,     /* Transformation Matrix (In/Out) */
 double       angle   /* Shear Angle in Degrees (In)    */
)
{
 /*
  * Concatenates a Shear Transformation of the Z-Axis parallel to
  * the Y-Axis into the Transformation Matrix (tm).
  */

 double tangent;
 int    r;

 tangent = tan(angle * g3d_DtoR);

 for (r = 0; r <= 3; r++)
  tm[r][1] += tm[r][2] * tangent;

} /* g3d_shear_z_parallel_to_y */

/*-----------------------------------------------------------------*/

void g3d_aim_x_axis
(
 g3d_matrix_t tm,  /* Transformation Matrix (In/Out) */
 g3d_vector_t V,   /* Aim Direction Unit Vector (In) */
 g3d_vector_t U    /* Up  Direction Unit Vector (In) */
)
{
```

```
 /*
  * Concatenates into the Transformation Matrix (tm) a Global Aim
  * transformation which causes the Global X-axis to be aligned to
  * the aim in the direction of the unit vector (V). The (U) vector
  * represents the relative up direction unit vector with respect to
  * the (V) vector.
  */

 g3d_matrix_t A;  /* Aim transformation matrix */
 g3d_vector_t X,  /* Local Axes               */
              Y,
              Z;

 /* Compute the new orientation of the transformed local axes */

 Mg3d_vector_copy(V, X);
 Mg3d_vector_normalise(X);

 Mg3d_vector_cross(U, X, Y);
 Mg3d_vector_normalise(Y);

 Mg3d_vector_cross(X, Y, Z);
 Mg3d_vector_normalise(Z);

 /* Construct the aim transformation matrix */

 A[0][0] = X[0];  A[0][1] = X[1];  A[0][2] = X[2];  A[0][3] = 0;
 A[1][0] = Y[0];  A[1][1] = Y[1];  A[1][2] = Y[2];  A[1][3] = 0;
 A[2][0] = Z[0];  A[2][1] = Z[1];  A[2][2] = Z[2];  A[2][3] = 0;
 A[3][0] = 0;     A[3][1] = 0;     A[3][2] = 0;     A[3][3] = 1;

 /* Concatenate the aim transformation with the (tm) transformation
    matrix */

 g3d_concatenate(tm, A);

} /* g3d_aim_x_axis */

/*-----------------------------------------------------------*/

void g3d_aim_y_axis
(
 g3d_matrix_t tm,  /* Transformation Matrix (In/Out) */
 g3d_vector_t V,   /* Aim Direction Unit Vector (In) */
 g3d_vector_t U    /* Up  Direction Unit Vector (In) */
)
{
 /*
  * Concatenates into the Transformation Matrix (tm) a Global Aim
  * transformation which causes the Global Y-axis to be aligned to
  * the aim in the direction of the unit vector (V). The (U) vector
  * represents the relative up direction unit vector with respect to
  * the (V) vector.
  */
```

```
g3d_matrix_t A;  /* Aim transformation matrix */
g3d_vector_t X,  /* Local Axes                */
             Y,
             Z;

/* Compute the new orientation of the transformed local axes */

Mg3d_vector_copy(V, Y);
Mg3d_vector_normalise(Y);

Mg3d_vector_cross(Y, U, X);
Mg3d_vector_normalise(X);

Mg3d_vector_cross(X, Y, Z);
Mg3d_vector_normalise(Z);

/* Construct the aim transformation matrix */

A[0][0] = X[0];  A[0][1] = X[1];  A[0][2] = X[2];  A[0][3] = 0;
A[1][0] = Y[0];  A[1][1] = Y[1];  A[1][2] = Y[2];  A[1][3] = 0;
A[2][0] = Z[0];  A[2][1] = Z[1];  A[2][2] = Z[2];  A[2][3] = 0;
A[3][0] = 0;     A[3][1] = 0;     A[3][2] = 0;     A[3][3] = 1;

/* Concatenate the aim transformation with the (tm) transformation
   matrix */

g3d_concatenate(tm, A);

} /* g3d_aim_y_axis */

/*-----------------------------------------------------------------*/
void g3d_aim_z_axis
(
 g3d_matrix_t tm,   /* Transformation Matrix (In/Out) */
 g3d_vector_t V,    /* Aim Direction Unit Vector (In) */
 g3d_vector_t U     /* Up  Direction Unit Vector (In) */
)
{
 /*
  * Concatenates into the Transformation Matrix (tm) a Global Aim
  * transformation which causes the Global Z-axis to be aligned to
  * the aim in the direction of the unit vector (V). The (U) vector
  * represents the relative up direction unit vector with respect to
  * the (V) vector.
  */

 g3d_matrix_t A;  /* Aim transformation matrix */
 g3d_vector_t X,  /* Local Axes                */
              Y,
              Z;

 /* Compute the new orientation of the transformed local axes */
```

```
 Mg3d_vector_copy(V, Z);
 Mg3d_vector_normalise(Z);

 Mg3d_vector_cross(u, Z, Y);
 Mg3d_vector_normalise(Y);

 Mg3d_vector_cross(Y, Z, X);
 Mg3d_vector_normalise(X);

/* Construct the aim transformation matrix */

 A[0][0] = X[0]; A[0][1] = X[1]; A[0][2] = X[2]; A[0][3] = 0;
 A[1][0] = Y[0]; A[1][1] = Y[1]; A[1][2] = Y[2]; A[1][3] = 0;
 A[2][0] = Z[0]; A[2][1] = Z[1]; A[2][2] = Z[2]; A[2][3] = 0;
 A[3][0] = 0;    A[3][1] = 0;    A[3][2] = 0;    A[3][3] = 1;

 /* Concatenate the aim transformation with the (tm) transformation
    matrix */

 g3d_concatenate(tm, A);
}  /* g3d_aim_z_axis */

/*-----------------------------------------------------------------*/

void g3d_reflect_about_arbitrary_point
(
 g3d_matrix_t tm,  /* Transformation Matrix (In/Out) */
 g3d_vector_t C    /* Centre of Reflection (In)      */
)
{
 /*
  * Concatenates a Global Reflection Transformation about an
  * arbitrary point (C) into the Transformation Matrix (tm).
  *
  * Done in three steps:
  *
  * 1. Translate the centre to the origin.
  *
  * 2. Reflect about the origin.
  *
  * 3. Translate the centre to its original position.
  */

 g3d_translate(tm, -C[0], -C[1], -C[2]);
 g3d_scale(tm, -1.0, -1.0, -1.0);
 g3d_translate(tm, C[0], C[1], C[2]);

}  /* g3d_reflect_about_arbitrary_point */

/*-----------------------------------------------------------------*/

void g3d_scale_about_arbitrary_point
(
 g3d_matrix_t tm,  /* Transformation Matrix (In/Out) */
```

```
 g3d_vector_t C,   /* Centre of Scale (In)          */
 double       sx,  /* Scale Factors (In)            */
 double       sy,
 double       sz
)
{
 /*
  * Concatenates a Global Scale Transformation about an arbitrary
  * point (C) into the Transformation Matrix (tm).
  *
  * Done in three steps:
  *
  * 1. Translate the centre to the origin.
  *
  * 2. Scale about the origin.
  *
  * 3. Translate the centre to its original position.
  */

 g3d_translate(tm, -C[0], -C[1], -C[2]);
 g3d_scale(tm, sx, sy, sz);
 g3d_translate(tm, C[0], C[1], C[2]);

}  /* g3d_scale_about_arbitrary_point */

/*---------------------------------------------------------------------*/

void g3d_reflect_about_arbitrary_axis
(
 g3d_matrix_t tm,   /* Transformation Matrix (In/Out)              */
 g3d_vector_t P,    /* Point on the Arbitrary Axis (In)            */
 g3d_vector_t D     /* Direction Unit Vector of the Arbitrary
                       Axis (In)                                   */
)
{
 g3d_matrix_t A;   /* Aim transformation matrix          */
 g3d_vector_t V,   /* Aim Direction Unit Vector          */
              U,   /* Relative Up Direction Unit Vector */
              X,   /* Local Axes                         */
              Y,
              Z;

  /*
   * Concatenates a Global Reflection Transformation about an
   * arbitrary Axis into the Transformation Matrix (tm).
   *
   * The Arbitrary Axis is defined by a point (P) and a direction
   * unit vector (D).
   *
   * This transformation is constructed by the following five steps:
   *
   * Step 1. Translate the point (P) to the origin.
   */
```

```
g3d_translate(tm, -P[0], -P[1], -P[2]);

/*
 * Step 2. Align the unit vector (V) with the with the X-axis, using
 *         the inverse of the aiming transformation that would align
 *         the X-axis with vector (V).
 */

/* Copy and normalise the Direction Vector of the arbitrary Axis */

 Mg3d_vector_copy(D, V);
 Mg3d_vector_normalise(V);


/*
 * Determine the relative up direction unit vector (U). No need to
 * worry about gimbal lock here as step 4 reverses step 2.
 */

if(
   (fabs(V[0])      <= g3d_IEEE_small_single) &&
   (fabs(V[1])      <= g3d_IEEE_small_single) &&
   (fabs(V[2] -1.0) <= g3d_IEEE_small_single)
  )
 {
  /* When (V == Z), set U = -X */

  Mg3d_vector_set(U, -1.0, 0.0, 0.0);
 }
else
if(
   (fabs(V[0])       <= g3d_IEEE_small_single) &&
   (fabs(V[1])       <= g3d_IEEE_small_single) &&
   (fabs(V[2] + 1.0) <= g3d_IEEE_small_single)
   )
 {
  /* When (V == -Z), set U = X */

  Mg3d_vector_set(U, 1.0, 0.0, 0.0);
 }
else
 {
  /* Otherwise, set U = Z */

  Mg3d_vector_set(U, 0.0, 0.0,1.0);
 }

/* Compute the new orientation of the transformed local axes */

Mg3d_vector_copy(V, X);
Mg3d_vector_normalise(X);

Mg3d_vector_cross(U, X, Y);
Mg3d_vector_normalise(Y);
```

```
 Mg3d_vector_cross(X, Y, Z);
 Mg3d_vector_normalise(Z);

/* Construct the aim transformation matrix */

 A[0][0] = X[0];  A[0][1] = X[1];  A[0][2] = X[2];  A[0][3] = 0;
 A[1][0] = Y[0];  A[1][1] = Y[1];  A[1][2] = Y[2];  A[1][3] = 0;
 A[2][0] = Z[0];  A[2][1] = Z[1];  A[2][2] = Z[2];  A[2][3] = 0;
 A[3][0] = 0;     A[3][1] = 0;     A[3][2] = 0;     A[3][3] = 1;

 /* Concatenate the aim transformation with the (tm) transformation
    matrix */

 g3d_concatenate_transpose(tm, A);

 /*
  * Step 3. Reflect about the X-axis.
  */

 g3d_scale(tm, 1.0, -1.0, -1.0);

 /*
  * Step 4. Return vector (V) to its original orientation, by
  *         applying the inverse of the transformation from
  *         step 2.
  */

 g3d_concatenate(tm, A);

 /*
  * Step 5. Translate point (P) to its original position, by
  *         applying the inverse of the transformation from
  *         step 1.
  */

 g3d_translate(tm, P[0], P[1], P[2]);

} /* g3d_reflect_about_arbitrary_axis */

/*-----------------------------------------------------------------*/

void g3d_rotate_about_arbitrary_axis
(
 g3d_matrix_t tm,     /* Transformation Matrix (In/Out)              */
 g3d_vector_t P,      /* Point on the Arbitrary Axis (In)            */
 g3d_vector_t D,      /* Direction Unit Vector of the Arbitrary
                         Axis (In)                                   */
 double       angle   /* Rotation Angle in Degrees (In)              */
)
{
 g3d_matrix_t A;   /* Aim transformation matrix           */
 g3d_vector_t V,   /* Aim Direction Unit Vector           */
              U,   /* Relative Up Direction Unit Vector */
```

```
            X,  /* Local Axes                              */
            Y
            Z;

/*
 * Concatenates a Global Rotation Transformation about an arbitrary
 * Axis into the Transformation Matrix (tm).
 *
 * The Arbitrary Axis is defined by a point (P) and a direction unit
 * vector (D).
 *
 * This transformation is constructed by the following five steps:
 *

 * Step 1. Translate the point (P) to the origin.
 */

g3d_translate(tm, -P[0], -P[1], -P[2]);

/*
 * Step 2. Align the unit vector (V) with the with the X-axis, using
 *         the inverse of the aiming transformation that would align
 *         the X-axis with vector (V).
 */

/* Copy and normalise the Direction Vector of the arbitrary Axis */

Mg3d_vector_copy(D, V);
Mg3d_vector_normalise(V);

/*
 * Determine the relative up direction unit vector (U). No need to
 * worry about gimbal lock here as step 4 reverses step 2.
 */

if(
    (fabs(V[0])       <= g3d_IEEE_small_single) &&
    (fabs(V[1])       <= g3d_IEEE_small_single) &&
    (fabs(V[2] -1.0)  <= g3d_IEEE_small_single)
  )
  {
   /* When (V == Z), set U = -X */

   Mg3d_vector_set(U, -1.0, 0.0, 0.0);
  }
 else
 if(
    (fabs(V[0])        <= g3d_IEEE_small_single) &&
    (fabs(V[1])        <= g3d_IEEE_small_single) &&
    (fabs(V[2] + 1.0)  <= g3d_IEEE_small_single)
    )
  {
   /* When (V == -Z), set U = X */
```

```
    Mg3d_vector_set(U, 1.0, 0.0, 0.0);
   }
  else
   {
    /* Otherwise, set U = Z */

    Mg3d_vector_set(U, 0.0, 0.0,1.0);
   }

/* Compute the new orientation of the transformed local axes */

 Mg3d_vector_copy(V, X);
 Mg3d_vector_normalise(X);

 Mg3d_vector_cross(U, X, Y);
 Mg3d_vector_normalise(Y);

 Mg3d_vector_cross(X, Y, Z);
 Mg3d_vector_normalise(Z);

/* Construct the aim transformation matrix */

 A[0][0] = X[0]; A[0][1] = X[1]; A[0][2] = X[2]; A[0][3] = 0;
 A[1][0] = Y[0]; A[1][1] = Y[1]; A[1][2] = Y[2]; A[1][3] = 0;
 A[2][0] = Z[0]; A[2][1] = Z[1]; A[2][2] = Z[2]; A[2][3] = 0;
 A[3][0] = 0;    A[3][1] = 0;    A[3][2] = 0;    A[3][3] = 1;

 /* Concatenate the aim transformation with the (tm) transformation
    matrix */

 g3d_concatenate_transpose(tm, A);

 /*
  * Step 3. Rotate about the X-axis.
  */

 g3d_rotate_x(tm, angle);

 /*
  * Step 4. Return vector (V) to its original orientation, by
  *         applying the inverse of the transformation from
  *         step 2.
  */

 g3d_concatenate(tm, A);

 /*
  * Step 5. Translate point (P) to its original position, by
  *         applying the inverse of the transformation from
  *         step 1.
  */

 g3d_translate(tm, P[0], P[1], P[2]);
```

```
}  /* g3d_rotate_about_arbitrary_axis */

/*-----------------------------------------------------------------*/

  void g3d_scale_along_arbitrary_axis
(
 g3d_matrix_t tm,   /* Transformation Matrix (In/Out)              */
 g3d_vector_t P,    /* Point on the Arbitrary Axis (In)            */
 g3d_vector_t D,    /* Direction Unit Vector of the Arbitrary      */
                       Axis (In)                                   */
 double       sf    /* Scale Factor (In)                           */
)
{
 g3d_matrix_t A;  /* Aim transformation matrix          */
 g3d_vector_t V,  /* Aim Direction Unit Vector          */
              U,  /* Relative Up Direction Unit Vector */
              X,  /* Local Axes                         */
              Y,
              Z;

 /*
  * Concatenates a Global Scaling Transformation along an arbitrary
  * Axis into the Transformation Matrix (tm).
  *
  * The Arbitrary Axis is defined by a point (P) and a direction
  * unit vector (D).
  *
  * This transformation is constructed by the following five steps:
  *
  * Step 1. Translate the point (P) to the origin.
  */

 g3d_translate(tm, -P[0], -P[1], -P[2]);

 /*
  * Step 2. Align the unit vector (V) with the X-axis, using
  *         the inverse of the aiming transformation that would
  *         align the X-axis with vector (V).
  */


 /* Copy and normalise the Direction Vector of the arbitrary Axis */

 Mg3d_vector_copy(D, V);
 Mg3d_vector_normalise(V);

 /*
  * Determine the relative up direction unit vector (U). No need to
  * worry about gimbal lock here as step 4 reverses step 2.
  */

 if(
    (fabs(V[0])      <= g3d_IEEE_small_single) &&
    (fabs(V[1])      <= g3d_IEEE_small_single) &&
```

```
    (fabs(V[2] -1.0) <= g3d_IEEE_small_single)
  )
 {
  /* When (V == Z), set U = -X */

  Mg3d_vector_set(U, -1.0, 0.0, 0.0);
 }
else
if(
    (fabs(V[0])       <= g3d_IEEE_small_single) &&
    (fabs(V[1])       <= g3d_IEEE_small_single) &&
    (fabs(V[2] + 1.0) <= g3d_IEEE_small_single)
    )
 {
  /* When (V == -Z), set U = X */

  Mg3d_vector_set(U, 1.0, 0.0, 0.0);
 }
else
 {
  /* Otherwise, set U = Z */

  Mg3d_vector_set(U, 0.0, 0.0,1.0);
 }

/* Compute the new orientation of the transformed local axes */

 Mg3d_vector_copy(V, X);
 Mg3d_vector_normalise(X);

 Mg3d_vector_cross(U, X, Y);
 Mg3d_vector_normalise(Y);

 Mg3d_vector_cross(X, Y, Z);
 Mg3d_vector_normalise(Z);

 /* Construct the aim transformation matrix */

 A[0][0] = X[0]; A[0][1] = X[1]; A[0][2] = X[2]; A[0][3] = 0;
 A[1][0] = Y[0]; A[1][1] = Y[1]; A[1][2] = Y[2]; A[1][3] = 0;
 A[2][0] = Z[0]; A[2][1] = Z[1]; A[2][2] = Z[2]; A[2][3] = 0;
 A[3][0] = 0;    A[3][1] = 0;    A[3][2] = 0;    A[3][3] = 1;

 /* Concatenate the aim transformation with the (tm) transformation
    matrix */

 g3d_concatenate_transpose(tm, A);

 /*
  * Step 3. Scale along the X-axis.
  */

 g3d_scale(tm, sf, 1.0,1.0);
```

```
 /*
  * Step 4. Return vector (V) to its original orientation, by
  *         applying the inverse of the transformation from
  *         step 2.
  */

 g3d_concatenate(tm, A);

 /*
  * Step 5. Translate point (P) to its original position, by
  *         applying the inverse of the transformation from
  *         step 1.
  */

 g3d_translate(tm, P[0], P[1], P[2]);

} /* g3d_scale_along_arbitrary_axis */

void g3d_translate_along_arbitrary_axis
(
 g3d_matrix_t tm,   /* Transformation Matrix (In/Out)               */
 g3d_vector_t P,    /* Point on the Arbitrary Axis (In)             */
 g3d_vector_t D,    /* Direction Unit Vector of the Arbitrary
                       Axis (In)                                    */
 double       dd    /* Displacement (In)                            */
)
{
 g3d_matrix_t A;  /* Aim transformation matrix          */
 g3d_vector_t V,  /* Aim Direction Unit Vector          */
              U,  /* Relative Up Direction Unit Vector */
              X,  /* Local Axes                         */
              Y,
              Z;

 /*
  * Concatenates a Global Translation Transformation along an
  * arbitrary Axis into the Transformation Matrix (tm).
  *
  * The Arbitrary Axis is defined by a point (P) and a direction
  * unit vector (D).
  * This transformation is constructed by the following five steps:
  *
  * Step 1. Translate the point (P) to the origin.
  */

 g3d_translate(tm, -P[0], -P[1], -P[2]);

 /*
 * Step 2. Align the unit vector (V) with the with the X-axis, using
 *         the inverse of the aiming transformation that would align
 *         the X-axis with vector (V).
 */

 /* Copy and normalise the Direction Vector of the arbitrary Axis */
```

```
Mg3d_vector_copy(D, V);
Mg3d_vector_normalise(V);

/*
 * Determine the relative up direction unit vector (U). No need to
 * worry about gimbal lock here as step 4 reverses step 2.
 */

if(
   (fabs(V[0])       <= g3d_IEEE_small_single) &&
   (fabs(V[1])       <= g3d_IEEE_small_single) &&
   (fabs(V[2] -1.0) <= g3d_IEEE_small_single)
  )
 {
  /* When (V == Z), set U = -X */

  Mg3d_vector_set(U, -1.0, 0.0, 0.0);
 }
else
if(
   (fabs(V[0])        <= g3d_IEEE_small_single) &&
   (fabs(V[1])        <= g3d_IEEE_small_single) &&
   (fabs(V[2] + 1.0) <= g3d_IEEE_small_single)
   )
 {
  /* When (V == -Z), set U = X */

  Mg3d_vector_set(U, 1.0, 0.0, 0.0);
 }
else
 {
  /* Otherwise, set U = Z */

  Mg3d_vector_set(U, 0.0, 0.0,1.0);
 }

/* Compute the new orientation of the transformed local axes */

Mg3d_vector_copy(V, X);
Mg3d_vector_normalise(X);

Mg3d_vector_cross(U, X, Y);
Mg3d_vector_normalise(Y);

Mg3d_vector_cross(X, Y, Z);
Mg3d_vector_normalise(Z);


/* Construct the aim transformation matrix */

A[0][0] = X[0];  A[0][1] = X[1];  A[0][2] = X[2];  A[0][3] = 0;
A[1][0] = Y[0];  A[1][1] = Y[1];  A[1][2] = Y[2];  A[1][3] = 0;
A[2][0] = Z[0];  A[2][1] = Z[1];  A[2][2] = Z[2];  A[2][3] = 0;
```

```
 A[3][0] = 0;      A[3][1] = 0;      A[3][2] = 0;      A[3][3] = 1;

 /* Concatenate the aim transformation with the (tm) transformation
    matrix */

 g3d_concatenate_transpose(tm, A);

/*
 * Step 3. Translate along the X-axis.
 */

 g3d_translate(tm, dd, 0.0, 0.0);

 /*
  * Step 4. Return vector (V) to its original orientation, by
  *         applying the inverse of the transformation from
  *         step 2.
  */

 g3d_concatenate(tm, A);

 /*
  * Step 5. Translate point (P) to its original position, by
  *         applying the inverse of the transformation from
  *         step 1.
  */
 g3d_translate(tm, P[0], P[1], P[2]);

}  /* g3d_translate_along_arbitrary_axis */

/*------------------------------------------------------------------*/

void g3d_reflect_about_arbitrary_plane
(
 g3d_matrix_t tm,   /* Transformation Matrix (In/Out)                */
 g3d_vector_t P,    /* Point on the Arbitrary Plane (In)             */
 g3d_vector_t N     /* Unit Normal Vector of the Arbitrary
                       Plane (In)                                    */
)
{
 g3d_matrix_t A;  /* Aim transformation matrix            */
 g3d_vector_t V,  /* Aim Direction Unit Vector            */
              U,  /* Relative Up Direction Unit Vector */
              X,  /* Local Axes                           */
              Y,
              Z;

 /*
  * Concatenates a Global Reflection Transformation about an arbitrary
  * Plane into the Transformation Matrix (tm).
  *
  * The Arbitrary Plane is defined by a point (P) and the direction
  * unit vector of its normal (N).
  *
```

```
 * This transformation is constructed by the following five steps:
 *
 * Step 1. Translate the point (P) to the origin.
 */

g3d_translate(tm, -P[0], -P[1], -P[2]);

/*
 * Step 2. Align the unit vector (V) with the with the X-axis, using
 *         the inverse of the aiming transformation that would align
 *         the X-axis with vector (V).
 */

/* Copy and normalise the Normal Vector of the arbitrary Plane */

Mg3d_vector_copy(N, V);
Mg3d_vector_normalise(V);


/*
 * Determine the relative up direction unit vector (U). No need to
 * worry about gimbal lock here as step 4 reverses step 2.
 */
if(
   (fabs(V[0])      <= g3d_IEEE_small_single) &&
   (fabs(V[1])      <= g3d_IEEE_small_single) &&
   (fabs(V[2] -1.0) <= g3d_IEEE_small_single)
  )
 {
  /* When (V == Z), set U = -X */

  Mg3d_vector_set(U, -1.0, 0.0, 0.0);
 }
else
if(
   (fabs(V[0])       <= g3d_IEEE_small_single) &&
   (fabs(V[1])       <= g3d_IEEE_small_single) &&
   (fabs(V[2] + 1.0) <= g3d_IEEE_small_single)
   )
 {
  /* When (V == -Z), set U = X */

  Mg3d_vector_set(U, 1.0, 0.0, 0.0);
 }
else
 {
  /* Otherwise, set U = Z */

  Mg3d_vector_set(U, 0.0, 0.0,1.0);
 }

/* Compute the new orientation of the transformed local axes */

Mg3d_vector_copy(V, X);
```

```
  Mg3d_vector_normalise(X);

  Mg3d_vector_cross(U, X, Y);
  Mg3d_vector_normalise(Y);

  Mg3d_vector_cross(X, Y, Z);
  Mg3d_vector_normalise(Z);

  /* Construct the aim transformation matrix */

  A[0][0] = X[0];  A[0][1] = X[1];  A[0][2] = X[2];  A[0][3] = 0;
  A[1][0] = Y[0];  A[1][1] = Y[1];  A[1][2] = Y[2];  A[1][3] = 0;
  A[2][0] = Z[0];  A[2][1] = Z[1];  A[2][2] = Z[2];  A[2][3] = 0;
  A[3][0] = 0;     A[3][1] = 0;     A[3][2] = 0;     A[3][3] = 1;

  /* Concatenate the aim transformation with the (tm) transformation
     matrix */

  g3d_concatenate_transpose(tm, A);

  /*
   * Step 3. Reflect about the YZ-plane.
   */
  g3d_scale(tm, -1.0,1.0,1.0);

  /*
   * Step 4. Return vector (V) to its original orientation, by
   *         applying the inverse of the transformation from
   *         step 2.
   */

  g3d_concatenate(tm, A);

  /*
   * Step 5. Translate point (P) to its original position, by
   *         applying the inverse of the transformation from
   *         step 1.
   */

  g3d_translate(tm, P[0], P[1], P[2]);

} /* g3d_reflect_about_arbitrary_plane */

/*-------------------------------------------------------------*/
/*!!!!!!!!!!!!!!!!!!!!!!!!!!!!!!!!!!!!!!!!!!!!!!!!!!!!!!!!!!!!!*/
/*!             3D Local Transformation Routines              !*/
/*!!!!!!!!!!!!!!!!!!!!!!!!!!!!!!!!!!!!!!!!!!!!!!!!!!!!!!!!!!!!!*/
/*-------------------------------------------------------------*/

void g3d_local_translate(g3d_matrix_t tm,  /* Transformation Matrix
                                              (In/Out)          */
                         double       dx,  /* Displacements (In) */
                         double       dy,
                         double       dz
                        )
```

```
{
 /*
  * Concatenates a Local Translation Transformation into the
  * Transformation Matrix (tm).
  */

 int c;

 for (c = 0; c <= 3; c++)
  tm[3][c] = tm[0][c] * dx + tm[1][c] * dy + tm[2][c] * dz + tm[3][c];

} /* g3d_local_translate */

/*-------------------------------------------------------------*/

void g3d_local_scale(g3d_matrix_t tm,  /* Transformation Matrix
                                           (In/Out)           */
                     double       sx,  /* Scale Factors (In)  */
                     double       sy,
                     double       sz
                    )
{
 /*
  * Concatenates a Scale relative the Local Origin Transformation
  * into the Transformation Matrix (tm).
  */

 int c;

 for (c = 0; c <= 3; c++)
  {
   tm[0][c] *= sx;
   tm[1][c] *= sy;
   tm[2][c] *= sz;
  }
} /* g3d_local_scale */

/*-------------------------------------------------------------*/

void g3d_local_rotate_x(g3d_matrix_t tm,    /* Transformation
                                                Matrix (In/Out)  */
                        double       angle  /* Rotation Angle in
                                                Degrees (In)     */
                       )
{
 /*
  * Concatenates a Rotation about the Local X-Axis Transformation
  * into the Transformation Matrix (tm).
  */

 g3d_matrix_t m;  /* Temporary Matrix */
 double       cosine, sine;
 int          c;
```

```
 /* Copy Matrix (tm) into Matrix (m) */

 memcpy(m, tm, sizeof(g3d_matrix_t));

 cosine = cos(angle * g3d_DtoR);
 sine   = sin(angle * g3d_DtoR);

 for (c == 0; c <= 3; c++)
  {
   tm[1][c] = m[1][c] * cosine - m[2][c] * sine;
   tm[2][c] = m[1][c] * sine   + m[2][c] * cosine;
  }
} /* g3d_local_rotate_x */

/*-------------------------------------------------------------------*/

void g3d_local_rotate_y(g3d_matrix_t tm,     /* Transformation
                                                Matrix (In/Out)    */
                        double       angle   /* Shear Angle in
                                                Degrees (In)       */
                       )
{
 /*
  * Concatenates a Rotation about the Local Y-Axis Transformation
  * into the Transformation Matrix (tm).
  */

 g3d_matrix_t m;  /* Temporary Matrix */
 double       cosine, sine;
 int          c;

 /* Copy Matrix (tm) Into Matrix (m) */

 memcpy(m, tm, sizeof(g3d_matrix_t));

 cosine = cos(angle * g3d_DtoR);
 sine   = sin(angle * g3d_DtoR);

 for (c = 0; c <= 2; c++)
  {
   tm[0][c] = m[0][c] * cosine + m[2][c] * sine;
   tm[2][c] = m[2][c] * cosine - m[0][c] * sine;
  }
} /* g3d_local_rotate_y */

/*-------------------------------------------------------------------*/

void g3d_local_rotate_z(g3d_matrix_t tm,     /* Transformation
                                                Matrix (In/Out)    */
                        double       angle   /* Shear Angle in
                                                Degrees (In)       */
                       )
{
```

```
 /*
  * Concatenates a Rotation about the Local Z-Axis Transformation
  * into the Transformation Matrix (tm).
  */

 g3d_matrix_t m;  /* Temporary Matrix */
 double       cosine, sine;
 int          c;

 /* Copy Matrix (tm) into Matrix (m) */

 memcpy(m, tm, sizeof(g3d_matrix_t));

 cosine = cos(angle * g3d_DtoR);
 sine   = sin(angle * g3d_DtoR);

 for (c = 0; c <= 2; c++)
  {
   tm[0][c] = m[0][c] * cosine - m[1][c] * sine;
   tm[1][c] = m[0][c] * sine   + m[1][c] * cosine;
  }
} /* g3d_local_rotate_z */
/*-----------------------------------------------------------*/

void g3d_local_shear_x_parallel_to_y
(
 g3d_matrix_t tm,    /* Transformation Matrix (In/Out) */
 double       angle  /* Shear Angle in Degrees (In)    */
)
{
 /*
  * Concatenates a Local Shear Transformation of the X-Axis parallel
  * to the Y-Axis into the Transformation Matrix (tm).
  */

 double tangent;
 int    c;

 tangent = tan(angle * g3d_DtoR);

 for (c = 0; c <= 3; c++)
  tm[0][c] -= tm[1][c] * tangent;

} /* g3d_local_shear_x_parallel_to_y */

/*-----------------------------------------------------------*/

void g3d_local_shear_x_parallel_to_z
(
 g3d_matrix_t tm,    /* Transformation Matrix (In/Out) */
 double       angle  /* Shear Angle in Degrees (In)    */
)
{
```

```
  /*
   * Concatenates a Local Shear Transformation of the X-Axis parallel
   * to the Z-Axis into the Transformation Matrix (tm).
   */

  double tangent;
  int    c;
  tangent = tan(angle * g3d_DtoR);

  for (c = 0; c <= 3; c++)
   tm[0][c] += tm[2][c] * tangent;

} /* g3d_local_shear_x_parallel_to_z */

/*------------------------------------------------------------------*/

void g3d_local_shear_y_parallel_to_x
(
 g3d_matrix_t tm,     /* Transformation Matrix (In/Out) */
 double       angle   /* Shear Angle in Degrees (In)    */
)
{
  /*
   * Concatenates a Local Shear Transformation of the Y-Axis parallel
   * to the X-Axis into the Transformation Matrix (tm).
   */

  double tangent;
  int    c;

  tangent = tan(angle * g3d_DtoR);

  for (c = 0; c <= 3; c++)
   tm[1][c] += tm[0][c] * tangent;

} /* g3d_local_shear_y_parallel_to_x */

/*------------------------------------------------------------------*/

void g3d_local_shear_y_parallel_to_z
(
 g3d_matrix_t tm,     /* Transformation Matrix (In/Out) */
 double       angle   /* Shear Angle in Degrees (In)    */
)
{
  /*
   * Concatenates a Local Shear Transformation of the Y-Axis parallel
   * to the Z-Axis into the Transformation Matrix (tm).
   */

  double tangent;
  int    c;
```

```
  tangent = tan(angle * g3d_DtoR);

  for (c = 0; c <= 3; c++)
   tm[1][c] -= tm[2][c] * tangent;

} /* g3d_local_shear_y_parallel_to_z */

/*-----------------------------------------------------------------*/

void g3d_local_shear_z_parallel_to_x
(
 g3d_matrix_t tm,    /* Transformation Matrix (In/Out) */
 double       angle  /* Shear Angle in Degrees (In)    */
)
{
 /*
  * Concatenates a Local Shear Transformation of the Z-Axis parallel
  * to the X-Axis into the Transformation Matrix (tm).
  */

 double tangent;
 int    c;

 tangent = tan(angle * g3d_DtoR);

 for (c = 0; c <= 3; c++)
  tm[2][c] -= tm[0][c] * tangent;

} /* g3d_local_shear_z_parallel_to_x */

/*-----------------------------------------------------------------*/

void g3d_local_shear_z_parallel_to_y
(
 g3d_matrix_t tm,    /* Transformation Matrix (In/Out) */
 double       angle  /* Shear Angle in Degrees (In)    */
)
{
 /*
  * Concatenates a Local Shear Transformation of the Z-Axis parallel
  * to the Y-Axis into the Transformation Matrix (tm).
  */

 double tangent;
 int    c;

 tangent = tan(angle * g3d_DtoR);

 for (c =; 0; c <= 3; c++)
  tm[2][c] += tm[1][c] * tangent;

} /* g3d_local_shear_z_parallel_to_y */

/*-----------------------------------------------------------------*/
```

```
void g3d_local_aim_x_axis
(
 g3d_matrix_t tm, /* Transformation Matrix (In/Out) */
 g3d_vector_t V,  /* Aim Direction Unit Vector (In) */
 g3d_vector_t U   /* Up  Direction Unit Vector (In) */
)
{
 /*
  * Concatenates into the Transformation Matrix (tm) a Local Aim
  * transformation which causes the Local X-axis to be aligned to
  * the aim in the direction of the unit vector (V). The (U)
  * vector represents the relative up direction unit vector with
  * respect to the (V) vector.
  */

 g3d_matrix_t A;  /* Aim transformation matrix */
 g3d_vector_t X,  /* Local Axes                */
              Y,
              Z;

 /* Compute the new orientation of the transformed local axes */

 Mg3d_vector_copy(V, X);
 Mg3d_vector_normalise(X);

 Mg3d_vector_cross(U, X, Y);
 Mg3d_vector_normalise(Y);

 Mg3d_vector_cross(X, Y, Z);
 Mg3d_vector_normalise(Z);

/* Construct the aim transformation matrix */

 A[0][0] = X[0];  A[0][1] = X[1];  A[0][2] = X[2];  A[0][3] = 0;
 A[1][0] = Y[0];  A[1][1] = Y[1];  A[1][2] = Y[2];  A[1][3] = 0;
 A[2][0] = Z[0];  A[2][1] = Z[1];  A[2][2] = Z[2];  A[2][3] = 0;
 A[3][0] = 0;     A[3][1] = 0;     A[3][2] = 0;     A[3][3] = 1;

 /* Concatenate the aim transformation with the (tm) transformation
    matrix */

 g3d_local_concatenate(tm, A);

}  /* g3d_local_aim_x_axis */

/*-----------------------------------------------------------------*/


void g3d_local_aim_y_axis
(
 g3d_matrix_t tm,  /* Transformation Matrix (In/Out) */
 g3d_vector_t V,   /* Aim Direction Unit Vector (In) */
 g3d_vector_t U    /* Up  Direction Unit Vector (In) */
```

```
)
{
 /*
  * Concatenates into the Transformation Matrix (tm) a Local Aim
  * transformation which causes the Local Y-axis to be aligned to
  * the aim in the direction of the unit vector (V). The (U)
  * vector represents the relative up direction unit vector
  * with respect to the (V) vector.
  */

 g3d_matrix_t A;  /* Aim transformation matrix */
 g3d_vector_t X,  /* Local Axes                */
              Y,
              Z;

 /* Compute the new orientation of the transformed local axes */

 Mg3d_vector_copy(V, Y);
 Mg3d_vector_normalise(Y);

 Mg3d_vector_cross(Y, U, X);
 Mg3d_vector_normalise(X);

 Mg3d_vector_cross(X, Y, Z);
 Mg3d_vector_normalise(Z);

 /* Construct the aim transformation matrix */

 A[0][0] = X[0];  A[0][1] = X[1];  A[0][2] = X[2];  A[0][3] = 0;
 A[1][0] = Y[0];  A[1][1] = Y[1];  A[1][2] = Y[2];  A[1][3] = 0;
 A[2][0] = Z[0];  A[2][1] = Z[1];  A[2][2] = Z[2];  A[2][3] = 0;
 A[3][0] = 0;     A[3][1] = 0;     A[3][2] = 0;     A[3][3] = 1;

 /* Concatenate the aim transformation with the (tm) transformation
    matrix */

 g3d_local_concatenate(tm, A);

} /* g3d_local_aim_y_axis */

/*-------------------------------------------------------------------*/

void g3d_local_aim_z_axis
(
 g3d_matrix_t tm,  /* Transformation Matrix (In/Out) */
 g3d_vector_t V,   /* Aim Direction Unit Vector (In) */
 g3d_vector_t U    /* Up  Direction Unit Vector (In) */
)
{
 /*
  * Concatenates into the Transformation Matrix (tm) a Local Aim
  * transformation which causes the Local Z-axis to be aligned to
  * the aim in the direction of the unit vector (V). The (U)
  * vector represents the relative up direction unit vector
```

```
  * with respect to the (V) vector.
  */

 g3d_matrix_t A;  /* Aim transformation matrix */
 g3d_vector_t X,  /* Local Axes                */
              Y,
              Z;

  /* Compute the new orientation of the transformed local axes */

 Mg3d_vector_copy(V, Z);
 Mg3d_vector_normalise(Z);

 Mg3d_vector_cross(U, Z, Y);
 Mg3d_vector_normalise(Y);

 Mg3d_vector_cross(Y, Z, X);
 Mg3d_vector_normalise(X);

/* Construct the aim transformation matrix */

 A[0][0] = X[0]; A[0][1] = X[1]; A[0][2] = X[2]; A[0][3] = 0;
 A[1][0] = Y[0]; A[1][1] = Y[1]; A[1][2] = Y[2]; A[1][3] = 0;
 A[2][0] = Z[0]; A[2][1] = Z[1]; A[2][2] = Z[2]; A[2][3] = 0;
 A[3][0] = 0;    A[3][1] = 0;    A[3][2] = 0;    A[3][3] = 1;

 /* Concatenate the aim transformation with the (tm) transformation
    matrix */

 g3d_local_concatenate(tm, A);

}  /* g3d_local_aim_z_axis */

/*-------------------------------------------------------------------*/

void g3d_local_reflect_about_arbitrary_point
(
 g3d_matrix_t tm,  /* Transformation Matrix (In/Out) */
 g3d_vector_t C    /* Centre of Reflection (In)      */
)
{
 /*
  * Concatenates a Local Reflection Transformation about an
  * arbitrary point (C) into the Transformation Matrix (tm).
  *
  * Done in three steps:
  *
  * 1. Translate the centre to the origin.
  *
  * 2. Reflect about the origin.
  *
  * 3. Translate the centre to its original position.
  */
```

```
 g3d_local_translate(tm, -C[0], -C[1], -C[2]);
 g3d_local_scale(tm, -1.0, -1.0, -1.0);
 g3d_local_translate(tm, C[0], C[1], C[2]);

} /* g3d_local_reflect_about_arbitrary_point */

/*---------------------------------------------------------------------*/

 void g3d_local_scale_about_arbitrary_point
 (
 g3d_matrix_t tm,   /* Transformation Matrix (In/Out) */
 g3d_vector_t C,    /* Centre of Scale (In)           */
 double       sx,   /* Scale Factors (In)             */
 double       sy,
 double       sz
)
{
 /*
  * Concatenates a Local Scale Transformation about an arbitrary
  * point (C) into the Transformation Matrix (tm).
  *
  * Done in three steps:
  *
  * 1. Translate the centre to the origin.
  *
  * 2. Scale about the origin.
  *
  * 3. Translate the centre to its original position.
  */

  g3d_local_translate(tm, -C[0], -C[1], -C[2]);
  g3d_local_scale(tm, sx, sy, sz);
  g3d_local_translate(tm, C[0] C[1], C[2]);

 } /* g3d_local_scale_about_arbitrary_point */

 /*---------------------------------------------------------------------*/

 void g3d_local_reflect_about_arbitrary_axis
 (
  g3d_matrix_t tm,   /* Transformation Matrix (In/Out)                  */
  g3d_vector_t P,    /* Point on the Arbitrary Axis (In)                */
  g3d_vector_t D     /* Direction Unit Vector of the Arbitrary
                        Axis (In)                                       */
  )
 {
  g3d_matrix_t A;  /* Aim transformation matrix          */
  g3d_vector_t V,  /* Aim Direction Unit Vector          */
               U,  /* Relative Up Direction Unit Vector */
               X,  /* Local Axes                         */
               Y,
               Z;
```

```
/*
 * Concatenates a Local Reflection Transformation about an arbitrary
 * Axis into the Transformation Matrix (tm).
 *
 * The Arbitrary Axis is defined by a point (P) and a direction unit
 * vector (D).
 *
 * This transformation is constructed by the following five steps:
 *
 * Step 1. Translate the point (P) to the origin.
 */

g3d_local_translate(tm, -P[0], -P[1], -P[2]);

/*
 * Step 2. Align the unit vector (V) with the with the X-axis,
 *         using the inverse of the aiming transformation that
 *         would align the X-axis with vector (V).
 */

/* Copy and normalise the Direction Vector of the arbitrary Axis */

Mg3d_vector_copy(D, V);
Mg3d_vector_normalise(V);

/*
 * Determine the relative up direction unit vector (U). No need to
 * worry about gimbal lock here as step 4 reverses step 2.
 */

if(
    (fabs(V[0])       <= g3d_IEEE_small_single) &&
    (fabs(V[1])       <= g3d_IEEE_small_single) &&
    (fabs(V[2] -1.0) <= g3d_IEEE_small_single)
  )
 {
  /* When (V == Z), set U = -X */

  Mg3d_vector_set(U, -1.0, 0.0, 0.0);
 }
else
if(
    (fabs(V[0])        <= g3d_IEEE_small_single) &&
    (fabs(V[1])        <= g3d_IEEE_small_single) &&
    (fabs(V[2] + 1.0) <= g3d_IEEE_small_single)
    )

 {
  /* When (V == -Z), set U = X */

  Mg3d_vector_set(U, 1.0, 0.0, 0.0);
 }
else
 {
```

```
  /* Otherwise, set U = Z */

  Mg3d_vector_set(U, 0.0, 0.0,1.0);
 }

/* Compute the new orientation of the transformed local axes */

Mg3d_vector_copy(V, X);
Mg3d_vector_normalise(X);

Mg3d_vector_cross(U, X, Y);
Mg3d_vector_normalise(Y);

Mg3d_vector_cross(X, Y, Z);
Mg3d_vector_normalise(Z);

/* Construct the aim transformation matrix */

A[0][0] = X[0];  A[0][1] = X[1];  A[0][2] = X[2];  A[0][3] = 0;
A[1][0] = Y[0];  A[1][1] = Y[1];  A[1][2] = Y[2];  A[1][3] = 0;
A[2][0] = Z[0];  A[2][1] = Z[1];  A[2][2] = Z[2];  A[2][3] = 0;
A[3][0] = 0;     A[3][1] = 0;     A[3][2] = 0;     A[3][3] = 1;

/* Concatenate the aim transformation with the (tm) transformation
   matrix */

g3d_local_concatenate_transpose(tm, A);


/*
 * Step 3. Reflect about the X-axis.
 */

g3d_local_scale(tm, 1.0, -1.0, -1.0);

/*
 * Step 4. Return vector (V) to its original orientation,
 *         by applying the inverse of the transformation
 *         from step 2.
 */

g3d_local_concatenate(tm, A);

/*
 * Step 5. Translate point (P) to its original position,
 *         by applying the inverse of the transformation
 *         from step 1.
 */

g3d_local_translate(tm, P[0], P[1], P[2]);

} /* g3d_local_reflect_about_arbitrary_axis */

/*-----------------------------------------------------------*/
```

```
void g3d_local_rotate_about_arbitrary_axis
(
 g3d_matrix_t tm,     /* Transformation Matrix (In/Out)                     */
 g3d_vector_t P,      /* Point on the Arbitrary Axis (In)                   */
 g3d_vector_t D,      /* Direction Unit Vector of the Arbitrary
                         Axis (In)                                          */
 double       angle   /* Rotation Angle in Degrees (In)                     */
)
{
 g3d_matrix_t A;   /* Aim transformation matrix          */
 g3d_vector_t V,   /* Aim Direction Unit Vector          */
              U,   /* Relative Up Direction Unit Vector */
              X,   /* Local Axes                         */
              Y,
              Z;

 /*
  * Concatenates a Local Rotation Transformation about an arbitrary
  * Axis into the Transformation Matrix (tm).
  *
  * The Arbitrary Axis is defined by a point (P) and a direction unit
  * vector (D).
  *
  * This transformation is constructed by the following five steps:
  *
  * Step 1. Translate the point (P) to the origin.
  */

 g3d_local_translate(tm, -P[0], -P[1], -P[2]);

 /*
  * Step 2. Align the unit vector (V) with the with the X-axis,
  *         using the inverse of the aiming transformation that
  *         would align the X-axis with vector (V).
  */

 /* Copy and normalise the Direction Vector of the arbitrary Axis */

 Mg3d_vector_copy(D, V);
 Mg3d_vector_normalise(V);

 /*
  * Determine the relative up direction unit vector (U). No need to
  * worry about gimbal lock here as step 4 reverses step 2.
  */

 if(
    (fabs(V[0])      <= g3d_IEEE_small_single) &&
    (fabs(V[1])      <= g3d_IEEE_small_single) &&
    (fabs(V[2] -1.0) <= g3d_IEEE_small_single)
   )
  {
   /* When (V == Z), set U = -X */
```

```
  Mg3d_vector_set(U, -1.0, 0.0, 0.0);
 }
else

if(
   (fabs(V[0])       <= g3d_IEEE_small_single) &&
   (fabs(V[1])       <= g3d_IEEE_small_single) &&
   (fabs(V[2] + 1.0) <= g3d_IEEE_small_single)
   )
 {
  /* When (V == -Z), set U = X */

  Mg3d_vector_set(U, 1.0, 0.0, 0.0);
 }
else
 {
  /* Otherwise, set U = Z */

  Mg3d_vector_set(U, 0.0, 0.0,1.0);
 }

/* Compute the new orientation of the transformed local axes */
Mg3d_vector_copy(V, X);
Mg3d_vector_normalise(X);

Mg3d_vector_cross(U, X, Y);
Mg3d_vector_normalise(Y);

Mg3d_vector_cross(X, Y, Z);
Mg3d_vector_normalise(Z);

/* Construct the aim transformation matrix */

A[0][0] = X[0]; A[0][1] = X[1]; A[0][2] = X[2]; A[0][3] = 0;
A[1][0] = Y[0]; A[1][1] = Y[1]; A[1][2] = Y[2]; A[1][3] = 0;
A[2][0] = Z[0]; A[2][1] = Z[1]; A[2][2] = Z[2]; A[2][3] = 0;
A[3][0] = 0;    A[3][1] = 0;    A[3][2] = 0;    A[3][3] = 1;

/* Concatenate the aim transformation with the (tm) transformation
   matrix */

g3d_local_concatenate_transpose(tm, A);

/*
 * Step 3. Rotate about the X-axis.
 */

g3d_local_rotate_x(tm, angle);

/*
 * Step 4. Return vector (V) to its original orientation,
 *         by applying the inverse of the transformation
 *         from step 2.
```

```
  */

 g3d_local_concatenate(tm, A);

 /*
  * Step 5. Translate point (P) to its original position,
  *         by applying the inverse of the transformation
  *         from step 1.
  */

 g3d_local_translate(tm, P[0], P[1], P[2]);

} /* g3d_local_rotate_about_arbitrary_axis */

/*---------------------------------------------------------------*/

void g3d_local_scale_along_arbitrary_axis
(
 g3d_matrix_t tm,  /* Transformation Matrix (In/Out)             */
 g3d_vector_t P,   /* Point on the Arbitrary Axis (In)           */
 g3d_vector_t D,   /* Direction Unit Vector of the Arbitrary
                      Axis (In)                                  */
 double       sf   /* Scale Factor (In)                          */
)
{
 g3d_matrix_t A;  /* Aim transformation matrix         */
 g3d_vector_t V,  /* Aim Direction Unit Vector         */
              U,  /* Relative Up Direction Unit Vector */
              X,  /* Local Axes                        */
              Y,
              Z;

 /*
  * Concatenates a Local Scaling Transformation along an arbitrary
  * Axis into the Transformation Matrix (tm).
  *
  * The Arbitrary Axis is defined by a point (P) and a direction
  * unit vector (D).
  *
  * This transformation is constructed by the following five steps:
  *
  * Step 1. Translate the point (P) to the origin.
  */

 g3d_local_translate(tm, -P[0], -P[1], -P[2]);

 /*
  * Step 2. Align the unit vector (V) with the with the X-axis,
  *         using the inverse of the aiming transformation that
  *         would align the X-axis with vector (V).
  */

 /* Copy and normalise the Direction Vector of the arbitrary Axis */
```

```
Mg3d_vector_copy(D, V);
Mg3d_vector_normalise(V);

/*
 * Determine the relative up direction unit vector (U). No need to
 * worry about gimbal lock here as step 4 reverses step 2.
 */

if(
   (fabs(V[0])      <= g3d_IEEE_small_single) &&
   (fabs(V[1])      <= g3d_IEEE_small_single) &&
   (fabs(V[2] -1.0) <= g3d_IEEE_small_single)
  )
 {
  /* When (V == Z), set U = -X */

  Mg3d_vector_set(U, -1.0, 0.0, 0.0);
 }
else
if(
   (fabs(V[0])       <= g3d_IEEE_small_single) &&
   (fabs(V[1])       <= g3d_IEEE_small_single) &&
   (fabs(V[2] + 1.0) <= g3d_IEEE_small_single)
   )
 {
  /* When (V == -Z), set U = X */

  Mg3d_vector_set(U, 1.0, 0.0, 0.0);
 }
else
 {
  /* Otherwise, set U = Z */

  Mg3d_vector_set(U, 0.0, 0.0,1.0);
 }

/* Compute the new orientation of the transformed local axes */

Mg3d_vector_copy(V, X);
Mg3d_vector_normalise(X);

Mg3d_vector_cross(U, X, Y);
Mg3d_vector_normalise(Y);

Mg3d_vector_cross(X, Y, Z);
Mg3d_vector_normalise(Z);

/* Construct the aim transformation matrix */

A[0][0] = X[0];  A[0][1] = X[1];  A[0][2] = X[2];  A[0][3] = 0;
A[1][0] = Y[0];  A[1][1] = Y[1];  A[1][2] = Y[2];  A[1][3] = 0;
A[2][0] = Z[0];  A[2][1] = Z[1];  A[2][2] = Z[2];  A[2][3] = 0;
A[3][0] = 0;     A[3][1] = 0;     A[3][2] = 0;     A[3][3] = 1;
```

```
  /* Concatenate the aim transformation with the (tm) transformation
     matrix */

  g3d_local_concatenate_transpose(tm, A);

  /*
   * Step 3. Scale along the X-axis.
   */

  g3d_local_scale(tm, sf, 1.0,1.0);

  /*
   * Step 4. Return vector (V) to its original orientation,
   *         by applying the inverse of the transformation
   *         from step 2.
   */

  g3d_local_concatenate(tm, A);

  /*
   * Step 5. Translate point (P) to its original position,
   *         by applying the inverse of the transformation
   *         from step 1.
   */

  g3d_local_translate(tm, P[0], P[1], P[2]);

} /* g3d_local_scale_along_arbitrary_axis */

/*------------------------------------------------------------------*/

void g3d_local_translate_along_arbitrary_axis
(
 g3d_matrix_t tm,  /* Transformation Matrix (In/Out)              */
 g3d_vector_t P,   /* Point on the Arbitrary Axis (In)            */
 g3d_vector_t D,   /* Direction Unit Vector of the Arbitrary
                      Axis (In)                                   */
 double       dd   /* Displacement (In)                           */
)
{
 g3d_matrix_t A;  /* Aim transformation matrix          */
 g3d_vector_t V,  /* Aim Direction Unit Vector          */
              U,  /* Relative Up Direction Unit Vector */
              X,  /* Local Axes                         */
              Y,
              Z;

 /*
 * Concatenates a Local Translation Transformation along an
 * arbitrary Axis into the Transformation Matrix (tm).
 *
 * The Arbitrary Axis is defined by a point (P) and a direction unit
 * vector (D).
 *
```

```
* This transformation is constructed by the following five steps:
*
* Step 1. Translate the point (P) to the origin.
*/

g3d_local_translate(tm, -P[0], -P[1], -P[2]);

/*
 * Step 2. Align the unit vector (V) with the with the X-axis,
 *         using the inverseof the aiming transformation that
 *         would align the X-axis with vector (V).
 */

/* Copy and normalise the Direction Vector of the arbitrary Axis */

Mg3d_vector_copy(D, V);
Mg3d_vector_normalise(V);

/*
 * Determine the relative up direction unit vector (U). No need to
 * worry about gimbal lock here as step 4 reverses step 2.
 */

if(
    (fabs(V[0])       <= g3d_IEEE_small_single) &&
    (fabs(V[1])       <= g3d_IEEE_small_single) &&
    (fabs(V[2] -1.0)  <= g3d_IEEE_small_single)
  )
 {
  /* When (V == Z), set U = -X */

  Mg3d_vector_set(U, -1.0, 0.0, 0.0);
 }
else
if(
    (fabs(V[0])        <= g3d_IEEE_small_single) &&
    (fabs(V[1])        <= g3d_IEEE_small_single) &&
    (fabs(V[2] + 1.0)  <= g3d_IEEE_small_single)
    )
 {
  /* When (V == -Z), set U = X */

  Mg3d_vector_set(U, 1.0, 0.0, 0.0);
 }
else
 {
  /* Otherwise, set U = Z */

  Mg3d_vector_set(U, 0.0, 0.0,1.0);
 }

/* Compute the new orientation of the transformed local axes */

Mg3d_vector_copy(V, X);
```

```
 Mg3d_vector_normalise(X);

 Mg3d_vector_cross(U, X, Y);
 Mg3d_vector_normalise(Y);

 Mg3d_vector_cross(X, Y, Z);
 Mg3d_vector_normalise(Z);

 /* Construct the aim transformation matrix */

 A[0][0] = X[0];  A[0][1] = X[1];  A[0][2] = X[2];  A[0][3] = 0;
 A[1][0] = Y[0];  A[1][1] = Y[1];  A[1][2] = Y[2];  A[1][3] = 0;
 A[2][0] = Z[0];  A[2][1] = Z[1];  A[2][2] = Z[2];  A[2][3] = 0;
 A[3][0] = 0;     A[3][1] = 0;     A[3][2] = 0;     A[3][3] = 1;

 /* Concatenate the aim transformation with the (tm) transformation
    matrix */

 g3d_local_concatenate_transpose(tm, A);

 /*
  * Step 3. Translate along the X-axis.
  */

 g3d_local_translate(tm, dd, 0.0, 0.0);

 /*
  * Step 4. Return vector (V) to its original orientation,
  *         by applying the inverse of the transformation
  *         from step 2.
  */

 g3d_local_concatenate(tm, A);

 /*
  * Step 5. Translate point (P) to its original position,
  *         by applying the inverse of the transformation
  *         from step 1.
  */

 g3d_local_translate(tm, P[0], P[1], P[2]);

} /* g3d_local_translate_along_arbitrary_axis */

/*--------------------------------------------------------------------*/

void g3d_local_reflect_about_arbitrary_plane
(
 g3d_matrix_t tm,  /* Transformation Matrix (In/Out)             */
 g3d_vector_t P,   /* Point on the Arbitrary Plane (In)          */
 g3d_vector_t N    /* Unit Normal Vector of the Arbitrary
                      Plane (In)                                 */
)
{
```

```
g3d_matrix_t A;   /* Aim transformation matrix           */
g3d_vector_t V,   /* Aim Direction Unit Vector           */
             U,   /* Relative Up Direction Unit Vector */
             X,   /* Local Axes                          */
             Y,
             Z;

/*
 * Concatenates a Local Reflection Transformation about an arbitrary
 * Plane into the Transformation Matrix (tm).
 * The Arbitrary Plane is defined by a point (P) and the direction
 * unit vector of its normal (N).
 *
 * This transformation is constructed by the following five steps:
 *
 * Step 1. Translate the point (P) to the origin.
 */

g3d_local_translate(tm, -P[0], -P[1], -P[2]);

/*
 * Step 2. Align the unit vector (V) with the with the X-axis,
 *         using the inverse of the aiming transformation that
 *         would align the X-axis with vector (V).
 */

/* Copy and normalise the Normal Vector of the arbitrary Plane */

Mg3d_vector_copy(N, V);
Mg3d_vector_normalise(V);

/*
 * Determine the relative up direction unit vector (U). No need to
 * worry about gimbal lock here as step 4 reverses step 2.
 */

if(
   (fabs(V[0])       <= g3d_IEEE_small_single) &&
   (fabs(V[1])       <= g3d_IEEE_small_single) &&
   (fabs(V[2] -1.0)  <= g3d_IEEE_small_single)
  )
 {
  /* When (V == Z), set U = -X */

  Mg3d_vector_set(U, -1.0, 0.0, 0.0);
 }
else
if(
   (fabs(V[0])        <= g3d_IEEE_small_single) &&
   (fabs(V[1])        <= g3d_IEEE_small_single) &&
   (fabs(V[2] + 1.0)  <= g3d_IEEE_small_single)
   )
 {
  /* When (V == -Z), set U = X */
```

```
  Mg3d_vector_set(U, 1.0, 0.0, 0.0);
 }
else
 {
  /* Otherwise, set U = Z */

  Mg3d_vector_set(U, 0.0, 0.0,1.0);
 }

/* Compute the new orientation of the transformed local axes */

Mg3d_vector_copy(V, X);
Mg3d_vector_normalise(X);

Mg3d_vector_cross(U, X, Y);
Mg3d_vector_normalise(Y);

Mg3d_vector_cross(X, Y, Z);
Mg3d_vector_normalise(Z);

/* Construct the aim transformation matrix */

A[0][0] = X[0];  A[0][1] = X[1];  A[0][2] = X[2];  A[0][3] = 0;
A[1][0] = Y[0];  A[1][1] = Y[1];  A[1][2] = Y[2];  A[1][3] = 0;
A[2][0] = Z[0];  A[2][1] = Z[1];  A[2][2] = Z[2];  A[2][3] = 0;
A[3][0] = 0;     A[3][1] = 0;     A[3][2] = 0;     A[3][3] = 1;

/* Concatenate the aim transformation with the (tm) transformation
   matrix */

g3d_local_concatenate_transpose(tm, A);

/*
 * Step 3. Reflect about the YZ-plane.
 */

g3d_local_scale(tm, -1.0,1.0,1.0);

/*
 * Step 4. Return vector (V) to its original orientation,
 *         by applying the inverse of the transformation
 *         from step 2.
 */

g3d_local_concatenate(tm, A);

/*
 * Step 5. Translate point (P) to its original position,
 *         by applying the inverse of the transformation
 *         from step 1.
 */

g3d_local_translate(tm, P[01, P[1], P[2]);
```

```
} /* g3d_local_reflect_about_arbitrary_plane */

/*------------------------------------------------------------------*/
/*!!!!!!!!!!!!!!!!!!!!!!!!!!!!!!!!!!!!!!!!!!!!!!!!!!!!!!!!!!!!!!!!!!*/
/*!            Apply Transformation to Point Routines              !*/
/*!!!!!!!!!!!!!!!!!!!!!!!!!!!!!!!!!!!!!!!!!!!!!!!!!!!!!!!!!!!!!!!!!!*/
/*------------------------------------------------------------------*/

void g3d_transform_3D_point(g3d_matrix_t tm, /* Transformation
                                                Matrix (In)      */
                            g3d_vector_t P   /* Point (In/Out)   */
                           )
{
 /* Apply the Transformation Matrix (tm) to the Point P=[x, y, z] */
 double x, y, z;

 x = P[0];
 y = P[1];
 z = P[2];

 P[0] = x * tm[0][0] + y * tm[1][0] + z * tm[2][0] + tm[3][0];
 P[1] = x * tm[0][1] + y * tm[1][1] + z * tm[2][1] + tm[3][1];
 P[2] = x * tm[0][2] + y * tm[1][2] + z * tm[2][2] + tm[3][2];

} /* g3d_transform_3D_point */

/*------------------------------------------------------------------*/

void g3d_transform_4D_point(g3d_matrix_t tm,  /* Transformation
                                                 Matrix (In)     */
                            g4d_vector_t P    /* Point (In/Out)  */
                           )
{
 /* Apply the Transformation Matrix (tm) to the Homogeneous Point
    P=[x, y, z, w] */

 double x, y, z, w;

 x = P[0];
 y = P[1];
 z = P[2];
 w = P[3];

 P[0] = x * tm[0][0] + y * tm[1][0] + z * tm[2][0] + w * tm[3][0];
 P[1] = x * tm[0][1] + y * tm[1][1] + z * tm[2][1] + w * tm[3][1];
 P[2] = x * tm[0][2] + y * tm[1][2] + z * tm[2][2] + w * tm[3][2];
 P[3] = x * tm[0][3] + y * tm[1][3] + z * tm[2][3] + w * tm[3][3];

} /* g3d_transform_4D_point */

/*------------------------------------------------------------------*/
/*!!!!!!!!!!!!!!!!!!!!!!!!!!!!!!!!!!!!!!!!!!!!!!!!!!!!!!!!!!!!!!!!!!*/
/*!              3D Viewing Transformation Routines                !*/
```

```
/*!!!!!!!!!!!!!!!!!!!!!!!!!!!!!!!!!!!!!!!!!!!!!!!!!!!!!!!!!!!!!!!!!*/
/*---------------------------------------------------------------*/

void g3d_viewing_transformation
(
 g3d_matrix_t tm,  /* Transformation Matrix (In/Out) */
 g3d_vector_t Vp,  /* Viewing Point (In)             */
 g3d_vector_t Cp,  /* Centre Point (In)              */
 g3d_vector_t U    /* Up Direction Unit Vector (In)  */
)
{
 /*
  * Concatenates the Viewing Transformation into the Transformation
  * Matrix(tm).
  *
  * The Viewing Transformation is defined by the Viewing Point
  * Vp=[vx, vy, vz], the Centre Point Cp=[cx, cy, vz] and the
  * relative up direction unit vector U=[ux, uy, uz]. This form
  * of the viewing transformation prevents gimbal lock.
  *
  * Done in two steps:
  *
  * 1. Apply the inverse of the transformation that would translate
  *    the origin of the new eye-space axes to the Viewing Point
  *    V=[vx, vy, vz]. This is equivalent translating the Viewing
  *    point to the origin of the object-space.
  *
  * 2. Re-arrange these new axes so that the z-axis points in the
  *    direction of the viewing axis and that they form a left-
  *    handed basis. Apply the inverse of this transformation.
  *   (See section 7.2)
  */

 g3d_matrix_t A;  /* Aim transformation matrix      */
 g3d_vector_t V,  /* Aim Direction Unit Vector      */
              X,  /* Local Axes                     */
              Y,
              Z;

 /*
  * Step 1.
  */

 g3d_translate(tm, -Vp[0], -Vp[1], -Vp[2]);

 /*
  * Step 2.
  */

 /* Compute the unit vector V that points in the direction of the
    Viewing Axis */

 V[0] = Cp[0] - Vp[0];
 V[1] = Cp[1] - Vp[1];
```

```
V[2] = Cp[2] - Vp[2];

if (Mg3d_vector_magnitude(V) <= g3d_IEEE_small_single)
 {
  printf("g3d_viewing_transformation: Distance between the Vp and",
         "Cp is too small! \n");
  return;
 }

Mg3d_vector_normalise(V);

/* Set-up the original orientations of the X and Y axes */

Mg3d_vector_set(X, 1.0, 0.0, 0.0);
Mg3d_vector_set(Y, 0.0, 1.0, 0.0);

/* Aim the local Z-axis along the V vector */

Mg3d_vector_copy(V, Z);

Mg3d_vector_cross(U, Z, X);
Mg3d_vector_normalise(X);

Mg3d_vector_cross(Z, X, Y);
Mg3d_vector_normalise(Y);

/* Reverse the direction of the local X-axis */

X[0] = -X[0];
X[1] = -X[1];
X[2] = -X[2];

/* Set-up the transpose (in this case also the inverse) of the
   aim transformation */

A[0][0] = X[0];  A[0][1] = Y[0];  A[0][2] = Z[0];  A[0][3] = 0;
A[1][0] = X[1];  A[1][1] = Y[1];  A[1][2] = Z[1];  A[1][3] = 0;
A[2][0] = X[2];  A[2][1] = Y[2];  A[2][2] = Z[2];  A[2][3] = 0;
A[3][0] = 0;     A[3][1] = 0;     A[3][2] = 0;     A[3][3] = 1;

} /* g3d_viewing_transformation */

/*-----------------------------------------------------------------*/

void g3d_viewing_transformation_V2
(
 g3d_matrix_t tm,  /* Transformation Matrix (In/Out) */
 g3d_vector_t Vp,  /* Viewing Point (In)             */
 g3d_vector_t Cp   /* Centre Point (In)              */
)
{
 /*
  * Concatenates the Viewing Transformation defined by the Viewing
  * Point Vp=[vx, vy, vz] and the Centre Point C=[cx, cy, cz] into
```

```
 * the Transformation Matrix (tm).
 *
 * Done in four steps:
 *
 * 1. Translate the Viewing Point Vp=[vx, vy, vz] to the origin of
 *    the Object Space.
 *
 * 2. Rearrange the new axes so that we have a left-handed
 *    coordinate system.
 *
 * 3. Rotate the new axes about the Ye-axis by an angle <theta>,
 *    so that the Ze-axis points towards the point Cp=[cx, cy, vz].
 *
 * 4. Rotate the new axes about the xe-axis by an angle <phi>,
 *    so that the Ze-axis points towards the Centre Point
 *    Cp=[cx, cy, cz].
 */

g3d_matrix_t R, Rx, Ry;
double       dx, dy, dz, d2d, d3d, cos_theta, sin_theta, cos_phi,
             sin_phi;

/*
 * Step 1.
 */

dx = Vp[0] - Cp[0];
dy = Vp[1] - Cp[1];
dz = Vp[2] - Cp[2];

d2d = sqrt(dx * dx + dy * dy);
d3d = sqrt(dx * dx + dy * dy + dz * dz);

if (d3d <= g3d_IEEE_small_single)
 {
  printf("g3d_viewing_transformation: Distance between the Vp and",
         "Cp is too small! \n");
  return;
 }

g3d_translate(tm, -Vp[0], -Vp[1], -Vp[2]);

/*
 * Step 2.
 */

R[0][0] = -1.0;  R[0][1] = 0.0;  R[0][2] =  0.0;  R[0][3] = 0.0;
R[1][0] =  0.0;  R[1][1] = 0.0;  R[1][2] = -1.0;  R[1][3] = 0.0;
R[2][0] =  0.0;  R[2][1] = 1.0;  R[2][2] =  0.0;  R[2][3] = 0.0;
R[3][0] =  0.0;  R[3][1] = 0.0;  R[3][2] =  0.0;  R[3][3] = 1.0;

g3d_concatenate(tm, R);

/*
```

```
  * Step 3.
  */

 if (d2d > g3d_IEEE_small_single)
  {
   cos_theta = dy / d2d;
   sin_theta = dx / d2d;

   g3d_initialize_matrix(Ry);
   Ry[0][0] =  cos_theta;  Ry[0][2] = sin_theta;
   Ry[2][0] = -sin_theta;  Ry[2][2] = cos_theta;

   g3d concatenate(tm, Ry);
  }

 /*
  * Step 4.
  */

 cos_phi = d2d / d3d;
 sin_phi = dz  / d3d;

 g3d_initialize_matrix(Rx);

 Rx[1][1] = cos_phi;  Rx[1][2] = -sin_phi;
 Rx[2][1] = sin_phi;  Rx[2][2] =  cos_phi;

 g3d_concatenate(tm, Rx);

}  /* g3d_viewing_transformation_V2 */

/*-------------------------------------------------------------------*/

void g3d_polar_viewing_transformation
(
 g3d_matrix_t tm,    /* Transformation Matrix (In/Out)          */
 double       vcd,   /* Viewing to Centre Point Distance (In)   */
 double       aa,    /* Azimuth Angle (In)                      */
 double       ia,    /* Incidence Angle (In)                    */
 double       ta,    /* Twist Angle (In)                        */
 g3d_vector_t Cp     /* Centre Point (In)                       */
)
{
 /*
  * Concatenates the Polar Viewing Transformation defined by:
  *
  * 1. the Viewing to Centre Point distance,
  * 2. the Azimuth angle (aa),
  * 3. the Incidence angle (ia),
  * 4. the Twist angle (ta) and
  * 5. The Centre Point (Cp)
  *
  * into the Transformation Matrix (tm).
  *
```

```
 * The Azimuth angle lies in the x-y plane is measured from
 * the y-axis.
 * The Incidence angle lies in a plane perpendicular to the x-y
 * plane is measured from the z-axis.
 * The Twist angle rotates the camera around the Viewing Axis using
 * a right-handed rule.
 *
 * This transformation is done in seven steps:
 *
 * 1. Translate the Centre Point (Cp) to the origin of the
 *    Object Space.
 *
 * 2. Compute the coordinates of the new Viewing Point
 *    [vx, vy, vz],
 *
 * 3. Translate the Viewing Point [vx, vy, vz] to the origin of
 *    the Object Space.
 *
 * 4. Rearrange the new axes so that we have a left-handed
 *    coordinate system.
 *
 * 5. Rotate the new axes about the Ye-axis by the azimuth angle,
 *    so that the new negative Ze-axis points towards point
 *    [0, 0, vz].
 *
 * 6. Rotate the new axes about the Xe-axis by the complement of
 *    the incidence angle (i.e. 90-ia), so that the new negative
 *    Ze-axis points towards the origin.
 *
 * 7. Rotate the new axes about the Ze-axis by the twist angle.
 */

g3d_matrix_t R;
double       vx, vy, vz, d2d;

/*
 * Step 1.
 */

if (vcd <= g3d_IEEE_small_single)
 {
  printf("g3d_polar_viewing_transformation: (vcd) distance is",
         "too small!\n");
  return;
 }

g3d_translate(tm, -Cp[0] -Cp[1], -Cp[2]);

/*
 * Step 2.
 */

d2d = vcd * sin(ia * g3d_DtoR);
```

```
vz = vcd * cos(ia * g3d_DtoR);
vx = d2d * sin(aa * g3d_DtoR);
vy = d2d * cos(aa * g3d_DtoR);

/*
 * Step 3.
 */

g3d_translate(tm, -vx, -vy, -vz);

/*
 * Step 4.
 */

R[0][0] = -1.0;  R[0][1] = 0.0;  R[0][2] =  0.0;  R[0][3] = 0.0;
R[1][0] =  0.0;  R[1][1] = 0.0;  R[1][2] = -1.0;  R[1][3] = 0.0;
R[2][0] =  0.0;  R[2][1] = 1.0;  R[2][2] =  0.0;  R[2][3] = 0.0;
R[3][0] =  0.0;  R[3][1] = 0.0;  R[3][2] =  0.0;  R[3][3] = 1.0;

g3d_concatenate(tm, R);

/*
 * Step 5.
 */

g3d_rotate_y(tm, aa);

/*
 * Step 6.
 */

g3d_rotate_x(tm, 90 - ia);

/*
 * Step 7.
 */

g3d_rotate_z(tm, ta);

} /* g3d_polar_viewing_transformation */

/*-------------------------------------------------------------------*/
/*!!!!!!!!!!!!!!!!!!!!!!!!!!!!!!!!!!!!!!!!!!!!!!!!!!!!!!!!!!!!!!!!!!!*/
/*!              3D Projection Transformation Routines               !*/
/*!!!!!!!!!!!!!!!!!!!!!!!!!!!!!!!!!!!!!!!!!!!!!!!!!!!!!!!!!!!!!!!!!!!*/
/*-------------------------------------------------------------------*/

void g3d_orthographic_projection
(
 g3d_matrix_t tm  /* Transformation Matrix (In/Out) */
)
{
 /*
  * Concatenates the Orthographic Projection Transformation into the
```

```
  * Transformation Matrix (tm).
  */

 tm[0][2] = 0.0;
 tm[1][2] = 0.0;
 tm[2][2] = 0.0;
 tm[3][2] = 0.0;

} /* g3d_orthographic_projection */

/*-----------------------------------------------------------------*/

void g3d_oblique_projection
(
 g3d_matrix_t tm,  /* Transformation Matrix (In/Out)            */
 double       a1,  /* Oblique Projection Angles in Degrees (In) */
 double       a2
)
{
 /*
  * Concatenates the Oblique Projection Transformation into the
  * Transformation Matrix (tm).
  */

 g3d_matrix_t h;
 double       l, a, b;
 int          r;


 l = 1.0 / tan(a1 * g3d_DtoR);
 a = l * cos(a2 * g3d_DtoR);
 b = l * sin(a2 * g3d_DtoR);

 memcpy(h, tm, sizeof(g3d_matrix_t));

 for (r = 0; r <= 3; r++)
  {
   tm[r][0] = h[r][0] + h[r][2] * a;
   tm[r][1] = h[r][1] + h[r][2] * b;
   tm[r][2] = 0.0;
  }
 } /* g3d_oblique_projection */

/*-----------------------------------------------------------------*/

void g3d_perspective_projection
(
 g3d_matrix_t tm,  /* Transformation Matrix (In/Out) */
 double       vd   /* Viewing Distance (In)          */
)
{
 /*
  * Concatenates the Perspective Projection Transformation into the
  * Transformation Matrix (tm).
```

```
  */

 int r;

 if (vd > g3d_IEEE_small_single)
  {
   for (r = 0; r <= 3; r++)
    tm[r][3] = tm[r][2] / vd;
  }
 else
  printf("g3d_perspective_projection: The Viewing Distance is",
                             "too small!\n");

}  /* g3d_perspective_projection */

/*------------------------------------------------------------*/

void g3d_perspective_projection_fv
(
 g3d_matrix_t tm,    /* Transformation Matrix (In/Out)          */
 double       vay,   /* View Angle in the y direction (In)      */
 double       xya,   /* x/y Aspect Ratio which determines
                        fov_x (In)                              */
 double       nz,    /* Near Clipping Plane Z (In)              */
 double       fz     /* Far Clipping Plane Z (In)               */
)
{
 /*
  * Concatenates the Perspective Projection Transformation into the
  * Transformation Matrix (tm).
  *
  * In this transformation the Viewing Pyramid is truncated by the
  * near and far clipping planes to form a Frustum of Vision.
  *
  * This version of the Perspective Projection transformation
  * incorporates the Eye-Space to Clip-Space transformation.
  */

 g3d_matrix_t P;
 double       cvay,   /* Cotangent of the View Angle in the
                         y direction */
              dz;     /* Depth in z */

 if (xya < g3d_IEEE_small_single)
  {
   printf("g3d_perspective_projection_fv: The x to y aspect ratio
          is too small!\n");
   return;
  }

 if ((fz - nz) < g3d_IEEE_small_single)
 {
  printf("g3d_perspective_projection_fv: The near and far planes",
         "are too close!\n");
```

```
  return;
 }

 /*
  * Compute the cotangent of the Field of View angle.
  */

 cvay = 1.0 / tan(vay / 2.0 * g3d_DtoR);
 dz   = fz - nz;

 /*
  * Set-up the Perspective Projection Matrix.
  */

 P[0][0] = cvay / xya;  P[0][1] = 0.0;   P[0][2] =  0.0;
 P[1][0] = 0.0;         P[1][1] = cvay;  P[1][2] =  0.0;
 P[2][0] = 0.0;         P[2][1] = 0.0;   P[2][2] =  fz / dz;
 P[3][0] = 0.0;         P[3][1] = 0.0;   P[3][2] = -fz * nz / dz;

 P[0][3] = 0.0;
 P[1][3] = 0.0;
 P[2][3] = 1.0;
 P[3][3] = 0.0;

 g3d_concatenate(tm, P);

}  /* g3d_perspective_projection_fv */

/*-----------------------------------------------------------------*/
/*!!!!!!!!!!!!!!!!!!!!!!!!!!!!!!!!!!!!!!!!!!!!!!!!!!!!!!!!!!!!!!!!!*/
/*!                     3D Clipping Routines                      !*/
/*!!!!!!!!!!!!!!!!!!!!!!!!!!!!!!!!!!!!!!!!!!!!!!!!!!!!!!!!!!!!!!!!!*/
/*-----------------------------------------------------------------*/

boolean_t g3d_clip_point(double *x,  /* Test Point (In/Out) */
                         double *y,
                         double *z
                        )
{
 boolean_t inside;

 inside = (
           (-(*z) <= (*x)) &&
           ( (*x) <= (*z)) &&
           (-(*z) <= (*y)) &&
           ( (*y) <= (*z))
          );

 /*
  * If the point is visible add a small displacement to its
  * z-coordinate in order to prevent a potential zero divide
  * in the perspective division.
  */
```

```
 if (inside) (*z) += g3d_z_displacement;

 return (inside);

} /* g3d_clip_point */

/*---------------------------------------------------------------*/

void g3d_set_clip_code(double            x,
                       double            y,
                       double            z,
                       g3d_region_code_t *c
                      )
{
 *c=0;

 if (x < -z) *c = g3d_left_plane; else
 if (x >  z) *c = g3d_right_plane;

 if (y < -z) *c |= g3d_bottom_plane; else
 if (y >  z) *c |= g3d_top_plane;

} /* g3d_set_clip_code */

/*---------------------------------------------------------------*/

 boolean_t g3d_clip_line_cs(double *x1,  /* Test Line (In/Out) */
                            double *y1,
                            double *z1,
                            double *x2,
                            double *y2,
                            double *z2
                           )
 {
  /*
   * Clip a 3D line segment in the clip-space coordinate system.
  */

 boolean_t         result;
 g3d_region_code_t c, c1, c2, done;
 double            x, y, z, t;

 result = False;

 done = 0;

 g3d_set_clip_code(*x1, *y1, *z1, &c1);
 g3d_set_clip_code(*x2, *y2, *z2, &c2);

 while ((c1 !== 0) || (c2 != 0))
  {
   if (c1 & c2) return(result);  /* Trivial Rejection */

   /*
```

```
 * Line segment is at least partially outside the clipping pyramid.
 */

if (c1 == 0) c = c2;
else         c = c1;

if (c & g3d_left_plane)
 {
  /*
   * Compute the intersection with y = -z clipping plane.
   */

  if (done & g3d_left_plane) return(result);
  else                       done |= g3d_left_plane;

  t = *x1 - *x2 - *z2 + *z1;

  if (fabs(t) < g3d_small_t) t = g3d_small_t;

  t = (*z1 + *x1) / t;

  x = -z;
  y = t * (*y2 - *y1) + *y1;
  z = t * (*z2 - *z1) + *z1;
 }
else

if (c & g3d_right_plane)
 {
  /*
   * Compute the intersection with y = +z clipping plane.
   */

  if (done & g3d_right_plane) return(result);
  else                        done |= g3d_right_plane;

  t = *x2 - *x1 - *z2 + *z1;

  if (fabs(t) < g3d_small_t) t = g3d_small_t;

  t = (*z1 - *x1) / t;

  x = z;
  y = t * (*y2 - *y1) + *y1;
  z = t * (*z2 - *z1) + *z1;
 }
else

if (c & g3d_bottom_plane)
 {
  /*
   * Compute the intersection with y = -z clipping plane.
   */
```

```
    if (done & g3d_bottom_plane) return(result);
    else                         done | = g3d_bottom_plane;

    t = *y1 - *y2 - *z2 + *z1;

    if (fabs(t) < g3d_small_t) t = g3d_small_t;

    t = (*z1 + *y1) / t;

    x = t * (*x2 - *x1) + *x1;
    y = -z;
    z = t * (*z2 - *z1) + *z1;
   }
  else
  if (c & g3d_top_plane)
   {
    /*
     * Compute the intersection with y = +z clipping plane.
     */

    if (done & g3d_top_plane) return(result);
    else                      done |= g3d_top_plane;

    t = *y2 - *y1 - *z2 + *z1;

    if (fabs(t) < g3d_small_t) t = g3d_small_t;

    t = (*z1 - *y1) / t;

    x = t * (*x2 - *x1) + *x1;
    y = Z;
    z = t * (*z2 - *z1) + *z1;
   }

  /*
   * Recomputed the code.
   */

  if (c == c1)
   {
    *x1 = x;
    *y1 = y;
    *z1 = Z;

    g3d_set_clip_code(x, y, z, &c1);
   }
  else
   {
    *x2 = x;
    *y2 = y;
    *z2 = z;

    g3d_set_clip_code(x, y, z, &c2);
   }
```

```
    }  /* while loop */

  /*
   * Add a small displacement to the z-coordinates of the line
   * segment in order to prevent a potential zero divide in the
   * perspective division.
   */

*z1 += g3d_z_displacement;
*z2 += g3d_z_displacement;

 return(True);

}  / * g3d_clip_line_cs */
```

Index

A

B

D

E

F

G

H

I

K

L

M

S

T

U

V

Symbols

2D

3D